T0236130

Lecture Notes in Artificial Intelligence 9983

Subseries of Lecture Notes in Computer Science

LNAI Series Editors

Randy Goebel
University of Alberta, Edmonton, Canada
Yuzuru Tanaka
Hokkaido University, Sapporo, Japan
Wolfgang Wahlster
DFKI and Saarland University, Saarbrücken, Germany

LNAI Founding Series Editor

Joerg Siekmann
DFKI and Saarland University, Saarbrücken, Germany

Franz Lehner · Nora Fteimi (Eds.)

Knowledge Science, Engineering and Management

9th International Conference, KSEM 2016
Passau, Germany, October 5–7, 2016
Proceedings

 Springer

Editors
Franz Lehner
University of Passau
Passau
Germany

Nora Fteimi
University of Passau
Passau
Germany

ISSN 0302-9743   ISSN 1611-3349 (electronic)
Lecture Notes in Artificial Intelligence
ISBN 978-3-319-47649-0   ISBN 978-3-319-47650-6 (eBook)
DOI 10.1007/978-3-319-47650-6

Library of Congress Control Number: 2016954191

LNCS Sublibrary: SL7 – Artificial Intelligence

Printed on acid-free paper

This Springer imprint is published by Springer Nature
The registered company is Springer International Publishing AG
The registered company address is: Gewerbestrasse 11, 6330 Cham, Switzerland

Preface

The 9th International Conference on Knowledge Science, Engineering, and Management 2016 (KSEM 2016) was the latest in the KSEM series, building on the success of eight previous events held in Guilin, China (KSEM 2006); Melbourne, Australia (KSEM 2007); Vienna, Austria (KSEM 2009); Belfast, UK (KSEM 2010); Irvine, USA (KSEM 2011); Dalian, China (KSEM 2013), Sibiu, Romania (KSEM 2014), and Chongqing, China (KSEM 2015). The series was initiated in 2006 by Prof. Ruqian Lu from the Chinese Academy of Sciences, with the aim of providing a forum for researchers in the broad areas of knowledge science, knowledge engineering, and knowledge management to exchange ideas and to report state-of-the-art research results.

KSEM 2016 was held in Passau, an ancient city located almost exactly in the geographic center of Europe, close to the metropolitan area of Munich. Passau looks back to more than 2,000 years of history and according to a statement attributed to Alexander von Humboldt, Passau was referred to as one of the seven most beautiful towns of the world. Passau is located at the confluence of three rivers, one of them being the Danube, the second longest of Europe. The appearance of the town itself is characterized by medieval, narrow allies and cobblestone streets, with buildings originating from the medieval times. The conference was hosted by the University of Passau, founded in 1978 and being the youngest university in Bavaria. Nevertheless, it has developed into one of the best academic addresses in Germany due to its high academic standard as regularly confirmed by rankings.

Today, knowledge management is from a scientific perspective a highly heterogeneous field combining the influences and ideas of various disciplines. Therefore, the main orientation of KSEM 2016 was influenced by the interdisciplinary character of knowledge management. The conference provided space for discussions on scientific as well as practice-oriented contributions and promoted the exchange of experiences, ideas, and opinions among all participants. The topics covered a broad field of theoretical and practical contributions ranging from classification and clustering, text mining and content analysis, knowledge management and systems, semantics and ontologies, to neural networks, artificial intelligence, and recommendation systems.

KSEM 2016 attracted a total of 116 submissions from 28 countries all over the world. The Program Committee members together with external reviewers contributed 238 reviews. As a result, 49 full papers were selected to be included in the proceedings with a very competitive acceptance rate of 42 %.

Moreover, we were honored to have two prestigious scholars giving keynote speeches at the conference, Prof. Bernd Krieg-Brückner, German Research Center for Artificial Intelligence DFKI and University of Bremen, and Prof. Eric Tsui, Hong Kong Polytechnic University. The abstracts of their talks are included in this volume.

KSEM 2016 would not have been possible without the contributions and efforts of a large scientific community. We thank our authors for being willing to submit their work to KSEM. We sincerely appreciate the large amount of valuable and timely reviews from the members of the Program Committee, the members of three special session committees, and helpful external reviewers. Moreover, we would like to express our gratitude to the conference general chair, Prof. Martin Wirsing (University of Munich, Germany), as well as to the conference general co-chairs, Prof. Zili Zhang (Southwest University, China) and Prof. Dimitris Karagiannis (University of Vienna, Austria). We are also grateful to the team at Springer led by Alfred Hofmann for publication of this volume.

Many thanks also go to our sponsors, msg systems AG and 4process AG for their kind support of KSEM 2016. The conference management system EasyChair was used to handle the submissions, conduct the electronic Program Committee meetings, and assist with the assembly of the proceedings.

The conference proceedings are dedicated to Prof. Ruqian Lu, Chinese Academy of Sciences, Beijing, who founded the KSEM conference series. We feel very honored that Prof. Ruqian Lu personally attended the conference and would like to cordially thank him for his extraordinary contributions and the continuous support over the years.

August 2016   Franz Lehner
                                                               Nora Fteimi

Organization

KSEM 2016 was hosted by the Chair of Information Systems at the University of Passau in Germany. The conference was held during October 5–7, 2016.

Organizing Committee

General Chair

Martin Wirsing   Ludwig Maximilians University Munich, Germany

Conference General Co-chairs

Zili Zhang	Southwest University, China
Dimitris Karagiannis	University of Vienna, Austria

Program Committee Co-chairs

Franz Lehner   University of Passau, Germany

Local Organizing Committee

Franz Lehner	University of Passau, Germany
Nora Fteimi	University of Passau, Germany
Aleksandra Dzepina	University of Passau, Germany

Steering Committee

Yaxin Bi	Ulster University, Belfast, UK
Cungen Cao	Chinese Academy of Sciences, China
Zhi Jin	Peking University, China
Dimitris Karagiannis	University of Vienna, Austria
Claudio Kifor	Sibiu University, Romania
Franz Lehner	University of Passau, Germany
Ruqian Lu	Chinese Academy of Sciences, China
Martin Wirsing	Ludwig Maximilians University Munich, Germany
Chengqi Zhang	University of Technology, Sydney, Australia
Zili Zhan	Southwest University, China
Hui Xiong (Chair)	State University of New Jersey, Rutgers, USA

Program Committee

Andreas Albrecht	Middlesex University, UK
Klaus-Dieter Althoff	DFKI/University of Hildesheim, Germany
Andreia Gabriela Andrei	Alexandru Ioan Cuza University of Iasi, Romania
Albena Antonova	Sofia University, Bulgaria
Nekane Aramburu	University of Deusto, Spain
Serge Autexier	DFKI Bremen, Germany
Kerstin Bach	Norwegian University of Science and Technology, Norway
Dirk Basten	University of Cologne, Germany
Salem Benferhat	Université d'Artois, France
Philippe Besnard	CNRS/IRIT, France
Markus Bick	ESCP Berlin, Germany
Freimut Bodendorf	University of Erlangen-Nuremberg, Germany
Ettore Bolisani	University of Padova, Italy
Remus Brad	Lucian Blaga University of Sibiu, Romania
Constantin Bratianu	Bucharest Academy of Economic Studies, Romania
Krysia Broda	Imperial College, UK
Robert Andrei Buchmann	Babes-Bolyai University, Romania
Cungen Cao	Chinese Academy of Sciences, China
Melisachew Chekol	University of Mannheim, Germany
Paolo Ciancarini	University of Bologna, Italy
Ireneusz Czarnowski	Gdynia Maritime University, Poland
Richard Dapoigny	LISTIC/Polytech'Savoie, France
Juan Manuel Dodero	Universidad de Cádiz, Spain
Josep Domenech	Universitat Politècnica de València, Spain
Dieter Fensel	University of Innsbruck, Austria
Josep Domingo-Ferrer	Universitat Rovira i Virgili, Spain
Sebastian Floerecke	University of Passau, Germany
John Edwards	Aston Business School, Birmingham, UK
Hans-Georg Fill	University of Vienna, Austria
Tatiana Gavrilova	St. Petersburg University, Russia
Stefan Güldenberg	University of Liechtenstein, Liechtenstein
Meliha Handzic	International Burch University, Bosnia
Jose Eduardo Munive Hernandez	University of Bradford, UK
Jun Hong	Queen's University Belfast, UK
Zhisheng Huang	Vrije University Amsterdam, The Netherlands
Alexander Keller	University of Passau, Germany
Gabriele Kern-Isberner	Technische Universität Dortmund, Germany
Zhi Jin	Peking University, China
Claudiu Kifor	Lucian Blaga University of Sibiu, Romania
Michael Langbauer	University of Passau, Germany
Ge Li	Peking University, China
Li Li	Southwest University, China

Li Liu   Chongqing University, China
Weiru Liu   Queen's University Belfast, UK
Xudong Luo   Sun Yat-sen University, China
Markus Manhart   University of Innsbruck, Austria
Pierre Marquis   CRIL-CNRS and Université d'Artois, France
Stewart Massie   Robert Gordon University, UK
John-Jules Meyer   Utrecht University, The Netherlands
Mihaela Muntean   West University of Timisoara, Romania
Oleg Okun   Kreditech Holding SSL GmbH, Germany
Dantong Ouyang   Jilin University, China
Maurice Pagnucco   The University of New South Wales, Australia
René Peinl   Hof University, Germany
Stavros Ponis   National Technical University Athens, Greece
Ulrich Reimer   University of Applied Sciences St. Gallen, Switzerland
Eric Schoop   TU Dresden, Germany
Stefan Smolnik   University of Hagen, Germany
Christian Stary   JK University of Linz, Austria
Stefan Thalmann   University of Innsbruck, Austria
Eduardo Tomé   Universidade Europeia, Portugal
Madalina Vatamanescu   SNSPA, Romania
Lucian Vintan   Lucian Blaga University of Sibiu, Romania
Carl Vogel   Trinity College Dublin, Ireland
Daniel Volovici   Lucian Blaga University of Sibiu, Romania
Martin Wirsing   Ludwig Maximilians University Munich, Germany
Robert Woitsch   BOC Asset Management, Austria
Tong Xu   University of Science and Technology of China, China
Slawomir Zadrozny   Polish Academy of Sciences, Poland
Chunxia Zhang   Beijing Institute of Technology, China
Songmao Zhang   Chinese Academy of Sciences, China

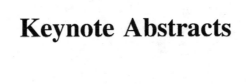

Keynote Abstracts

Generic Ontology Design Patterns: Qualitatively Graded Configuration

Bernd Krieg-Brückner

German Research Center for Artificial Intelligence DFKI,
University of Bremen, Bremen, Germany
Bernd.Krieg-Brueckner@dfki.de

Abstract. Formal Methods for provably correct systems have, alas, not (yet?) achieved widespread application today. This is partly due to their inherent sophistication that, apparently, can only be conquered by theoretically trained scientists; but it is also due to a lack of methodological support, adequate development tools, in particular for verification, and digestible training material for the uninitiated. There should be better academic incentives for theoreticians to develop tools and beginners' training materials showing pitfalls and how to overcome them.

For abstract modelling, ontologies seem to be a good compromise between formality, with deduction as added value, and accessibility to the layman, provided that sufficient methodological support and development tools are available.

The objectives of this talk are to introduce Generic Ontology Design Patterns as a methodological tool in abstract semantic modelling, and to illustrate their use for qualitatively graded configuration in two application domains.

A Generic Ontology Design Pattern abstracts away from particular domains. It serves initially as guideline for a design with a particular form or semantic content. As a constrained development context, it only allows particular operations, with the possibility to automatically check for violation of semantic constraints.

Several patterns are devoted to various kinds of abstraction. The methodology for qualitatively graded configuration supports grading at several levels of qualitative abstraction. It is illustrated by configuration of mobility devices to provide suitable assistance for persons with individual age-related impairments, and configuration of food for persons with individual diet restrictions.

Re-wiring our Brain in the Cloud: Opportunities and Challenges in the Big Data era

Eric Tsui

Department of Industrial and Systems Engineering,
The Hong Kong Polytechnic University, Hong Kong, China
Eric.Tsui@polyu.edu.hk

Abstract. This talk will discuss the power of cloud computing and how a knowledge cloud, which consists not only IT infrastructure but also connections, people power and artificial intelligence, can support product and services innovations via community building, crowdsourcing, ideation, micro-tasking and human-machine cooperative problem solving. With its high connectivity, massive repositories and scalability, the Cloud is naturally the home to Big Data. The second half of the talk will showcase a range of Big Data achievements from various industries including automobile, healthcare. transportation and retail. Success in Big Data not only requires an analytical and intuitive mind but often one also needs to let go exactness for approximations. Among other things, one may need to bypass casual explanations and pattern matching to fully exploit the power of Big Data. Despite the reported successes, many of the existing enterprise applications and computer algorithms are not suitable for processing data in the Big Data era; new algorithms need to be developed. This talk will conclude with a sensible roadamp for organisations to embark on their Big Data journey.

Contents

Knowledge Enrichment and Visualization

Knowledge Management

Knowledge Retrieval

Knowledge Systems and Security

Neural Networks and Artificial Intelligence

Ontologies

Recommendation Algorithms and Systems

Clustering and Classification

BOWL: Bag of Word Clusters Text Representation Using Word Embeddings

Weikang Rui[✉], Kai Xing, and Yawei Jia

School of Computer Science, University of Science and Technology of China,
Hefei, Anhui, China
{jasonrui,ywjia}@mail.ustc.edu.cn, kxing@ustc.edu.cn

Abstract. The text representation is fundamental for text mining and information retrieval. The Bag Of Words (BOW) and its variants (e.g. TF-IDF) are very basic text representation methods. Although the BOW and TF-IDF are simple and perform well in tasks like classification and clustering, its representation efficiency is extremely low. Besides, word level semantic similarity is not captured which results failing to capture text level similarity in many situations. In this paper, we propose a straightforward Bag Of Word cLusters (BOWL) representation for texts in a higher level, much lower dimensional space. We exploit the word embeddings to group semantically close words and consider them as a whole. The word embeddings are trained on a large corpus and incorporate extensive knowledge. We demonstrate on three benchmark datasets and two tasks, that BOWL representation shows significant advantages in terms of representation accuracy and efficiency.

Keywords: Text representation · Word embeddings · Text classification

1 Introduction

One of most important works in text mining and information retrieval is text representation. The text representation problem is to convert each document into a feature vector. The Bag Of Words (**BOW**) represents each document by a sparse vector with high dimensionality where each dimension corresponds to the term frequency. An extension to BOW is the Term Frequency-Inverse Document Frequency (**TF-IDF**) [17], where each dimension is weighted by term's inverse document frequency. Although the BOW and TF-IDF approaches are commonly used and effective to some degree, its representation efficiency is low. The dimensionality of each feature vector is the size of dictionary which can be up to tens of thousands. Curse of dimensionality problem can arise while representing each document with a high dimensional vector as most algorithms used to classify or cluster texts do not scale well to high-dimensional data.

Besides the representation efficiency, the BOW and TF-IDF representations do not capture word level semantic similarity, that is to say the words with similar meaning are considered independently. Those representations are incapable

© Springer International Publishing AG 2016
F. Lehner and N. Fteimi (Eds.): KSEM 2016, LNAI 9983, pp. 3–14, 2016.
DOI: 10.1007/978-3-319-47650-6_1

of handling some situations. To exemplify this, we use a same example that is given by Kusner et al. [12]. Considering two sentences in two documents: *Obama speaks to the media in Illinois* and: *The President greets the press in Chicago*, in BOW based representations, they have no similarity although they express almost the same thing which is illustrated in Fig. 1. In this case, word pairs: (Obama, President); (speaks, greets); (media, press); and (Illinois, Chicago) are semantically close words. They should not be considered independently. It is natural to group those semantically close words and consider them as a whole. In Fig. 1, if we consider each word pair as a whole and represent the two documents based on word pairs, the two documents are represented by two same 4-dimensional vectors while the BOW represents the 2 documents by two totally different 8-dimensional vectors.

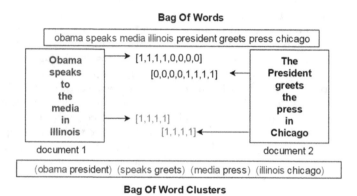

Fig. 1. A comparison of Bag Of Words representation and Bag Of Word Clusters representation. All stop words are removed for both representations. All words are lowercased.

Considering the importance of text representation, there have been numerous works that attempt to represent texts in a low-dimensional space. The Latent Semantic Indexing (**LSI**) [5] finds the low-dimensional feature space by applying singular value decomposition to the BOW features. The Latent Dirichlet Allocation (**LDA**) [2] and Probabilistic Latent Semantic Indexing (**PLSI**) [9] learn words distribution over topics and topics distribution over the documents given a collection of texts. Although they provide a more coherent text representation, they often do not improve empirical performance of BOW on tasks like classification and clustering and their parameters require fine-tuning [12,13].

In this paper, we propose a Bag Of Word cLusters (**BOWL**) representation for texts. As illustrated in Fig. 1, we group the semantically close terms and consider those terms as a whole so that we can reduce dimensionality of BOW and capture word level semantic similarity. To group the semantically close terms, we exploit advantages of word embeddings which project each word into a low dimensional feature space relative to the size of one-hot representation. The *word2vec* model [14] attracts much attention for its simplicity and effectiveness. Empirical result shows that distance between two words reflects their semantic similarity to a large extent. We perform clustering algorithm on the word embeddings to group semantically close words.

We represent each text as a projection to the word clusters, namely bag of word clusters. In BOW, each dimension is assigned with the corresponding term frequency. In our BOWL representation, we propose a dynamic k-max weighting method which borrows idea of pooling operation in convolutional neural network [3]. The pooling in convolutional neural network performs a downsampling operation along the spatial dimensions while our k-max weighting operation downsamples each word cluster. Besides, we dynamically adjust value of k.

The empirical results show that our BOWL representation is effective in text representation in terms of representation accuracy and efficiency. In text classification and clustering tasks, our method outperforms LSI and LDA representations in all cases and outperforms BOW and TF-IDF representations in most cases while using a much lower dimensional space.

The BOWL representation has several attractive properties.

- It represents documents in low dimensional vectors effectively and takes word level semantic similarity into consideration in a straightforward and high interpretable way.
- It utilizes the word embeddings to group semantically close words. The word embeddings are trained from a large corpus and incorporate extensive knowledge naturally.
- Its parameter, number of word clusters, is more robust than parameters of LSI and LDA. It means that our method requires much less work to choose the parameter.

The remainder of this paper is organized as follows. We introduce related work next. A more detailed explanation of our method is given in Sect. 3. We conduct experiments and give analysis in Sect. 4. The last section is conclusion of this paper.

2   Related Work

Many works consider the text representation as feature selection problem. They select a subset of the term features based on some criteria such as information gain, χ^2 statistic (CHI) and mutual information [19]. In addition to single criterion based feature selection, there are more sophisticated optimization methods like Bayesian network-based selection [10] and genetic algorithms [15] to find optimal feature subset. In contrast to feature selection methods where the original features are preserved, feature extraction methods find correlation between different words and combine existing features to create new features. The LSI and LDA [2,5] and their variations [7,16] are typical feature extraction methods for text representation. Jiang et al. [11] use graphs to represent documents and extract frequent subgraphs to produce feature vectors for classification.

There are new approaches to learn document representations. Stacked Denoising Autoencoders (**SDA**) [6] and the faster **mSDA** [4] learn a high level representations for text by deep neural networks. Xu et al. [18] use frequent words to represent rare words which map the whole vocabulary to a smaller frequent used word set. Kusner et al. [12] use word embeddings' distances to measure the documents' distances. The difference between it and our method is that it only measures distances between documents without representing documents with vectors which can only be used by algorithms like KNN.

The idea of using word clusters to represent text was introduced by Bekkerman [1] for text classification. It uses Information Bottleneck clustering to generate document representation in a word cluster space where each cluster is a distribution over document classes. Like many feature selection and extraction methods, it uses class labels to find features that are highly indicative of categories. Those representations require much label information and can only be used in a specific classification task. Our method requires no supervision information which makes it a more general representation and independent of tasks.

3   Bag Of Word Clusters

The Fig. 2 gives architecture of our BOWL text representation which consists of two parts. The left part is the word clusters finding and the right part is weighting. Next, those two parts will be introduced respectively.

3.1   Word Clusters

To group words that are semantically close, it is necessary to learn their relations from a large corpus. The *word2vec* [14] uses a simple neural network architecture which consists of an input layer, a projection layer and an output layer to learn the word vector representations that can predict the nearby words well.

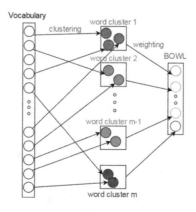

Fig. 2. Architecture of bag of word clusters representation.

The *word2vec* is extremely efficient due to its simple skip-gram model and hierarchical softmax so that we can train word embeddings on a large corpus. Relations between words are encoded in the vectors e.g. $vec(Czech) + vec(currency) \approx vec(koruna)$ and $vec(Berlin) - vec(Germany) \approx vec(Paris) - vec(France)$. Also, the distance between two word embeddings indicates their semantic closeness to a large degree. The Table 1 gives 8 most similar words of 4 words including noun, adjective and verb in the learned word embeddings. It is feasible to group semantically close words by clustering on word embeddings.

Table 1. Words with their 8 closest words

Beautiful	Car	Write	Money
Lovely	Cars	Read	Cash
Enchanting	Truck	Writing	Funds
Gorgeous	Driver	Publish	Monies
Pretty	Driving	Reread	Bankroll
Delightful	Chassis	Wrote	Embezzle
Charming	Automobile	Compose	Repay
Alluring	Vehicle	Script	Buy
Wonderful	Toyotas	Writes	Debt

The K-Means algorithm is one of most common clustering methods due to its efficiency and good performance. The K-Means algorithm requires a parameter, number of clusters, which becomes the parameter of our BOWL representation. The Table 2 shows 4 word clusters of 500 word clusters on the Amazon dataset, which will be introduced in experiments section. Words within a cluster are semantically close and cause the redundancy in indicating what a text is talking

Table 2. Words clusters

Cluster 1	Dog hounds terk kittens puppies puppy tiger pig dogs rabbit sled shaggy yummy fur pony terrier pet monkey cats poodle leash raccoon giraffe pets gorilla bear bull cat collie wolf grizzly poacher paw monkeyshines
Cluster 2	Section nearest lines parallel platform underground connects connect terminal main connecting link accessible elevated via extension linking connected layout loop tunnel line connection branch connections
Cluster 3	Avail mistakes remorse hesitation excuses intentions qualms regrets mistake regret excuse
Cluster 4	Italian italians monte marino gaggia rome mussolini sicily ronco leone venice mezzogiorno bologna italy milan crema grado bellagio

about. Also, considering words within a cluster as a whole can take word level similarity into consideration when measuring text level similarity. For example, two texts talking about Italy, one uses word *Italy* mostly and another uses *Rome* mostly, will have far distance while they are close content. It is more severe in short texts as semantically close words rarely appear together in one short text.

3.2 Word Clusters Weighting

We consider each word cluster as one dimension in BOWL representation. Unlike the BOW where each dimension corresponds to one word, the BOWL has to consider frequencies of a number of words to weight each dimension. Like convolutional neural network, we use the pooled feature to weight each word cluster. For each word cluster, we perform pooling operation on word features (term frequencies). Typical pooling operations are max pooling and mean pooling. The max pooling uses max word frequency to weight each dimension while the mean pooling uses mean word frequency to weight each dimension. Texts contain noise, which can be incorporated by mean pooling as it takes all terms into consideration. So, the max pooling is adopted by us.

We found that the max pooling has good performance when word clusters are not large. When word cluster is large, i.e. the number of clusters is small, frequent words can appear in most word clusters and dominate max pooling features. Those frequent terms like *many, great, play* etc. contain less information. So, in this paper, we propose a dynamic k-max pooling method to weight word cluster feature, that is to say, to increase sample features when word clusters are large. The summation of k maximum term features is used to weight cluster feature, thereby reducing the dominance of frequent terms. As the size of word clusters becomes small, the k should be small either to avoid incorporating noise. We use the following empirical formula to determine the k by average size of clusters:

$$k = \lceil \frac{|v|}{p * \sqrt{n}} \rceil \tag{1}$$

where the $|v|$ is size of vocabulary and n is the number of clusters. The p is a parameter and is set to 1200 in our experiment. The k decreases as clusters go smaller. The k reduces to 1, that is max pooling, when we use enough word clusters to represent a text. The Fig. 3 shows classification accuracy of dynamic k-max weighting compared with max weighting on 2 datasets. The classification results get improved on both classifiers when sampling features according to average size of word clusters. As the number of clusters increases, the k reduces to 1 and two weighting methods become the same. It indicates that the dynamic k-max weighting improves the text representation accuracy compared with max weighting on very low dimensional BOWL representation.

4   Experiments

To evaluate the effectiveness of our word clusters based text representation, Bag Of Words cLusters (**BOWL**), we use the BOWL representation in classification

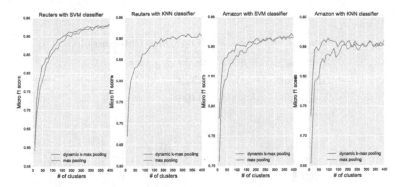

Fig. 3. Classification results of two weighting methods on 2 text datasets using 2 classifiers

and clustering tasks on three datasets. Also, we compare our text representation with 4 commonly used text representations:

1. Bag-Of-Words (**BOW**). A vector of counts of terms that in the vocabulary.
2. Term Frequency-Inverse Document Frequency (**TF-IDF**) [17]. Weight the bag-of-words representation by each word's inverse document frequency.
3. Latent Semantic Indexing (**LSI**) [5]. Perform singular value decomposition on the BOW representation to project the BOW representation to a semantic feature space.
4. Latent Dirichlet Allocation (**LDA**) [2]. Posit that each text is a mixture of a small number of topics and represent each text as distribution over those topics. We use the LDA model in Gensim[1] which is based on online stochastic optimization [8] and use default parameters except the number of topics.

For classification task, we use both KNN and SVM algorithms on each representation. For LSI and LDA, the parameter is the number of topics and for BOWL, the parameter is the number of word clusters. We use part of the training data as validation set to find the parameter that can result in best classification in both KNN classifier and SVM classifier and retrain on the whole training data with the best parameter.

For clustering task, we use K-Means as the clustering algorithm. To evaluate clustering result, we use labels of the data. As clustering is an unsupervised task, we do not use the validation set to find the best parameters.

We also evaluate robustness of parameters of LSI, LDA and BOWL.

4.1 Datasets and Preprocessing

All approaches are evaluated in 3 datasets. The Table 3 lists those datasets and their characteristics including the size of the dataset n, the dimensionality of bag-of-words representation and the number of categories.

---

[1] https://radimrehurek.com/gensim/.

Table 3. Datasets

Name	n	BOW Dim	Categories
Reuters-21578	8609	25059	20
BBCSport	737	12328	5
Amazon	8000	19400	4

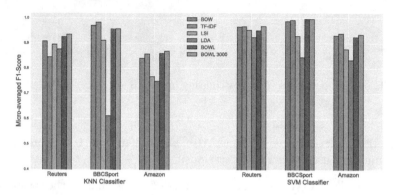

Fig. 4. Classification results on 3 text datasets using 2 classifiers, compared with 4 text representation methods

The Reuters-21578[2] is widely used test collection for text categorization labeled by news topics. We use ModApte train/test split and keep documents that belong to 20 major categories. The BBCSports dataset[3] consists of 737 sports news articles from the BBC Sport website in five topical areas from 2004–2005. The Amazon dataset[4] is a set of customers' reviews towards four product categories, *books, dvd, electronics and kitchen*. For BBCSports and Amazon datasets, we randomly split the whole dataset into train and test parts.

For all datasets, we remove stopwords and digits, and lowercase all words. To avoid oversize BOW dimensionality and reduce noise, we additionally remove words that appear in less than 2 texts in Amazon dataset. The LSI and LDA take BOW representations of texts as input. For all representations, the vectors are L2 normalized at last.

4.2   Word Embeddings

We use *word2vec* tool[5] to train word embeddings on English Wikipedia dump which includes more than 2 billion words. All words are lowercased. We use the skipgram model with window size 5 and filter words with frequency less than 5.

---

[2] http://www.daviddlewis.com/resources/testcollections/reuters21578/.
[3] http://mlg.ucd.ie/datasets/bbc.html.
[4] https://www.cs.jhu.edu/~mdredze/datasets/sentiment/index2.html.
[5] https://code.google.com/p/word2vec/.

After training, we got word vectors with dimensionality of 200 for more than 2 million words. For our BOWL representation, words that do not appear in the *word2vec* model are discarded. For BBCSport and Amazon datasets, there are less than 5 % words are discarded. There are about 10 % words are discarded in Reuters-21578 dataset. It is worth mentioning that the corpus have some influence on the final result. Generally, large corpus result in more accurate word embeddings. Impure corpus can have some benefits. For instance, texts that are mixed with different languages make the word embeddings capture the similar semantic words in different languages like *Milano* and *Milan*.

4.3   Results

Classification. One of applications of text representation is text classification and goodness of text representation determines classification accuracy to a large extent. As the datasets contain multiple categories with unbalanced data, we use Micro-averaged F1-score as the classification evaluation metric. The Fig. 4 gives classification results on those datasets using KNN and SVM classifier where the k is set to 5 for KNN classifier. Parameters, number of topics and number of clusters, are chosen by using a validation set and are limited to 800. In the Reuters dataset with KNN classifier, BBCSport dataset with SVM classifier and Amazon dataset with KNN classifier, our representation achieves best performances. In three other cases, the BOWL representation has competitive classification results compared with BOW and TF-IDF. Meanwhile, the BOWL representation uses less than 800 features where 95 % features are reduced relative to BOW and TF-IDF which makes BOWL much more efficient. In terms of text representation for classification, the LSI and LDA representations do not perform well and BOWL representation outperforms them significantly. When we increase the number of clusters to 3000, our method has better performance. The 3000 dimensional BOWL representation has highest micro f1-score in almost all cases. Although we use 3000 features, the size of feature space is only about 12 % of the size of BOW feature space.

Clustering. A good text representation should have good performances on different tasks. We also evaluate those representations on clustering task by K-Means algorithm. As the K-Means cannot handle clusters with different sizes, we select same amount texts for different categories. To speed up running, we only select a subset of texts for each category. The result is average value on 5 runs of K-Means due to its slight randomness. We use the Adjusted Rand Index (ARI) as the evaluation metric. The Rand Index (RI) is calculated by:

$$RI = \frac{TP + TN}{TP + FP + FN + TN} = \frac{TP + TN}{\binom{n}{2}} \tag{2}$$

where n is the number of texts. The TP is the number of text pairs that belong to same category and assigned with same cluster label. The TN is the number of text pairs that belong to different categories and assigned with different cluster

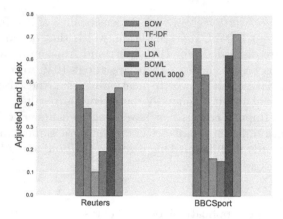

Fig. 5. Clustering results on 2 text datasets

labels. The Adjusted Rand Index is adjusted RI to make sure have a value close to 0 for random labeling and exactly 1 when clusterings are identical.

The Fig. 5 is the ARI scores for those representations on two datasets. Scores for LSI, LDA and BOWL are best scores with dimensions limited to 800. Similar with classification results, the BOWL representation has better results than LSI and LDA, and competitive results with BOW. Unlike classification, our representation outperforms TF-IDF significantly. When we use 3000-dimensional vectors to represent texts, BOWL representation gets better ARI scores and exceeds BOW significantly on BBCSport dataset. Besides, the representation efficiency is much higher than BOW and TF-IDF.

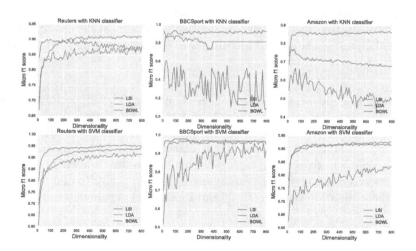

Fig. 6. Classification results with different dimensionality of representation vectors

Connection with LSI and LDA. BOWL shares some common elements with LSI and LDA. They all project the BOW features to a latent feature space. For LDA, latent features are distributions over topics. For BOWL, the feature space is semantically close words which is less latent than LSI and LDA. We prefer it *concepts* which is higher level than *words* and lower level than *topics*. So, for BOWL it performs worse than LSI and LDA when dimensionality of feature space is very limited. But when increase the dimensionality, the BOWL representation has best result as in Fig. 6.

Also in Fig. 6, the parameter of BOWL is much more robust than LSI and LDA. As the dimensionality increases, the performance of BOWL increases and becomes stable gradually with little fluctuation while the fluctuation is more intensive and the performance even decreases for LSI and LDA. It means it is much easier for BOWL to choose the parameter. We assume the reason for this is that the BOWL learns the feature space from a very large corpus by word embeddings which is independent of current texts while LSI and LDA learn it from current small size texts.

5   Conclusion

In this paper, we propose a straightforward Bag Of Word Clusters (BOWL) text representation which groups semantically close words and considers them as one dimension in the representation space. For each dimension, we use dynamic k-max pooling method to weight. Words are grouped by performing clustering algorithm on word embeddings which is trained from a large corpus. Unlike LSI and LDA, our method finds the new feature space from much larger and independent texts. Experiments on classification and clustering tasks show significant advantages of BOWL in terms of representation accuracy and efficiency.

References

1. Bekkerman, R., El-Yaniv, R., Tishby, N., Winter, Y.: Distributional word clusters vs. words for text categorization. J. Mach. Learn. Res. **3**, 1183–1208 (2003)
2. Blei, D.M., Ng, A.Y., Jordan, M.I.: Latent dirichlet allocation. J. Mach. Learn. Res. **3**, 993–1022 (2003)
3. Blunsom, P., Grefenstette, E., Kalchbrenner, N., et al.: A convolutional neural network for modelling sentences. In: Proceedings of the 52nd Annual Meeting of the Association for Computational Linguistics. Proceedings of the 52nd Annual Meeting of the Association for Computational Linguistics (2014)
4. Chen, M., Xu, Z., Weinberger, K., Sha, F.: Marginalized denoising autoencoders for domain adaptation. arXiv preprint arXiv:1206.4683 (2012)
5. Deerwester, S.C., Dumais, S.T., Landauer, T.K., Furnas, G.W., Harshman, R.A.: Indexing by latent semantic analysis. JAsIs **41**(6), 391–407 (1990)
6. Glorot, X., Bordes, A., Bengio, Y.: Domain adaptation for large-scale sentiment classification: a deep learning approach. In: Proceedings of the 28th International Conference on Machine Learning (ICML-11), pp. 513–520 (2011)

7. Griffiths, T.L., Steyvers, M., Blei, D.M., Tenenbaum, J.B.: Integrating topics and syntax. In: Advances in Neural Information Processing Systems, pp. 537–544 (2004)
8. Hoffman, M., Bach, F.R., Blei, D.M.: Online learning for latent dirichlet allocation. In: Advances in Neural Information Processing Systems, pp. 856–864 (2010)
9. Hofmann, T.: Probabilistic latent semantic indexing. In: Proceedings of the 22nd Annual International ACM SIGIR Conference on Research and Development in Information Retrieval, pp. 50–57. ACM (1999)
10. Inza, I., Larrañaga, P., Etxeberria, R., Sierra, B.: Feature subset selection by Bayesian network-based optimization. Artif. Intell. **123**(1), 157–184 (2000)
11. Jiang, C., Coenen, F., Sanderson, R., Zito, M.: Text classification using graph mining-based feature extraction. Knowl. Based Syst. **23**(4), 302–308 (2010)
12. Kusner, M.J., Sun, Y., Kolkin, N.I., Weinberger, K.Q.: From word embeddings to document distances. In: Proceedings of The 32nd International Conference on Machine Learning, pp. 957–966 (2015)
13. Lu, Y., Mei, Q., Zhai, C.: Investigating task performance of probabilistic topic models: an empirical study of PLSA and LDA. Inf. Retrieval **14**(2), 178–203 (2011)
14. Mikolov, T., Sutskever, I., Chen, K., Corrado, G.S., Dean, J.: Distributed representations of words and phrases and their compositionality. In: Advances in Neural Information Processing Systems, pp. 3111–3119 (2013)
15. Oh, I.S., Lee, J.S., Moon, B.R.: Hybrid genetic algorithms for feature selection. IEEE Trans. Pattern Anal. Mach. Intell. **26**(11), 1424–1437 (2004)
16. Petterson, J., Buntine, W., Narayanamurthy, S.M., Caetano, T.S., Smola, A.J.: Word features for latent dirichlet allocation. In: Advances in Neural Information Processing Systems, pp. 1921–1929 (2010)
17. Salton, G., Buckley, C.: Term-weighting approaches in automatic text retrieval. Inf. Process. Manag. **24**(5), 513–523 (1988)
18. Xu, Z.E., Chen, M., Weinberger, K.Q., Sha, F.: From sbow to dCoT marginalized encoders for text representation. In: Proceedings of the 21st ACM International Conference on Information and Knowledge Management, pp. 1879–1884. ACM (2012)
19. Yang, Y., Pedersen, J.O.: A comparative study on feature selection in text categorization. In: ICML, vol. 97, pp. 412–420 (1997)

Clustering Categorical Sequences with Variable-Length Tuples Representation

Liang Yuan[1], Zhiling Hong[2(✉)], Lifei Chen[3], and Qiang Cai[4]

[1] Network Operation Maintenance Center,
University of Electronic Science and Technology of China,
Chengdu 611731, China
[2] Software School, Xiamen University, Xiamen 361005, China
hongzl@xmu.edu.cn
[3] School of Mathematics and Computer Science,
Fujian Normal University, Fuzhou 350117, Fujian, China
[4] Technique Department, Xiamen Customs, Xiamen 361000, China

Abstract. Clustering categorical sequences is currently a difficult problem due to the lack of an efficient representation model for sequences. Unlike the existing models, which mainly focus on the fixed-length tuples representation, in this paper, a new representation model on the variable-length tuples is proposed. The variable-length tuples are obtained using a pruning method applied to delete the redundant tuples from the suffix tree, which is created for the fixed-length tuples with a large memory-length of sequences, in terms of the entropy-based measure evaluating the redundancy of tuples. A partitioning algorithm for clustering categorical sequences is then defined based on the normalized representation using tuples collected from the pruned tree. Experimental studies on six real-world sequence sets show the effectiveness and suitability of the proposed method for subsequence-based clustering.

Keywords: Sequence clustering · Representation model · Variable-length tuples · Pruning method · Entropy-based measure

1 Introduction

Data clustering has a wide range of applications and has been one of the essential methods used in knowledge systems. In the past decades, it was studied extensively in the statistics, machine learning and data mining communities, and a number of clustering algorithms have been proposed [1,2]. However, most of them are principally designed for attribute-value data, say, vector data. Currently, categorical sequences, such as speech sequences in natural language processing, are widely used in real-world applications. Clustering such sequences is a difficult problem due to the fact that the chronological order of symbols (categories) that compose the sequences is very important for clustering tasks; this remains a major obstacle in applying the traditional clustering algorithms [3–5].

© Springer International Publishing AG 2016
F. Lehner and N. Fteimi (Eds.): KSEM 2016, LNAI 9983, pp. 15–27, 2016.
DOI: 10.1007/978-3-319-47650-6_2

A widely accepted solution to the problem is using a tuples-based representation for sequences, which effectively equates to project each sequence onto the new pattern space spanned by a set of tuples [4,6]. Roughly speaking, the tuple of sequences is one kind of short subsequences; thus, the locally chronological order of symbols can be preserved to some extent. Such a representation model is somewhat similar to the Vector Space Model (VSM) for representing documents in text mining [7], where each term in the documents is considered as a dimension and each document is typically represented as a vector of the term frequencies. With the tuples-based representation, sequences are viewed as "documents" with the tuples representing their "terms", and can thus be clustered like the common vector data.

Obviously, the ability of the representation model to capture the structural features hidden within sequences depends on the tuples chosen for the model. The common method is the n-tuples (alternatively known as n-grams) approach [4,8,9], which is the set of all possible tuples (grams) with their length fixed at n. As choosing an appropriate tuple length is currently a difficult problem, generally, one tends to use a large n, because small n likely breaks long sequence patterns into small segments [10]. However, a large n would result in a huge number of tuples which is exponential in the length. More importantly, with a fixed length, all tuples of length n are collected without distinguishing between significant and non-significant tuples [5], which challenges the traditional clustering algorithms by the existence of many noisy features (tuples) or redundant features (tuples) that do not contribute to clustering.

The popular approach adapting the algorithms to the high-dimensional data is to eliminate these features by combining feature selection techniques, for example, by removing those tuples whose frequency are less than the user-defined threshold [10]. Clearly, such a threshold is difficult to determine. Another approach is to perform subspace clustering on the high-dimensional data: examples include entropy-weighting K-means (EWKM) [11], model-based projective clustering (MPC) [12], etc. Note that such feature-weighting-based algorithms are designed on the assumption that each of the dimensions (here, the tuples) spanning the new pattern space is independent of the others, which hardly holds in the n-tuples representation for sequences: tuples that share the same preceding subsequence may be highly correlated with each other.

In this paper, a new method is proposed to produce the variable-length tuples, with a large number of redundant tuples removed. We propose a pruning method for the purpose, by organizing the original n-tuples into a suffix tree, on which those leaves corresponding to the redundant tuples are iteratively deleted in terms of the information gain provided to their parent (i.e., the preceding tuples). The remaining tuples in the pruned tree are then collected to create a normalized representation model, with which a partitioning algorithm is defined for the clustering task. We conducted a series of experiments on real-world categorical sequences. The results show that the proposed method significantly outperforms other mainstream methods.

The remainder of this paper is organized as follows. Section 2 presents some preliminaries and related work. Section 3 describes the new representation model and the clustering algorithm. Experimental results are presented in Sect. 4. Finally, Sect. 5 gives our conclusion and discusses directions for future work.

2   Preliminaries and Related Work

A categorical sequence is a linear chain made up of symbols, containing some structural features. Figure 1 gives an example, where two sequences denoted by s_1 and s_2 are shown. Both s_1 and s_2 are made up of 3 symbols "A", "B" and "C", but they have different lengths, saying, 14 and 12, respectively. Clustering such sequences is a challenging problem due to the difficulties in defining a meaningful distance measure for sequences [1,4]. The existing measures fall in two groups: alignment-based and alignment-free measures [6,13]. In the first group, the distance is computed by an alignment algorithm, such as the well-known edit distance and its approximate algorithms [14]. Generally, they have a high time complexity. The alignment-free measures in the second group calculate the distance between sequences based on statistical models [3,10], information theory [9] or subsequences [5], without identifying the similar regions of sequences; thus, they are computationally efficient.

s_1: **ABAABBACACBACB**

s_2: **ACBABAABBACB**

Fig. 1. An example of categorial sequences made up of 3 symbols "A", "B" and "C".

To define an alignment-free distance measure for sequences, the tuple-based representation has been widely used due to its simplicity [4,6]. Using the model, each sequence can be transformed into a vector of tuple frequencies. Table 1 illustrates the 3-tuples representation for the sequences of Fig. 1, where each column corresponds to an unique tuple comprising 3 symbols and the digit in each cell indicates the number of the tuple appearing in the sequence. Based on the table, distance between sequences can be easily computed using Euclidean distance [9], Mahalanobis distance [15], etc. The common clustering methods can also be easily applied to categorical sequences, such as the hierarchical clustering algorithms [9,10] aimed at organizing sequences into a tree of clusters and the partitioning methods including the well-known K-means and its numerous variants [1,11,12]. Since the aim is to generate flat clusters in this paper, we will focus on the latter, that is, grouping sequences according to the occurrences of the tuples given the number of clusters K.

As discussed previously, the number of tuples (i.e., the data dimensionality) would be huge in practice, when the n-tuples representation is employed with the length fixed at a large n. For example, the number of symbols composing a speech sequence typically reaches 20 (see Sect. 4.1); therefore, there is a set of 20^n

Table 1. 3-tuples representation for the sequences shown in Fig. 1.

	AAB	ABA	ABB	ACA	ACB	BAA	BAB	BAC	BBA	CAC	CBA
s_1	1	1	1	1	2	1	0	2	1	1	1
s_2	1	1	1	0	2	1	1	1	1	0	1

possible tuples. To cluster such high-dimensional data, one has to resort to the unsupervised feature-selection techniques, implemented by the filter methods or the built-in methods [1,12]. Subspace clustering, aimed at grouping data objects into clusters projected in some subspaces, is one of the popular methods using the built-in mechanism for feature selection. Examples include PROCLUS and its variants [16], the entropy-weighting algorithm EWKM [11], etc. The goal of a filter method is to choose an appropriate subset of the original features in the preprocessing step before clustering, where some heuristic criteria are defined to evaluate the significance of features [17]. Due to the huge number of admissible subsets which is exponential in the data dimensionality, usually, both methods choose the subset based on the assumption that the attributes are independent of each other. In the conditional probability model (CPD) for sequences [10], for instance, only the frequent subsequences (corresponding to the tuples) are chosen, given a threshold defining the minimal frequency of the resulting tuples.

In the variable-length representation for the tuples proposed in this paper, however, we focus on the identification of possibly redundant tuples. With the redundant tuples removed, the correlations between features (tuples) are thus reduced. Our efficient method for producing the new representation model is based on the pruning strategy, while surmounting the independent assumption, as described in the next section.

3   Sequences Clustering with Variable-Length Tuples

In this section, we propose a variable-length tuples representation for categorical sequences, followed by a new K-means-type algorithm for clustering the sequences. We begin by introducing the notation used throughout the paper.

3.1   Basic Notation

In what follows, the sequence set is denoted by $S = \{s_i | i = 1, 2, \ldots, N\}$ from which $K(1 < K < N)$ clusters are searched for. Here, s_i stands for the ith sequence and N the number of sequences to be clustered. Let s be a categorical sequence of length L, where each of the L symbols is one of the categories $\forall x \in X$, with X being the set of symbols and $|X|$ the number of symbols. Moreover, the K clusters are denoted by $c_1, \ldots, c_k, \ldots, c_K$, each consisting of a disjoint subset of S; therefore, $S = \cup_{k=1}^{K} c_k$. The set of K clusters is denoted by $C = \{c_k | k = 1, 2, \ldots, K\}$.

A n-length subsequence, also called n-tuples, of sequence $s \in S$ is a segment of n consecutive symbols in s. Note that the length n is also referred to as *memory length* in the case of Markov chain model for sequences [10]. Letting t be a n-length tuple and $\#_S(t)$ the number of t appearing in S, we denote the set of n-tuples by $T = \{t|\#_S(t) > 0\}$; in other words, each of the tuples should appear in at least one of the sequences in S. Based on the definitions, the cardinality of T, i.e., $|T|$, is precisely the dimensionality of the data using the tuples-based representation model. We denote $D = |T|$ and t_d the dth n-tuples of S, where $d = 1, 2, \ldots, D$.

Each n-length tuple with $n > 1$ can be viewed as a combination of the preceding $(n-1)$-length subsequence and the ending symbol. According to this view, the tuple t can be rewritten as $t = \delta x$, where δ denotes the preceding subsequence of the ending symbol x. We denote the conditional probability of x given its preceding subsequence by $p(\mathcal{X} = x|\mathcal{Y} = \delta)$, where $\mathcal{X}$ and $\mathcal{Y}$ are the random variables associated with the symbols and the preceding subsequences. To simplify the representation, we will use $p(x|\delta)$ to denote $p(\mathcal{X} = x|\mathcal{Y} = \delta)$ in the following pages.

3.2 Variable-Length Tuples Representation

Given a large memory-length n, for example, $n = 10$, the resulting n-tuples representation for sequences is generally in high dimensionality (recall that $D \approx |X|^n$). In this subsection, we aim at reducing the dimensionality by removing the redundant tuples from T. For the purpose, we first organize the n-tuples into a suffix tree, where each path (from the root to one of the leaf nodes of the tree) corresponds to a tuple of length n. Here, the root of the tree is a virtual symbol indicating the beginning of each tuple; thus, the height of the tree is $n+1$, and the number of children for each node except the leaves is at most $|X|$. Figure 2(b) shows a subtree created for the 3-tuples of the sequences s_1, s_2 in Fig. 1; the tuples used have the same preceding symbol "A", as Fig. 2(a) shows.

During the creation of the tree, each node (except the root) is attached a value indicating its conditional probability with regard to its preceding

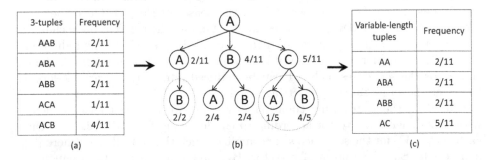

Fig. 2. Illustration of the pruning method for generating the variable-length tuples (preceded "A") from the sequences shown in Fig. 1

subsequence. The conditional probability is estimated by the frequency estimator. For example, according to Fig. 2(a), the number of the subsequence "AC" appearing in the sequence set $\{s_1, s_2\}$ is $1 + 4 = 5$ and the total number of tuples preceded "A" is 11; then, we estimate the conditional probability by $p(C|A) = 5/11$, as the value attached to the node labeled "C" shows in Fig. 2(b). Likewise, the conditional probabilities for the tuples "ACA" and "ACB" can be estimated, i.e., $p(A|AC) = 1/5$ and $p(B|AC) = 4/5$, respectively.

The variable-length tuples representation for sequences can be derived based on the n-tuples tree. This is achieved by pruning the tree using a post-pruning method similar to that applied to decision-tree induction [1]. As Figs. 2(b) and (c) show, with some leaves removed, the tree changes to accommodate both 2-tuples and 3-tuples. Since we aim at eliminating the redundant tuples, a criterion should be defined to measure the redundancy of a subtree, and subsequently to determine whether the subtree need to prune or not. Taking for example the tree shown in Fig. 2(b) again, obviously, the leaf labeled "B" for the 3-length tuple "AAB" can be deleted, because there is no information loss if it is replaced with the shorter tuple "AA". However, it is not the case for "AB": lengthening "AB" to "ABA" and "ABB" is able to obtain considerable information gain. This observation suggests an entropy-based judgement for the redundancy evaluation. Formally, we first compute the entropy for the tuples having the same preceding subsequence δ by

$$H(\delta) = -\sum_{x \in X} p(x|\delta) \times \log_2 p(x|\delta) \tag{1}$$

with

$$p(x|\delta) = \frac{\#_S(\delta x)}{\#_S(\delta)}.$$

Then, the redundant tuples are identified based on the following Definition 1.

Definition 1 (Redundant tuples). The tuples δx for $\forall x \in X$ are redundant if $H(\delta) < \tau$, where $\tau \geq 0$ is a threshold defining the minimal information gain.

Given the n-tuples tree, the pruning process begins by examining the leaf nodes according to Definition 1; here, the symbols corresponding to the leaf and its siblings are considered as $\forall x \in X$ in the senses of Eq. (1). Once they are identified being redundant, the leaves are deleted and their parent node changes to the new leaf. The new leaves are then re-scanned to search for the redundant subsequences of shorter length. With such an iterative pruning method, the 3-tuples for the sequences in Fig. 1 can be reduced into a set of variable-length tuples as shown in Table 2, by setting $\tau = 1$.

Based on the D resulting tuples $t_1, t_2, \ldots, t_D$, we represent each sequence with a D-dimensional vector according to the following Definition 2. Note that each vector $\boldsymbol{V}_s$ for the sequence s is normalized such that $\| \boldsymbol{V}_s \| = 1$, where $\| \cdot \|$ denotes the Euclidean norm of a vector. By the normalization, the frequencies are smoothed to counteract the effect induced by the varying lengths of the sequences as well as the tuples.

Table 2. Variable-length tuples representation for the sequences shown in Fig. 1.

	AA	ABA	ABB	AC	BAA	BAB	BAC	BB	CA	CB
s_1	0.224	0.224	0.224	0.670	0.224	0.000	0.447	0.224	0.224	0.224
s_2	0.289	0.289	0.289	0.577	0.289	0.289	0.289	0.289	0.000	0.289

Definition 2 (Variable-length tuples representation). The variable-length tuples representation for each sequence $s \in S$ is the vector V_s, given by

$$V_s = \; < f_s(t_1), f_s(t_2), \ldots, f_s(t_d), \ldots, f_s(t_D) > \tag{2}$$

where $f_s(t_d) = \#_s(t_d) \times (\sum_{d'=1}^{D}[\#_s(t_{d'})]^2)^{-\frac{1}{2}}$ with $\#_s(t_d)$ being the number of t_d appearing in s.

Now, the only pending factor of the representation model is the setting for τ. Intuitively, the desired value of τ connects to both the number of symbols composing the sequences ($|X|$) and the number of clusters K. In particular, τ should be enlarged with a large $|X|$ and a small K; thus, an obvious setting for τ could be

$$\tau = \max\{\log_2 \tfrac{|X|}{K}, 0\}. \tag{3}$$

According to Eq. (3), $\tau > 0$ when $|X| > K$, which is often the case in practice. In the case where $|X| \le K$, $\tau = 0$ which means that it is not necessary to prune the tuples tree. In this case, the variable-length representation degenerates to the traditional n-tuples representation.

3.3   Clustering Algorithm

In this subsection, a partitioning algorithm is presented for clustering categorical sequences based on the variable-length tuples representation discussed in the previous subsection. The algorithm named *KM-NVLT* (for K-Means with Normalized Variable-Length Tuples), as outlined in Algorithm 1, starts clustering from transforming the sequences in S into vectors using the new representation (steps (1) $\sim$ (3)). Then, the algorithm groups the sequences in an iterative process like the K-means clustering (Step (5)).

The aim of Step (4) in *KM-NVLT* is to build a robust condition for the coming iterative process, by choosing a set of well-scattered sequences as the initial cluster centers. The first two centers I_1 and I_2 are chosen according to the following rule:

$$(I_1, I_2) = \operatorname{argmax}_{(s,s') \in S \times S} \| V_s - V_{s'} \|^2. \tag{4}$$

Then, the remaining $K - 2$ centers are selected based on the maximum-minimum principle [9], i.e.,

$$I_{k+1} = \operatorname{argmax}_{s \in S \setminus \{I_i | i=1,\ldots,k\}} \min_{i=1,\ldots,k} \| M_s - M_{I_i} \|^2 \tag{5}$$

where $k \in [2, K - 1]$.

Input: the sequence set S, the number of clusters K and the memory-length n
Output: the set of resulting clusters $C = \{c_k | k = 1, 2, \ldots, K\}$
begin

    (1) Generate n-tuples from the sequences in S, and create the n-tuples tree using the method described in Sect. 3.2;

    (2) Determine τ according to Eq. (3) and prune the n-tuples tree based on Definition 1;

    (3) Collect the variable-length tuples from the pruned tree, and represent each sequence $s \in S$ by the vector $\boldsymbol{V}_s$ according to Eq. (2) and Definition 2;

    (4) Choose K vectors for the initial cluster centers using Eqs. (4) and (5);

    (5) repeat

        (5.1) Generate $c_1, c_2, \ldots, c_K$ by assigning each sequence $s \in S$ to its closest cluster center, in terms of the Euclidean distance between $\boldsymbol{V}_s$ and each cluster center;

        (5.2) Recompute the center for each cluster c_k by averaging the vectors belonging to c_k, where $k = 1, 2, \ldots, K$.

    until *C is not changed.*;

end

Algorithm 1. Outline of the *KM-NVLT* algorithm.

The time complexity for generating the variable-length tuples is $O(n \times |X| \times \mathcal{L})$, where $\mathcal{L}$ is the total lengths of the sequences in S. The time complexities of the K-means-type clustering and the centers selection method are $O(KND)$ and $O(N^2D)$, respectively. Generally, $|X| > K$ and $n \times \mathcal{L} > ND$; thus, for *KM-NVLT*, the time complexity can be finally given as $O(\max\{n \times |X| \times \mathcal{L}, N^2D\})$.

4   Experimental Evaluation

In this section, we evaluate the performance of *KM-NVLT* on real-world categorical sequences, and also experimentally compare the variable-length tuples representation with a few other methods.

4.1   Sequence Sets and Experimental Setup

Six sequence sets for speech recognition are used. We obtained the sequence sets from [18], namely *locmelovoy, locmrlovoy, locmslovoy, locfjlavoy, locflauvoy* and *locffpevoy*, abbreviated to S1 ∼ S6, respectively. Each sequence in the sets is generated from the pronunciation of one of the five French vowels ("a", "e", "i", "o" and "u") by binning the sound wave. So, the true number of clusters for the sequence sets are known, i.e., $K = 5$. Details of the data sets are summarized in Table 3. The average length of sequences in these sets varies from 560 to approximately 1900, in order to evaluate the capability of different representation models and clustering methods.

We used the K-means algorithm on the common n-tuples representation as the baseline in the experiments. The algorithm is abbreviated KM-NT (for K-Means with Normalized Tuples), which was implemented as a special case of our

Table 3. Details of the real-world sequence sets

Dataset	#clusters(K)	#symbols($|X|$)	#sequences(N)	Length(L)	Average length
S1	5	20	50	[203, 1035]	560
S2	5	18	50	[226, 1253]	737
S3	5	20	50	[506, 1564]	924
S4	5	19	50	[405, 1986]	1088
S5	5	19	50	[581, 2382]	1570
S6	5	18	50	[701, 3753]	1899

KM-NVLT by fixing τ at 0. Based on the normalized n-tuples representation, we also chose EWKM [11], a feature-weighting-based subspace clustering algorithm, as the competing method. Its weighting parameter γ was set to the author-recommended value 0.5.

Frequency-based feature selection method was also used to provide a reference point for comparison. For the purpose, the frequent n-tuples were collected to create a reduced tuples set for each sequence set. Since it is difficult to determine the threshold for the frequencies, we selected those tuples whose frequency is larger than 1 for the resulting representation. The vectors were finally normalized using the same method to that of Definition 2. We will denote the K-means algorithm applied to such a representation by KM-NFT (for K-Means with Normalized Frequent Tuples).

Bisection K-means [9] was also chosen for comparison. For this algorithm, we created a TF-IDF representation model for each sequence set, where IDF is the abbreviation for the *inverse document frequency* popularly used in the text mining community [7]. After assigning the n-tuples with the IDF weights, the vectors were normalized. The bisection K-means with the TF-IDF representation will be denoted as BKM-NIDF. The initial cluster centers for all the competing algorithms were selected using the same approach as that of *KM-NVLT* (see Step (4) of Algorithm 1).

4.2   Experimental Results

The performance of the algorithms is evaluated in terms of *clustering accuracy* (CA), which is computed as $CA = \frac{1}{N}\sum_{k=1}^{K} a_k$, where a_k is the number of sequences in the majority group corresponding to c_k. Clearly, this measure requires that the ground truth of the datasets be known, which is the case in our experiments. Table 4 shows the clustering accuracy obtained by the five algorithms on the six sequence sets, with the memory-length n set to 10 (the performance with regard to different n will be examined in Fig. 3). The best results are marked in bold typeface.

From Table 4, we can see that our *KM-NVLT* is able to achieve high-quality overall results, outperforming the competing algorithms on all the six sequence sets. In fact, only KM-NFT obtains comparable results on S1, S3 and S6, whereas

Table 4. Comparisons of clustering accuracy (CA) obtained by different algorithms ($n = 10$)

Dataset	KM-NVLT	KM-NFT	EWKM-NT	BKM-NIDF	KM-NT
S1	**1.000**	0.980	0.620	0.720	0.620
S2	**0.980**	0.600	0.600	0.700	0.600
S3	**1.000**	**1.000**	0.660	0.640	0.660
S4	**0.860**	0.640	0.680	0.720	0.680
S5	**1.000**	0.780	0.620	0.680	0.620
S6	**0.980**	0.900	0.680	0.660	0.700

EWKM-NT and KM-NT perform poorly (note that both are based on the n-tuples representation). We also observe that the clustering accuracies of EWKM-NT and KM-NT are close, indicating that the built-in feature-selection scheme used in the soft subspace clustering methods fails in distinguishing between significant and non-significant tuples. BKM-NIDF, which makes use of the TF-IDF representation, yields higher accuracy than EWKM-NT on most of the data sets. This indicates that the IDF weighting method is able to identify significant tuples to some extent.

Comparisons of the clustering accuracy with varying memory-length of tuples are given in Fig. 3. It can be seen from the figures that our *KM-NVLT* achieves robust performance accompanied by high clustering accuracy, along with the increment of the memory-length n. Except the cases of KM-NFT on S1 and S3, the competing algorithms yield instable results that are sensitive to the setting of n. The good performance of *KM-NVLT* owes to the use of the pruning method in producing the variable-length tuples with redundant tuples deleted, which, in effect, improve the performance of the remaining features (tuples). Another gain of the pruning method is to reduce the number of features for the clustering algorithms, as Fig. 4 shows.

Figure 4 illustrates the number of resulting tuples in the traditional n-tuples representation (used by EWKM-NT and KM-NT), the frequent n-tuples representation used by KM-NFT and our variable-length representation. One can see that the number of tuples are significantly reduced by using our pruning method. The number can also be reduced using the frequency-based selection method; however, the results are clearly dependent on the user-defined threshold, which is difficult to estimate. The figures also show that, when the memory-length n goes from 6 to 14, the numbers of resulting variable-length tuples remain approximately unchanged on the six sequence sets. This result suggests that the optimal length of tuples for the speech sequences is about 6. As the memory length substantially connects to the order of Markov chain model [10], our pruning method might be helpful in estimating the order for such models.

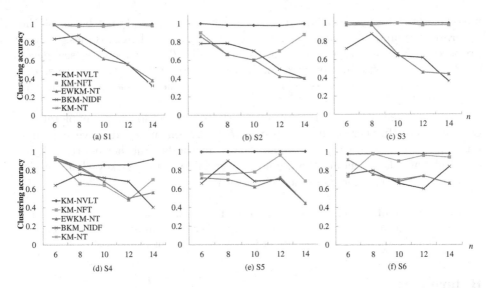

Fig. 3. Clustering accuracy of the algorithms with various memory-lengths of tuples.

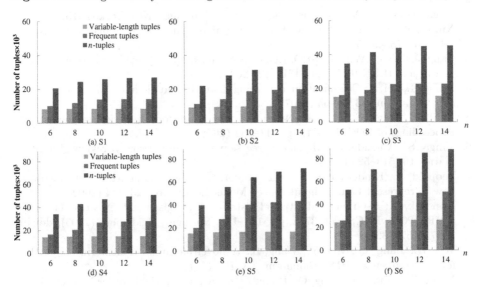

Fig. 4. Comparisons of the number of tuples using different representation models.

5   Conclusion and Perspectives

In this paper, we proposed a variable-length tuples representation for categorical sequence clustering, unlike the existing methods, which are generally based on the fixed-length tuples (n-tuples) representation. We proposed to organize the original n-tuples into a tree, in order to derive a pruning method to obtain the variable-length tuples from the pruned tree. We defined an entropy-based measure to evaluate the redundance of tuples and to provide the basis for the removal of redundant tuples. Using the resulting variable-length tuples, we also proposed a K-means-type algorithm, call *KM-NVLT*, for categorical sequences clustering. The experiments were conducted on six speech sequence sets, the results show the effectiveness of the new representation for sequences and the new algorithm for clustering.

There are many directions that are clearly of interest for future exploration. One avenue of further study is to test *KM-NVLT* on more extensive sequence sets, and to compare with other mainstream methods. Another efforts will be directed towards extending the method to variable-order Markov chain model for model-based categorical sequence clustering.

Acknowledgments. This work was supported by the National Natural Science Foundation of China under Grant No. 61175123, and partially supported by the Natural Science Foundation of Fujian Province of China under Grant No. 2015J01238.

References

1. Aggarwal, C.C.: Data Mining: The Textbook. Springer, New York (2015)
2. Xu, R., Wunsch, D.C.: Survey of clustering algorithms. IEEE Trans. Neural Netw. **16**, 645–678 (2005)
3. Yang, J., Wang, W.: CLUSEQ: Efficient and effective sequence clustering. In: Proceedings of IEEE ICDE, pp. 101–112 (2003)
4. Dong, G., Pei, J.: Classification, clustering, features and distances of sequence data. Seq. Data Min. **33**, 47–65 (2007)
5. Kelil, A., Wang, S.: SCS: a new similarity measure for categorical sequences. In: Proceedings of IEEE ICDM, pp. 343–352 (2008)
6. Vinga, S., Almeida, J.: Alignment-free sequence comparison: a review. Bioinformatics **19**, 513–523 (2003)
7. Leopold, E., Kindermann, J.: Text categorization with support vector machines: how to represent texts in input space? Mach. Learn. **46**, 423–444 (2002)
8. Kondrak, G.: N-Gram similarity and distance. In: Consens, M., Navarro, G. (eds.) SPIRE 2005. LNCS, vol. 3772, pp. 115–126. Springer, Heidelberg (2005). doi:10. 1007/11575832_13
9. Wei, D., Jiang, Q., Wei, Y., Wang, S.: A novel hierarchical clustering algorithm for gene sequences. BMC Bioinform. **13**, 174 (2012)
10. Xiong, T., Wang, S., Jiang, Q., Huang, J.Z.: A novel variable-order Markov model for clustering categorical sequences. IEEE Trans. Knowl. Data Eng. **26**, 2339–2353 (2014)
11. Jing, L., Ng, M.K., Huang, J.Z.: An entropy weighting k-means algorithm for subspace clustering of high-dimensinoal sparse data. IEEE Trans. Knowl. Data Eng. **19**, 1–16 (2007)
12. Chen, L., Jiang, Q., Wang, S.: Model-based method for projective clustering. IEEE Trans. Knowl. Data Eng. **24**, 1291–1305 (2012)
13. Herranz, J., Nin, J.: Solé M.: optimal symbol alignment distance: a new distance for sequences of symbols. IEEE Trans. Knowl. Data Eng. **23**, 1541–1554 (2011)
14. Chen, L.: EM-type method for measuring graph dissimilarity. Int. J. Mach. Learn. Cybern. **5**, 625–633 (2014)

15. Wu, T.J., Burke, J.P., Davison, D.B.: A measure of DNA sequence dissimilarity based on Mahalanobis distance between frequencies of words. Biometrics. **53**, 1431–1439 (1997)
16. Wu, T., Fan, Y., Hong, Z., Chen, L.: Subspace clustering on mobile data for discovering circle of friends. In: Zhang, S., Wirsing, M., Zhang, Z. (eds.) KSEM 2015. LNCS (LNAI), vol. 9403, pp. 703–711. Springer, Heidelberg (2015). doi:10.1007/978-3-319-25159-2_64
17. Peng, H., Long, F., Ding, C.: Feature selection based on mutual information: criteria of max-dependency, max-relevance, and min-redundancy. IEEE Trans. Pattern Anal. Mach. Intell. **27**, 1226–1238 (2005)
18. Loiselle, S., Rouat, J., Pressnitzer, D., Thorpe, S.: Exploration of rank order coding with spiking neural networks for speech recognition. Proc. IEEE IJCNN **4**, 2076–2080 (2005)

Cellular Automata Based on Occlusion Relationship for Saliency Detection

Hao Sheng[1,2(✉)], Weichao Feng[1], and Shuo Zhang[1]

[1] State Key Laboratory of Software Development Environment,
School of Computer Science and Engineering, Beihang University, Beijing, China
{shenghao,fengwc,zhangshuo}@buaa.edu.cn
[2] Research Institute of Beihang University in Shenzhen,
Shenzhen 518057, People's Republic of China

Abstract. Different from the traditional images, 4D light field images contain the scene structure information and have been proved that can better obtain the saliency. Instead of estimating depth or using the unique refocusing capability, we proposed to obtain the occlusion relationship from the raw image to calculate saliency detection. The occlusion relationship is calculated using the Epipolar Plane Image (EPI) from the raw light field image which can distinguish a region is most likely a foreground or background. By analyzing the occlusion relationship in the scene, true edges of objects can be selected from the surface textures of objects, which is effective to segment the object completely. Moreover, we assume that objects which are non-occluded are more likely to be the foreground and objects that are occluded by lots of objects are background. Then the occlusion relationship is integrated into a modified saliency detection framework to obtain the saliency regions. Experiment results demonstrate that the occlusion relationship can help to improve the saliency detection accuracy, and the proposed method achieves significantly higher accuracy and robustness in comparison with state-of-the-art light field saliency detection methods.

Keywords: Light field · Saliency detection · Occlusion relationship · Raw image · EPI image

1 Introduction

Saliency detection has been an important research direction in computer vision which is aiming at locating salient pixels or regions in an image. Precise and trustworthy saliency detection plays an important preprocessing role in many computer vision tasks, including image segmentation [1], object detection [2], image thumbnailing [3] and retargeting [4]. Due to the absence of the high-level knowledge, almost all the 2D methods based on traditional images rely on assumptions that appearance contrasts between objects and their surrounding regions are high. This assumptions are called contrast prior and are used in many saliency detection methods [5–10]. Besides contrast prior, boundary prior [8]

© Springer International Publishing AG 2016
F. Lehner and N. Fteimi (Eds.): KSEM 2016, LNAI 9983, pp. 28–39, 2016.
DOI: 10.1007/978-3-319-47650-6_3

has been used in many approaches [6,8,9]. Image boundary regions are mostly identified as backgrounds and boundary priors are further used to enhance the saliency detection. Although many methods suggest that the boundary prior is effective, it has its own drawbacks.

We can't easily assume that the salient object is not in the boundary and the background is in the boundary. In many cases, this assumption is wrong. In this paper, a more reliable prior, which consider the occlusion information, is introduced to obtain the background and foreground measurement.

Saliency detection for light field images is first proposed in [9], which has proved that light field image can obtain more information than the image captured by traditional cameras. Traditional cameras only capture color (RGB) information, whereas the light field camera has capability of refocusing and obtain the depth of the scene. In this paper, we explore the salient object detection problem by using a different input: the raw picture of the Lytro camera. Lytro camera is a light field camera which mounts a lenslet array in front of the sensor. It is the first commercial light field cameras and has $328 * 328$ spatial resolution and $10 * 10$ angular resolution. A light field image can be viewed as an array of images captured by a camera array. Specifically, it captures the scene from different continues angels and we can obtain the parallax image. Like the binocular images and RGBD images, parallax images can calculate the depth information. However, the disparity estimation from parallax images has been a difficult problem for a long time and various advanced methods have been proposed. Moreover, light field images captured by Lytro camera are much more difficult to obtain the accurate depth because the microlens are closely arranged and the parallax images have a very small difference. Although the light field cameras are difficult to obtain the accurate depth, it can benefit saliency detection in numbers of ways, like the capability of post-capture refocusing [11].

In this paper, we obtain a novel information from light field image called occlusion relationship. The occlusion relationship is calculated using the Epipolar Plane Image (EPI) from the raw light field image. Then we try to utilize the occlusion relationship as a new feature to judge the foreground regions and background regions. In other words, we utilize the occlusion relationship to obtain a more reliable background measure instead of the boundary. Moreover, we modified the cellular automata methods [12] and combined the color priors and background priors to calculate the salience object with our occlusion relationship features from light field images and achieve comparable results with the state-of-the-art methods.

2   Related Work

Saliency detection can be categorized into two categories, bottom-up approaches [13–16] and top-down approaches. Bottom-up approaches usually exploit low-level cues such as feature, colors and spatial distances to construct saliency maps. Itti [17] propose a saliency model based on the neural network which integrates three feature. Harel [18] propose a graph-based saliency measure. In [19–21], a saliency models based on Bayesian has been proposed which is exploiting low and mid level cues. Sun et al. [22] improve the Xie's model by introducing boundary and soft segmentation. Perrazzi [23] propose a contrast-based saliency filter and calculate saliency detection by uniqueness and spatial distribution of the patch on the image. Shen and Wu [24] combine lower-level feature and higher-level feature to construct a model based on the low-rank matrix recovery. Cheng et al. [5] utilize a soft abstraction method to remove the image useless details and calculate salient regions. Goferman et al. [25] propose a context-aware saliency algorithm to detect the image regions which represent the scene based on the four principles of the human's visual attention. [26] compute the contrast of the center and surround distribution of features which is based on the Kullback-Leibler divergence for salient object detection. In [27], a graph-based bottom-up method is proposed using manifold ranking. Zhu [28] construct a saliency object detection method based on boundary connectivity. Compared to bottom-up approaches, top-down approaches are task-driven and require supervised learning with manually labeled ground truth. Zhang [29] construct a Bayesian-based top-down methods. Judd [30] learned a top-down saliency model object detectors such as faces, humans, animals, and text. [31] propose a saliency model by jointly learning a conditional Random Field and a dictionary. The bottom-up approaches of saliency detection are data-driven and usually based on low-level visual information and more interesting in detection the local feature rather than a global feature. On the other hands, top and down approaches of saliency detection are interesting in detect object easy to obtain visual attention like human face and body. Consider the time complexity problem, bottom-up approaches usually efficient than top-down approaches. Our approach belongs to bottom-up approaches where we add an occlusion cues. Early saliency detection method exploits center priors which expected to exhibit a high contrast in the center. Koch and Itti [17] are the first to use the center prior to detection saliency. After that, many methods computer the center prior locally or globally. Local methods computer the difference within some nearby pixels or super pixels. Global methods computer the difference into the whole images by computer the power spectrum [32], color histogram [33], and so on. Even if the use of center priors are proved effective, Wei [8] use background priors to calculate saliency detection and have been proved that background priors are also effective methods. Zhu [28] observed the boundary connectivity as a robust background measure and calculate the relationship between foreground and background. In most approaches, they use the color character to judge foreground or background. However, most images in the LF have similar color so that these approaches can easily fail. Our approach calculates each EPI image to obtain the occlusion relationship and combine the color contrast, background prior to obtaining the saliency detection.

3   Occlusion Relationship Analysis Algorithm

In this paper, we present the 4D light field with $L(s, t, u, v)$ representation where (s, t) is the coordinate of the image in different views and (u, v) is the coordinate

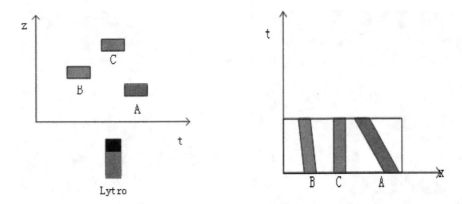

Fig. 1. Principle chart: the slope of the mutual occluded objects is different, the larger slope is occluded by the smaller slope

of the main lens plane. ST plane and UV plane are two parallel planes and $L(s, t, u, v)$ represent the intersection of incoming light from a different view perspective. Using $L(s, t, u, v)$ representation is time-saving in computing raw data and occlusion relation.

3.1   Hierarchical Relationship Model

Epipolar Plane Image (EPI) is constructed by collating multiple images taken from the equidistant location along a line. EPI slice can be taken from parallax image in equidistant location along a line. The point in parallax image maps to a straight line in EPI slice. The slope of EPI line is proportional to its depth in real space as Fig. 1 and adjacent EPI lines with different slope has an occlusion relationship in real space. In Fig. 1, there are three objects A, B, C in front of the Lytro camera, object A is the closest object to the camera which is more possible to be the salience object and C is the farthest object to the camera which has a little possible to be the salience object. Object A and B are closer to the camera than C, so object C is occluded by the object A and B. As a result, object A and B may be more salience because they shelter C. In EPI slice, there are three trails and the largest slope represented to C and A has the smallest slope. In other words, the smaller slope of the trail represents that the point is closer to the camera and the larger slope of the trail represent that the point is far from the camera. So, the object has smaller slope occluded the object which has larger slope.

Based on the observations, we construct a specific feature to present whether a pixel is occluded by other or shelter others. After obtained the EPI image, we first apply Canny edge detection on the EPI slice. Then we curve fit all the point on the line and obtain the slope of each line. Because of the texture, adjacent trails may belong to the same object, so we fuse the adjacent trails which have the same color feature as follows:

(a) RGB image   (b) occlusion relationship

Fig. 2. The scene image and the result of occlusion relationship image. The red point represents the occlusion point and the green represents the occluded point (Color figure online)

$$t_i = ||c_i - c_j||, \tag{1}$$

where $||c_i - c_j||$ is the Euclidean Distance between the superpixel i and j in CIELAB color space. We consider the superpixel i and j belongs to the same object when $t_i < 10$. Then we calculate the adjacent lines in the EPI lines using their slope and the equation as:

$$T_i = s_i - s_j, \tag{2}$$

where s_i is the slope the line i. If the $T_i < -0.15$, we considered that the line i occluded line $i + 1$. using this equation, we can obtain the occlusion relationship. Then collect all the relationship of each EPI slice and obtain the all occlusion relationship of the origin image as Fig. 2.

3.2   Hierarchical Relationship Based on Lytro Camera

Images photoed by Lytro camera are stored in the .lfp file format as their RAW data and we can not handle to extract parallax images directly. So we must convert it into the files we can compute. In [34], we know the .lfp file format contains a lot of information such as focal length and gamma parameter in the file header. The RAW image file is a gray-scale image with BGGR Bayer pattern to store different RGB channels values. We first demosaic the RAW image and then estimate the rotation from the sensor and the micro-lens. After rotated the image, we need to obtain each micro-lens image and calculate the epipolar image. Due to the rotation, the total resolution of the image is less than $3280 * 3280$ and

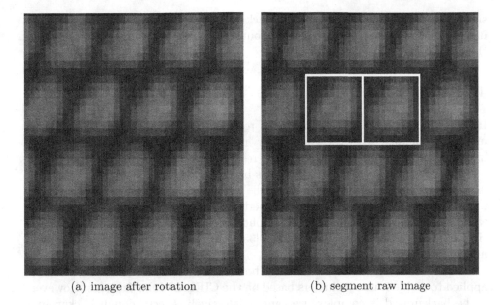

(a) image after rotation   (b) segment raw image

Fig. 3. Obtain the micro-len image

the total number of micro-lens is $328 * 328$, so some micro-lens are $10 * 10$ and some are less than $10 * 10$. We utilize the characteristics that there are a much darker line between the adjacent micro-lens. We obtain each micro-lens image using this method and then obtain the EPI image. The left image in Fig. 3 is the image after rotation and the right image extracts the middle two micro-lens image.

Because of each miro-lens has about $10 * 10$ pixels, we can obtain about $10 * 10$ parallax pictures. When we use the 10 pictures in the x direction, we can obtain the EPI in x direction. Similarly, when we use the 10 pictures in the y direction, we can obtain the EPI in y direction. So we can obtain the occlusion relationship in both x and y direction. Traditional camera array can obtain occlusion relationship in only one direction and we can obtain the occlusion relationship in two directions easily.

3.3   Background Measure

We observe that foreground and background regions in natural images and datasets are quite different. Foreground region tends to be in a small depth and in the front of the image, on the contrary, background regions tend to be in a high depth value and far from the image. It is intuitive cognition to the foreground and background and we considered the occluded region as a background point and occlude another region as a foreground point. To accurately measure background regions and improve computational efficiency, an input image is segmented into N small superpixels by the simple linear iterative clustering (SLIC) algorithm [35]. The number background points and foreground points in each superpixel are accumulated and the background measure is defined as: Where indicate the region belongs to a background and indicate the region belongs to the foreground.

4   Saliency Detection

In this section, we detect the salience region of the image using background measure combined with the RGB color information. We show how to use our information to obtain the final salience result and compare with another methods. The rest salience detection frame is based on the existed cellular automate method worked by [12] and we add the occlusion relationship cues into the method in the first section to obtain a better salience result. To improve computational efficiency, we segmented the input image into N small superpixels using simple linear iterative clustering(SLIC) [35] algorithm. We utilize the superpixel instead of pixels to calculate saliency.

In [12], background seeds are image boundary and the K-means algorithm is applied to divide it into 3 clusters based on the CIELAB color features. However, if the background is complex, we cannot effectively select enough background seeds and cannot obtain a reasonable result for global color distinctionm (GCD) maps. We utilize the background pixels as background seed and then apply the K-means algorithm to obtain the GCD map. GCD map is computed as: where is defined as in [12], after calculated global color distinction maps, we calculated global spatial distance (GSD) matrix.

Based on the GCD map and GSD matrix, the single-layer cellular automate (SCA) method is used to calculate the salient regions. Each cell denotes a superpixel and the saliency value of each superpixel denotes its state and it is discrete. Updating rule is the main point of the cellular automata and is defined as:

$$S^{t+1} = C^* \cdot S^t + (I - C^*) \cdot F^* \cdot S^t, \tag{3}$$

as in [12], the saliency map is computed after N1 times.

5   Experiment

We evaluate the proposed method on Light field (LF) datasets [9], which are captured by Lytro camera and include 100 light field images with the estimated depth map and salient region ground truth. We compare our method with state-of-the-art saliency detection methods, including LF [9], MR [27], GS [8], SF [23], wCtr [28]. LF [9] is based on the light field camera and MR [27] which is used on Manifold Ranking, GS [8] which is used the geodesic saliency using background priors, SF [23] which is used saliency filters to obtain the salient region and wCtr [28] which is used on the background priors are all based on the RGB space. In order to prove the effectiveness of the proposed occlusion relationship, we also evaluate the saliency results using depth cues instead of the occlusion relationship. We use depth map from select the most far from the screen which has more great depth value. We sort all the depth values and select

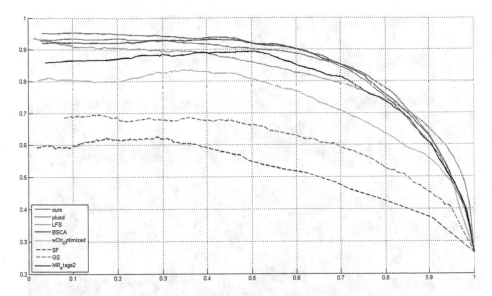

Fig. 4. Precision recall curve comparation with state-of-the-art methods

the maximum 20 % depth pixels as background points and use cellular automate to obtain saliency detection.

We evaluate our saliency result with occlusion relationship cues and depth cues to show the effectiveness of our light field cues. We can observe that the depth cues is useful in the saliency detection because the depth cues help the background priors more precisely. By contrast, the occlusion relationship is able to detect the background more precise than the depth cues. The reason is the occlusion relation can find the relatively background and is more precise than the background in the all image.

The visual examples are showed in Fig. 5, we can see that the occlusion relationship cues are able to improve the cellular automate method and highlight the salience parts.

We follow the canonical precision-recall curve (PRC) to evaluate the accuracy of the detected saliency on databases of different dimension. For details about PRC methods, we refer readers to [17] and the parameters setting in our implement is the same as [36]. Figure 4 shows the result of the PRC comparison architecture. Our method achieves the highest precision in most of the recall range on all datasets. Occlusion relationship and depth cues can improve the salience and our feature achieves a higher precision and recall rate compared with using depth cues. The reason is that the depth cues always located at a corner and our cues are around all the map.

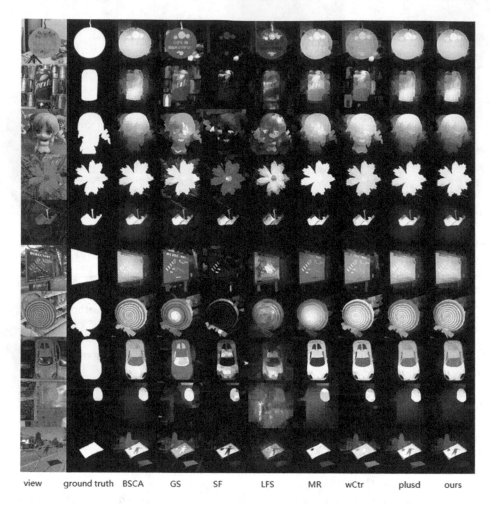

view   ground truth   BSCA   GS   SF   LFS   MR   wCtr   plusd   ours

Fig. 5. Comparison of our saliency maps with state-of-the-art methods

6   Conclusion

We propose a bottom-up method to detect salient regions in light field images through cellular automata method, which incorporates occlusion relationship cues and background and foreground priors. The occlusion relationship is calculated on the raw images and is showed that it is simple and effective. Based on the EPI images, we can obtain the occlusion relationship and use this cues to calculate the background and foreground. The information then extracts the location of the possible background and foreground in the image. We evaluate the proposed algorithm on large datasets and demonstrate promising results with comparisons to several state-of-the-art methods. Furthermore, the proposed algorithm is efficient. Our future work will focus on integration of multiple features with applications to other vision problems in light field.

References

1. Rother, C., Kolmogorov, V., Blake, A.: GrabCut: interactive foreground extraction using iterated graph cuts. ACM Trans. Graph. **23**(3), 307–312 (2004)
2. Borji, A., Sihite, D.N., Itti, L.: Salient object detection: a benchmark. IEEE Trans. Image Process. **24**(12), 414–429 (2015)
3. Sun, J., Ling, H.: Scale and object aware image retargeting for thumbnail browsing. 2011 IEEE International Conference on Computer Vision (ICCV), pp. 1511–1518. IEEE (2011)
4. Ding, Y., Xiao, J., Yu, J.: Importance filtering for image retargeting. In: 2011 IEEE Conference on Computer Vision and Pattern Recognition (CVPR), pp. 89–96. IEEE (2011)
5. Cheng, M.-M., Warrell, J., Lin, W.-Y., Zheng, S., Vineet, V., Crook, N.: Efficient salient region detection with soft image abstraction. In: Proceedings of the IEEE International Conference on Computer Vision, pp. 1529–1536 (2013)
6. Jiang, H., Wang, J., Yuan, Z., Yang, W., Zheng, N., Li, S.: Salient object detection: a discriminative regional feature integration approach. In: Proceedings of the IEEE Conference on Computer Vision and Pattern Recognition, pp. 2083–2090 (2013)
7. Jiang, Z., Davis, L.S.: Submodular salient region detection. In: Proceedings of the IEEE Conference on Computer Vision and Pattern Recognition, pp. 2043–2050 (2013)
8. Wei, Y., Wen, F., Zhu, W., Sun, J.: Geodesic saliency using background priors. In: Fitzgibbon, A., Lazebnik, S., Perona, P., Sato, Y., Schmid, C. (eds.) ECCV 2012. LNCS, vol. 7574, pp. 29–42. Springer, Heidelberg (2012). doi:10.1007/978-3-642-33712-3_3
9. Li, N., Ye, J., Ji, Y., Ling, H., Yu, J.: Saliency detection on light field. In: IEEE Conference on Computer Vision and Pattern Recognition, pp. 2806–2813 (2014)
10. Yan, Q., Xu, L., Shi, J., Jia, J.: Hierarchical saliency detection. In: Proceedings of the IEEE Conference on Computer Vision and Pattern Recognition, pp. 1155–1162 (2013)
11. Ng, R., Levoy, M., Brédif, M., Duval, G., Horowitz, M., Hanrahan, P.: Light field photography with a hand-held plenoptic camera. Computer Science Technical Report CSTR, vol. 2, no. 11, pp. 1–11 (2005)
12. Qin, Y., Lu, H., Xu, Y., Wang, H.: Saliency detection via cellular automata. In: Proceedings of the IEEE Conference on Computer Vision and Pattern Recognition, pp. 110–119 (2015)
13. Achanta, R., Estrada, F., Wils, P., Süsstrunk, S.: Salient region detection and segmentation. In: Gasteratos, A., Vincze, M., Tsotsos, J.K. (eds.) ICVS 2008. LNCS, vol. 5008, pp. 66–75. Springer, Heidelberg (2008). doi:10.1007/978-3-540-79547-6_7
14. Achanta, R., Hemami, S., Estrada, F., Susstrunk, S.: Frequency-tuned salient region detection. In: IEEE Conference on Computer Vision and Pattern Recognition, CVPR 2009, pp. 1597–1604. IEEE (2009)
15. Bruce, N., Tsotsos, J.: Saliency based on information maximization. In: Advances in Neural Information Processing Systems, pp. 155–162 (2005)
16. Zhai, Y., Shah, M.: Visual attention detection in video sequences using spatiotemporal cues. In: Proceedings of the 14th ACM International Conference on Multimedia, pp. 815–824. ACM (2006)
17. Itti, L., Koch, C., Niebur, E., et al.: A model of saliency based visual attention for rapid scene analysis. IEEE Trans. Pattern Anal. Mach. Intell. **20**(11), 1254–1259 (1998)

18. Harel, J., Koch, C., Perona, P.: Graph-based visual saliency. In: Advances in Neural Information Processing Systems, pp. 545–552 (2006)
19. Rahtu, E., Kannala, J., Salo, M., Heikkilä, J.: Segmenting salient objects from images and videos. In: Daniilidis, K., Maragos, P., Paragios, N. (eds.) ECCV 2010. LNCS, vol. 6315, pp. 366–379. Springer, Heidelberg (2010). doi:10.1007/978-3-642-15555-0_27
20. Xie, Y., Lu, H.: Visual saliency detection based on bayesian model. In: 2011 18th IEEE International Conference on Image Processing (ICIP), pp. 645–648. IEEE (2011)
21. Xie, Y., Huchuan, L., Yang, M.-H.: Bayesian saliency via low and mid level cues. IEEE Trans. Image Process. **22**(5), 1689–1698 (2013)
22. Sun, J., Lu, H., Li, S.: Saliency detection based on integration of boundary and soft-segmentation. In: 2012 19th IEEE International Conference on Image Processing, pp. 1085–1088. IEEE (2012)
23. Perazzi, F., Krähenbühl, P., Pritch, Y., Hornung, A.: Saliency filters: contrast based filtering for salient region detection. In: 2012 IEEE Conference on Computer Vision and Pattern Recognition (CVPR), pp. 733–740. IEEE (2012)
24. Shen, X., Wu, Y.: A unified approach to salient object detection via low rank matrix recovery. In: 2012 IEEE Conference on Computer Vision and Pattern Recognition (CVPR), pp. 853–860. IEEE (2012)
25. Goferman, S., Zelnik-Manor, L., Tal, A.: Context-aware saliency detection. IEEE Trans. Pattern Anal. Mach. Intell. **34**(10), 1915–1926 (2012)
26. Klein, D.A., Frintrop, S.: Center-surround divergence of feature statistics for salient object detection. In: 2011 International Conference on Computer Vision, pp. 2214–2219. IEEE (2011)
27. Yang, C., Zhang, L., Lu, H., Ruan, X., Yang, M.-H.: Saliency detection via graph-based manifold ranking. In: Proceedings of the IEEE Conference on Computer Vision and Pattern Recognition, pp. 3166–3173 (2013)
28. Zhang, X., Wang, Z., Yan, C., Zou, H., Peng, Q., Jiang, X., Dan, W.U.: Animal Nutrition Institute, and Sichuan Agricultural University, "Saliency optimization from robust background detection". In: IEEE Conference on Computer Vision and Pattern Recognition, pp. 2814–2821 (2014)
29. Zhang, L., Tong, M.H., Marks, T.K., Shan, H., Cottrell, G.W.: Sun: a bayesian framework for saliency using natural statistics. J. Vis. **8**(7), 32–32 (2008)
30. Judd, T., Ehinger, K., Durand, F., Torralba, A.: Learning to predict where humans look, vol. 30, no. 2, pp. 2106–2113 (2009)
31. Yang, J.: Top-down visual saliency via joint CRF and dictionary learning. In: IEEE Computer Society Conference on Computer Vision and Pattern Recognition Proceedings/CVPR. IEEE Computer Society Conference on Computer Vision and Pattern Recognition, vol. 157, no. 10, pp. 1–1 (2012)
32. Hou, X., Zhang, L.: Saliency detection: a spectral residual approach. In: IEEE Conference on Computer Vision and Pattern Recognition, pp. 1–8 (2007)
33. Cheng, M., Zhang, G., Mitra, N.J., Huang, X., Hu, S.: Global contrast based salient region detection. IEEE Trans. Pattern Anal. Mach. Intell. **37**(3), 409–416 (2015)
34. Cho, D., Lee, M., Kim, S., Tai, Y.W.: Modeling the calibration pipeline of the Lytro camera for high quality light-field image reconstruction. In: Proceedings of the 2013 IEEE International Conference on Computer Vision, pp. 3280–3287 (2013)

35. Radhakrishna, A., Shaji, A., Smith, K., Lucchi, A., Fua, P., Susstrunk, S.: Slic superpixels, Dept. School Comput. Commun. Sci., EPFL, Lausanne, Switzerland. Technical report 149300 (2010)
36. Cheng, M.-M., Mitra, N.J., Huang, X., Torr, P.H.S., Hu, S.-M.: Global contrast based salient region detection. IEEE Trans. Pattern Anal. Mach. Intell. **37**(3), 569–582 (2015)

Text Mining and Lexical Analysis

A Practical Method of Identifying Chinese Metaphor Phrases from Corpus

Jianhui Fu[1,2(✉)], Shi Wang[1], Ya Wang[1], and Cungen Cao[1]

[1] Key Laboratory of Intelligent Information Processing,
Institute of Computer Technology, Chinese Academy of Sciences, Beijing, China
{fujianhui,wangshi,wangya,cgcao}@ict.ac.cn
[2] University of Chinese Academy of Sciences, Beijing, China

Abstract. Research of linguistic metaphors is an important branch of natural language processing. Applications (e.g. semantic understanding, machine translation, and information retrieval) are affected if metaphors can not be identified appropriately. This paper presents a three-phase method for recognizing Chinese metaphor phrases from a large-scale corpus. First, we acquire the context of every candidate phrase. Then hierarchical clustering is used to cluster the phrases based on their contextual information. Finally, heuristic rules are used on the clustering result to determine whether a candidate phrase is a metaphor phrase. Experimental results show the method achieves a satisfactory performance.

Keywords: Metaphor phrase recognition · Chinese metaphor phrase · Phrases clustering

1 Introduction

Metaphors generally exist in human language, and they are an important research topic in disciplines such as rhetorics and cognitive science. Metaphor processing has been taken seriously as an important branch of natural language processing in recent years. Metaphor processing has important applications in machine translation [1], effective computing [2], text entailment [3] and so on.

There are a number of definitions of metaphors [4–6]. There mainly exist two opinions behind the definitions about metaphors. The first idea is to treat a metaphor as a kind of rhetoric, and the other idea is to regard metaphors as a cognitive phenomena. In this paper, like most computer researchers, we analyzes metaphors from the perspective of cognitive linguistics. In cognitive linguistics, the metaphor is defined as the figure of speech without a marking word. For example, "马路杀手" (road killer) refers to the drivers who harm others life without driving skill on the road. The phase doesn't hint at the killer but the drivers who killed persons. "职业杀手" (professional killer) refers to the persons who take a job as a killer. A metaphor contains two parts: target domain (or target for short) and source domain (or source for short). The target is the subject for which some attributes are to be borrowed from the source [7], The source is the object, from which the target borrows some attributes. The word which appears as

F. Lehner and N. Fteimi (Eds.): KSEM 2016, LNAI 9983, pp. 43–54, 2016.
DOI: 10.1007/978-3-319-47650-6_4

the source domain in a metaphor phrase is referred to as the source word, such as "海洋" (ocean), "杀手" (killer), "大军" (army).

Many metaphors tend to appear in the form of phrases in Chinese. When "海洋" (ocean) and "知识" (knowledge) are combined into a phrase as "知识海洋" (knowledge ocean), the phrase contains a metaphorical meaning. "知识海洋" (knowledge ocean) means that the number of knowledge is huge like the volume of the ocean.

We find that the last word of a metaphor phrase is often the source word through an analysis of a large number of Chinese metaphor phrases. The last word of a metaphor phrase is called the tail word for short in our work. For example, "海洋" (ocean) in "知识海洋" (knowledge ocean) is the tail word. Our previous experiments show that more than 97 % tail words of Chinese metaphor phrases are source words. Therefore, we focus on the case that the source word appears at the end of the metaphor phrase in this paper.

This paper presents a method for acquiring metaphor phrases from a Chinese phrase corpus. Taking the existing massive Chinese phrases as input, the goal is to recognize the metaphor phrases. First, we acquire the context of every candidate phrase. Then a clustering method is used to cluster the phases based on their context. Clustering results show that most of the similar phrases are clustered together. Finally, heuristic rules are used to filter out the clustering results to determine whether a phrase is metaphorical. The method presented in this paper is based on Chinese noun phase. Whether the method is applicable to other languages needs further research.

2　Related Work

Birke and Sarkar [8] presented a computational model called Trope Finder in order to classify verbs in sentences as either literal or non-literal. Krishnakumaran and Zhu [9] proposed three algorithms of "A is B", "verb+noun" and "adj+noun" for distinguishing between live and dead metaphors, and conditional probabilities combined with WordNet are adopted in the algorithms [10].

Neuman et al. presented three rule-based algorithms to identify three expression types of metaphors, i.e. "A is B", "verb+noun" and "adj+noun" [11]. Based on negating literalness they defined a set of rules for determining whether a particular expression is literal. If an expression not literal, it is considered to be a metaphor.

Shutova and Korhonen [12] focused on the metaphors of the form "verb+noun" based on semi-supervised learning. The method is based on the principle of clustering by association. Shutova et al. also presented an approach to identifying "verb+noun" metaphors [13]. The method first identifies metaphorical expressions in text and then paraphrases them with their literal paraphrases.

Turney et al. presented an algorithm called Concrete-Abstract for the metaphor classification problem [14]. By considering the abstractness level of words in a given expression, they discovered that the nouns in a wide range of "adj+noun" metaphoric expressions tend to be abstract and the adjectives tend to be concrete. Their method ranks the abstractness level of a given word by comparing it to 20 abstract words and 20 concrete words. The abstract words are used as paradigms of abstractness while the concrete words are used as paradigms of concreteness.

Hovy et al. used SVMs with tree kernels to metaphor classification [15]. The features contain the semantic aspects of each word and different tree representations of each sentence.

Using translation dictionaries, a trained English model was applied to three other languages (Spanish, Russian, and Farsi) with the features of abstractness and image ability, word super senses and unsupervised vector-space word representations [16]. 3,737 annotated sentences extracted from the Wall Street Journal domain are included in the English training dataset.

Shlomo and Last presented a supervised learning approach to identifying three major types of metaphoric expressions without using any knowledge resources or handcrafted rules [17]. The method is called MIL (Metaphor Identification by Learning). A set of statistical features are learned from a corpus of a given domain. They also used an annotated set of sentences, which contain candidate expressions labeled as 'metaphoric' or 'literal' by native English speakers.

Navarrocolorado and Tomás [18] presented a new unsupervised approach to metaphor identification based on LDA. Assuming a correlation between topic models and conceptual domains, the topics of each word are used to identify the semantic inconsistency between a word and its context.

In the field of Chinese metaphor recognition, Wang presented a machine learning model to identify the "noun+noun" type of metaphors based on the CCD dictionary [19]. The model adopts the method of reasoning by similarity calculation. Zhao integrated the synonymy information of Synonymy Thesaurus and the semantic information of HowNet into CRF and ME in order to solve the problem of metaphor recognition [20]. Xu [21] and Wang [22] built a statistical model of ME to identify the "noun+noun" type of metaphors on the basis of artificial annotation corpus, respectively. Xu presents the specific methods of feature extraction and model training. Huang suggested tree pattern matching algorithms based on various dependency patterns of metaphorical sentences acquired by a dependency parser [23].

It is observed that different source words of metaphor phrases have different context. Therefore, it is difficult to build the classification model of "海洋" (ocean) by utilizing the context of "杀手" (killer). Training corpora of classification methods must be large enough so that it can cover all the source words. Meanwhile, it is difficult to construct a large enough corpus because of the dynamic increase of source words.

3 An Empirical Analysis of Chinese Metaphor Phrases

We use the source word "杀手" (killer) as an example to explain the thought of our method. "政府杀手" (governmental killer) is not a metaphor. "政府杀手" (governmental killer) itself is a "killer". The hypernym of "政府杀手" (governmental killer) is "杀手" (killer). A governmental killer means a person or organization hired by a government who carries out the activity of killing some people. However, the "电脑杀手" (computer killer), which is not a "killer", refers to viruses or other harmful things to the computer. Through the analysis of the corpus for "政府杀手" (government killer) and "电脑杀手" (computer killer) we find that "政府杀手" (governmental killer) is always collocated with words such as "谋杀" (murder), "枪杀" (shooting), or "犯人"

(criminals). This kind of collocations reflects the original meaning of "杀手" (killer). We refer to those collocated terms as words originally related to "杀手" (killer). However, the context of a metaphor phrase may not contain any originally related word. An example is shown in Fig. 1.

"这种'电脑杀手'能够在十秒内'杀死'你的台式机，并......"。
Translation:
This "Computer killer" could "kill" your desktop in ten seconds, and...

Fig. 1. An example of the context of "电脑杀手" (computer killer)

In the context of the example, terms related to "computer" frequently appear; that is to say, the source word often co-occurs with the word set related to its original meaning when it does not show the metaphorical meaning in Chinese phrases and the context has some certain commonality. Therefore, we try to identify the metaphor phrases or non-metaphor phrase by using the commonality of the context.

First, we retrieve every phrase with our search engine and collect its context information. Second, we cluster the phrases with the context information, and try to identify the metaphor phrases and none-metaphor phrases, which relied on a few heuristic rules to be presented shortly.

10 million phrases were randomly extracted from our Chinese phrases corpus, and examples are "知识海洋" (knowledge ocean), "政府杀手" (government killer) and "电脑杀手" (computer killer). After word segmentation, we manually picked out 1021 head words which could be used as source words, some of which are shown in Table 1.

Table 1. Tail words

安全带 (safety belt)	长廊 (long corridor)	东西 (thing)	共同体 (community)	画廊 (art gallery)
顾客 (customer)	平台 (platform)	冬瓜 (wax gourd)	平原 (plain)	画面 (frame)
城楼 (gate tower)	冠军 (champion)	旅馆 (hotel)	键盘 (keyboard)	开水 (boiled water)
巴士 (bus)	都市 (metropolis)	豆腐 (bean curd)	古墓 (ancient tomb)	话剧 (modern drama)
海洋 (ocean)	沙漠 (desert)	大军 (army)	杀手 (killer)	天空 (sky)
长度 (length)	花瓶 (vase)	间谍 (spy)	天堂 (heaven)	宫殿 (palace)

4 Clustering of Chinese Metaphor Phrases

4.1 Search of Phrases

We generally search a vocabulary K by Google. Meanwhile, in order to retrieve more original meaning information related to source word in the context as much as possible, we not only directly search for the source word itself, but also its synonyms. For example, when we retrieve "撒哈拉大沙漠" (Sahara), the retrieval terms should be: "撒哈拉大沙漠"(Sahara) + "沙漠"(desert) + "戈壁" (gobi).

We extract the webpage fragments of the top 100 search results by our search engine. Every fragment contains a or two sentences about the item. Then we combine the 100 fragments into a document. At the same time, a stop words table is used to filter out some words, such as "网页快照" (snapshot), "图片" (images), "网页" (web) and so on. The final document is called the original meaning document of vocabulary K. We will use this document to replace vocabulary K and acquire the similarity of these documents using clustering method.

In addition, we construct a query with the source word itself and its synonyms, and its retrieved original meaning document is also used to cluster. For example, the original meaning document acquired by retrieving "desert" + "gobi" also participates in the cluster.

4.2 Similarity Computation

We already prepare the original meaning document and the next problem is how to calculate the similarity which is used to cluster the source word. We use the $sim(D_1,D_2)$ to calculate the similarity. $sim(D_1,D_2)$ is the cosine distance between two documents which is often used in text computation tasks. The definition of formula is as follows.

$$sim(D_1, D_2) = COS(D_1, D_2) = \frac{d_1 \cdot d_2}{||d_1|| \cdot ||d_2||} = \frac{\sum_{i=0}^{k} (w(t_i) \cdot w(t_j))}{\sqrt{\sum_{i=0}^{n} w(t_i)^2} \cdot \sqrt{\sum_{i=0}^{m} w(t_j)^2}} \quad (1)$$

$$w(t) = tf_t \cdot idf_t = tf_t \cdot \log(N/n_t) \quad (2)$$

$||d_1||$ denotes the length of documents D_1. $||d_2||$ denotes the length of documents D_2. $w(t)$ calculated by the TF-IDF formula denotes the feature weights of the vector. tf_t denotes the frequency of t. N denotes the number of the whole documents. n_t denotes the number of the documents which contain word t.

This measure is not empirical in evaluating the similarity. So we prepared another two scores as follows, and compared their performance with $sim(D_1,D_2)$. Experimental results in Sect. 6 demonstrate that $sim(D_1,D_2)$ outperforms others evidently.

$$sim_1(D_1, D_2) = D_1 \cap D_2 / D_1 \cup D_2 \quad (3)$$

$$sim_2(D_1, D_2) = D(D_1 || D_2) + D(D_2 || D_1)$$
$$= \sum_{w_i} p(w_i | D_1) \times \log p(w_i | D_1)/p(w_i | D_2) + p(w_i | D_2) \times \log p(w_i | D_2)/p(w_i | D_1) \quad (4)$$

$sim_1(D_1,D_2)$ is the Jaccard distance. $|D_1 \cap D_2|$ denotes the number of the words both in D_1 and D_2 without repetition. $|D_1 \cup D_2|$ denotes the number of the words in D_1 or D_2

without repetition. In $sim_2(D_1,D_2)$, $D(D_1||D_2)$ is the Kullback-Leibler divergence of D_1 and D_2.

4.3   Clustering Algorithm and Evaluation Criteria

We use the bottom-up hierarchical clustering in order to respond to the problem that we do not know the proper number of the clusters. When the distance between the clusters is greater than or equal to a threshold value, the clustering is terminated.

We use the purity of the clustering result in [24] to evaluate the effect of clustering. The definition is as follows. Given a clustering result C and a class-label system A,

$$p_{ca} = f(c,a)/f(c,*) \tag{5}$$

The symbol a is a class label in A. The symbol c is a cluster in C. $f(c,a)$ is the number of the elements of c which is labeled a. The symbol '*' is the wildcard. Then the entropy of c is computed as follows.

$$E_c = -\sum_{a\in A} p_{ca} \times \log(p_{ca}) \tag{6}$$

The formula to calculate the final purity is as follows.

$$E(C) = \sum_{c\in C} (|c| \times E_c)/|C| \tag{7}$$

If the number of elements in a cluster is equal to 1, it is meaningless; it will not be used to calculate the purity. This work only considers the clusters in which the number of elements is greater than 5 (called big clusters). Meanwhile, we give another evaluation indicator in order to get better evaluation of the clustering result, called the big-cluster probability, as shown below.

$$BCP(C) = |BC|/|C| \tag{8}$$

where $|BC|$ denotes the number of big clusters in a clustering result, and $|C|$ denotes the number of all clusters.

5   Heuristic Rules for Metaphor Phrase Recognition

After the clustering, we have to automatically label all clusters as "metaphor" or "non-metaphor". Comprehensive analysis of the clustering results indicated that a cluster may contain non-metaphor phrases, and correct metaphor phrases often get together. We introduce several heuristic rules to further recognize phrases in clusters and to label clusters as "metaphor" or "non-metaphor".

First, we introduce attribute rules. Attribute rules take advantage of the attributes of the source word to recognize a metaphor phrase. We find that a Chinese phrase is usually the hyponym of its tail word, and the hyponym inherits the attributes from the hypernym. A metaphor phrase does not have all the attributes of the source word because the metaphor phrase is not the hyponym of its source word. The metaphor phrase and its source word generally do not have common attributes.

We used our search engine to extract attributes of a phrase. We have accumulated a lot of attribute words about the source word [25]. Table 2 shows some attribute words of the five experimental source words. Given a phrase P and an attribute word B of the source word of P, we construct two queries "P的b" (b of P) and "P的*b" (*b of P), and then put the queries to our search engine. If the average number of the returned results is greater than a threshold T^*, we think P has the attribute b. Experiments show that the best value of T^* is 30. The average number of the returned results is computed by the following formula.

$$T(P,B) = \frac{\sum\limits_{b_i \in B} fre(P,b_i)}{|B|} \qquad (9)$$

where B is the attribute word set of the source word of P, $fre(P,b_i)$ denotes the number of the returned search results using the queries constructed with P and b_i.

Table 2. Some attribute words of the five experimental source words

Source word	Attributes of the objects denoted by the source word
海洋 (ocean)	面积 (area), 水深 (depth of water), 深度 (depth), 宽度 (width), 水温 (water temperature), 地理位置 (geographical location)
大军 (army)	战场 (battlefield), 敌人 (enemy), 战斗力 (fighting capacity), 伤亡人数 (number of casualties), 任务 (task), 使命 (mission)
杀手 (killer)	动机 (motivation), 武器 (lethal weapon), 任务 (task), 长相 (appearance), 身高 (stature), 特征 (characteristic)
沙漠 (desert)	面积 (area), 气候 (climate), 降雨量 (rainfall capacity), 分布 (distribution), 主要植被 (main vegetation types), 生态特征 (ecological characteristics), 植被 (vegetation), 生态环境 (ecological environment),
天空 (sky)	云层 (clouds), 监测 (monitoring), 特征 (feature), 预报 (forecast)

The attribute rule says: Given a P, and B is the attribute word set of the source word of P, if $T(P, B) < T^*$, P is a metaphor.

For example, as shown in Table 2, the attribute word set of "沙漠"(desert) includes "面积" (area), "温度" (temperature), and "降雨量" (rainfall capacity). "撒哈拉大沙漠" (Sahara) is the hyponym of "沙漠"(desert). We constructed the queries "撒哈拉大沙漠的面积" (area of Sahara) and "撒哈拉大沙漠的温度" (temperature of Sahara) based on the attribute words "面积" (area) and "温度" (temperature) of the source word "沙漠"(desert). Then we put the queries to our search engine: "撒哈拉大沙漠的面积" (area of Sahara) returned 1030 results. "撒哈拉大沙漠的温度" (temperature of Sahara) returned 647 results. "爱情沙漠" (love desert) is a metaphor phrase, whose source word is "沙漠" (desert). "爱情沙漠" (love desert) is not the hyponym of "沙漠"

(desert). It does not have the attributes "面积" (area) and "温度" (temperature). Queries of "爱情沙漠的面积" (area of love desert) nor "爱情沙漠的温度" (temperature of love desert) returned no results. This means "爱情沙漠" (love desert) is a metaphor phrase.

Now, we introduce other useful heuristic rules for metaphor recognition and cluster labeling. The heuristic rules are acquired by observation of a large number of experimental results. Through the results of 30403 Chinese phrases' word segmentation and POS, we present the POS rules and Discontinuous-character rules. Through the observation of the clustering result in Sect. 4, we introduce the cluster labeling rules.

First of all, we present the microscopic-structure of phrases. We define a phrase as **phrase** = $w_1 w_2 ... w_n$, where n denotes the number of words segmented by our segmentation tool. We introduce more heuristic rules as follows.

(1) **POS rules.** If the POS of w_i is not a noun, verb, adjective or discriminative, the phrase is not a metaphor.
(2) **Discontinuous-character rules.** If there are consecutive discontinuous characters in the phrase, then the phrase is not a metaphor.
(3) **Cluster labeling rules.**
    a. If 50 % or 8 of candidate phrases in a cluster are metaphorical, all the phrases in the cluster are metaphorical; that is, the cluster is labeled "metaphor". The others are labeled "non-metaphor".
    b. If a cluster contains the source word itself, all the phrases in the cluster are not metaphorical; that is, the cluster is labeled "non-metaphor".

6 Experimental Results

In our experiments, we chose "海洋" (ocean), "杀手" (killer), "大军" (army), "沙漠" (desert) and "天空" (sky) to evaluate our method. We randomly selected 315 phrases which ends with one of the five source words. The set of phrases contains 160 metaphor phrases and 155 non-metaphor phrases, as shown in Table 3.

Table 3. The distribution of metaphor phrases and non-metaphor phrases

Source word	Number of metaphor phrases	Number of non-metaphor phrases
海洋 (ocean)	35	24
大军 (army)	40	39
杀手 (killer)	37	23
沙漠 (desert)	33	29
天空 (sky)	15	40

Tables 4, 5, 6, 7 and 8 present the evaluation results of the five clustering results. Experimental results demonstrate that $sim(D_1, D_2)$ is much better than other scores.

Table 4. The evaluation of the clustering result of "大军"

	|C|	E(C)	BCP(C)
$sim(D_1,D_2)$	12	0.15	67.7 %
$sim_1(D_1,D_2)$	14	1.56	42.5 %
$sim_2(D_1,D_2)$	17	−1.35	40.6 %

Table 5. The evaluation of the clustering result of "海洋"

	|C|	E(C)	BCP(C)
$sim(D_1,D_2)$	13	1.12	74.5 %
$sim_1(D_1,D_2)$	15	1.57	67.5 %
$sim_2(D_1,D_2)$	15	1.43	69.1 %

Table 6. The evaluation of the clustering result of "天空"

	|C|	E(C)	BCP(C)
$sim(D_1,D_2)$	35	1.35	34.8 %
$sim_1(D_1,D_2)$	40	1.37	30.4 %
$sim_2(D_1,D_2)$	37	1.52	32.3 %

Table 7. The evaluation of the clustering result of "沙漠"

	|C|	E(C)	BCP(C)
$sim(D_1,D_2)$	17	0.54	68.3 %
$sim_1(D_1,D_2)$	19	1.21	60.3 %
$sim_2(D_1,D_2)$	18	1.19	65.8 %

Table 8. The evaluation of the clustering result of "杀手"

	|C|	E(C)	BCP(C)
$sim(D_1,D_2)$	23	0.18	29.1 %
$sim_1(D_1,D_2)$	25	0.54	20.2 %
$sim_2(D_1,D_2)$	27	1.13	25.7 %

Figure 2 shows the 12 clusters of the clustering result of "大军" (army). The clustering result indicates that the metaphor phrases and non-metaphor phrases can be well separated.

In cluster 1 which is labeled as "non-metaphor" using the heuristic rules, most phrases are non-metaphorical phrases, and two exceptions are "记者大军" (army of journalists) and "皇马大军"(army of Real Madrid Club de Fútbol). In cluster 2 which is labeled "metaphor", all phrases are metaphor ones, which are commonly used in China.

It is interesting to note that one-phrase clusters tends to be metaphorical, and the exceptions are "项羽大军" (The Army of Xiang Yu) and "古罗马大军" (The Army of Ancient Rome).

The following performance measures are calculated: precision, recall, and F-measure. Table 9 shows the performance of our method. "大军" (army) and "沙漠" (desert) get the best two F-measure. All the source words have higher precision. But they have relatively low recall. The reason is that, when we put the method into practice, we only consider the clusters whose number of elements is greater than 5, and discard smaller clusters which are generally difficult to be labeled with the rules. This method is based on the context of the source word. If there are more context, the recall will be higher. Nevertheless, the results demonstrate the effectiveness of our method.

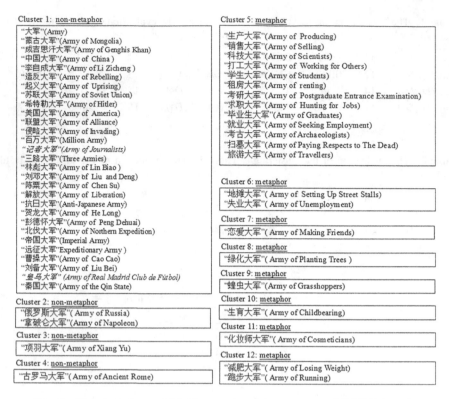

Fig. 2. The clustering and labeling results of the source word "大军" (army)

Table 9. Performance of metaphor phrase recognition

Source word	Precision	Recall	F-measure
海洋 (ocean)	87.1 %	75.6 %	80.9 %
天空 (sky)	76.3 %	70.3 %	73.2 %
沙漠 (desert)	92.4 %	83.1 %	87.5 %
大军 (army)	91.8 %	81.1 %	86.1 %
杀手 (killer)	83.3 %	70.1 %	76.1 %

7   Conclusion

We observed that a Chinese phrase is usually the hyponym of its tail word and the phrases with the same tail word have similar context. With the features we have observed, we presented a metaphor phrase recognition method based on phrase clustering and heuristic rules. A hierarchical clustering method was used to cluster the phases based on their context, and heuristic rules were used to label the clustering results to determine whether a phrase is metaphorical. The results showed our method is more effective and satisfactory.

There exist some problems in our work. First, we need an automatic source word extraction method, and the number of attribute words is still small. Metaphor phrases are easier to be distinguished from the phrases with more attribute words. Also, there are some names of people and brands in the phrase library which have a negative influence on the metaphor recognition process. All these problems will be solved in the future.

Acknowledgments. This work is supported by the National Science Foundation of China (under grant Nos. 91224006 and 61173063) and the Ministry of Science and Technology (under grant No. 201303107).

References

1. Yun, J.X.: A semantic approach to English-Chinese machine translation, 基于语义语言的英汉机器翻译研究. University of Dalin Science and Technology, MS thesis (2011)
2. Huang, X., Yang, Y., Zhou, C.-L.: Emotional metaphors for emotion recognition in chinese text. In: Tao, J., Tan, T., Picard, R.W. (eds.) ACII 2005. LNCS, vol. 3784, pp. 319–325. Springer, Heidelberg (2005)
3. Agerri, R: Metaphor in textual entailment. In: International Conference on Computational Linguistics, Posters Proceedings, Manchester, UK, pp. 18–22 (2008)
4. Martinich, A.P.: The Philosophy of Language. Oxford University Press, Oxford (1998)
5. Lakoff, G., Johnson, M.: Metaphors we live by. J. Aesthetics Art Criticism, 75–96 (1980)
6. Shu, D.F.: Studies in Metaphor, 隐喻学研究. Shanghai Foreign Language Education Press, Shanghai (2000)
7. Wang, Z.M.: Research on Chinese Noun Phrase Metaphor Recognition, 汉语名词短语隐喻识别研究. Beijing Language and Culture University Press, Beijing (2010)
8. Birke, J., Sarkaar, A.: A clustering approach for the nearly unsupervised recognition of nonliteral language. In: 11th Conference of the European Chapter of the Association for Computational Linguistic, Trento, pp. 329–336 (2006)
9. Krishnakumaran, S., Zhu, X.: Hunting elusive metaphors using lexical resources. In: Proceedings of the Workshop on Computational Approaches to Figurative Language, Rochester, NY, pp. 13–20 (2007)
10. Leacock, C., Chodorow, M.: Wordnet an Electronic Lexical Database. MIT Press, Boston (1998)
11. Neuman, Y., Assaf, D., Cohen, Y., et al.: Metaphor identification in large texts corpora. Plos One **8**(4), e62343 (2013)
12. Shutova, E., Korhonen, A.: Metaphor identification using verb and noun clustering. In: Proceedings of the 23rd International Conference on Computational Linguistics, Beijing, pp. 1002–1010 (2010)
13. Shutova, E., Teufel, S., Korhonen, A.: Statistical metaphor processing. Comput. Linguist. **39**(2), 301–353 (1974)
14. Turney, P.D., Neuman, Y., Assaf, D., Cohen, Y.: Literal and metaphorical sense identification through concrete and abstract context. In: EMNLP 2011 Proceedings of the Conference on Empirical Methods in Natural Language Processing, Edinburgh, Scotland, UK, pp. 680–690 (2011)

15. Hovy, D., Srivastava, S., Jauhar, S.K., et al.: Identifying metaphorical word use with tree kernels. In: Proceedings of the First Workshop on Metaphor in NLP, Atlanta, Georgia, pp. 52–57 (2013)
16. Tsvetkov, Y., Boytsov, L., Gershman, A., et al.: Metaphor detection with cross-lingual model transfer. In: Proceedings of the 52nd Annual Meeting of the Association for Computational Linguistics, Baltimore, Maryland, USA, pp. 248–258 (2014)
17. Ben, Y., Last, M.: MIL: automatic metaphor identification by statistical learning. In: The Workshop on Interactions Between Data Mining and Natural Language Processing, pp. 19–29, Nancy, France (2015)
18. Navarrocolorado, B., Tomás, D.: A fully unsupervised topic modeling approach to metaphor identification. In: XXXI Congreso de la Sociedad Española para el Procesamiento del Lenguaje Natural, Alicante (2015)
19. Wang, Z.M.: A study on metaphorical similarity and metaphorical inference identification, 名词隐喻相似度及推理识别研究. J. Chin. Inf. Process. 22(3), 37–43 (2008)
20. Zhao, H., Qu, W., Zhang, F., Zhou, J.: Chinese verb metaphor recognition based on machine learning and semantic knowledge, 基于机器学习与语义知识的动词隐喻识别. Nanjing Normal Univ. (Eng. Technol.) 11(3), 59–64 (2011)
21. Xu, Y.: Recognition of the Chinese metaphor phenomena based on the maximum entropy model, 基于最大熵模型的汉语隐喻现象识别. Comput. Eng. Sci. 29(4), 95–103 (2007)
22. Wang, Z., Wang, H., Yu, S.: Chinese nominal metaphor recognition based on machine learning, 基于机器学习方法的汉语名词隐喻识别. High Technol. Lett. 17(6), 575–580 (2007)
23. Huang, X.X.: Research on some key issues of metaphor computation, 隐喻机器理解的若干关键问题研究. Zhejiang University, MS thesis (2009)
24. Steinbach, M., Karypis, G., Kumar, V.: A comparison of document clustering techniques (2000)
25. Wang, P., Cao, C., Wang, S.: An interative approach to automatic attribute acquisition, 一种迭代式的概念属性名称自动获取方法. J. Chin. Inf. Process. 28(4), 58–67 (2014)

Knowledge Extraction from Chinese Records of Cyber Attacks Based on a Semantic Grammar

Fang Fang[1,2], Ya Wang[1], Luchen Zhang[3(✉)], and Cungen Cao[1]

[1] Key Laboratory of Intelligent Information Processing,
Institute of Computer Technology, Chinese Academy of Sciences, Beijing, China
fangfang900703@163.com
[2] University of Chinese Academy of Sciences, Beijing, China
[3] National Computer Network Emergency Response Technical
Team/Coordination Center of China, Beijing, China
zhangluchen2007@163.com

Abstract. Knowledge acquisition from text is an important research of artificial intelligence. In this paper, we present a method of acquiring knowledge from Chinese records of events of cyber attacks based on a semantic grammar. In order to parse the sentences in the records, the method first identifies Chinese noun phrases in the records, and then use the semantic grammar of the cyber-attack domain to parse the records. Finally, knowledge is extracted from the parsing trees. Experimental results show that our method for noun phase identification has a good performance, and the precision of knowledge acquisition reaches a high level of 90 %.

Keywords: Cyber attack records · Knowledge extraction · Noun phrase identification · Semantic grammar

1 Introduction

Knowledge acquisition from text (KAT) is an important research of artificial intelligence. It automatically or semi-automatically transforms natural language text into computer-understandable knowledge.

As we know, cyber attacks occur every second all over the world, and severe attacks are formally recorded in text for further analysis. In this paper, we present a method for extracting knowledge from text records of cyber attacks – a special of knowledge acquisition from text.

At present, methods for acquiring knowledge from text are mainly either based on machine learning methods or lexico-syntactic patterns. In the line of machine learning methods, there are a great number of famous systems [6], e.g. TextRunner [4], NELL [5] and OMCS [7]. But all of them deal with English text. And our work is similar to them, but deals with Chinese text.

Another important point that we want to highlight is that we are pursuing a method for high-precision knowledge extraction, and we also rely on lexico-syntactic patterns as well.

© Springer International Publishing AG 2016
F. Lehner and N. Fteimi (Eds.): KSEM 2016, LNAI 9983, pp. 55–68, 2016.
DOI: 10.1007/978-3-319-47650-6_5

However, in Chinese text processing, methods based on lexico-syntactic patterns are have two limitations. First, incorrect identification of noun phrases (NP) causes pattern matching to output wrong structured information [11–13]. This is a common problem! Second, lexico-syntactic patterns have little semantic information, and therefore often cause the pattern matcher to output mistaken structures (e.g. trees or key-value pairs).

The main aim of this work is to achieve a knowledge extraction method with a high precision. Therefore, our method mainly relies on a manually designed semantic grammar, in addition to heuristic and statistical analysis. The method is in the same spirit of lexico-syntactic patterns, but those patterns are specially designed, called semantic patterns [1].

A semantic pattern is different from a syntactic one in several important aspects. First, in a semantic pattern, each unit is associated with a thematic role and a type as well, which is similar to FrameNet [2, 3, 8–10]. This will help the parser determine whether the pattern can match an input sentence. Second, the semantic category of patterns has a few knowledge templates or predicates which will help the parser output the knowledge in a formal way.

We have developed a preliminary knowledge extractor called NkiExtractor to parse the input text using the semantic patterns, and to generate the knowledge from the parsing trees [1].

The outline of the paper is as follows. Section 2 describes a hybrid method of Chinese noun phrase identification (which is an important part of text parsing). Section 3 presents the process of extracting knowledge from records of cyber attacks. Finally, Sect. 4 concludes our work.

2 Chinese Noun Phrase Identification Before Knowledge Extraction

As we all know, Chinese is very complex. The sentence should be segmented into Word/POS pairs at first. But the inaccuracy of segmentation generally makes the sentences difficult to be parsed correctly. Therefore, noun phrase (NP) identification is an inevitable process before parsing, and NP identification has a critical influence on the speed and correctness of parsing. So, in this section, we solely focus on the problem of NP identification.

In Chinese noun phrase identification, the main difficulties include:

1. *Word Segmentation Problems.*
   Word segmentation tools often misbehave in some special situations, and this causes a lot of problems in our practice, as will be presented shortly in Sect. 3.1.
2. *Boundary Problem.*
   Some Chinese noun phrases contain collocation information, and these collocation can be serve for extracting boundaries. For example, in the phrase "在/p 后期/n 维护/v 方面/n (in the later maintenance)", "在/p" is the left boundary of the NP, and "方面/n" is the right boundary of the phrase.

3. *Nominality Problems of VN and NV.*
Some Chinese noun phrases are in the structure of N+V or V+N. It is often hard to determine whether they are NPs or whether the verb V is the main verb. For example, "还行/v 内存/n" could mean RAM, *or* literally "executes memory" in which *executes* is a verb. This problem represents one of major ambiguities in Chinese text.

2.1 Postprocessing of Chinese Word Segmentation

Chinese word segmentation is itself a research direction, and there are a number of tools on the Web [14, 15]. But these tools still have quite a number of flaws or limitations which sometime severely affect identification of Chinese noun phrases. In this section, introduce three common postprocessing tasks.

Firstly, Chinese has a special kind of words called idioms. These words' POS tag is "/i". But unfortunately, Chinese idioms are sometime mistakenly segmented. For example, the Chinese idiom "身临其境 (you didn't be one place but you seem to have experienced the same experience by yourself)" is often mistakenly segmented as "身/Ng 临/v 其/r 境/Ng". We solve this task with an idiom dictionary of 50286 idioms.

Secondly, brackets may often cause problems in Chinese processing. Typical brackets of Chinese are " ", 《 》, < >, { } and []. However, the words in the brackets are usually segmented. For example, " 《玩偶之家》 " (Doll's House) is the name of a book. It often mistakenly segmented as " 《/w 玩偶/n 之/u 家/n 》 /w". Our postprocessing task is to merge the whole words.

Finally, Chinese sentences may contain long English phrases, ASCII strings, numbers and so on, and these are wrongly segmented during word segmentation. For example, the long ASCII string "TCLL40F3320-3D (a machine model)" is wrongly segmented as "TCLL/nx 40F/n 3320-3D/nz". For such a problem, we merge the segmented string together to its original.

2.2 Identifying Boundaries of Noun Phrases

As we have mentioned above, an effective way to identify a noun phrase is to identify its left and right boundaries; that is, if we can determine the boundaries, the noun phrase can be easily determined. In Chinese processing, there are several work about NP identification, but the precision is rather low [16]. In this paper, we introduce a hybrid method which handles the NP identification from three different perspective.

2.2.1 Collocation-Based Boundary Identification

In Chinese, there are a lot of fixed collocations, which are good boundaries for noun phrases. Because we use the FSTD framework (i.e. Framework of Semantic Taxonomy and Description, which will be introduced in the next chapter) to parse the sentences. We extract a lot of fixed collocations from the grammatical rules of FSTD, as partly shown in Table 1.

Table 1. Part of collocations extracted from FSTD

Collocation	Means in English	Collocation	Means in English
按…方法	According to … method	在…情况下	in…situation(case)
作为…回报	in return for…	用…方式	Use…(method)

2.2.2 Suffix/Prefix-Word Based Boundary Identification

Some noun phrases contain headwords or suffix words. A headword is a word that presents a kind of things. Using these words, some NPs can be identified precisely.

In our work, we use more than 5 million Wiki entries of Baidu Baike [17] and a dictionary of synonyms developed by Harbin Institute of Technology [18]. Every Baidu Baike entry is viewed as a phrase. The dictionary of synonyms is used to extend the extracted headwords.

The procedure is shown in Table 2 as follows.

Table 2. The algorithm of extracting prefix/suffix

Algorithm: Extracting Prefix/Suffix
Step 1: From the 5 million segmented Baidu Baike entries, we count the frequencies of prefix and suffix words. Then we set the words whose frequency is more than 500 as candidate words.
Step 2: From the candidate words, we find the synonym for each candidate word based on the synonym dictionary. Then put synonyms into the candidate words whose similarity is more than 0.8.
Step 3: Output all the candidate words.

In step 2, we calculate the similarity by using the coding schema of the synonym dictionary. A code is made up of five levels and eight bits which contains number or letter. The coding schema is shown in Table 3.

Table 3. The coding schema of the synonym dictionary

Coding bit	1	2	3	4	5	6	7	8
Coding examples	D	a	1	5	B	0	2	=\# \@
Level	Level 1	Level 2	Level 3	Level 4	Level 5			

In the table, "=" denotes "equal"; "#" denotes "unequal", but the relevant; "@" denotes "independence", meaning that there are no synonyms and relevant words in the dictionary.

The similarity between W_1 and W_2 is defined as follows.

$$sim(W_1, W_2) = \frac{differLevel(W_1, W_2) - 1}{5} \times Weight(W_1, W_2)$$

In the formula, $differLevel(W_1, W_2)$ is the level at which W_1 and W_2 begin to have different coding bits. "5" is the maximum of level. We denote the longest common prefix of the codes of W_1 and W_2 as $LCPC(W_1, W_2)$.

$$Weight(W_1, W_2) = \begin{cases} 0, & if\ W_1\ or\ W_2\ is\ end\ with\ '@' \\ w_{differLevel(W_1, W_2)} \times \frac{n-k+1}{n}, & other \end{cases}$$

In the formula, n denotes the number of codes which are prefixed with LCPC (W_1, W_2). k denotes the difference between the code bit of W_1 at the $differLevel(W_1, W_2)$ and the code bit of B at $differLevel(W_1, W_2)$. $w_{differLevel(W_1, W_2)}$ denotes the weight of a level. $w_{differLevel(A,B)}$ is calculated mainly rely on manual and experience.

Part of prefix and suffix words are shown in Table 4.

Table 4. The portion of prefix/suffix

Prefix	Means in English	Suffix	Means in English
中国	China	公司	Company
北京	Beijing	学校	School

2.2.3 Rule-Based Noun Phrase Identification

In addition to the two methods above, we rely on the semantic patterns in our FSTD to help identify noun phrases. In a semantic pattern, there are some collocations.

Noun phrases identified by both boundaries are precise, as our observation has shown. But one single boundary is often uncertain. So we determine the effectiveness of a single boundary using Baidu Baike entries. We define the evaluation formula as

$$S = \frac{frquent(B)}{N_m}$$

N_m denotes the most frequent word that Baidu Baike entries contain, $frquent(B)$ denotes the frequence of boundary B.

Table 5 lists a few more boundaries in our FSTD.

Table 5. Four cases of boundaries in FSTD

Boundaries in semantic patterns	Literal translation in English
<把将介词词类 > N<叫作中心词类>	<Name>N<As>
RN < 被介词词类 > <称为中心词类>	RN<Be><Called>

2.3 Handling Problems of N+V and V+N Phrases

Although the methods mentioned above can effectively identify noun phrases, the most cases of noun phrases are phrases of the form N+V or V+N. In Chinese, there are some researches about the N+V and V+N structures. In their research, they listed some word pairs that have the structures. Our work is extending word pairs. We extract word pairs of the N+V structure and V+N structure from Baidu Baike entries to a candidate set.

For the whole process, the step of obtaining the candidate set is relatively straightforward: it just extracts the data which have satisfy the two structures. And the verification process includes two methods. One is based on the synonym dictionary [18], and the other is based on Baidu Baike entries.

The verification process based on Baidu Baike entries uses the Witten-Bell back off method [19]. We define the word pair is "$w_1 w_2$", then

$$P(w_2|w_1) = \frac{c(w_1 w_2)}{c(w_1) + n(w_1)} + \frac{n(w_1)}{c(w_1) + n(w_1)} \frac{c(w_2)}{N}$$

$$Pb(w_2|w_1) = \frac{n(w_1)}{c(w_1) + n(w_1)} \frac{c(w_2)}{N}$$

$c(w_1 w_2)$ denotes the total times of "$w_1 w_2$" in the corpus (Baidu Baike entries), $c(w_1)$ denotes the times of w_1 in the corpus, $c(w_2)$ denotes the times of w_2 in the corpus, N is the number of entries, and $n(w_1)$ denotes the number of the different words after w_1.

If "$w_1 w_2$" satisfies the condition

$$\log(P(w_2|w_1)) - \log(Pb(w_2|w_1)) < K$$

$w_1 w_2$ will be deleted.

Through all of the above process, we finally retain the result of V+N structure and N+V structure as shown in Table 6.

Table 6. The results of the V+N structure and N+V structure

	V+N structure	N+V structure
Number	78264	53507

2.4 Experiments

Because some of the intermediate results have been mentioned above, here we directly give the results of the recognition in sentences.

We randomly selected 150 sentences from the corpus, including news, film and finance. At first, we evaluate the results of noun phrase identification using accuracy, recall and F_1 values. Our results are all calculated manually. The result is shown in the Table 7.

Table 7. The result of noun phrase identification

Type	Accuracy	Recall	F₁ values
Value	89.72 %	87.97 %	88.84 %

Note that we pay more attention to the results of the sentences that have multiple verbs. So we selected 80 matching sentences. The result is shown in Table 8.

Table 8. The result of noun phrase identification

Type	Error number	Error rate	Accuracy
Value	13	16.25 %	83.75 %

It should be noted here that some of the results are also correct even though the sentence contains multiple words. Because in the Chinese exist the sentence contains multiple (pivotal sentences and serial-verb sentence).

In the result, the most errors are modal verbs and a complete word is divided into a number of words during segmentation.

At present, the existing NP identification methods are divided into two categories. One is statistical method (such as the maximum entropy, the longest phrase identification, etc.). The other is method based rules (our method). Statistical methods are used in a mathematical way. These methods have highly interpretability, but their accuracy is not high. Especially when the data is sparse. Our method based on rules has high accuracy, and we also use entries of Baidu Baike and a dictionary of synonyms. The shortcoming is that we need artificial intervention.

3 Parsing Chinese Records of Cyber Attacks Using the FSTD

In recent years, our research group has been developing a Framework of Semantic Taxonomy and Description (FSTD) [1]. It aims to extract knowledge from text with a high precision. In this section, we will introduce the FSTD from the perspective of extracting knowledge from cyber attack records.

3.1 The Cyber Attack FSTD

3.1.1 Basic FSTD

A key concept of FSTD is semantic category. And there are two semantic categories, i.e. *event* and *state*. A semantic category is represented as a frame with slots as follows in Fig. 1 [1]:

So far, we have 5120 semantic categories with about 31100 predicates in FSTD.

- **Definition**: Informal interpretation of semantic category in natural language.
- **Rules**: A rule is a semantic pattern. A semantic category may have multiple patterns for parsing sentences whose meaning is the same as that of the category. Semantic patterns are used to parse those sentences. The left-hand side of a pattern is a non-terminal which represents the semantic category. The right-hand side is a sequence of units, which may be a non-terminal or a synset(i.e. A set of words with the same meaning), and some of the units can be optional.
- **Predicate**: Predicates are relations, and they can be of the is-a, part-of, and temporal relations, and organize various categories into a network.
- **Template**: Templates lead to the parser to encode the knowledge after parttern matching.
- **Example**: illustrative sentences contain the semantic category word.
- **Preconditions**: Conditions for an event of the category to occur to or a state to be true.
- **Effects**: Something produced after an event of the category occurs or a state of the category is true.
- **Axioms**: Formal formulae which constrain the current and other semantic categories.

Fig. 1. Notions in FSTD

3.1.2 Some Semantic Categories for Cyber Attack Analysis

We apply FSTD to cyber attack text analysis. This can attain better results. FSTD has a few specially-designed semantic categories, and they are used to parse and extract knowledge from records of cyber attacks. Before we present examples of such categories, we introduce an attack model which are used in designing the FSTD templates, as shown in Table 9.

Table 9. The attack model

Contents	Explanation
timeOfAttack	The time of the attack happened
attacker	Agent who launched the attack, including individual and organization
targetOfAttack	Attack target that the attacker chose, e.g. a system, website, or network
attentionOfAttack	Attacker's intention, e.g. protesting against a government and information interception
typeOfAttack	E.g. DDOS, data leakage, and webpage tampering.
effectOfAttack	effect, e.g. information lose and system paralysis…
scopeOfAttack	The scope impacted by the attack
processOfAttack	The process of the attack

Based on the attack model, we design a few templates for representing extracted knowledge, as shown in Figs. 2 and 3.

defcategory 篡改语义类(CategoryOfTampering)

{

Definition：The cognitive subject or the organization change or distort other things(classics, theory, policy, web pages, etc.) with "hypocritical" means.

Predicates：

{篡改，修改}

{tamper, modify}

Rule:

<篡改类语句>::= <当事:实体><遭受词类> [<施事:实体>] [<范围:数据>]<篡改中心词类>

<SentOfTampering>::=<Essive:Entity><SynsetOfSuffering>[<Agnet:Entity>] [<Scope:Data>] <TamperPredicate>

Templates：

攻击者:施事

attacker: Agent

攻击目标:当事

targetOfAttack: Essive

攻击类型:篡改信息

typeOfAttack: data tamper

}

Fig. 2. The content of "CategoryOfTampering"

defcategory 攻击语义类(CategoryOfAttack)

{

Definition：Agent attacks network system hardware and software by making use of network deficiencies and security flaws.

Predicates：

{攻击，网络攻击}

{attack, attack through network}

Rule:

<攻击类语句>::=<施事:实体><攻击词类><受事:实体>[<意图>]

<SentOfAttack>::=<Agnet:Entity><AttackPredicate><Patient:Entity>[<Attention>]

Templates：

攻击者:施事

attacker: Agent

攻击目标:受事

targetOfAttack: Patient

攻击意图:意图

attentionOfAttack: Attention

}

Fig. 3. The content of "CategoryOfAttack"

3.2   The Procedure of Knowledge Extraction

We parsed the attack sentences by using the NkiExtractor[1]. In this work about extracting knowledge from Chinese records of cyber attacks, we identify noun phrases in segmented records at first. This process will help the parsing process. It can improve the parsing speed and accuracy. The knowledge extraction process consists of two parts, the text parsing process and the extraction process. The procedure is shown as Fig. 4.

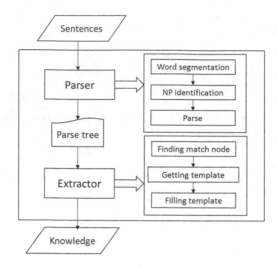

Fig. 4. Procedure of knowledge extraction

3.2.1 The Procedure of Parsing

The main purpose of the parsing process is to parse the cyber attack records by the network attack FSTD. The cyber attack FSTD has been created before and we have introduced it in Sect. 3.1.

The whole parsing procedure mainly includes the following four steps. The procedure is shown as Fig. 5.

— First of all, we segment the cyber attack texts.
— Then we identify the noun phrase in the texts after the segmentation processing. The main method about NP identification we used has been mentioned in the second chapter. After analyses for the preliminary experiment results, we find out that the word segmentation errors have a great influence on the parsing results. So in order to ensure the speed and accuracy of parsing, NP identification before parsing is an inevitable process. Besides, the NP identification results also have an influence on the final parsing results. Therefore, we will also focus on the NP identification.
— Next, we use the improved robust Earley algorithm [20] to parse the texts on the basis of the cyber attack FSTD.
— Finally, we will get the parsing trees from the sentences.

Fig. 5. Procedure of parsing

3.2.2   The Procedure of Extraction

After generating parsing trees in the parsing process. The next step is to extract knowledge from parsing trees. Of course, the knowledge extraction is also rely on the definition of templates in the FSTD.

The whole extraction procedure mainly includes three steps. The procedure is shown as Fig. 6.

— We first search a parse tree for the nodes that correspond to semantic categories.
— Then, we get the rules used in the parse tree and retrieve the knowledge templates corresponding to the rule.
— At last, we fill knowledge templates with the words in the sentence which match the tree nodes, yielding all knowledge tuples.

Fig. 6. Procedure of extraction

3.3   Experiments

The cyber texts are parsed by NkiExtractor. An example of a specific text and the acquired knowledge is shown below in Fig. 7.

内容：黑客组织"匿名者"在"绿色权益行动"中对XX官网实施DoS攻击，旨在抗议YY海军在ZZ市修建"WW系统"
Content: Hacker group "anonymous" attacked XX website using XX DoS in the "green Rights Act". Purpose is to protest the YY Navy to build the WW system in ZZ city

结果：   Result:
  攻击者：黑客组织"匿名者"   Attacker: Hacker group "anonymous"
  攻击目标：XX官网   Attack target: XX website
  攻击类型：DoS攻击   Attack type: DoS
  攻击意图：旨在抗议YY海军在ZZ市   Attack intention: protest the YY Navy
            修建"WW系统"   to build the WW system in ZZ city

Fig. 7. Example of knowledge extraction (because we use the corpus is secret data provided by one department, the example hidden the name).

In the cyber attack FSTD, we have 91 semantic categories with 527 rules. We used the corpus about cyber attack one department provided. We selected 201 paragraph with 1518 sentences as training data to design grammar. And we left 32 paragraph with 195 sentences as test data to calculate the accurate rate. We propose three metrics for result:

— *Accuracy of Entity's name:* The accuracy for the extraction of the attacker, the attack target and the attack type.

$$\frac{\text{the number of correct entity names}}{\text{total number of entity names}}$$

– *Accuracy of entity's attribute:* The accuracy for the extraction of the attack time, the attack intention, attack process and so on.

$$\frac{\text{the number of correct entity } attributes}{\text{total number of entity } attributes}$$

– *Accuracy of entities' relationship:* The accuracy for extracting sentences about attack from all text.

$$\frac{\text{the number of extracted sentences}}{\text{total number of sentecne}}$$

The result is shown below in Table 10.

Table 10. Result of knowledge extraction

Type	Total	Correct numbers	Error numbers	Accuracy
Accuracy of Entity's name	129	122	7	94.57 %
Accuracy of entity's attribute	111	105	6	94.59 %
Accuracy of entities' relationship	130	121	9	93.08 %

The experimental results show that our method can extract knowledge from the attack records with high accuracy. The efficiency is also high. Although the experimental results show that the accuracy rate is high, but there still are some errors.

The main reasons of the errors include: (1) errors caused by word segmentation; (2) errors caused by noun phase identification, and predicate verbs; (3) errors caused by lack of rules in the categories.

4   Conclusion

In this paper, we presented a method to parse Chinese records of cyber attacks to extract knowledge from them. Because of the complexity of the Chinese language, and in order to better parse the records, we first performed Chinese noun phrase identification, and used the NkiExtractor to parse the records to knowledge templates.

Through experiments, we can see that our method about noun phrase identification has a good performance, and the record parsing and knowledge extraction have a high precision.

Of course, there are some problems in our works. (1) The noun phrase identification may pass the errors to the knowledge extraction; (2) In noun phrase identification, we do not consider the pivotal sentences and serial-verb sentences; and (3) our rules (or semantic patterns) in FSTD can not cover all the sentences of the records and machine learning is required to automatically enrich the rules. All these problems are our future research tasks.

Acknowledgments. This word is supported by the National Science Foundation of China (under grant No. 91224006 and 61173063) and the Ministry of Science and Technology (under grant No. 201307107).

References

1. Zang, L., et al.: A Chinese framework of semantic taxonomy and description: preliminary experimental evaluation using web information extraction. In: Zhang, S., et al. (eds.) KSEM 2015. LNCS, vol. 9403, pp. 275–286. Springer, Heidelberg (2015). doi:10.1007/978-3-319-25159-2_25
2. Framenet. https://framenet.icsi.berkeley.edu/fndrupal/home
3. Fillmore, C.J., Lee-Goldman, R., Rhodes, R.: Sign-based construction grammar and the framenet constructicon. Boas/Sag (Hg.) (2012)
4. Banko, M., Cafarella, M.J., Soderland, S., Broadhead, M., Etzioni, O.: Open information extraction from the web. Commun. ACM **51**(12), 68–74 (2008)
5. Carlson, A., Betteridge, J., Kisiel, B., Settles, B., Hruschka Jr., E.R., Mitchell, T.M.: Toward an architecture for never-ending language learning. In: AAAI, vol. 5, p. 3 (2010)
6. Lenat, D.B.: CYC: a large-scale investment in knowledge infrastructure. Commun. ACM **38** (11), 33–38 (1998)
7. Singh, P., et al.: The public acquisition of commonsense knowledge. In: Proceedings of AAAI Spring Symposium: Acquiring (and Using) Linguistic (and World) Knowledge for Information Access (2002)
8. Baker, C.F., Fillmore, C.J., Lowe, J.B.: The Berkeley framenet project. In: Proceedings of the 17th International Conference on Computational Linguistics, vol. 1, pp. 86–90. Association for Computational Linguistics (1998)
9. Fillmore, C.J., Johnson, C.R., Petruck, M.R.L.: Background to framenet. Int. J. Lexicography **16**(3), 235–250 (2003)
10. Chinese framenet. http://sccfn.sxu.edu.cn/portal-en/home.aspx
11. Guo, H., Ying, M.: Application study of hidden Markov model based on genetic algorithm in noun phrase identification. Comput. Sci. **36**(10), 244–247 (2009)
12. Li, R.: Noun phrase identification based on genetic algorithm and hidden Markov model. Int. J. Syst. Control **2**(3), 221–227 (2007)
13. Kong, L., Ren, F., Sun, X., Quan, C.: Word frequency statistics model for Chinese base noun phrase identification. In: Huang, D.-S., Jo, K.-H., Wang, L. (eds.) ICIC 2014. LNCS, vol. 8589, pp. 635–644. Springer, Heidelberg (2014). doi:10.1007/978-3-319-09339-0_64
14. Stanford Word Segmenter. http://nlp.stanford.edu/software/segmenter.shtml

15. ICTCLAS (Institute of Computing Technology, Chinese Lexical Analysis System). http://www.nlp.org.cn/project/project.php?proj_id=6
16. Yuming, W.U., Luo, X., Yang, Z.: Semantic separator learning and its applications in unsupervised Chinese text parsing. Front. Comput. Sci. 7(1), 55–68 (2013). Selected Publications from Chinese Universities
17. Baidu Baike. http://baike.baidu.com/
18. Synonym dictionary. http://ir.hit.edu.cn/demo/ltp/Sharing_Plan.htm
19. Witten, I.H., Bell, T.C.: The zero-frequency problem: estimating the probabilities of novel events in adaptive text compression. IEEE Trans. Inf. Theory 37(4), 1085–1094 (1991)
20. Earley, J.: An Efficient Context-Free Parsing Algorithm. Morgan Kaufmann Publishers Inc., San Francisco (1986)

Increasing Topic Coherence
by Aggregating Topic Models

Stuart J. Blair$^{(\boxtimes)}$, Yaxin Bi, and Maurice D. Mulvenna

School of Computing and Mathematics, Ulster University,
Newtownabbey BT37 0QB, UK
Blair-S4@email.ulster.ac.uk, {y.bi,md.mulvenna}@ulster.ac.uk

Abstract. In this paper, we introduce a novel method for aggregating multiple topic models to produce an aggregate model that contains topics with greater coherence than individual models. When generating a topic model a number of parameters must be specified. Depending on the parameters chosen the resulting topics can be very general or very specific. In this paper the process of aggregating multiple topic models generated using different parameters is investigated; the hypothesis being that combining the general and specific topics can increase topic coherence. The aggregate model is created using cosine similarity and Jensen-Shannon divergence to combine topics which are above a similarity threshold. The model is evaluated using evaluation methods to calculate the coherence of topics in the base models against those of the aggregated model. The results presented in this paper show that the aggregated model outperforms standard topic models at a statistically significant level in terms of topic coherence when evaluated against an external corpus.

Keywords: Topic models · Semantic coherence · Ensemble methods

1 Introduction

In the modern era of computing and with the advent of big data we have unrivalled access to massive data that would have been unimaginable in the past. A challenge faced is extracting the underlying information from massive data. Of particular interest to researchers is how to extract this information from textual data. Textual data is hard to analyse due to their varied syntactical structures and semantics, one method that attempts to identify the underlying topical structure of textual data is topic models such as latent Dirichlet allocation (LDA) [3].

Topic models are a type of statistical model that can quickly find the latent topics within large collections of documents. When generating a topic model some parameters need to be set, effecting on the topics output by the model. If there is no prior knowledge of the corpus being modelled, setting the number of topics to generate too low will result in very broad topics containing mostly

© Springer International Publishing AG 2016
F. Lehner and N. Fteimi (Eds.): KSEM 2016, LNAI 9983, pp. 69–81, 2016.
DOI: 10.1007/978-3-319-47650-6_6

common words and stopwords. If the number of topics is set too high, they will be very granular and overfitted to the corpus. Other parameters specified when using some topic models such as alpha and beta Dirichlet prior values in LDA are crucial to the model and can have a noticeable effect on the quality of the model [18].

In order to help avert the very general or very specific topics that could be generated using non-optimal initial parameters, we propose a novel method of aggregating topic models. Under this method, a user defines a set of different parameters and multiple topic models are generated using these parameters; the topics which are found to be similar amongst these models are then aggregated. The main contribution of this proposed approach is the post-processing of topic models to increase topic coherence. This has the advantage of allowing very granular topics that may only be produced in a model with many topics to have a presence in a topic model which is more representative of the corpus as a whole. The proposed method is also advantageous as it requires no changes to the underlying topic model's generative method.

The overall coherence of the topics produced from the proposed aggregate model (AGGM) will be assessed using both intrinsic and extrinsic measures. The use of an intrinsic coherence measure allows us to verify that the models topics accurately represent the contents of the corpus used to generate the model based on word co-occurrences (WCO). Using extrinsic coherence measures allows for the checking of the produced topics against a corpus of general English documents to make sure that the topics are generally coherent in everyday language, similar to how a human observing the topics would decide if they are coherent or not. Additionally, the statistical significance of the AGGM will also be calculated with regards to the base model.

2   Related Work

There has been little research in the area of combining multiple topic models and topic model ensembles. One ensemble method for topic models is not related to textual data but is used on medical data to predict risk of disease [13]. In this ensemble, each model is trained with a dataset such as poverty level, gender, age and race. The topics discovered in each model are then combined into a single matrix. This combined topic matrix is then used to infer a topic distribution over new patients to predict their disease risk.

Another piece of research on topic model ensembles uses the co-occurrence of document-word tuples and their assigned topic [14]. This method assumes a vector of document-word tuple topic assignments. The corpus is then split into sub-corpora and the topic assignment vector for each sub-corpus is combined into one vector. This is then used to infer topic distributions over documents. This method is used on both LDA and latent semantic analysis (LSA) ensembles; in the tests on both real and synthetic data, the LDA ensemble outperforms LSA in terms of perplexity, however, the LSA ensemble outperforms the LDA ensemble in terms of efficiency.

The application of classic ensemble methods to topic models has been researched little. The boosting ensemble method has been applied to topic models with good results and generalisation ability due to LDA mapping features to topic space rather than word space [20]. The boosting method for topic models was enhanced by integrating a supervised hidden Markov model, improving the accuracy [21].

A method for generating a pachinko allocation model (PAM) has also been proposed [9]. This method utilises correlations between topics to create a directed acyclic graph. This results in a tree-like structure where leaves are words from the vocabulary and interior nodes show correlation between the child nodes. Other similar work in this area includes hierarchical topic models [2]. In this model a document is generated by creating a route from the root node to a leaf node. Where the root is very general words and the further the tree is traversed, the more specific words get. This method assumes that the corpus falls into a hierarchical structure of topics. This also has the advantages of being able to easily accommodate growing corpora and having nonparametric priors. It is similar to the work presented in [7].

The idea of using similarity measures to compare topic models has also been studied in the past. Cosine similarity has been used to compare the top words in topics [19]. Jensen-Shannon (JS) divergence has also been used to assess the similarity of documents based on their topic distributions [6]. These works use the similarity measures for evaluation. A method for modelling consensus topics across a range contexts has also been proposed [15]. This method implements a co-regularisation framework to create pseudo-documents using a centroid-based regularisation to make topics from different contexts agree with each other based on a set of general topic distributions. This allows for context specific topics to be bridged together using the general topics.

The closest piece of previous work to this paper uses the process of self-aggregation of short text during the model generation stage of a topic model [12]. This process integrates clustering of text before the topic modelling process begins. This assumes that each document only comes from one latent topic. This allows for the creation of larger pseudo-documents to be used in the topic models generative procedure.

The main difference between the work presented in this paper and the work discussed in this section is that the proposed method focuses on post-processing of topics after the models have been generated. This has the advantage of not relying on arbitrary manipulations of the underlying topic model structure, resulting in a less complex model structure requiring no context specific information. This method also allows for leniency in model parameter selection. As stated previously, a set of parameters needs to be defined before the models are generated. If one of these parameters results in topics that are very different to other models' topics, they are unlikely to be aggregated with other topics; resulting in them not having an effect on the aggregate topics.

3 Topic Modelling

Topic modelling originates with LSA. However, when applied to an information retrieval task, it is often known as latent semantic indexing [5]. LSA utilises a document-term matrix and singular value decomposition to find similar documents, making the assumption that words which frequently appear together are related. Two notable disadvantages of LSA are that the model is based around the bag-of-words method and that it struggles with polysemy. This means that word order in documents is abandoned and that it cannot distinguish between the different meanings of a single word. For example, *crane* can refer to both a bird as well as a piece of construction machinery.

LDA has helped to eliminate the polysemy difficulties by introducing a probabilistic element to the model but it continues to struggle with the bag-of-words assumption, abandoning all sentence structure when creating the model [3].

In this paper, the topic model that will be utilised when creating the base models in experiments is LDA. LDA is a generative probabilistic model that finds latent topics in a collection of documents by learning the relationship between words (w_j), documents (d_i), and topics (z_j). The data used by an LDA model is input in bag-of-words form, word counts are preserved but the ordering of the words is lost. The generative process for document d_i assumes the following:

- There is a fixed number of topics K.
- Each topic z has a multinomial distribution over vocabulary ϕ_z drawn from Dirichlet prior $Dir(\beta)$.
- $i \in \{1, \ldots, M\}$ where M is the number of documents in the corpus.
- $Dir(\alpha)$ is the document-topic Dirichlet distribution.

The following is the generative process for document d_i:

1. Choose $\theta_i \sim Dir(\alpha)$.
2. For word $w_j \in d_i$:
   (a) Draw a topic $z_j \sim \theta_i$.
   (b) Draw a word $w_j \sim \phi_{z_j}$.

3.1 Topic Coherence Measures

Topic coherence can be defined as how interpretable a topic is based on the relationships between the words within the topic itself. Topic coherence measures aim to quantify the quality of the latent topics discovered by a topic model from a human-like perspective in order to ensure a high degree of semantic coherence. Much research has been performed in statistical topic model evaluation [16, 17], however, it has been found that these methods are not always reflective of how humans view the topics [4]. It was shown that metrics based on word co-occurrences and mutual information approaches are more representative of how humans would evaluate a topic [11]; for this reason, these approaches are used in this paper.

Intrinsic topic coherence measures utilise word co-occurrences in documents from the corpus used to train the model [10]. The intrinsicness of this method allows for better judgement of the coherence of a topic based on the training documents. It is measured on a scale of $0-1$, a result closer to 1 means that the model correctly identified words that occur together frequently in documents as a topic. However, this does not guarantee that they make semantic sense or that they are interpretable by a human; it means that the topics represent the corpus they were generated from.

Extrinsic topic coherence measures require the use of an external corpus to calculate the observed coherence. One approach to calculating extrinsic topic coherence utilises pointwise mutual information (PMI) [11]. This metric is extrinsic in nature as it calculates the PMI between the top N words in a sliding window using an external corpus such as a Wikipedia dump file (a plain text version of every article present on Wikipedia). The calculation can be seen in (1). Where $p(w_i, w_j)$ is the probability of two words co-occurring in a document, $p(w_i)$ and $p(w_j)$ is the probability of word w_i and word w_j occurring in the document, respectively.

$$PMI(w_i, w_j) = log\frac{p(w_i, w_j)}{p(w_i)\,p(w_j)} \tag{1}$$

Research into using the PMI has found that it has a high degree of correlation with a human's evaluation of topic coherence, however, a normalized version of the PMI approach results in an even higher level of correlation with human judgement (NPMI) [8]. Using PMI with an external corpus allows for the calculation of how frequently words in the topic occur together; because an external corpus is used, it can be seen as how humans would interpret the topics from everyday language.

4 Aggregating Topic Model Outputs

One problem with trying to aggregate topic model outputs in an ensemble-like style is that unlike conventional classifier ensembles that have a finite set of possible classes $(C_1, \ldots, C_n)$, topic models have an infinite number of outputs that are unknown until the model has been created. For this reason, the proposed aggregation method must adjust to the multitude of outputs it may face.

In the field of information retrieval there exists many methods for measuring similarity; these methods include cosine similarity, Pearson correlation coefficient, and the Jaccard index. The Jaccard index could be used to find similar topics by simply calculating the similarity coefficient between the top N words in two given topics. Gaining a high value from the result of the Jaccard index shows that there is indeed some similarity between the topics. However, the downside is that a threshold for similarity needs to be set via introspection as there is no fool proof method of statistically assessing the similarity threshold. Previously, there has been research regarding using JS divergence and Kolmogorov-Smirnov divergence to assess the similarity of topic probability distributions [1] within the topic model's the phi (topic-word) distribution.

To perform the aggregation cosine similarity and JS divergence will be used to assess the similarity of topics' Phi distributions. Both methods will then be evaluated for performance. The cosine similarity allows for more flexibility in setting the similarity threshold and is also not invariant to shifts in input as opposed to measures such as Pearson correlation coefficient which is invariant to input shifts. The upper and lower bounds for cosine similarity are 1 for complete similarity and 0 for complete dissimilarity. The process for aggregating topics will now be described. Firstly, the user will define a set of parameters that will be used to generate the base models for this technique. A threshold for similarity will then need to be set using grid search with development sets of data to see which threshold produces the most coherent topics on average; although this may be seen as computationally expensive, on modern hardware these tuning experiments using a subset of data are relatively quick to execute and easy to run in parallel. Then each topic from each of the other models will be compared to the base topics in a pairwise method for similarity. If the cosine similarity of the two topics is above the threshold then they will be combined, the method for combining the similar topics is to simply calculate the mean probability of each word in the Phi distributions. It should be noted that the number of topics in the base model does not increase or decrease, nor does the number of words in the topic as the alphabet for each model is the same. Equation (2) shows the combination process for cosine similarity, where $\hat{\varphi}_k$ is an aggregated topic, n is the number of similar topics, M is the number of models, T_i is the number of topics in a model, $\varphi_{(i,j)}$ is the phi distribution for topic T_j in model M_i, φ_x is the xth phi distributions from the base model, and γ is the similarity threshold.

$$\hat{\varphi}_k = \frac{\sum_{i=1}^{M} \sum_{j=1}^{T_i}}{n} \begin{cases} \varphi_{(i,j)}, & if \quad \frac{\varphi_{(i,j)} \cdot \varphi_x}{\|\varphi_{(i,j)}\| \|\varphi_x\|} \geq \gamma \\ 0, & otherwise \end{cases} \tag{2}$$

JS divergence allows for the symmetric measurement of similarity between distributions. Using the base 2 logarithm, the JS divergence has the bounds $0 \leq D_{JS}(P \parallel Q) \leq 1$ where 0 indicates complete similarity and 1 is complete dissimilarity. The JS divergence is a symmetrised and smoothed version of Kullback-Leibler (KL) divergence, using the average KL divergence for each Phi distribution to the average of each Phi distribution. It is shown in (3) where P and Q are distributions and M is the average of distributions P and Q, that is $M = \frac{1}{2}(P + Q)$.

$$D_{JS}(P \parallel Q) = \frac{1}{2} D_{KL}(P \parallel M) + \frac{1}{2} D_{KL}(Q \parallel M) \tag{3}$$

The process for using JS divergence to create aggregate topics can be seen in (4) using the same notation as (2). The main difference is that the JS divergence result should be $\leq \gamma$ as opposed to cosine similarity where the result should be $\geq \gamma$.

$$\hat{\varphi}_k = \frac{\sum\limits_{i=1}^{M} \sum\limits_{j=1}^{T_i}}{n} \begin{cases} \varphi_{(i,j)}, & if \quad D_{JS}(\varphi_{(i,j)} \parallel \varphi_x) \leq \gamma \\ 0, & otherwise \end{cases} \qquad (4)$$

4.1 Choosing Similarity Threshold

An important aspect of this method for aggregating topic models is the choice of similarity threshold. The overall problem attempting to be solved can be viewed as optimising the semantic coherence of topics by finding the optimal similarity threshold and sliding window size. The sliding window size is directly related to calculating coherence as it sets the window size for calculating word probabilities. For example, if the word co-occurrence probability for word w_i and word w_j was calculated using a sliding window of size 50 words, then as long as the words occur at least once in the 50 word window it will count as the words having co-occurred, irrelevant as to whether they are in different sentences or paragraphs. However, if a lower window size such as 10 is used, it is stricter as it limits where the words can co-occur. This allows for more confidence that the words actually occurred together in the same context.

In this paper a grid search will be used over a set of similarity thresholds and a set of sliding window sizes. A small subset of the full dataset will be used for finding the optimal values. The grid search will then allow for topics to be aggregated and the coherence calculated using the Cartesian product of the set of similarity thresholds and set of sliding window sizes. Although methods such as Bayesian optimisation can be used to optimise parameters, it is unnecessary for this task which can be easily parallelised despite suffering from the curse of dimensionality. This makes grid search a feasible option without overcomplicating the problem by using more complex methods.

5 Experimental Setup

The following section contains experiments conducted to assess the performance of the proposed aggregated topic model. Each experiment will use LDA with 2000 iterations of Gibbs sampling. The method for deciding the similarity threshold for each experiment will be detailed, and comparisons of how the aggregated model compares to the base models used to create the aggregated model will be presented. The topic coherence test consists of extrinsic PMI. The reference corpus used for the extrinsic test will be the English Wikipedia. This corpus was chosen as Wikipedia has over five million articles, therefore it will be a good reference of general English language; the average document length of the Wikipedia dump was 133.07 words. An intrinsic coherence test will also be performed to measure how well topics capture the structure of the underlying corpus.

The data used to generate the models in the following experiments is a set of Associated Press articles supplied with David Blei's lda-c package[1]. The corpus contains 2246 documents and 34977 tokens after removal of stopwords.

---

[1] Available at: http://www.cs.princeton.edu/~blei/lda-c/.

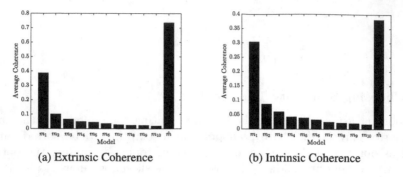

(a) Extrinsic Coherence   (b) Intrinsic Coherence

Fig. 1. Average coherences for topics in the base models and AGGM in experiment 1

The first experiment will consist of creating models of different topic numbers and aggregating the similar topics; this will show how the model performs aggregating general topics and specific topics. The second experiment consists of aggregating models with different Dirichlet priors; this experiment will show what effect different priors have on aggregating multiple models.

5.1 Models Created with Different Numbers of Topics

This experiment consists of creating ten models each with a different number of topics $T = \{10, 20, 30, 40, 50, 60, 70, 80, 90, 100\}$ and an alpha prior of $50/T$. Using different numbers of topics allows for representations of the corpus at different granularity levels, for example, 10 topics provides a very general overview and 100 provides a very specific set of topics. An aggregated model that combines similar topics from multiple models using different numbers of topics could produce a more coherent set of topics due to the output containing a mix of general topics, as well as specific topics.

The first step in generating the AGGM is to choose the similarity threshold at which similar topics will be combined. Grid search will be used on small development sets of data at different similarity thresholds, starting at 0.1 and increasing to 0.9 in increments of 0.1. The sliding window size will also be changed at intervals of 10 words from 10 words to 100 words, as well as using the whole document as the window. The grid search found the optimal similarity threshold according to PMI is >0.7 for cosine similarity and <0.5 for JS divergence. Note that cosine similarity thresholds are in the form $>n$ as the value for complete similarity is 1, whereas JS divergence thresholds are $<n$ as complete similarity is 0. This means that JS divergence is more lenient in the topics that it aggregates, resulting in many more similar topics being combined, allowing for the combination of general and specific topics. Cosine similarity has a higher similarity threshold meaning that not as many topics will be combined, however, JS divergence achieves a higher topic coherence.

Following the tuning experiments, the full experiment was run using JS divergence as the similarity measure as the grid search found it to have the highest coherence; and a similarity threshold of <0.5, the results of this experiment for extrinsic and intrinsic coherence tests can be seen in Figs. 1a and b, respectively. It should be noted that when the coherence of base models and AGGMs are compared, the same sliding window size is used for each model. In this experiment m_1 is the base model, $m_2 - m_{10}$ are the other models to be compared, and $\hat{m}$ is the AGGM. Any model can be the base model, m_1 was chosen in this experiment arbitrarily. Also, if a different model was chosen as the base, the same topic similarity comparisons would be made; the only difference of using m_1 over the other models in this case is the number of topics in the final AGGM. As Fig. 1a shows, the AGGM has an extrinsic PMI value of 0.75, this is higher than any of the model used to create it. This shows that the AGGM's topics are more coherent based on general English language. The AGGM also has the highest intrinsic coherence. This means the AGGM's topics have been complemented with additional relevant topic words leading to topics that are more representative of the corpus.

This experiment results in some differences between the base model topics' top words and the AGGM's top words. A comparison between the base model and AGGM is visible in Tables 1a and b, respectively. In t_1 the AGGM has additional words including "Nicaragua" and "Contra"; this supplemented the words from the base model, "united" and "states". It would be logical to connect these words through the Nicaraguan Revolution, when the United States supported the Contra forces in a campaign against the Sandinista National Liberation Front. Another major change can be seen in t_7 where the AGGM contains more words about medical research and disease, whereas the base model includes some less relevant words such as "children", "percent" and "space". Additionally, t_8 sees the addition of the words "index" and "exchange", this makes it more obvious that this topic is about stock markets and finance. The AGGM also allowed for subtle changes such as the addition of Jackson in t_6, which refers to Jesse Jackson, the opponent of Michael Dukakis in the 1988 Democratic presidential primaries.

5.2 Models Created with Different Parameters

This experiment consists of creating ten models each with a different alpha Dirichlet prior value $\alpha = \{1.0, 2.0, 3.0, 4.0, 5.0, 6.0, 7.0, 8.0, 9.0, 10.0\}$ and fixed number of topics $T = 10$. Using different alpha Dirichlet priors will have a noticeable effect on topic distribution. A high alpha value means that documents are likely to have a mixture of many topics with no single topic being dominant. A low alpha value results in very few and in some cases, only one, topics being in the document.

As with experiment 1, the first step in generating the AGGM is to set the similarity threshold at which similar topics will be combined. The same method of grid search will be used for this experiment. This grid search found the optimal similarity threshold according to PMI to be >0.9 for cosine similarity and <0.1 for JS divergence. This grid search is more stringent in the similarity of topics to be aggregated. Experiment 1 was more lenient in the topics it would combine

Table 1. Comparison of topics from the base model and AGGM with noticeable changes

(a) Base Model			
t1	t6	t7	t8
government	bush	health	percent
united	president	people	market
states	house	children	year
military	dukakis	percent	million
aid	campaign	study	prices
panama	bill	report	billion
china	state	program	dollar
president	senate	aids	rose
year	democratic	years	oil
rights	congress	space	stock

(b) AGGM			
t1	t6	t7	t8
government	bush	aids	percent
aid	dukakis	health	market
military	campaign	disease	dollar
rebels	president	drug	stock
states	house	study	year
united	jackson	medical	prices
nicaragua	bill	virus	trading
panama	senate	research	index
contra	republican	blood	rose
president	democratic	hospital	exchange

by using lower similarity thresholds. The results in this experiment mean that only very similar topics will be combined; this may lead to less of a change in the AGGM topics as less will be aggregated.

Following the tuning experiments, the full experiment was run using JS divergence as the similarity measure and a similarity threshold of <0.1, the results of this experiment for extrinsic and intrinsic coherence tests can be seen in Figs. 2a and b, respectively. It should be noted that when the coherence of base models and AGGMs are compared, the same sliding window size is used for each model. In this experiment m_1 is the base model, $m_2 - m_{10}$ are the other models to be compared, and $\hat{m}$ is the AGGM. As Fig. 1a shows, the AGGM has an extrinsic PMI value of 0.7, this is much higher than any of the model used to create it. This shows that the AGGM's topics are much more coherent based on general English language. The AGGM also has the highest intrinsic coherence. This means the AGGM's topics have gained additional relevant topic words leading to topics that are more representative of the corpus.

In terms of how the underlying topics changed in the AGGM, there were not as many changes as in experiment 1, however, the few changes that did occur improved topic coherence. For example, in m_1 there is a topic about Mikhail Gorbachev, the Soviet Union and the United States; in the AGGM this topic was supplemented with the words "east" and "Germany"; this makes the topic more clearly about the Berlin wall and tensions between the West and East towards the end of the Cold War. The other major difference between base model topics and aggregate topics was in one about finances. The base model contained units of money such as "million" and "billion"; as well as words to do with the workforce, such as "workers" and "business". The AGGM's equivalent topic also contained the words "industry" and "company". Experiment 2 is interesting as its topics seen less changes than experiment 1, but the few changes resulted in increases in topic coherence. This could be because some topics in the base model were quite specific, but were generalised more in the AGGM.

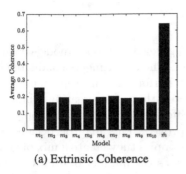

(a) Extrinsic Coherence

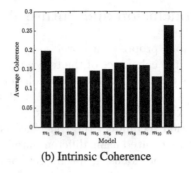

(b) Intrinsic Coherence

Fig. 2. Average coherences for topics in the base models and AGGM in experiment 2

6   Results

The results show that aggregating models does increase the coherence of topics. Figures 1a, b, 2a and b show that the model with the lowest number of topics or highest alpha prior (m_1 from both experiments) are normally the most coherent topic but after aggregation, the aggregate topic is the most coherent. This could be because m_1 is usually the most general model, therefore when evaluated extrinsically the words would have a high probability of co-occurring as they are not specific. What is also interesting is the fact that the AGGM also has the highest intrinsic coherence, meaning that combining elements of more specific models into the general model allows for a greater representation of the modelled corpus.

It was found that to maximise coherence the sliding window size had to be set to the size of the document being analysed. Using the full document size is not detrimental to results as the average document length is 133.07 words, which is only 33.07 words more than the second highest average coherence sliding window size of 100.

Additionally, it was found that the results of both extrinsic experiments had a correlation of 0.98 and the results of the intrinsic experiments had a correlation of 0.94; this suggests that both methods of aggregation have a similar effect on the final result.

A paired t-test was performed to compare the topic coherence for the base model and best AGGM from each experiment; this is advantageous as it takes into account the difference between base models and AGGMs, as well as the mean values to assess if there is a statistically significant difference in topic coherence after aggregating the similar models. The significance level is 0.05. In the first experiment there is a significant difference in the topic coherence for the base model ($\mu = 0.386$, $\sigma = 0.179$) and AGGM ($\mu = 0.736$, $\sigma = 0.224$); $t(9) = 7.173$, $P = 0.0000523$. In the second experiment there is also significant difference in the topic coherence for the base model ($\mu = 0.253$, $\sigma = 0.17$) and AGGM ($\mu = 0.64$, $\sigma = 0.229$); $t(9) = 12.03$, $P = 0.00000075$. These results show that aggregating the output of multiple topic models can increase the topic coherence.

7   Conclusion and Future Work

This paper proposed a novel solution for aggregating topic models that can improve the coherence of the topics produced. The experiments conducted show that it is possible to improve the coherence by performing this topic model aggregation. Through the experiments we discovered that after creating a number of models with different numbers of topics or different parameters and applying the aggregation technique, the coherence is improved. This may be because when models are created using different numbers of topics they create a mix of general; as well as more focused, specific sets of topics as the number of topics increases. The advantage of this is that the AGGMs have more general topics which leads to the AGGM being more representative of the corpus it was generated from as shown by the intrinsic coherence results.

The results of experiment 2 were also interesting as despite having less changes in the AGGM than experiment 1; there was a noticeable difference in coherence. This suggests that aggregation allows for more general topics, and that any form of generalisation results in a higher topic coherence. An important point to note is that although the top N words in a topic may not appear to change much in some cases; the underlying word distribution in the topics will change after the AGGM is formed.

The proposed aggregation technique shows that it outperforms standard topic models in topic coherence but it can still be improved. For example, by clustering the corpus into subsets and generating models using these subsets. The topics generated from these subsets when aggregated could provide a good mix of general topics, as well as specific topics. This work could also be furthered by creating aggregated topics from different types of topic models. Another important area of further work is to present the base models and AGGMs to humans and have them to rank the topics. This will allow for the correlation of the coherence of the aggregated model with human opinion.

References

1. Aletras, N., Stevenson, M.: Measuring the Similarity between automatically generated topics. In: Proceedings of the 14th Conference of the European Chapter of the Association for Computational Linguistics, pp. 22–27, Gothenburg (2014)
2. Blei, D.M., Griffiths, T.L., Jordan, M.I., Tenenbaum, J.B.: Hierarchical topic models and the nested Chinese restaurant process. Adv. Neural Inf. Process. Syst. **16**, 17–24 (2004)
3. Blei, D.M., Ng, A., Jordan, M.I.: Latent Dirichlet allocation. J. Mach. Learn. Res. **3**, 993–1022 (2003)
4. Chang, J., Gerrish, S., Wang, C., Boyd-Graber, J.L., Blei, D.M.: Reading tea leaves: how humans interpret topic models. In: Advances in Neural Information Processing Systems, pp. 288–296 (2009)
5. Deerwester, S., Dumais, S., Furnas, G., Landauer, T., Harshman, R.: Indexing by latent semantic analysis. J. Am. Soc. Inf. Sci. **41**(6), 391–407 (1990)

6. Hall, D., Jurafsky, D., Manning, C.D.: Studying the history of ideas using topic models. In: Proceedings of the Conference on Empirical Methods in Natural Language Processing, pp. 363–371 (2008)
7. Hofmann, T.: The cluster-abstraction model: unsupervised learning of topic hierarchies from text data. IJCAI **99**, 682–687 (1999)
8. Lau, J.H., Newman, D., Baldwin, T.: Machine reading tea leaves: automatically evaluating topic coherence and topic model quality. In: Proceedings of the 14th Conference of the European Chapter of the Association for Computational Linguistics, pp. 530–539, Gothenburg (2014)
9. Li, W., McCallum, A.: Pachinko allocation: DAG-structured mixture models of topic correlations. In: Proceedings of the 23rd International Conference on Machine Learning, pp. 577–584 (2006)
10. Mimno, D., Wallach, H.M., Talley, E., Leenders, M., McCallum, A.: Optimizing semantic coherence in topic models. In: Proceedings of the Conference on Empirical Methods in Natural Language Processing, pp. 262–272. Association for Computational Linguistics, july 2011
11. Newman, D., Lau, J.H., Grieser, K., Baldwin, T.: Automatic evaluation of topic coherence. In: Human Language Technologies: The 2010 Annual Conference of the North American Chapter of the ACL, pp. 100–108, Los Angeles (2010)
12. Quan, X., Kit, C., Ge, Y., Pan, S.J.: Short and sparse text topic modeling via self-aggregation. In: Proceedings of the 24th International Joint Conference on Artificial Intelligence, pp. 2270–2276 (2015)
13. Rider, A.K., Chawla, N.V.: An ensemble topic model for sharing healthcare data and predicting disease risk. In: Proceedings of the International Conference on Bioinformatics, Computational Biology and Biomedical Informatics, pp. 333–340 (2013)
14. Shen, Z., Luo, P., Yang, S., Shen, X.: Topic modeling ensembles. In: 2010 IEEE International Conference on Data Mining. pp. 1031–1036 (2010)
15. Tang, J., Zhang, M., Mei, Q.: One theme in all views: modeling consensus topics in multiple contexts. In: Proceedings of the 19th ACM SIGKDD International Conference on Knowledge Discovery and Data Mining, pp. 5–13 (2013)
16. Wallach, H.M.: Structured topic models for language. Ph.D. thesis, Universty of Cambridge (2008)
17. Wallach, H.M., Mimno, D., Mccallum, A.: Rethinking LDA: why priors matter. In: NIPS (2009)
18. Wallach, H.M., Murray, I., Salakhutdinov, R., Mimno, D.: Evaluation methods for topic models. In: Proceedings of the 26th Annual International Conference on Machine Learning - ICML 2009, New York, USA, pp. 1–8. ACM, New York, June 2009
19. Yan, X., Guo, J., Lan, Y., Cheng, X.: A biterm topic model for short texts. In: Proceedings of the 22nd International Conference on World Wide Web, pp. 1445–1456 (2013)
20. Yongliang, W., Qiao, G.: Multi-LDA hybrid topic model with boosting strategy and its application in text classification. In: 33rd Chinese Control Conference, pp. 4802–4806 (2014)
21. Yongliang, W., Qiao, G.: Modeling texts in semantic space and ensemble topic-models via boosting strategy. In: 34th Chinese Control Conference, pp. 3838–3843 (2015)

Learning Chinese-Japanese Bilingual Word Embedding by Using Common Characters

Jilei Wang[1], Shiying Luo[2], Yanning Li[3], and Shu-Tao Xia[1(✉)]

[1] Tsinghua University, Beijing 100084, China
wangjl15@mails.tsinghua.edu.cn, xiast@sz.tsinghua.edu.cn
[2] Northeastern University, Shenyang 110819, China
[3] Renmin University of China, Beijing 100872, China

Abstract. Bilingual word embedding, which maps word embedding of two languages into one vector space, has been widely applied in the domain of machine translation, word sense disambiguation and so on. However, no model has been universally accepted for learning bilingual word embedding. In this work, we propose a novel model named CJ-BOC to learn Chinese-Japanese word embeddings. Given Chinese and Japanese share a large portion of common characters, we exploit them in our training process. We demonstrated the effectiveness of such exploitation through theoretical and also experimental study. To evaluate the performance of CJ-BOC, we conducted a comprehensive experiment, which reveals its speed advantage, and high quality of acquired word embeddings as well.

Keywords: Bilingual word embedding · Distributed representation · Common characters · Chinese-Japanese

1 Introduction

Due to the boost of social network, massive text data are generated every day, reaching an enormous bulk beyond human's reading ability. Therefore, there is now an urgent need of high-quality knowledge extraction from text, which is often associated with natural language processing (NLP) techniques. Word embedding, originally referred as distributed representation, was proposed by Hinton [1], whose basic idea is to denote word as low-dimension and real-valued vector. A favorable feature of word embedding is that the trained vectors can reflect the similarity among words. Such merit led to the popularity and thus broad applications of word embedding, in which Word2Vec[1] by Google has been extensively used in text mining, text segmentation, synonym discovery, etc.

Bilingual word embedding, as a specific form of word embedding, maps vectors of two different languages into the same vector space. Bilingual word embedding directly depicts the internal relatedness among words of two different languages, and is therefore believed able to facilitate many NLP tasks. Actually

---

[1] https://code.google.com/archive/p/word2vec, accessed date: March 17, 2016.

© Springer International Publishing AG 2016
F. Lehner and N. Fteimi (Eds.): KSEM 2016, LNAI 9983, pp. 82–93, 2016.
DOI: 10.1007/978-3-319-47650-6_7

many research efforts have strengthened such belief, as bilingual word embedding has been successfully applied in machine translation [2], word sense disambiguation [3] and so on. Despite its advantages, bilingual word embedding is still at its initial stage, without commonly accepted approaches.

In this paper, we propose a model CJ-BOC for learning Chinese-Japanese word embedding, based on CBOW (short for continuous bag-of-words), one of the two models given in [2]. The exploitation of common Chinese characters shared by Chinese and Japanese is the main difference distinguishing CJ-BOC from CBOW, whose validity is well demonstrated according to both theoretical and empirical study. In our experiment, CJ-BOC significantly outperformed sentence-aligned approaches in terms of the quality of word embedding.

The contributions of our work are: 1. We analyze and demonstrate why common characters shared by Chinese and Japanese are effective for learning word embedding, from a view of information theory; 2. Based on the exploitation of common characters, we design a novel model for learning Chinese-Japanese word embedding; 3. We conduct a comprehensive experiment to verify the effectiveness and efficiency of our approach.

The rest of this paper is structured as follows: Sect. 2 reviews the related work, Sect. 3 introduces common characters, and Sect. 4 elaborately introduces our model CJ-BOC. In Sect. 5, we present our experimental results. Finally in Sect. 6, we conclude this paper and give the future work.

2 Related Work

Distributed representation, now commonly referred as word embedding, was firstly proposed by Hinton [1]. Compared with the one-hot representation in early works of NLP and information retrieval, distributed representation can better reflect the relatedness among words. Due to this unique merit, recent years have seen vast research efforts on word embedding.

One typical framework for learning word embeddings is neural network language models (NNLM) proposed by Bengio et al. [4], which uses back propagation to train each word embedding and parameter of the model. NNLM has been widely applied [5]. As an open source framework, Word2Vec provides multiple models, including H-Softmax, NEG, CBOW, and Skip-Gram, and is known for its fast training speed, as well as the quality word embeddings learned from it. By exploiting the advantages of both global matrix factorization and local context window methods, Pennington et al. [7] proposed a model named GloVe.

In 2013, Mikolov et al. presented a work [2] which obtained English and Spanish word embedding separately, and then generated English-Spanish embeddings using a linear mapping between vector spaces of both languages. Compared with traditional monolingual word embedding, bilingual word embedding is still at its initial stage. On the other hand, bilingual word embedding directly depicts the internal relatedness among words of two different languages; theoretically, such merit enables bilingual word embedding to facilitate machine translation, or even fully replace word mapping matrix or dictionary in traditional machine translation approaches. The experiment result of [2] provided solid proof, which achieved an accuracy of around 90 % for English-Spanish word translation, with the help of bilingual word embedding. In light of this, bilingual word embedding has recently been applied successfully to machine translation [2], named entity recognition, word sense disambiguation [3], ctc.

BilBOWA by Gouws et al. [8] is a model for learning bilingual word embeddings. An outstanding advantage of BilBOWA is it eliminates the need of word alignments or dictionaries; and in an English-German cross-lingual classification task, BilBOWA achieved a speedup as high as three orders of magnitude.

As for non-alphabet-based languages like Chinese and Japanese, word embedding (and other NLP research as well) is also basically studied at the word level. But the individual character of a word actually often has semantical meaning itself. By taking the rich internal information of composing characters into consideration, Chen et al. [9] proposed a new model called character-enhanced word embedding model (CWE), and outperformed approaches that neglect such internal information. Another interesting fact of Chinese and Japanese is: both languages share a large portion of common characters. Through these common characters, Chu et al. [10] successfully constructed a Chinese-Japanese parallel corpus, which is highly accurate.

3   Common Character

3.1   Chinese and Japanese Characters

Chinese characters were invented over 4000 years ago, and have been spread with Chinese culture to neighboring countries like Korea, Japan, Vietnam, etc. Combined with the native languages of these countries, Chinese characters have become part of their writing systems. Early Japanese writing system was copied from Chinese characters, which made both countries able to understand the literature of each other. Through the years, Chinese characters were gradually integrated with Japanese native languages, and later formed a writing system with Chinese characters (kanji), hiragana and katakana combined.

There are three Chinese character systems currently, including traditional Chinese (in Hong Kong, Macau, Taiwan), simplified Chinese (in mainland China, Singapore, Malaysia, etc.), and Japanese Kanji. The latter two were independently simplified from traditional Chinese. Most Chinese characters in these three systems can be reciprocally corresponded, with the same shape or slight variances only. In Table 1 we can see 4 groups of corresponding Chinese characters with basically the same meanings. Therefore in real-world applications, most Chinese characters in a specific system can find their corresponding characters in other systems. For example, Chu et al. [11] proposed a Chinese character table involving traditional Chinese, simplified Chinese and Japanese.

Japanese vocabulary is categorized into several types: kango, wago, gairaigo, etc. Kango was either borrowed from Chinese or constructed from Chinese roots. As for wago, some of them are purely made up of Chinese characters, and there are also some mixed by Chinese characters and kana (hiragana and katakana),

Table 1. Examples of common characters in Simplified Chinese, Traditional Chinese, and Japanese Kanji.

Simplified	Traditional	Japanese	Meaning
王	王	王	King
图	圖	図	Picture
云	雲	雲	Cloud
国	國	国	Country

also a few wholly formed by kana. According to the Shinsen Kokugo Jiten Japanese dictionary, kango comprises 49.1 % vocabulary and wago makes up 33.8 %.

Hence roughly 80 % of Japanese words contain Chinese characters. A plausible conjecture is that the meanings of these Chinese characters are equivalent or somehow related to their meanings in Chinese. Sufficient works have been done to compare Chinese and Japanese words formed by common Chinese characters, which provide solid proof for this conjecture. Due to space limits, we skip these details here, but there can be totally 4 cases: $w_{zh} = w_{ja}$, $w_{zh} > w_{ja}$, $w_{zh} < w_{ja}$, and $w_{zh} \approx w_{ja}$, in which w_{zh} and w_{ja} denote the meaning count in Chinese and Japanese. An example for $w_{zh} \approx w_{ja}$ is that the word "意见" means "opinion" in both Chinese and Japanese. Moreover, it also means "dissatisfied" in Chinese, and "suggest" in Japanese.

3.2 From a View of Information Theory

From the above comparison between common characters in Chinese and Japanese, we can figure out the difference between Chinese/Japanese and western writing system. To be more specific, in Chinese or Japanese, a word can be constructed either by multiple Chinese characters, or by only one, since every Chinese character can solely make sense. Take the Chinese character "天" to illustrate, which can be either a word itself or part of a word in both Chinese and Japanese. In either case, "天" means one of "sky", "heaven", "day", and so on.

Word embedding essentially vectorizes the semantics of a word, and actually has been studied on Chinese characters. Consider a Chinese character C_{zh} in a Chinese article, and its corresponding character C_{ja} in Japanese; the semantic difference of C_{zh} and C_{ja} in their respective contexts can be depicted using mutual information, a concept measuring the dependency among random variables. Mutual information is defined as follows:

$$I(C_{zh}; C_{ja}) = \sum_{x \in C_{zh}} \sum_{y \in C_{ja}} p(x,y) log \frac{p(x,y)}{p(x)q(y)} \qquad (1)$$

Mutual information is non-negative and symmetric; given a condition Z about the distribution between C_{zh} and C_{ja}, conditional mutual information is:

$$I(C_{zh}; C_{ja}|Z) = E_{p(x,y|z)} log \frac{p(x,y|z)}{p(x|z)q(y|z)} \tag{2}$$

In this paper, we let the condition Z be $C_{zh} \in S_{zh}$, $C_{ja} \in S_{ja}$, in which S_{zh} and S_{ja} are a Chinese and Japanese sentence respectively in Chinese-Japanese translation. Through qualitative analysis, we can conclude that for common Chinese characters in corresponding Chinese and Japanese sentences, the probability of semantic equivalence is notably high. In subsequent experiments, to prove this conclusion, we will estimate the conditional mutual information of some Chinese characters within a certain amount of translated sentences.

4   Model

In this section, we first introduce the widely used model CBOW, and also some extensions on it. Then we present our own model named CJ-BOC.

4.1   CBOW

Mikolov et al. in [12] proposed continuous bag-of-words model (CBOW), where the optimization goal is to maximize a probabilistic language model as follows:

$$L(S) = \frac{1}{N} \sum_{i=1}^{N} p(w_i|Context(w_i)) \tag{3}$$

We use $Context(w_i)$ to denote the context of w_i here:

$$Context(w_i) = \{w_{i-K}, \ldots, w_{i-1}, w_{i+1}, \ldots, w_{i+K}\} \tag{4}$$

K is the window length here, which can be tuned in the model. The probability of each word in a given context can be computed using a softmax function:

$$p(w_i|Context(w_i)) = \frac{exp\{\bar{x}_i^T \cdot x_i\}}{\sum_{x_j \in W} exp\{\bar{x}_i^T \cdot x_j\}} \tag{5}$$

in which $\bar{x}_i$ is the average word embedding of the context:

$$\bar{x}_i = \frac{1}{2K} \sum_{x_j \in Context(x_i)} x_j \tag{6}$$

To better illustrate CBOW, we present its sketch as shown in Fig. 1. In the sentence "我们/的/祖国/像/花园" (Our homeland is like garden), we set the window length to 2; when learning the word "祖国", there are totally 4 words used for updating it, two of which following it and the other two before it.

In addition, CBOW can also be combined with Skip-gram, negative sampling and other methods for improvement. Such combination can be found in Word2Vec, and hence we do not elaborate here.

Fig. 1. An example of CBOW, in which the windows length is 2.

4.2 Bilingual CBOW

Now we consider CBOW in the scenario of bilingual word embedding. Generally, training bilingual word embedding requires a large corpus for each language, and also a relatively small parallel corpus acquired from strict translation and sentence-level alignment. The two monolingual corpora are trained to obtain the relatedness among words within a language, while the parallel corpus is used to calibrate the trained vectors between two languages.

Word-Aligned Method. The original CBOW can be extended to cater the bilingual scenario, with some extent of modifications. Two approaches are commonly used for such extension, one of them based on machine translation. This approach uses toolkit like GIZA++[2] to align parallel sentence-pairs on word level; by doing this, for each word, we can find its corresponding word. When two words reciprocally corresponded are trained, their respective contexts are used. For example, for sentence-pair $\langle S_{zh}, S_{ja} \rangle$, where $S_{zh} = \{w_{zh,1}, \ldots, w_{zh,N}\}$ and $S_{ja} = \{w_{ja,1}, \ldots, w_{ja,M}\}$, a pair of aligned words is $\langle w_{zh,i}, w_{ja,j} \rangle$, and the objective function for training $w_{zh,i}$ is:

$$L(S_{zh}) = \frac{1}{N} \sum_{i=1}^{N} \Big\{ p(w_{zh,i} | Context(w_{zh,i}, S_{zh})) + p(w_{zh,i} | Context(w_{ja,i}, S_{ja})) \Big\} \tag{7}$$

Using existing tools, the advantage of this approach is context can be optimally selected; this ensures an optimal training result, when data and other components of the model are fixed. However, a severe drawback of this approach is running the machine translation tools usually takes long time. Therefore, the training of model is unacceptably slow, leading to intractability in applications.

Sentence-Aligned Method. A simple modification can be made to avoid the above problem:

$$L(S_{zh}) = \frac{1}{N} \sum_{i=1}^{N} p(w_{zh,i} | Context(w_{zh,i}, S_{zh})) + p(w_{zh,i} | S_{ja}) \tag{8}$$

---

[2] http://www.statmt.org/moses/giza/GIZA++.html, accessed date: June 11, 2016.

In other words, all words in S_{ja} are introduced as the context of $w_{zh,i}$ in training, which obviously brings a large portion of noise in. However, based on this idea, some works have achieved satisfying results. One of them is BilBOWA [8], which reached a comparable level with state-of-the-art approaches after some modifications on the aggregation process. More importantly, the model is trained comparably fast with training monolingual word embedding.

4.3   Our Model: CJ-BOC

As has been mentioned, common Chinese characters of Chinese and Japanese are semantically similar or related. Inspired by the above extensions, we further exploit this feature and propose a CBOW-like model: Chinese-Japanese Bag of Characters model (CJ-BOC). In our proposed model, the objective function is:

$$L(S_{zh}) = \frac{1}{N} \sum_{i=1}^{N} \Big\{ p(w_{zh,i}|Context(w_{zh,i}, S_{zh})) \\ + \lambda \cdot p(w_{zh,i}|Context(w_{zh,i}, S_{ja})) \\ + \mu \cdot p(w_{zh,i}|S_{ja}) \Big\} \tag{9}$$

λ and ν here are both parameters of the model, and $Context(w_{zh,i}, S_{ja})$ can be acquired via common character matching:

$$Context(w_{zh,i}, S_{ja}) = \bigcup_{w_{ja} \in CC(w_{zh,i}, S_{ja})} Context(w_{ja}, S_{ja}) \tag{10}$$

$CC(\cdot)$ here is character matching, which means:

$$CC(w_{zh,i}, S_{ja}) = \{w_{ja}|w_{ja} \in S_{ja}, c \in w_{zh,i}, c \in w_{ja}\} \tag{11}$$

We use c to denote common character, and the function $CC(\cdot)$ is to find all words sharing common characters in S_{ja} for $w_{zh,i}$. Then the contexts of all these words are used for training $w_{zh,i}$. We illustrate an example in Fig. 2. "春天/到/了/樱花/开了" and "春/に/なる/と/桜/が/咲き/ます" are a parallel sentence-pair, meaning "Spring is coming, and cherry blossom is coming out". The model detects the Chinese character "樱" in "樱花 (cherry blossom)", which is identical to "桜(cherry blossom)" in Japanese. Therefore, these two characters share their respective context during training.

   Our method is apparently better than sentence-aligned method, which can be verified through qualitative analysis. Theoretically, our method introduces extra noise compared with word-aligned method, since multiple common Chinese characters with different meanings may co-exist even in the same sentence. But according to our subsequent experiments on mutual information, we can see that the noise ratio is actually minor, owing to the intrinsic nature of parallel sentence-pair. We also introduce the character information as auxiliary, which, to some extent, provides more information than word-aligned method.

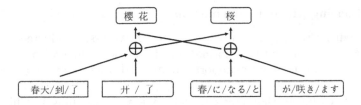

Fig. 2. An example of CJ-BOC, window length is 4, common character is "櫻(桜)".

5 Experiments and Analysis

5.1 Datasets and Experiment Settings

In our experiment, the parallel corpus is from [10], which includes Chinese-Japanese sentence-pair generated from wikipedia (both Chinese and Japanese version). We mainly used train.ja and train.zh in this dataset, both including 126811 lines of text. In both files, each line is a complete sentence, which is parallel to the sentence in its corresponding line of the other file. As preprocessing, we performed word segmentation on the two files using MeCab[3] for Japanese and Jieba[4] for Chinese before experiments. Parameters were tuned to ensure the segmentation on both languages were almost equally grained.

In our experiment, we took sentence-aligned method (Sen-AM) as the baseline, to comparably demonstrate the performance of our CJ-BOC. Both methods were implemented in C language[5]. Suppose there are N sentences in total, each of which containing k words, and each word consists of p characters; if the window length is w, and training a word costs $O(M)$ time, the theoretical worst-case time complexity of Sen-AM should be $O(k^2N + kMN)$, while our CJ-BOC has a worst-case time complexity of $O(p^2k^2wN + kMN)$. Since p, k and w are typically small, there is no notable difference between CJ-BOC and Sen-AM in terms of asymptotic worst-case complexity. To improve learning, we turned on both hierarchical softmax and 10-word negative sampling, which Mikolov proposed in Word2Vec. We set the window length to 5; λ and μ in CJ-BOC were 0.7 and 0.3 respectively. We used a Linux PC to run the experiment, which has a 2.6 GHz CPU and 8 G memory. Using Sen-AM and CJ-BOC to train the above parallel corpora, the time consumptions are 112.962 s and 178.022 s respectively; we can see that using CJ-BOC here only brought 63.5 % extra training time, compared with Sen-AM. Both methods finally generated 25219 word embeddings in Chinese and 22749 in Japanese, which will be used in subsequent experiments in Subsects. 5.3 and 5.4.

---

[3] http://taku910.github.io/mecab, accessed date: May 12, 2016.
[4] https://github.com/fxsjy/jieba, accessed date: August 2, 2015.
[5] Our source code is available at https://github.com/jileiwang/cjboc.

5.2 Estimating Mutual Information

Before evaluating the effect of word embedding, we first conducted an experiment to estimate the mutual information of common Chinese characters. We did this experiment, hoping to support our method from an information theory perspective. The process was as follows:

(1) Select a Chinese character ch, and retrieve its meaning in Chinese and Japanese according to Xinhua and Kojien dictionary respectively. Suppose ch has respectively n and m meanings in Chinese and Japanese, which are denoted as $(x_1, \ldots, x_n)$ and $(y_1, \ldots, y_m)$.

(2) Randomly choose M sentences with ch in both languages independently, and count the frequency of it with each meaning: $(a_1, \ldots, a_n)$ and $(b_1, \ldots, b_m)$. The meaning of ch in each sentence is acquired using manual annotation. If ch appears k times in a single sentence, each of the appearance is deemed to contribute a $\frac{1}{k}$ frequency to its corresponding meaning.

(3) Take the frequency of each meaning as the estimation of its appearance probability: $p(x_i) = \frac{a_i}{M}$ and $p(y_j) = \frac{b_j}{M}$.

(4) For the conditional probability $p(x_i|Z)$, $p(y_j|Z)$, and $p(x_i, y_j|Z)$, extract N parallel sentence pairs both of which include ch, and mark its semantics. Similar to steps 2 and 3, frequency is taken as an estimation of probability.

(5) Exploit the above information to calculate conditional mutual information:

$$I(X;Y|Z) = E_{p(x,y|z)} log \frac{p(x,y|z)}{p(x|z)q(y|z)} \tag{12}$$

(6) Given the unconditional joint probability:

$$p(x_i, y_j) = \frac{|Z| \cdot p(x_i, y_j|Z) + |\bar{Z}| \cdot p(x_i, y_j|\bar{Z})}{N^2} \tag{13}$$

where $|Z| = N$ and $|Z| + |\bar{Z}| = N^2$. We assume in this unconditional situation, a Chinese sentence may correspond to any Japanese sentence in the corpus, and vice versa. To be more specific, given this independency, we can thus compute $p(x_i, y_j|\bar{Z}) = p(x_i)p(y_j)$.

(7) The unconditional mutual information can be computed using the above information.

To illustrate, we selected two representative common characters "天(means sky, heaven, etc.)" with $M = 78$ and $N = 54$, and "地(means ground, earth, etc.)" with $M = 83$ and $N = 51$. The results are show in Table 2.

An intuitive conjecture, based on experiences, is that conditional mutual information should be greater than unconditional mutual information. Table 2 somehow provides evidence for this conjecture, as we can see that conditional mutual information of these two representative characters are significantly larger than the corresponding unconditional mutual information. This indicates in a sentence-pair of parallel corpus, a pair of common Chinese characters are likely to be equivalent semantically. Therefore, in our CJ-BOC, the extraction of context using common Chinese characters is rooted in solid theoretical base.

Table 2. Mutual information.

	天	地
Mutual information	0.3369	0.5804
Conditional mutual information	30.3057	87.4515

5.3 Cross-Lingual Synonym Comparison

In translation, some word in the source language can actually correspond to multiple words in the target language, and we can also find one-to-one correspondence of words between source and target language. To avoid ambiguity, we randomly selected 15 one-to-one corresponding word-pairs $\langle w_{zh}, w_{ja}\rangle$, and for each pair, we: (1) compute the cosine distance d between w_{zh} and w_{ja}; (2) compute the cosine distance between w_{zh} and every Japanese word in the corpus C_{ja}, and obtain the rank of d among them; (3) conduct similar operations as in (2) on w_{ja} and every Chinese word in the corpus C_{zh}.

Some of the selected words have common characters, while some do not. Two examples are given in Table 3.

Table 3. Examples of word-pairs and their experimental results.

Word-Pair			Cosine Distance		Rank $w_{zh} \to C_{ja}$		Rank $w_{ja} \to C_{zh}$	
w_{zh}	w_{ja}	Meaning	Sen-AM	CJ-BOC	Sen-AM	CJ-BOC	Sen-AM	CJ-BOC
巴黎	パリ	Paris	0.1271	0.7793	1422	1	729	1
东京	東京	Tokyo	0.1555	0.2029	910	1585	344	1007

For the rank of each word, we compute its relative rate among all words as below:

$$rate = (1 - \frac{rank - 1}{total\_word\_num}) \times 100\,\% \qquad (14)$$

None of the words in these word-pairs is ambiguous, which makes it apparent that both cosine distance and the rate are favorable to be large. The averaged results for all word-pairs are shown in Table 4. From Table 4, we can see obvious superiority of CJ-BOC compared with Sen-AM.

Table 4. Average of cosine distances and rates of 15 word-pairs.

Method	Cosine Distance	Rate $w_{zh} \to C_{ja}$	Rate $w_{ja} \to C_{zh}$
Sen-AM	0.1404	92.8\%	94.0\%
CJ-BOC	0.4682	97.6\%	96.3\%

5.4 Cross-Lingual Analogical Reasoning

Analogical reasoning [13,14] tries to figure out the potential analogy among word embeddings. Take a frequent question in IQ test as an example, what is the answer of "父亲(father):母亲(mother)::男孩(boy):X"? The expected answer is "女孩(girl)", which is deduced as: x=v(母亲)-v(父亲)+v(男孩) according to analogical reasoning. The closest word embedding to x is retrieved as the answer, which indicates a single successful task if it is v(女孩).

Given several related words from different languages, cross-lingual analogical reasoning works as follows: y=v(はは)-v(ちち)+v(男孩), we hope that the relatedness between Japanese words "はは(mother)" and "ちち(father)" could help us find the Chinese character "女孩(girl)" through "男孩(boy)".

Among the 4 words in cross-lingual analogy tasks, the number of Chinese word to the number of Japanese word can be 2 to 2, 1 to 3 or 3 to 1. For the 2 to 2 case, Mikolov in [2] mentioned that, even the word embeddings were trained independently within their own language (i.e., not in a cross-lingual manner), the trained embeddings could still capture good features. So we do not consider this case in our experiment. Instead we selected 72 groups of 1 to 3 or 3 to 1 cross-lingual analogy tasks, with the input format being $\{A : B :: C : D\}$. We computed the vector $V = B - A + C$, and just as we did in experiment 2, we calculated the cosine distance d between V and D; besides, we also figured out the cosine distance between every pair $\langle V, D' \rangle$ (D' is from the corpus of D), obtained the rank of d among them, and then transformed its rank to the corresponding rate. We averaged the results to make them comparable: the average Cosine distance of Sen-AM and CJ-BOC is 0.2309 and 0.3882 respectively; while Sen-AM has a rate of 85.8 %, roughly 10 % lower than that of CJ-BOC, which is 95.3. Based on these results, we can conclude that the quality of our word embedding is obviously higher than that of Sen-AM.

6 Conclusion and Future Work

In this work, we first reviewed the background of word embedding and bilingual word embedding, and then demonstrated the semantical equivalence and relatedness among common characters of Chinese and Japanese. In addition, we performed theoretical derivation and empirical study on the common character, from a perspective of information theory. In light of the feature of common character, we proposed our CJ-BOC model based on CBOW, which can learn Chinese-Japanese word embedding efficiently. According to our experiments on task of bilingual analogical reasoning, the word embeddings generated using CJ-BOC were obviously better than those of sentence-alignment methods.

As for future work, our research could possibly move on towards two directions: (1) the exploitation of common character should not be limited by the model. In other words, apart from the NNLM-based Word2Vec, we can also use methods like GloVe instead. (2) [9] actually provided a promising idea, which is to train word embeddings and character embeddings at the same time. And such combination could further verify the characteristics of common characters.

Acknowledgement. This research is supported in part by the Major State Basic Research Development Program of China (973 Program, 2012CB315803), the National Natural Science Foundation of China (61371078), and the Research Fund for the Doctoral Program of Higher Education of China (20130002110051).

References

1. Hinton, G.E.: Learning distributed representations of concepts. In: Proceedings of the Eighth Annual Conference of the Cognitive Science Society
2. Mikolov, T., Le, Q.V., Sutskever, I.: Exploiting similarities among languages for machine translation (2013). arXiv:1309.4168
3. Guo, J., Che, W., Wang, H., Liu, T.: Learning sense-specific word embeddings by exploiting bilingual resources. In: Proceedings of COLING, pp. 497–507 (2014)
4. Bengio, Y., Ducharme, R., Vincent, P., Janvin, C.: A neural probabilistic language model. J. Mach. Learn. Res. **3**, 1137–1155 (2003)
5. Mnih, A., Hinton, G.E.: A scalable hierarchical distributed language model. In: Advances in Neural Information Processing Systems, pp. 1081–1088 (2009)
6. Mikolov, T., Yih, W.T., Zweig, G.: Linguistic regularities in continuous space word representations. In: HLT-NAACL, pp. 746–751 (2013)
7. Pennington, J., Socher, R., Manning, C.: Glove: global vectors for word representation. In: Conference on Empirical Methods in Natural Language Processing (2014)
8. Gouws, S., Bengio, Y., Corrado, G.: BilBOWA: fast bilingual distributed representations without word alignments (2014). arXiv:1410.2455
9. Chen, X., Xu, L., Liu, Z., Sun, M., Luan, H.: Joint learning of character and word embeddings. In: International Conference on Artificial Intelligence. AAAI Press (2015)
10. Chu, C., Nakazawa, T., Kurohashi, S.: Constructing a Chinese-Japanese parallel corpus from Wikipedia. In: Proceedings of the Ninth Conference on International Language Resources and Evaluation (LREC 2014), Reykjavik, Iceland, May 2014
11. Chu, C., Nakazawa, T., Kurohashi, S.: Chinese characters mapping table of Japanese, traditional Chinese and simplified Chinese. In: Proceedings of the Eighth Conference on International Language Resources and Evaluation (LREC 2012), pp. 2149–2152, Istanbul, Turkey, May 2012
12. Mikolov, T., Chen, K., Corrado, G., Dean, J.: Efficient estimation of word representations in vector space. arXiv:1301.3781
13. Veale, T.: An analogy-oriented type hierarchy for linguistic creativity. Knowl. Based Syst. **19**(7), 471–479 (2006)
14. Veale, T., Li, G.: Analogy as an organizational principle in the construction of large knowledge-bases. In: Prade, H., Richard, G. (eds.) Computational Approaches to Analogical Reasoning: Current Trends. SCI, vol. 548, pp. 83–101. Springer, Heidelberg (2014). doi:10.1007/978-3-642-54516-0_4

Content and Document Analysis

Analyzing Topic-Sentiment and Topic Evolution over Time from Social Media

Yan Hu, Xiaofei Xu, and Li Li[(✉)]

School of Computer and Information Science, Southwest University, Chongqing
400715, China
guyuemuzhi@sina.cn, lily@swu.edu.cn

Abstract. Most online news websites have enabled users to annotate
their sentiments while reading the news. Different from traditional users'
feedbacks such as reviews or ratings, those annotations are more intuitive
to express the sentiment of the users. Topic model is proved more effective to analyze the text information, however, most existing topic models
focus on either extracting static topic sentiment or tracking topics over
time but ignoring sentiment analysis. In the paper, we propose a joint
topic-sentiment over time model (JTSoT) to detect the topic-sentiment
shift and track the topic evolution over time. The critical challenge is
how to balance the relationship among the topic, sentiment and time.
The topic is represented as a Beta distribution over time and a Dirichlet
distribution with respect to the sentiment. We evaluate our method on
the real-world news dataset. The experimental results show that we have
achieved high correlation between the topic and sentiment, better interpretable topic evolution, and higher document sentiment classification
result and perplexity.

Keywords: Topic models · Joint topic-sentiment over time model ·
Topic-sentiment shift · Topic evolution

1 Introduction

Topic model and sentiment analysis are two active research fields in natural
language processing as well as data mining. Some online news websites, such
as Baidu and Sina, allow users to annotate different types of sentiments after
reading news to express their personal sentiment. From these news and sentiment
annotations, we may get more accurate analysis about topic and sentiment. Some
useful works have been done in this aspect [1–5], but most of these works just
extract sentiments of topics in a static way, ignoring the dynamic property of
text data. Other works [6–8] focus on analyzing the evolution of topics without
considering the sentiment. Motivated by this observation, we propose a novel
LDA-based [9] approach to extract topic-sentiment from text as well as the
evolution of topics over time.

The proposed model mainly has 3-level output: document-topic, topic-
sentiment-word and topic evolution over time. The process of modeling topics

© Springer International Publishing AG 2016
F. Lehner and N. Fteimi (Eds.): KSEM 2016, LNAI 9983, pp. 97–109, 2016.
DOI: 10.1007/978-3-319-47650-6_8

is affected not only by the word co-occurrence but also the sentiment and time. Here the topic is presented as a Beta distribution over time and has a Dirichlet distribution with respect to sentiment. So it can get the relationship between topics and sentiments and the evolution of topics over time. All of these can be extracted simultaneously without any post-processing.

The contribution of this paper can be summarized as follows. First, time is jointly modeled with topics which can capture the evolution of topics over time by representing topic as a beta distribution over time. Generally speaking, we considered multi features including time factor, which makes the model more reliable in practice. Secondly, we can detect topic and sentiment simultaneously. Finally, we carry out extensive experiments to demonstrate the efficiency of our approach and analyze the association among topic, sentiment and time. Based on a ground-truth evaluation framework, we compare our model to other art-of-art models, such as TOT [6] and eToT [10].

The remainder of this paper is organized as follows. In Sect. 2, we briefly review several representative works related to our method. Section 3 mainly presents our approach in detail. We show the data preparation and discuss the experiment results in Sect. 4. Finally, Sect. 5 concludes this paper.

2   Related Work

We mainly analyze the existing work from two aspects containing topic models and topic sentiment analysis. We will briefly discuss these related works and the relationship between them and the proposed model in this paper.

2.1   Topic Models

Topic models are hierarchical Bayesian models of discrete data, becoming a widely used approach for exploratory and predictive analysis of text. There are many different kinds of topic models have been proposed [6,7,9,11,12], among which LDA [9] is a probabilistic extension of latent semantic indexing (LSI) [13] and probabilistic latent semantic indexing (pLSI) [14].

Documents are usually collected over time, such as online discussion, news, consequently these contents may evolve and change over time. To this end, several topic models have explored different topics and their evolvements over time. Topics over Time (TOT) [6] is a LDA-based model for capturing the topics evolvement over time and associates each topic with a Beta distribution over time. Besides, Dynamic Topic Model (DTM) [7] has used state space model to model the change of topic-word distribution over time. Wang et al. [15] replace state space model with Brownian motion which is actually a continuous generalization of state space model. Another dynamic topic model is Topic Tracking Model (TTM) [8] tracking consumer purchase behaviors. Different from DTM, it replaces Gaussian distribution with Dirichlet distribution to use Gibbs sampling to estimate the parameters.

2.2   Topic Sentiment Analysis

There exists many works about sentiment classification [16–18] without considering the mixture of topics in the text. Joint Sentiment Topic model (JST) [2,19] has been developed to extract topic under different sentiment label by inserting a sentiment layer before the topic layer based on LDA. Reverse-JST [2], JTS [20] and JTV [21] are a variant of JST where the order of sentiment and topic layers is inverted. There also exists some different models like Aspect-and-Sentiment Unification Model (ASUM) [22] and Sentiment-LDA [1].

Besides, the issue of modeling topic-sentiment evolution is explored recently. Topic sentiment trend model (TSTM) [23] extends a new layer to capture the temporal dimension based on TSM [3]. He et al. [24,25] introduce Dynamic-JST based on the previously proposed JST model to capture the qualitative topic evolution over time. Time-aware Topic-Sentiment (TTS) [26] is proposed to analyze topic-sentiment evolution over time by modeling time jointly with topics and sentiments. However, in our approach the topic is presented as a Beta distribution over time which is different from TTS. The emotion Topic Over Time (eToT) [10] can uncover the latent relationship among news, emotion and time directly. At the same time, emotion-based Dynamic Topic Model (eDTM) is proposed to explore the state space model for tracking the dynamics of topics where the model at each timestamp is derived from the model at the previous one.

From these observations, we mainly insert a sentiment layer after topic layer and regard time as an attribute of topic to avoid the dependence of time span. More details will be discussed subsequently.

3   Joint Topic Sentiment over Time Model

In this section, we will explain the details of joint topic sentiment over time (JTSoT) model to model topic-sentiment association as well as the topic evolution over time.

For social sentiment analysis, users' sentiments about a document are mainly decided by their sentiments towards to the topics of the document. We extend LDA model to a four layers model by adding sentiment layer and time layer to capture sentiment and time respectively as shown in Fig. 1.

Besides, documents from the news dataset must be annotated with time (day, month, year). Time is first discretized and each document d has a discrete

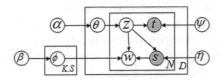

Fig. 1. Graphical representation of JTSoT model

Table 1. Notations in JSToT

D	Number of documents
V	Vocabulary size
K	Number of latent topics
S	Number of sentiments
α, β, η	Dirichlet parameters
ψ	Beta parameters
θ	Documents distribution for topics
ϕ	Topics distribution for words
n_d	Number of words in document d
$n_{d,k}$	Number of words in document d assigned to topic k
$n_{k,l}$	Number of words assigned to topic k and sentiment l
$n_{v,k,l}$	Number of word v assigned to topic k and sentiment l

timestamp t and a distribution over sentiment s. Next, we will explain our solution to model the topic-sentiment-time association. The notations in the rest of this paper are given in Table 1.

3.1 Generative Process

JTSoT is a generative model of words, sentiments and timestamp. Its generative process corresponding to the graphical model shown in Fig. 1 is as follows:

(1) Draw $K * S$ multinomials $\phi_{z,l} \sim Dir(\beta)$

(2) For each document d, draw a multinomial $\theta_d \sim Dir(\alpha)$, then for each word $w_{d,i}$ in d:

    (a) Draw a topic $z_{d,i} \sim Multi(\theta_d)$

    (b) Draw a multinomial distribution $s_{d,i} \sim Dir(\eta_{z_{d,i}})$

    (c) Draw a word $w_{d,i} \sim Multi(\phi_{z_{d,i},s_{d,i}})$

    (d) Draw a timestamp $t_{d,i} \sim Beta(\psi_{z_{d,i}})$

By examining the generative process of JTSoT model, we assume all words in a document share same sentiments distribution s and same timestamp t with the document.

3.2 Inference

The solution to estimate the parameters in topic models [2,5] is to use Gibbs sampling. The detailed inference via Gibbs sampling on LDA is presented in [27].

(a) Joint distribution: The joint probability of words, topics, sentiments and timestamp can be factored based on Bayes conditional independence rule as follows:

$$p(\mathbf{w}, \mathbf{t}, \mathbf{s}, \mathbf{z}|\alpha, \beta, \eta, \psi) = p(\mathbf{w}|\mathbf{s}, \mathbf{z}, \beta)p(\mathbf{t}|\mathbf{z}, \psi)p(\mathbf{s}|\mathbf{z}, \eta)p(\mathbf{z}|\alpha) \qquad (1)$$

The first term is obtained by integrating over ϕ.

$$p(\mathbf{w}|\mathbf{s},\mathbf{z},\beta) = \left(\frac{\Gamma(\sum_{v=1}^{V}\beta_v)}{\prod_{v=1}^{V}\Gamma(\beta_v)}\right)^{K\cdot S} \prod_{z=1}^{K}\prod_{l=1}^{S}\frac{\prod_{v=1}^{V}\Gamma(n_{v,z,l}+\beta_v)}{\Gamma\left(\sum_{v=1}^{V}n_{v,z,l}+\beta_v\right)} \quad (2)$$

The second term in Eq. (1) directly associates topics with time. Due to the versatile shapes of Beta distribution, we select it. The timestamp range of the data needs to be normalized to range from 0 to 1. Given a topic z, the probability of a observed timestamp t can be calculated as Eq. (3).

$$p(t|\mathbf{z},\psi) = \prod_{d=1}^{D}\prod_{i=1}^{N_d}\frac{(1-t_{d,i})^{\psi_{z_{d,i},1}-1}(t_{d,i})^{\psi_{z_{d,i},2}-1}}{B\left(\psi_{z_{d,i},1},\psi_{z_{d,i},2}\right)} \quad (3)$$

We assume each topic has a latent distribution with respect to sentiments. The number of each sentiment of each document in the news dataset can be observed. We assume that each topic z has a Dirichlet distribution η_z with respect to sentiment. Given a topic, the probability of a observed sentiment distribution $\mathbf{s}$ can be calculated as follows:

$$p(\mathbf{s}|\mathbf{z},\eta) = \prod_{d=1}^{D}\prod_{i=1}^{N_d}\frac{\Gamma\left(\sum_{s=1}^{S}\eta_{z_{d,i},l}\right)}{\prod_{s=1}^{S}\Gamma\left(\eta_{z_{d,i},l}\right)}\prod_{s=1}^{S}s_{d,i,s}^{\eta_{z_{d,i},s}-1} \quad (4)$$

where $\mathbf{s}$ is the vector of proportion of sentiments, S is the number of categories of sentiments which is the same with the size of vector η_z. The rest term of Eq. (1) is obtained in the same way by integrating over θ depicted in [26].

(b) Posterior distribution: In the Gibbs sampling procedure, what we need to calculate is the posterior distribution $p(s_{d,i} = l, z_{d,i} = k|\mathbf{w},\mathbf{t},\mathbf{s}_{\neg d,i},\mathbf{z}_{\neg d,i},\alpha,\beta,\eta,\psi)$ where $\neg d,i$ denotes the quantity of data that excludes the word w_i of current document d. Posterior probability can be derived from joint probability as follows:

$$p(s_{d,i}=l,z_{d,i}=k|\mathbf{w},\mathbf{t},\mathbf{s}_{\neg d,i},\mathbf{z}_{\neg d,i},\alpha,\beta,\eta,\psi) \propto$$

$$\frac{n_{d,k,\neg d,i}+\alpha_k}{\sum_{z=1}^{K}(n_{d,z}+\alpha_z)-1}\frac{n_{w_{d,i},k,l,\neg d,i}+\beta_{w_{d,i}}}{\sum_{v=1}^{V}(n_{v,k,l}+\beta_v)-1}$$

$$\frac{(1-t_{d,i})^{\psi_{k,1}-1}(t_{d,i})^{\psi_{k,2}-1}}{B\left(\psi_{k,1},\psi_{k,2}\right)}\left(\frac{\Gamma\left(\sum_{s=1}^{S}\eta_{k,s}\right)}{\prod_{s=1}^{S}\Gamma\left(\eta_{k,s}\right)}\cdot\prod_{s=1}^{S}s_{d,i,s}^{\eta_{k,s}-1}\right)\frac{\eta_{k,l}}{\sum_{s=1}^{S}\eta_{k,s}} \quad (5)$$

where $s_{d,i,s}$ is the s_{th} sentiment proportion of the word w_i of document d. After each iteration of Gibbs sampling, we update ψ and η as follows:

$$\psi_{k,1} = \overline{t_k}\left(\frac{\overline{t_k}(1-\overline{t_k})}{S^{t_k^2}}-1\right) \quad (6)$$

$$\psi_{k,2} = (1-\overline{t_k})\left(\frac{\overline{t_k}(1-\overline{t_k})}{S^{t_k^2}}-1\right) \quad (7)$$

$$\eta_{k,l} = \overline{s_{k,l}} \left(\frac{\overline{s_{k,l}}(1 - \overline{s_{k,l}})}{S^{s2}_{k,l}} - 1 \right) \tag{8}$$

where $\overline{t_k}$ and S^{t2}_k are the sample mean and the biased sample variance of the timestamps of words belonging to topic k, respectively. Similarly, $\overline{s_{k,l}}$ and $S^{s2}_{k,l}$ are the sample mean and the biased sample variance of the l_{th} sentiment of words belonging to topic k. After finishing iterations, we can get the distribution θ and ϕ as follows:

$$\theta_{d,k} = \frac{n_{d,k} + \alpha_k}{\sum_{z=1}^{K} (n_{d,z} + \alpha_z)} \tag{9}$$

$$\phi_{k,l,i} = \frac{n_{i,k,l} + \beta_i}{\sum_{v=1}^{V} (n_{v,k,l} + \beta_v)} \tag{10}$$

(c) Gibbs sampling algorithm: The Gibbs sampling procedure of JSToT is given in Algorithm 1.

Algorithm 1. Inference on JTSoT

Require: α, β, ψ, η
 1: Initialize matrices θ, ϕ
 2: **for** i=1 to max Gibbs sampling iterations **do**
 3:     **for** all documents $d \in [1, D]$ **do**
 4:         **for** all words $w \in [1, N_d]$ **do**
 5:             Exclude word w associated with l_{th} sentiment and topic k from document d and update count variables
 6:             Sample a topic and a sentiment for word w using Eq. (5)
 7:             Update count variables with new topic and sentiment
 8:         **end for**
 9:     **end for**
10:     Update the parameters ψ and η
11:     Update the matrices θ and ϕ
12: **end for**

---

4   Experiment Results

In this section, we evaluate the performance of our approach by comparing it with other methods in perplexity and document classification. The parameter settings for the whole experiments in this paper are consistent. Following with the settings in [10], the hyperparameters α and β are empirically set to $50/K$ and 0.01. Meanwhile, all results are obtained after 500 iterations when the algorithm achieves converge.

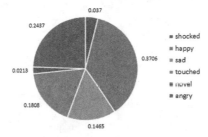

Fig. 2. Sentiment category of news

4.1 Data Preparation

The experiment dataset was collected from the Sina News[1] from August 21, 2012 to November 11, 2013 [10] which contains 7504 news with 4844594 sentiment annotations. After reading these news, readers can annotate their sentiments including "shocked", "happy", "sad", "touched", "novel" and "angry". For the process of word segmentation needed in Chinese, the JieBa Tokenization[2] is needed and all stopwords are removed. Besides, in order to express the results analysis, we translate the Chinese news expression into English [10].

Here we analyze the sentiments probability of each document and choose the max sentiment probability as the sentiment category of each document. Different sentiment categories document and their proportion distribution are shown in Fig. 2. We can find that "happy" and "angry" categories are more than others.

4.2 Analysis of Topic Sentiment Distribution and Topic Time Distribution

We have illustrated that the proposed model mainly can discover the relationship among topic, sentiment and time. Different topics have different sentiment distribution and time distribution. Here we choose several topics to analyze their distribution $p(s|z)$ over six sentiments shown in Fig. 3 and their distribution $p(t|z)$ over timestamp shown in Fig. 4.

From Fig. 3, we can find that topic 1, topic 2, topic 4 have similar distribution over sentiments, where the sentiment "touched" has the highest probability. However topic 3 has a more uniform distribution over sentiments. We can also get some information from the topic sentiment word distribution $\phi_{k,l,i}$. We mainly analyze the events that words of each topic can express and the time distribution of topic. Topic 1 involves a event report that a couple constantly saved the people trapped in the flood on April, 2013 and topic 3 mainly comes from some simple background reports. Topic 2 and topic 4 also illustrate social event reports about "return money found" and "offer help" respectively. However, the distributions of topic1, topic 2 and topic 4 over timestamp are very different and the distributions

---

[1] http://news.sina.com.cn/society.
[2] http://www.oschina.net/p/jieba.

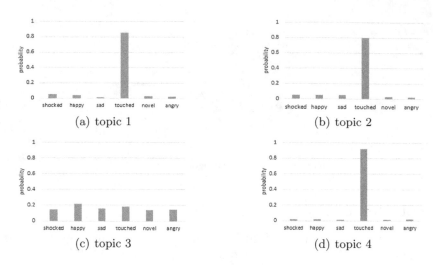

Fig. 3. The distributions of different sentiments in 4 different topics

of topic 2 and topic 3 are smoother in Fig. 4. The topic 1 mainly happened from January, 2013 (timestamp 0.4) to April, 2013 (timestamp 0.6) which reported some events containing words "accident", "save" and "hospital". The topic 4 and topic 6 also have higher probability during this period. So we can demonstrate that time has an important effect on topic analysis.

4.3   Document Sentiment Classification

Except analyzing the relationship among topic, sentiment and time, we can apply our model to solve the document sentiment classification problem. According to the above data preparation shown in Fig. 2, we conduct a six class-classification problem on the news dataset. Here we can predict the sentiment s* of each document d by calculating the highest generation probability as s* $= \arg\max_s p(s|d)$,

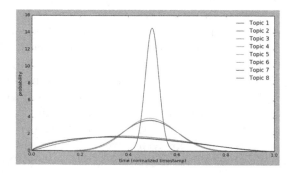

Fig. 4. The distribution of 8 topics over different time

Table 2. Classification accuracy

Method	Accuracy	Max accuracy	Min accuracy	SE
LDA-SVM	0.5993	0.6125	0.5875	0.008658
LDA-XGBoost	0.6378	0.6502	0.6283	0.007993
eToT	0.6835	0.7025	0.6575	0.015937
JTSoT	0.7037	**0.7205**	0.6825	0.014905

where $p(s|d) \propto \sum_z \prod_{w \in d} p(w|s, z)p(s|z)p(z)$. In this experiment, we mainly choose LDA-SVM, LDA-XGBoost, eToT as the baselines. We regard the document topic distribution as the features of document based on LDA when we set the topic number as 100 and use SVM and XGBoost [28] as the classifiers respectively. We run the classifiers five times using different 80−20% partitions of train-test sets. Besides, we run eToT and JTSoT at five different topic numbers, $k \in \{20, 30, 40, 50, 60\}$. The experiment results in terms of classification accuracy are shown in Table 2. The JTSoT can perform better than other baselines and the JTSoT and eToT can get max classification results when the topic number is set 50. Above, the standard error (SE) is defined as the standard deviation of the sampling distribution of the test data. The SE values in Table 2 are calculated according to their accuracy on five different results. The SE value presents the stability of classification accuracy. Because the results of eToT and JTSoT are influenced by the topic numbers, their stability is worse than other baselines. But eToT and JTSoT are weakly supervised classification methods rather than supervised SVM and XGBoost.

4.4   Perplexity Analysis

Here we use the perplexity criterion to measure the performance of the proposed topic model to fit a new held-out data. The lower the perplexity is, the less "perplexed" the model of the unseen data is, the better the generalization performance is. Perplexity is used to measure the likelihood of the test corpus given the proposed model parameters [29]. For a set of test document M, the perplexity is defined as:

$$perplexity(D_{test}) = \exp\left\{ -\frac{\sum_d \sum_{w \in d} \ln p(w)}{\sum_d |d|} \right\}. \tag{11}$$

Where D_{test} represents the test dataset and the $|d|$ is the length of document d. The probability of each word can be calculated as $p(w) = \sum_k \sum_l p(k, l, w) = \sum_k \sum_l p(w|k, l)p(l|k)p(k)$.

We computer the perplexity under estimated parameters of JTSoT model and compare it with those of eToT, ToT and LDA on news dataset. Here we run these models 5 times at different topics and select the average values to show in Fig. 5. The results show that the JTSoT outperforms better in the experiment. Besides, the change of perplexity of these methods is not too sensitive to the

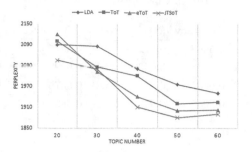

Fig. 5. Perplexity for different topics (lower is better)

Table 3. Example of topic 5 extracted by JSToT under different sentiments

Topic 5					
Shocked		Happy		Sad	
w	$p(w|k,l)$	w	$p(w|k,l)$	w	$p(w|k,l)$
Whu	0.005733	edible	0.001305	bag	0.002522
kill	0.004712	solatium	0.001287	kill	0.002091
capable	0.004366	relief	0.001205	comrade	0.001861
neighbor	0.004035	price	0.001186	Nanhai	0.001825
official	0.003825	fan	0.001136	muddy	0.001805
secretary	0.003505	homes	0.001132	partner	0.001765
lest	0.003435	writer	0.001087	sub-health	0.001672
embezzle	0.00331	space	0.001065	fisherman	0.001532
income	0.003325	drinks	0.000869	stunned	0.001386
Garnet	0.003015	interview	0.000652	wave	0.001261
Touched		Novel		Angry	
w	$p(w|k,l)$	w	$p(w|k,l)$	w	$p(w|k,l)$
Germany	0.001769	farmyard	0.022556	bridge	0.001704
safety	0.001676	apologize	0.015037	statue	0.001661
refugee	0.001632	Mansion	0.014255	arrest	0.001383
advanced	0.001585	sexy	0.013565	choke	0.001337
deeds	0.001532	drain	0.012986	cheongsam	0.001314
bomb	0.001415	populace	0.011346	scrap	0.001278
baby	0.001235	connotation	0.011146	powerless	0.001245
Jew	0.001147	in love	0.010625	ride	0.001199
Outpatient	0.1129	bed	0.009725	foraging	0.001129
exhaust	0.001061	relax	0.007518	hill	0.001083

change of topic number and the JTSoT reaches the minimal value when topic number is set to 50.

4.5 Case Study

In this subsection, we mainly extract topics from the dataset without subjectivity detection and evaluate the effectiveness of topic sentiment captured by the model. The distribution of words given topic and sentiment was estimated using Eq. (10).

A word is drawn from the distribution over words conditioned on both topics and sentiments unlike the LDA model that a word is drawn from the topic-word distribution. So we analyze the extracted topics under six different sentiments. Here we select to show the top 10 ranked words of topic 5 under six sentiments in Table 3.

We can observe that the "Whu (Wuhan university)" and "kill" can help us to target a social event on August, 2013, which was a young pregnant woman killed a university faculty member on the highway, so the most possible sentiment is "shocked" when users read the news. Besides, "Germany", "Jew", "bomb" and "refugee" mainly come from a report on October, 2013 that Daozhi Lin entrusted by a Jew and his family tried their best to protect the books during World War II. These results can illustrate the effectiveness of JTSoT in detecting the relationship between topic and sentiment.

5 Conclusion

In this paper, we proposed the joint topic sentiment over time model (JTSoT) which can track the evolution of the topic and discover the relationship between the topic and sentiment. It is demonstrated that JTSoT can outperform the state-of-the-art models in terms of perplexity and document sentiment classification. The proposed model can automatically and accurately detect the expression on the news dataset.

Acknowledgement. This work was supported by Natural Science Foundation of China (No. 61170192).

References

1. Li, F., Huang, M., Zhu, X.: Sentiment analysis with global topics and local dependency. In: AAAI 2010, pp. 1371–1376 (2010)
2. Lin, C., He, Y., Everson, R., Ruger, S.: Weakly supervised joint sentiment-topic detection from text. IEEE Trans. Knowl. Data Eng. **24**(6), 1134–1145 (2012)
3. Mei, Q., Ling, X., Wondra, M., Su, H., Zhai, C.: Topic sentiment mixture: modeling facets and opinions in weblogs. In: Sixteenth International Conference on World Wide Web, pp. 171–180. ACM (2007)

4. Jo, Y., Oh, A.H.: Aspect and sentiment unification model for online review analysis. In: Proceedings of the Fourth ACM International Conference on Web Search and Data Mining, pp. 815–824. ACM(2011)

5. Li, C., Zhang, J., Sun, J.T., Chen, Z.: Sentiment topic model with decomposed prior. In: SIAM International Conference on Data Mining (SDM 2013). Society for Industrial and Applied Mathematics (2013)

6. Wang, X., McCallum, A.: Topics over time: a non-Markov continuous-time model of topical trends. In: Proceedings of the 12th ACM SIGKDD International Conference on Knowledge Discovery and Data Mining, pp. 424–433. ACM (2006)

7. Blei, D.M., Lafferty, J.D.: Dynamic topic models. In: Proceedings of the 23rd International Conference on Machine Learning, pp. 113–120. ACM (2006)

8. Iwata, T., Watanabe, S., Yamada, T., Ueda, N.: Topic tracking model for analyzing consumer purchase behavior. IJCAI 10, 1427–1432 (2009)

9. Blei, D.M., Ng, A.Y., Jordan, M.I.: Latent Dirichlet allocation. J. Mach. Learn. Res. 3, 993–1022 (2003)

10. Zhu, C., Zhu, H., Ge, Y., Chen, E.: Tracking the evolution of social emotions: a time-aware topic modeling perspective. In: 2014 IEEE International Conference on Data Mining, pp. 697–706. IEEE (2014)

11. Hirschberg, J., Manning, C.D.: Advances in natural language processing. Science 349(6245), 261–266 (2015)

12. Liu, Q., Ge, Y., Li, Z., Chen, E.: Personalized travel package recommendation. In: 2011 IEEE 11th International Conference on Data Mining, pp. 407–416. IEEE (2011)

13. Deerwester, S., Dumais, S.T., Furnas, G.W., Landauer, T.K., Harshman, R.: Indexing by latent semantic analysis. J. Am. Soc. Inf. Sci. 41(6), 391 (1990)

14. Hofmann, T.: Probabilistic latent semantic indexing. In: Proceedings of the 22nd Annual International ACM SIGIR Conference on Research and Development in Information Retrieval, pp. 50–57.ACM (1999)

15. Wang, C., Blei, D.M., Heckerman, D.: Continuous time dynamic topic models. In: Proceedings of the 24th Conference in Uncertainty in Artificial Intelligence, Helsinki, Finland, 9-12 July 2008, pp. 579–586. AUAI Press (2008)

16. Wu, F., Huang, Y.: Collaborative multi-domain sentiment classification. In: 2015 IEEE International Conference on Data Mining, pp. 459 468. IEEE (2015)

17. Pang, B., Lee, L., Vaithyanathan, S.: Thumbs up? sentiment classification using machine learning techniques. In: Proceedings of the ACL 2002 Conference on Empirical Methods in Natural Language Processing, vol. 10, pp. 79–86. Association for Computational Linguistics (2002)

18. Blitzer, J., Dredze, M., Pereira, F.: Biographies, bollywood, boom-boxes and blenders: domain adaptation for sentiment classification. In: ACL 2007, pp. 40–447 (2007)

19. Lin, C., He, Y.: Joint sentiment/topic model for sentiment analysis. In: Proceedings of the 18th ACM Conference on Information and Knowledge Management, pp. 375–384. ACM (2009)

20. Dermouche, M., Kouas, L., Velcin, J., Loudcher, S.: A joint model for topic-sentiment modeling from text. In: Proceedings of the 30th Annual ACM Symposium on Applied Computing, pp. 819–824. ACM (2015)

21. Trabelsi, A., Zaiane, O.R.: Mining contentious documents using an unsupervised topic model based approach. In: 2014 IEEE International Conference on Data Mining, pp. 550–559. IEEE (2014)

22. Jo, Y., Oh, A.H.: Aspect and sentiment unification model for online review analysis. In: Proceedings of the Fourth ACM International Conference on Web Search and Data Mining, pp. 815–824. ACM (2011)
23. Zheng, M., Wu, C., Liu, Y., Liao, X.: Topic sentiment trend model: modeling facets and sentiment dynamics. In: 2012 IEEE International Conference on Computer Science and Automation Engineering (CSAE), vol. 3, pp. 651–657. IEEE (2012)
24. He, Y., Lin, C., Gao, W., Wong, K.F.: Dynamic joint sentiment-topic model. ACM Trans. Intell. Syst. Technol. (TIST) **5**(1), 6 (2013)
25. Lin, Y.H.C., Gao, W., Wong, K.F.: Tracking sentiment and topic dynamics from social media (2012)
26. Dermouche, M., Velcin, J., Khouas, L., Loudcher, S.: A joint model for topic-sentiment evolution over time. In: 2014 IEEE International Conference on Data Mining, pp. 773–778. IEEE (2014)
27. Heinrich, G.: Parameter estimation for text analysis. Technical report, University of Leipzig (2008)
28. Song, R., Chen, S., Deng, B., Li, L.: eXtreme gradient boosting for identifying individual users across different digital devices. In: Cui, B., Zhang, N., Xu, J., Lian, X., Liu, D. (eds.) WAIM 2016. LNCS, vol. 9658, pp. 43–54. Springer, Heidelberg (2016). doi:10.1007/978-3-319-39937-9_4
29. Wang, J., Li, L., Tan, F., Zhu, Y., Feng, W.: Detecting hotspot information using multi-attribute based topic model. PloS One **10**(10), e0140539 (2015)

An Unsupervised Framework Towards Sci-Tech Compound Entity Recognition

Yang Yan[1,2,3(✉)], Tingwen Liu[1,2(✉)], Li Guo[1,2],
Jiapeng Zhao[1,2], and Jinqiao Shi[1,2]

[1] National Engineering Laboratory for Information Security Technologies,
Beijing, China
[2] Institute of Information Engineering, Chinese Academy of Sciences, Beijing, China
{yanyang9021,liutingwen,guoli,liquangang,shijinqiao}@iie.ac.cn
[3] University of Chinese Academy of Sciences, Beijing, China

Abstract. Classifying sci-tech compound named entities, such as the names of patents and projects, plays an important role in enhancing many high-level applications. However, there are very little work on this novel and hard problem. Traditional sequence labeling strategies cannot apply on sci-tech compound entities due to heavy cost of human annotation and low data redundancy. This paper concludes three intrinsic characteristics of sci-tech compound entities, and further proposes a generic and unsupervised framework named SCSegVal to address the problem. Our SCSegVal consists of two components: text splitting and segment validating. We reduce the best split of a text to the problem of maximizing the stickiness sum of segments. The construction of indicative words used in segment validating is reduced to the classical minimum set cover problem. Experimental results on classifying real-world science-technology entities show that SCSegVal achieves a sharp increase comparing with the classical supervised HMM-based approach.

Keywords: Named entity recognition · Unsupervised algorithm · Segment stickiness · Minimum set cover problem

1 Introduction

As an classical information extraction task, Named Entity Recognition (NER) aims at locating and classifying elements in text into predefined categories such as the names of persons, organizations, locations, expressions of times, quantities, monetary values, percentages, *etc.* However, much less work have been done on classifying a novel and common type of named entity, referred to as sci-tech compound named entity, which is usually complicated professionalism that consists of multiple continuous words in text. Typical compound named entities are the name of drugs, chemicals, projects, papers.

Sci-tech compound NER, on one hand, plays an important role in many applications, such as intellectual property protection [29], open source intelligence analysis [30], scholar data discovery and machine translation. We believe

© Springer International Publishing AG 2016
F. Lehner and N. Fteimi (Eds.): KSEM 2016, LNAI 9983, pp. 110–122, 2016.
DOI: 10.1007/978-3-319-47650-6_9

that employing compound NER in higher-level applications will enhance performance.

We first make a detail observation of the structure of sci-tech compound entities, and conclude three unique characteristics. Based on our observations, we propose a novel unsupervised segmentation and validation framework, named SCSegVal to achieve the recognition of compound entities. We first split a given text into several segments, and then validate whether each segment is our interested compound entity. The split of texts and the validation of segments are two main technical challenges. The overview of SCSegVal framework is listed in Fig. 1.

To address the first challenge, we introduce word stickiness to quantify the tightness between two words. Based on word stickiness, we further introduce segment stickiness to quantify the compactness degree of a segment (essentially a text). We reduce the best split of a text to the problem of maximizing the stickiness sum of segments on dynamic programming algorithm. To address the second challenge, We reduce the construction of the indicative whitelist to the classical minimum set cover problem, and also design a dynamic programming algorithm to address the problem. Experimental results show that our approach achieves 82.6 % precision, 79.65 % recall and 81.1 % F_1-measure in average, which is much higher than the classical supervised HMM-based approach.

The rest of the paper is organized as follows. We review related work in Sect. 2. We explain our observed characteristics in Sect. 3. Sections 4 and 5 describes the text splitting and segment validating component of SCSegVal framework respectively. In Sect. 6, we present experimental results. Finally, We give conclusions in Sect. 7.

2   Related Work

Research on NER starts at 1990s. Prior NER work mainly focus on identifying the names of persons, organizations, locations *etc.*, which can be classified into two category: rule-based [2,18] and stochastic-based [1,21]. There exists burgeoning specific NER tasks wg such as query log NER [7], biomedical NER [6,33], chemical NER [24], social network events NER [10–12,12,16,17], which take advantage of "gregarious" or unique charismatic property of twitter to find the popular phrase. Those approach take advantage of manual annotation and data redundancy between n-gram collocations. They are unrealistic for sci-tech compound entities due to labor cost and low redundancy of collocated constituted entities.

Our work is also related to NLP task of quality phrase (multi-word semantic unit) mining [15] and word sequence segmentation. Extracting quality of multi-word semantic units from unstructured text [3–5,15] is diverse in applications [14, 26,27] and has been gaining increasing attention of NLP community. Different from NER, quality phrase mining focus on making segmentation on original sentence to get precise extracted phrases without caring its specific category. Earlier work of multiword phrase segmentation use predefine POS pattern or leverage annotated to learn rules on POS-tagged dataset [22,31]. Those types of methodology is not generic in domains and dependant on expensive manual notation, which make it challenging to their widespread application.

Despite great efforts have been made on NLP task of named entities recognition and multiword segmentation, most of them focus on shorter length of phrases and simple structure of single entities. The problem of compound entities consisting of collocated multiple entities, noun phrases, conjunctions and prepositions which has more complicated inner structure and has longer length, is still remain unsolved.

3   Characteristics of Compound Entity

We observe that compound entities have three novel characteristics that are significantly different from traditional named entities, such as person names.

Observation 1 (Multi-Constituent Compound): each compound entity consists of multiple different constituents, *e.g.* noun entities in many different fields (entities in the biology, chemical or computer field *etc.*) and even auxiliary words. Many noun entities, such as "hybrid rice" in "research and application in two-line hybrid rice", are Wikipedia terms.

Observation 2 (Limited Missing POSes): the number of Part-Of-Speech (POS) types in compound entities does not increase even when the number of compound entities is very large. As the total number of POS types is fixed, and some POSes are highly unlikely to appear in compound entities, such as be verbs (be/is/are/am/was/were *etc.*) and demonstrative pronouns (you/me/I/my/him/her *etc.*). We denote $\mathbb{P}$ as the set of all POS types in compound entities, and $p(w)$ as the POS type of its constituent word or tagged entities w. We also denote $\overline{\mathbb{P}}$ as the complementary set of $\mathbb{P}$, namely the set of POS types do not appears in any compound entity.

Observation 3 (Multiple Indicative Words): we note that people usually follow strong linguistic laws when naming compound entities. For example, project names usually have their own indicative words, such as "research", "system", "application" *etc.* Those indicative words occurs much more frequently than other non-indicative words. Moreover, most compound entities have more than one indicative words. And a text without any indicative word is very unlikely to contain any compound entity. Note that personal names and other traditional named entities do not have this characteristic. In this paper, we denote $\mathbb{W}$ as the set of indicative words extracted from compound entities.

4   Text Splitting

4.1   Problem Statement

Given a arriving raw text t, the problem of text splitting is to split t into a variable number of segments (denoted as m): $t = s_1 s_2 \cdots s_m$. Each segment

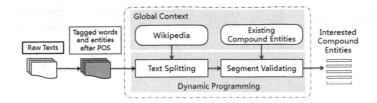

Fig. 1. Overview of GenricSegVal

consists of one or more continuous words or basic entities in t. As for our text splitting problem of compound NER, we add constraint conditions that (1) the maximum length of a segment is u, namely any segment has no more than u words; (2) each segment has no such two words that the POS type of one appears in some compound entities and that of the other does not appear in any compound entity. The main reason of the first constraint condition is that compound entities do not contain an infinite number of words. Adding the second condition is inspired by Observation 2.

To obtain the optimal segmentation, we use the following objective function, where C is the function that measures the stickiness of a segment or a text:

$$\max_{s_1, s_2, \ldots, s_m} C(t) = \sum_{i=1}^{m} C(s_i) \tag{1}$$

s.t.   C1. $1 \leq |s_i| \leq u, \forall i$ C2. $I(w_j) = I(w_k), \forall i, \forall w_j, w_k \in s_i$

where $t = s_1 s_2 \cdots s_m$

For any word w in s_i, $I(w) = 1$ if $p(w) \in \mathbb{P}$ and $I(w) = 0$ if $p(w) \in \overline{\mathbb{P}}$.

Segment stickiness is a quantification of the semantics compactness degree of a segment. Segment stickiness is composed by word stickiness. A high stickiness score of segment s indicates that it is not suitable to further split segment s as it is compact enough.

If the word length of text t is l, there exists $O(2^l)$ possible segmentations. It is inefficient and outweighes benefits to enumerate all of them and get the optimal segmentation. Inspired the work in [11–13], we design a dynamic programming algorithm to address the problem, as shown in Algorithm 1.

The main idea is to recursively conduct binary segmentations and then evaluate the stickiness of the resultant segments. Given any segment s from t (s can be t itself or a part of t) and suppose $s = w_1 w_2 \cdots w_n$, we try to conduct a binary segmentation by splitting s into two adjacent segments $s^1 = w_1 \cdots w_j$ and $s^2 = w_{j+1} \cdots w_n$ by satisfying:

$$\max_{s^1, s^2} C(s) = C(s^1) + C(s^2) \tag{2}$$

s.t.   C1. $n - u \leq j \leq n - 1$ C2. $I(w_k) = I(w_n), \forall j + 1 \leq k \leq n$

Algorithm 1. Text Splitting Algorithm

Input:
a text t of l words: $t = w_1 w_2 \cdots w_l$; u: the maximum length u of each segment;
v: set $\mathbb{S}(s)$ of top v segmentations for each segment s
Output:
the optimal segmentation $t = s_1 s_2 ... s_m$

1 set $\mathbb{S}(\varepsilon) = \{\varepsilon\}$ for a null text or segment ε;
2 $C(\varepsilon) = 0$;
3 **for** $i = 1 : l$ **do**
4   initialize set $\mathbb{S}(s_i) = \emptyset$ for $s_i = w_1 w_2 \cdots w_i$;
5   **for** $j = i - 1 : 0$ **do**
6   split s_i into two shorter sub segments $s_i^1 = w_1 \cdots w_j$ and $s_i^2 = w_{j+1} \cdots w_i$;
7   /* when $j = 0$, $s_i^1 = \varepsilon$ */
8   **if** $i - j \leq u$ *and* $I(w_{j+1}) == I(w_i)$ **then**
9   calculate $C(s_i^2)$;
10   **foreach** $S_j \in \mathbb{S}(s_i^1)$ **do**
11   /* S_j is a known good segmentation of s_i^1 */
12   concatenate S_j and s_i^2 to form a new segmentation S of s_i;
13   add S to $\mathbb{S}(s_i)$;
14                 $C(S) = C(S_j) + C(s_i^2)$;
15   **else**
16   **break**;
17   sort $\mathbb{S}(s_i)$ and keep only the top v segmentations of s_i;
18 **return** $S \in \mathbb{S}(t)$ with the highest score as the optimal segmentation;

To reduce the computation complexity and the memory cost, for each segment we only store the top v segmentations during the dynamic searching process (line 14 in Algorithm 1).

The complexity of Algorithm 1 is $O(luv \log(uv))$, where l is the average word length of texts, u is the upper bound of segment length, and v bounds top segmentations of a segment. We observed that in our data, $u = 30$ is a proper bound as the maximum length of a segment, which alleviate cost of the time-consuming iteration operation for solving all possible segmentations. We also set $v = 15$ so that the segmentation only focuses on high-quality segments and are not stuck by trivial ones, which leads to a complexity of $O(l)$.

4.2 Segment Stickiness

In Algorithm 1, an open question to be addressed is computing the segment stickiness function of C. Our segment stickiness is based on word stickiness, which is used to quantify the tightness between two words. Two words with high word stickiness should be put into the same segment. Another factor influence the stickiness of a segment is the word length of the segment. Two words with long distance should not be put into the same segment. And a segment with long word length should be further split. Thus, in this paper we define the stickiness of a segment s as follows:

$$C(s) = \frac{\sum\limits_{\forall w_1, w_2 \in s} C(w_1, w_2)}{|s|} \tag{3}$$

where $C(w_1, w_2)$ is the word stickiness between words w_1 and w_2.

4.3   Word Stickiness

As mentioned above, the stickiness between two words is a quantification of the tightness between them. Based on the intrinsic characteristics of compound entities and with the help of encyclopedia knowledge bases, we introduce three tools to describe that which two words should be tightly tied together: special marker based stickiness, Wikipedia term based stickiness and POS2Vector based stickiness.

For two words w_1 and w_2, the stickiness between them is defined as:

$$C(\mathrm{w}_1, \mathrm{w}_2) = (C_{\mathrm{sm}}(\mathrm{w}_1, \mathrm{w}_2) + C_{\mathrm{wt}}(\mathrm{w}_1, \mathrm{w}_2)) \times C_{\mathrm{p2v}}(\mathrm{w}_1, \mathrm{w}_2)$$

where $C_{\mathrm{sm}}(\mathrm{w}_1, \mathrm{w}_2)$, $C_{\mathrm{wt}}(\mathrm{w}_1, \mathrm{w}_2)$ and $C_{\mathrm{p2v}}(\mathrm{w}_1, \mathrm{w}_2)$ represent the special marker based stickiness, Wikipedia term based stickiness and POS2Vector based stickiness between w_1 and w_2 respectively.

Special Marker Based Stickiness. In most texts, special markers in pairs, such as quotation marks, are used to emphasize or note some special phrases that may consist of multiple words. Thus, two words in the same pair of special markers should be given high stickiness. In this paper, for a segment s and two words w_1 and w_2 in s, we use $D_s(w_1, w_2)$ to represents the word distance from one to the other in s, and $I_s^{\mathrm{sm}}(w_1, w_2)$ to indicates whether w_1 and w_2 are in the same pair of special markers in s. We define the special marker based stickiness between w_1 and w_2 as:

$$C_{\mathrm{sm}}(w_1, w_2) = e^{D_s(w_1, w_2)} \times I_s^{\mathrm{sm}}(w_1, w_2) \tag{4}$$

E.g., the text "Our | paper | at | IJCAI2015 | 'Bayesian | Active Learning | for | Posteriori Estimation' | studies | the | bayesian | setting | when | the likelihood | is | expensive | to | evaluate". "|" denotes the delimiters separator basic entities and words. The word "Active Learning" and "Posteriori Estimation" are both in quotation marks. Their connection should be stronger than other words pairs which is not both in quotation marks like "studies-setting" or "Bayesian-likelihood".

Wikipedia Term Based Stickiness. As compound entities usually contain Wikipedia terms, in this paper for a segment s and two words w_1 and w_2 in s, we introduce s' as segment of s that lies between w_1 and w_2 and function $I_{wt}(w)$

to indicates whether word w in a Wikipedia term. We define the Wikipedia term based stickiness between w_1 and w_2 as:

$$C_{\text{wt}}(w_1, w_2) = \frac{\sum\limits_{w \in s'} I_{wt}(w)}{|s'| - \sum\limits_{w \in s'} I_{wt}(w) + 1} \times I_{wt}(w_1) \times I_{wt}(w_2) \tag{5}$$

Here, we add one at denominator to avoid arithmetic overflow.

POS2Vector Based Stickiness. For compound entities, their POS (Part-Of-Speech) sequences also present some statistical characteristics. Moreover, the data on POS types are more dense than that on words, because the number of POS types is much less than that of words. As stated in Observation 2, some POSes rarely occurs in compound entities. In other word, these scarce POSes should have more distances with other POSes. The main idea of computing such POSes's distance is embedding each POS type into latent vector, and measuring the POS2Vector based stickiness between two words based on the cosine similarity for the vectors of their POS types.

In this paper, for any two words w_1 and w_2, we introduce vector of POS type (denoted as $\overrightarrow{p(w)}$) and define the POS2Vector based stickiness between w_1 and w_2 as:

$$C_{\text{p2v}}(w_1, w_2) = \frac{1 + S_c(\overrightarrow{p(w_1)}, \overrightarrow{p(w_2)})}{2} = \frac{1 + \frac{\overrightarrow{p(w_1)} \cdot \overrightarrow{p(w_2)}}{|\overrightarrow{p(w_1)}| \times |\overrightarrow{p(w_2)}|}}{2} \tag{6}$$

The vectors of POS types are obtained by using the POS sequences of texts as the inputs of Word2Vector tool [19].

5   Segment Validating

Based on our observed characteristics of compound entities, this paper adopts two main validating restriction to check whether segment s is a compound entity. First, the POS type of any word in segment s should not be in $\overline{\mathbb{P}}$. Second, segment s should contain at least one indicative word.

Note that given satisfying segmentation is not enough, although sci-tech compound entities are included in the segmentation result. Driven by the observation that sci-tech compound entities contains multiple indicative words, we use indicative words to determine whether given segments are sci-tech compound entities. Relaxing the restrictions on the construction of indicative words in $\mathbb{W}$ will cause over-matching for compound named entities, and tightening the restrictions will lead to low recall rate of compound entities. The reason we don't use tradition term frequency and tf-idf is that purely selecting word with high frequency words will recognize other unrelated words such as stop words, preposition, etc. as indicative words, and size of indicative words is dependent on manual setting. We define the problem of constructing indicative words as follows. Given a set of compound entities $R = \{r_1, r_2, \ldots r_m\}$, and word set $W = \{w_1, w_2, \ldots, w_n\}$ which consists of all the words in at least one member in R. We want to find a minimum subset of the indicative words set $\Psi \subseteq W$, which satisfies the following two limitations at the same time. First, Ψ covers R, implying that any compound entity in R should contain at least one word in Ψ. Second, $|\Psi|$ is the least.

Our problem of constructing indicative words can easily reduced to the classical minimum set cover problem. Thus it is NP-hard. In this paper, we design a dynamic programming algorithm to address the problem. Before describing the algorithm, We first introduce some useful symbols. We construct a $n \times m$ boolean matrix M, whose each row represents an element in W and each column represents an element in R. If the i-th word w_i appears in the j-th compound entity r_j, then $M_{ij} = 1$, otherwise $M_{ij} = 0$. We then introduce a binary operator $\oslash$ with a boolean matrix as the first argument and a word as the second argument, its result is also a boolean matrix. The operational result of $M \oslash w_j$ is removing corresponding row i and all the column j where $m_{ij} = 1$. For example, assuming compound entity phrase set $P = \{p_1, p_2, p_3\}$ get four entities after traditional POS, which is noted as $W = \{w_1, w_2, w_3, w_4\}$. And M is a 4×3 boolean matrix as shown in Fig. 2. The output $M \oslash w_2$ is a 3×2 boolean matrix, the w_2 row and the r_2 column from non-zero elements m_{22} are removed from M.

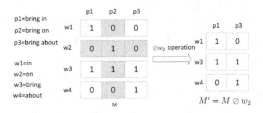

Fig. 2. Binary operator $\oslash$

We use the function F with a boolean matrix as the input to get the minimum number of indicative words, and the function G with a boolean matrix as the input to get the optimal set of indicative words. Then we can infer that:

$$w = \min_{w_i}\{F(M \oslash w_i) \mid 1 \leq i \leq n\} \tag{7}$$

$$F(M) = F(M \oslash w) + 1, \quad G(M) = G(M \oslash w) \cup \{w\}$$

Based on above equations, we can design a dynamic programming algorithm to address the problem of constructing indicative words sets.

6   Experimental Results

6.1   Experimental Setup

In our experiments, we compare our work with the classical HMM-based approach that widely-used in traditional NER. To our knowledge, there are neither handy solutions nor available benchmark datasets for our proposed compound entity recognition. We collect 1119 sci-tech compound entities supported by the National Natural Science Foundation of China in 2014, as existing compound entities to get approximate results of $\mathbb{P}$, $\mathbb{W}$ and obtain the embedding vector for each POS type. We collect 1869 web texts that each contains one or several sci-tech compound entities won the National Science and Technology Progress Award between 2005 and 2014 (referred to as NSTPA). Each of them contains one or several compound sci-tech entities. We select such dataset for the reason that it contains sci-tech compound entities from many fields, such as chemistry, physics, agriculture etc., which is rich in diversity of fields compared with previous work, such as gene and protein entity, [20,32], disease entity [9], chemical compound [28], etc. To compare with the HMM-based approach, we have to label the words in texts with our designed 40 hidden states Due to the heavy cost of manual annotation, we only label 681 texts of NSTPA (refer as 681-NSTPA) as the baseline dataset for comparison and utilize 10-fold cross-validation for the train/test split. Even so, we still employ 4 volunteers and spend 2 weeks on labeling the texts in 681-NSTPA. We also collect 766 web texts that each of them do not contain any compound named entity from Sougou Lab Data [25] (referred to as SOUGOU). All the 2635 texts constitute T as the training set to evaluate our work. All the word segmentation and POS tagging operations are conducted with the widely-used ICTCLAS [8] tool.

We evaluate our approach on the metrics of precision rate, recall rate and F-measure. For the i-th element t_i in T, the precision rate is $Pr(t_i) = \sum_j \frac{|L(e_{ij}, \widehat{e_{ij}})|}{|\widehat{e_{ij}}|}$ where e_{ij} is the j-th real compound entity in t_i, $\widehat{e_{ij}}$ is the j-th predicted compound entity for t_i, function L gives the longest common substring between e_{ij} and $\widehat{e_{ij}}$. Similarly, the recall rate is denoted as $Re(t_i) = \frac{|L(e_{ij}, \widehat{e_{ij}})|}{|e_{ij}|}$. Here, if e_{ij} and $\widehat{e_{ij}}$ are null, $Pr(t_i) = Re(t_i) = 1$; if e_{ij} is null and $\widehat{e_{ij}}$ is not null, then $Pr(t_i) = 0$ and $Re(t_i) = 1$; if e_{ij} is not null and $\widehat{e_{ij}}$ is null, then $Pr(t_i) = 1$ and $Re(t_i) = 0$. Then for text set T, its average precision rate is defined as the sum of precision rates of its elements divided by the size of T, namely $Pr(T) = \frac{\sum_{t_i \in T} Pr(t_i)}{|T|}$; its average recall rate is defined as the sum of total recall rates of its elements divided by the size of P, namely $Re(T) = \frac{\sum_{t_i \in T} Re(t_i)}{|T|}$. F-measure is the harmonic mean of precision and recall. In this paper, $F_1(T) = \frac{2 \times Pr(T) \times Re(T)}{Pr(T) + Re(T)}$.

6.2   Effects of Parameters in This Paper

Segment Maximum Length: As the name suggests, the maximum length of each segment, namely parameter u, is used to limit the upper bound of segment length. Small value of parameter u will split a text into too many segments, and large value of u will enlarge the searching space of Algorithm 1 and lead to high time cost.

Figures 3(a) and 4(a) show the precision rate, recall rate and F_1-measure of SCSegVal on classifying compound entities for different segment maximum lengths. From the two figures, we observe that the precision rate decreases and

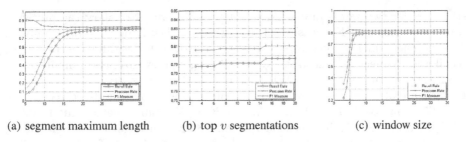

(a) segment maximum length   (b) top v segmentations   (c) window size

Fig. 3. Changes of precision, recall and F_1-measure on segment maximum length, top v segmentations and window size on 681-NSTPA

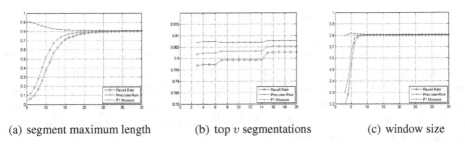

(a) segment maximum length   (b) top v segmentations   (c) window size

Fig. 4. Changes of precision, recall and F_1-measure on segment maximum length, top v segmentations and window size on NSTPA

the recall rate increases with u. When segment maximum length is 34, we both get the highest F1-measure on two dataset, 81.1 % on 681-NSTPA and 80.5 % on NSTPA When segment maximum length is 34, we get the highest F_1-measure on dataset 80.5 % on NSTPA and 81.1 % on 681-NSTPA. At the time, the precision rate and the recall rate are 82.6 % and 79.65 % on 681-NSTPA respectively, and 80.8 % and 80.2 % on NSTPA. We note that the changes is very little when $u \geq 30$ on both datasets, which implies that after reaching a certain threshold value, enlarging u will have little effectiveness on enhancing precision, recall and F_1-measure, but only increase the time and memory cost.

Top v Segmentations: During the dynamic searching process in Algorithm 1, we only keep the top v segmentations for each segment to reduce the time and space cost. Figures 3(b) and 4(b) show the precision rate, recall rate and F_1-measure of SCSegVal on classifying compound entities for different values of v. We observe that precision, recall and F_1-measure change simultaneously at only a few values of v, and remain stable in most of the time. In the overall trend, recall rate and F_1-measure increase with v, precision rate first declines slightly and then increases. The reason of this phenomenon is that Algorithm 1 gives a near optimal segmentation.

Sliding Window Size: In our description above, we assign the stickiness for any two words in a text no matter how great the distance between them is. In fact, we can set a sliding window, and add a limitation that only assigning the stickiness for two words in a window. Figures 3(c) and 4(c) show the precision rate, recall rate and F_1-measure of SCSegVal on classifying compound entities for different sliding window sizes. We observe that in the overall trend precision, recall and F_1-measure first increase with sliding window size, and remain stable after a certain threshold value. When sliding window size is 10, we get the highest F_1-measure of 81.10%, where the precision rate is 82.59% and the recall rate is 79.65% on 681-NSTPA. On NSTPA, our SCSegVal can achieve highest 80.5% in F_1-measure, 80.8% in precision rate and 80.2% in recall rate when sliding window size is 10. The results demonstrate the reasonableness of setting a sliding window.

6.3   Comparison with HMM-Based Approach

For HMM-based approach, we use widely-used features such as POS tag [23], context hint words and border indication as observation states and define 40 hidden states to capture the complicate structure of compound entities. Table 1 presents some representative hidden states. For SCSegVal, the values of parameters u and v are set to 30 and 15. We calculate the stickiness for any two words in a text no matter how great the distance between them is. The set of indicative words W contains 44 words.

Table 1. Comparing with HMM-based approach on experimental sets

Methodology	Recall	Precision	F_1-measure
HMM on 681-NSTPA	48.83%	47.76%	48.29%
SCSegVal on 681-NSTPA	79.65%	82.60%	81.10%
SCSegVal on NSTPA	80.2%	80.8%	80.5%
SCSegVal on SOUGOU	100%	55.22%	71.15%

Table 1 shows the experimental results of SCSegVal and HMM-based approach on our experimental sets. We can find that SCSegVal gives exciting results on classifying compound entities. Specifically, SCSegVal achieves 79.65% in recall rate, 82.6% in precision rate and 81.1% in F_1-measure on 681-NSTPA; 80.2% in recall rate, 80.8% in precision rate and 80.5% in F_1-measure on NSTPA; 100% in recall rate, 55.22% in precision rate and 71.15% in F_1-measure on SOUGOU. The results of our SCSegVal approach are much higher than that of the classical HMM-based approach. It is important to note that SCSegVal is unsupervised and we do not need to label training samples for SCSegVal.

7   Conclusion

This paper proposed an unsupervised framework named SCSegVal to address the problem of sci-tech compound NER. We note that sci-tech compound entities have distinct and intrinsic characteristics that are different from traditional named entities. SCSegVal consists of two components: text splitting and segment validating. We reduce the best split of a text to the problem of maximizing the stickiness sum of segments, and design a dynamic programming algorithm to address the maximizing problem. The construction of indicative words used in segment validating is reduced to the classical minimum set cover problem, and addressed also with a dynamic programming algorithm. Experimental results on classifying sci-tech entities show that unsupervised SCSegVal surpasses the classical supervised HMM-based approach on precision, recall and F_1-measure.

Acknowledgements. This work was supported in part by the National Natural Science Foundation of China under Grant No. 61303260; the Strategic Priority Research Program of the Chinese Academy of Sciences under Grant No. XDA06030200; Xinjiang Uygur Autonomous Region Science and Technology Project under Grant No. 201230123.

References

1. Bikel, D.M., Miller, S., Schwartz, R., Weischedel, R.: Nymble: a high-performance learning name-finder. In: Proceedings of ANLC, pp. 194–201 (1997)
2. Chiticariu, L., Krishnamurthy, R., Li, Y., Reiss, F., Vaithyanathan, S.: Domain adaptation of rule-based annotators for named-entity recognition tasks. In: Proceedings of the 2010 EMNLP, pp. 1002–1012 (2010)
3. Danilevsky, M., Wang, C., Desai, N., Ren, X., Guo, J., Han, J.: Automatic construction and ranking of topical keyphrases on collections of short documents. In: Proceedings of the SDM, pp. 398–406 (2014)
4. Deane, P.: A nonparametric method for extraction of candidate phrasal terms. In: Proceedings of the 43rd ACL, pp. 605–613 (2005)
5. El-Kishky, A., Song, Y., Wang, C., Voss, C.R., Han, J.: Scalable topical phrase mining from text corpora. VLDB J. **8**(3), 305–316 (2014)
6. Finkel, J., Dingare, S., Manning, C.D., Nissim, M., Alex, B., Grover, C.: Exploring the boundaries: gene and protein identification in biomedical text. BMC Bioinform. **6**(1), 1 (2005)
7. Guo, J., Xu, G., Cheng, X., Li, H.: Named entity recognition in query. In: Proceedings of SIGIR, pp. 267–274 (2009)
8. Zhang, H.: NLPIR/ICTCLAS (2012). http://ictclas.nlpir.org/
9. Yang, J., Qiubin, Y., Guan, Y., Jiang, Z.: An overview of research on electronic medical record oriented named entity recognition and entity relation extraction. Acta Automatica Sinica **40**(8), 1537 (2014)
10. Li, C., Sun, A., Datta, A.: Twevent: segment-based event detection from tweets. In: Proceedings of the 21st ACM CIKM, pp. 155–164 (2012)
11. Li, C., Sun, A., Weng, J., He, Q.: Exploiting hybrid contexts for tweet segmentation. In: Proceedings of ACM SIGIR, pp. 523–532 (2013)
12. Li, C., Sun, A., Weng, J., He, Q.: Tweet segmentation and its application to named entity recognition. IEEE TKDE **27**(2), 558–570 (2015)
13. Li, C., Weng, J., He, Q., Yao, Y., Datta, A., Sun, A., Lee, B.S.: Twiner: named entity recognition in targeted twitter stream. In: Proceedings of the ACM SIGIR, pp. 721 730 (2012)

14. Li, Y., Hsu, B.J.P., Zhai, C., Wang, K.: Unsupervised query segmentation using clickthrough for information retrieval. In: Proceedings of the 34th ACM SIGIR, pp. 285–294 (2011)
15. Liu, J., Shang, J., Wang, C., Ren, X., Han, J.: Mining quality phrases from massive text corpora. In: Proceedings of the 2015 ACM SIGMOD, pp. 1729–1744 (2015)
16. Liu, X., Wei, F., Zhang, S., Zhou, M.: Named entity recognition for tweets. ACM Trans. Intell. Syst. Technol. (TIST) 4(1), 3 (2013)
17. Liu, X., Zhang, S., Wei, F., Zhou, M.: Recognizing named entities in tweets. In: Proceedings of ACL, pp. 359–367 (2011)
18. Mikheev, A., Moens, M., Grover, C.: Named entity recognition without gazetteers. In: Proceedings of EACL, pp. 1–8 (1999)
19. Mikolov, T., Sutskever, I., Chen, K., Corrado, G.S., Dean, J.: Distributed representations of words and phrases and their compositionality. In: Proceedings of NIPS, pp. 3111–3119 (2013)
20. Mitsumori, T., Fation, S., Murata, M., Doi, K., Doi, H.: Gene/protein name recognition based on support vector machine using dictionary as features. BMC Bioinform. 6(1), 1 (2005)
21. Nadeau, D., Sekine, S.: A survey of named entity recognition and classification. Lingvisticae Investigationes 30(1), 3–26 (2007)
22. Punyakanok, V., Roth, D.: The use of classifiers in sequential inference. In: Proceedings of NIPS (2001)
23. Ratinov, L., Roth, D.: Design challenges and misconceptions in named entity recognition. In: Proceedings of CoNLL, pp. 147–155 (2009)
24. Rocktäschel, T., Weidlich, M., Leser, U.: ChemSpot: a hybrid system for chemical named entity recognition. Bioinformatics 28(12), 1633–1640 (2012)
25. Sogou Labs: Sogou text classification corpus (2008). http://www.sogou.com/labs/dl/c.html/
26. Tan, B., Peng, F.: Unsupervised query segmentation using generative language models and wikipedia. In: Proceedings of the 17th WWW, pp. 347–356 (2008)
27. Tjong Kim Sang, E.F., Buchholz, S.: Introduction to the CoNLL-2000 shared task: chunking. In: Proceedings of the 4th ConLL, pp. 127–132 (2000)
28. Usié, A., Alves, R., Solsona, F., Vázquez, M., Valencia, A.: Chener: chemical named entity recognizer. Bioinformatics 30(7), 1039–1040 (2014)
29. Wikipedia: Intellectual property protection (2015). https://en.wikipedia.org/wiki/Intellectual_property
30. Wikipedia: Open-source intelligence (2015). https://en.wikipedia.org/wiki/Open-source_intelligence
31. Xun, E., Huang, C., Zhou, M.: A unified statistical model for the identification of english baseNP. In: Proceedings of the 38th ACL, pp. 109–116 (2000)
32. Yeh, A., Morgan, A., Colosimo, M., Hirschman, L.: BioCreAtIvE task 1A: gene mention finding evaluation. BMC Bioinform. 6, S2 (2005)
33. Yoshida, K., Tsujii, J.: Reranking for biomedical named-entity recognition. In: Proceedings of the Workshop on BioNLP 2007: Biological, Translational, and Clinical Language Processing, pp. 209–216 (2007)

Identifying Helpful Online Reviews with Word Embedding Features

Jie Chen[1], Chunxia Zhang[2], and Zhendong Niu[1(✉)]

[1] School of Computer Science and Technology,
Beijing Institute of Technology, Beijing, China
{bit_chenjie,zniu}@bit.edu.cn
[2] School of Software, Beijing Institute of Technology, Beijing, China
cxzhang@bit.edu.cn

Abstract. The advent of Web 2.0 has enabled users to share their opinions via various social media websites. People's decision-making process is strongly influenced by online reviews. Predicting the helpfulness of reviews can help to save time and find helpful suggestions. However, most of previous works focused on exploring new features with external data source, such as user's profile, semantic dictionaries, etc. In this paper, we maintain that the helpfulness of an online review can be predicted by knowing only word embedding information. Word embedding information is a kind of word semantic representation computed with word context. We hypothesize that word embedding information would allow us to accurately predict the helpfulness of an online review. The experiments were conducted to prove this hypothesis and the results showed a substantial improvement compared with baselines of features previously used.

Keywords: Automatic helpfulness voting · User preference · Helpfulness classification

1 Introduction

The internet contains a wealth of reviews and opinions on any topics. User-generated contents come in various forms and sizes, objective opinions and subjective opinions. Postings in internet forums and user comments in websites are the important sources of information. The decision-making process of people is affected by the opinions of others in the information age [4]. When a person wants to change a job, he or she will start by searching for reviews and opinions written by the employees and former employees regarding the companies in his or her wish list. However, the number of reviews is often very large, which causes lots of reviews and opinions to be unnoticed, even though some of them are very helpful. As a result, predicting the helpfulness of a review is very important.

Many websites rank reviews by their published time, product rating, user voting, etc. Compared to sort by published time and product rating, the user voting method seems to be better and more helpful, since its results are cumulative from

© Springer International Publishing AG 2016
F. Lehner and N. Fteimi (Eds.): KSEM 2016, LNAI 9983, pp. 123–133, 2016.
DOI: 10.1007/978-3-319-47650-6_10

lots of visitors. For example, in Amazon.com, they employ a voting system to collect the feedback by asking "was this review helpful to you? Yes/no". It would be useful to rank reviews based on the quality as soon as these reviews are shown. This would save lots of time on surfing the web-pages and finding helpful reviews. However, user voting mechanisms are controversial, including the imbalance vote bias, the winner cycle bias and the early bird bias [14]. These kinds of bias show that voting system is not the best choice for ranking user-generated contents.

Previous works approximate the ground truth of helpfulness from users' voting results. If there are X of Y users who consider a review to be helpful, then the helpfulness score is X/Y. However, it is hard to collect the right value of Y. For example, when a user opens a product details page with many reviews, he just read the basic information about the product and leaves. It's hard to decide whether we should add 1 to all the reviews in this page. In addition, the review voting itself can be influenced by many factors, such as page structure adjustment, review recommendation, etc.

In this paper, we model the problem of predicting review helpfulness score as a regression problem and analysis performance of different features used in previous researches. Many researches [3,5,10,18,25,26] focus on exploring new features to model review sentences, then gain better results on the task. However, novel features are limited by the data resources, language of sentence, third-party tools, etc. In order to overcome these limitations, word embedding features are introduced to model sentences. Experimental results show that word embedding features outperform other features used in previous research. From the point view of dimensionality reduction, we also compared the Unigram features with Latent Semantic Analysis (LSA) to other features. Result showed that LSA technology with unigram features gain better performance.

The following section discusses related works about review helpfulness prediction. The definition of helpfulness prediction and the format of data used in our experiments are given in Sect. 3. Details about features used in our approach are introduced in Sect. 4. Experiments and evaluation metrics are described in Sect. 5. In Sect. 6 we discuss and analysis the results. We conclude and present directions for future research in the last section.

2   Related Works

Presenting the helpful content to visitors is an important component for any content-centric websites. Engineers of such kind of website have been committed to improve the click rate of reviews, either using normal ranking mechanism or carefully improved mechanism. Consequently, there has been plenty of researches on various aspects of ratings and the quality of review contents.

Some of them focus on finding the most helpful features for predicting the quality of review content [9,10,14,20,25]. Meanwhile, there are also some researches focus on exploring new algorithms [5,13,22,26,27].

In the research of Kim et al. [9], lexical, structural, syntactic, semantic and meta-data related features were used for automatic helpfulness prediction. Text surface features and unigrams are proved to be the most helpful features and widely used in later researches.

Zhang and Varadarajan [27] built a regression model by incorporating a diverse set of features, and achieved highly competitive performance of utility scoring on three real-world data sets. Their experiments also proved that the shallow syntactic features turned out to be the most influential predictors.

Liu [14] worked on how to detect low quality reviews. They introduced features to model the informativeness, subjectiveness and readability of a review and classified them into high or low qualities.

Yang et al. [25] hypothesized that helpfulness is an internal property of text and introduced LIWC and INQUIRER semantic features to model the review text. Their experiments showed that two semantic features could accurately predict helpfulness scores and greadly improve the performance compared with features previously used.

RevRank is an unsupervised algorithm to ranking helpfulness of online book reviews [22]. They first constructed a lexicon of dominant terms across reviews, then a virtual core review based on this lexicon was created. They used the distance between the virtual review and each real review to determine overall helpfulness ranking.

Hong et al. [5] developed a binary helpfulness classification system. The system used a set of novel features based on needs fulfillment, information reliability and sentiment divergence measure. Their system outperformed some earlier researches with the same dataset.

Lee and Choeh [13] proposed a helpfulness prediction neural network model and made use of products, review characteristics, and reviewer information as features. This is the first study to predict helpfulness using neural networks. The authors proved that their model outperform the conventional linear regression model analysis in predicting helpfulness.

Rong Zhang et al. [26] proposed a comment-based collaborative filtering approach which captures correlations between hidden aspects in review comments and numeric ratings. They also estimated the aspects of comments based on profiles of users and items, the model outperformed baseline system in Chinese review dataset.

Srikumar [10] proposed a predictive model extracts novel linguistic category features by analysing the textual content of review. He made use of review metadata, subjectivity and readability related features for helpfulness prediction. He proved that the proposed linguistic category features were better predictors of review helpfulness for experience goods.

3   Task Definition

In this section, we defined the task of review helpfulness prediction (RHP), and we introduce the data format of Amazon.com reviews. This data have been successfully used in related review helpfulness prediction tests. All the data analysis, illustrations and experiments are based on the dataset.

Table 1. An example of reviews in Amazon dataset

Tag	Value
Member id	A1004AX2J2HXGL
Product id	B00064LJVE
Date	January 13, 2005
Number of helpful feedbacks	5
Number of feedbacks	15
Rating	1.0
Title	Into the woods
Body	M. is a hack, a second-place magician in a high school talent show. He's drawn comparisons to Hitchcock and Spielberg - in the same sentence no less? Resting on the laurels of exactly ONE good movie, he manages to eek out a career for himself. Since THE SIXTH SENSE, his movies have gotten progressively worse. UNBREAKABLE was fair at best. An interesting idea with a dull, rumbling ride to the conclusion. SIGNS was a very rough movie to watch. The characters were cookie-cutter samples of human emotion and conflict - toss in a guy in an alien suit and you have what exactly?...

3.1   Task of RHP

The task of RHP aims to automatically predict the helpfulness score of a specific product's reviews. In this task, in order to eliminate the interference of external information, only text information is considered rather than any other human interaction information, such as user background, user level etc. The RHP should assign a high score to a review which gains a high manual voting score and assigns a low value to a review which gains a low manual voting score.

Therefore, given a set of reviews, the RHP should output a score list of each review's helpfulness score. We treat this as a regression task of reviews regarding their helpfulness.

3.2   Amazon.com Data Format

We use the Amazon review data which was prepared for Opinion Spam Detection [6]. This dataset provides 5.8 million reviews about products sold in Amazon. Each review contains product number, date, number of helpful feedback, number of feedbacks, rating, title and body. An example of reviews in this dataset was shown in Table 1.

In this paper, we only consider the body part of each review as the available local resources for RHP. The 'body' part gives the content of a review. Other items, such as 'title', 'ratings', are not totally available in this dataset for each

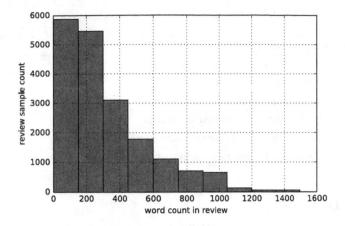

Fig. 1. Word count distribution in the corpus

product. In order to avoid dealing with missing information, we do not use the title and other fields which are optional in this experiment. The length of 'body' part of reviews in this dataset is various. Word count distribution about review sample in the corpus is given in Fig. 1.

4   Features

To make the experiment reproducible, only text-based features are used and discussed in this work. Text surface features [9,15,17,24], Unigram features [1,9, 24], Part-of-speech (POS) features [9,10,15] are widely used in previous research work, then we considered them as baselines.

4.1   Surface Features

Following previous researches [24,25], text surface features used are shown in Table 2. These features have proven effectiveness and are easy to implement for a new corpus.

4.2   Unigram Features

It is proved that the unigram feature is a reliable feature for review helpfulness prediction in previous work [25]. After removing all the stop words and word frequency lower than 10, we build a word dict. Each review is represented as a word vector, in which the value is *TF-IDF* weight.

In addition, for getting the semantic features and saving the training time, we also employ the LSA [11] technology to perform dimensionality reduction of vector space. Each review represented with unigram features is re represented in a lower dimension vector space.

Table 2. The description of surface features

Number	Feature description
1	The number of sentences in the review
2	The number of words in the review
3	The average length of sentences
4	The number of exclamation marks
5	The percentage of question sentences
6	The ratio of uppercase to lowercase characters in the review text

4.3 POS Features

The efficiency of part of speech (POS) features has been proved in previous research and there is not much difference among ways of implementing of POS features, which made it to be a reasonable feature in RHP. We use the following POS features: number of Noun words, number of Adjective words, number of Verb words, and number of Adverb words.

4.4 Word Embedding Features

We use the Genism tool[1] to learn the word embeddings from the provided 5.8 M Amazon product reviews, with the following settings:

1. we removed non-english reviews, which reduces the corpus to 5.5 M reviews.
2. we used the skip-gram model with window size 5 and filtered words with a frequency less than 10.

We use word embeddings of size 100, which means the dimension of output vector is 100. This setting is same with default settings of other tools, such as *word2vec*. The details of computing word embedding features are introduced in previous researches [16,19].

5 Experiments and Results

We empirically evaluate our approach, described in Sect. 4, by comparing the performance of different features combination. Below, we describe our experimental setup, choose evaluation metric, present our results and analyze different features' performance.

5.1 Evaluation Setup and Evaluation Metrics

In order to predict the helpfulness score of reviews, we focus on reviews with helpful feedback voting in the Amazon dataset. For removing duplicate reviews in the dataset, we use Hong's [5] deduplication method to filter the redundant reviews. There are too many reviews without voting information or feedback information. For this, we filter out the reviews with feedbacks count lower than 100.

---

[1] http://radimrehurek.com/gensim.

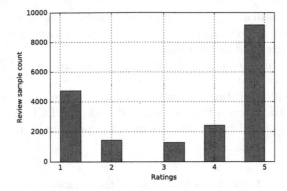

Fig. 2. Distribution of review ratings

The final dataset involves 19,030 reviews on 9805 products. The distribution of review ratings is shown in Fig. 2. To obtain the helpfulness voting score, we follow the annotation of review quality defined by Liu [14]. On the basis, we tested each group of feature combination on the whole dataset.

In the training process, we use three regression methods including Linear Regression (LR), Linear Support Vector Regression (LSVR) and Support Vector Regression (SVR)[21]. In the evaluation process, we run 10-fold cross validation. The original Amazon ratings are not used as ground truth, because the ratings are stared by their author for the product not for the review text.

In our experiment, we use the Root Mean Square Error (RMSE) metric to evaluate the performance.

5.2   Results

In this experiment, we test the performance with single feature groups described in Sect. 4 and results are shown in Table 3. Different combinations of features are also tested and results are shown in Table 4.

Table 3. RMSE of single feature

Features	LR	LSVR	SVR
Surface features (SF)	0.314	0.591	0.341
POS features (PF)	0.323	0.349	0.337
Unigram features (UF)	0.376	0.283	0.343
LSA + Unigram features (LUF)	**0.245**	0.252	0.285
Embedding features (EF)	**0.248**	0.254	0.250

Feature Performance. The first group of results is baselines of this experiment. As described in previous researches, SF features focus on statistics information and they are used as the baseline.

The second group of results is about Unigram features with LSA technology and word embedding features. From the results, LUF gains better performance than word embedding features. However, the difference between them is not large. Compared to Unigram features without LSA, the LUF improves the performance a lot. The word embedding features also perform better than Unigram features.

Table 4. RMSE of feature combinations

Features	LR	LSVR	SVR
SF + PF	**0.305**	0.324	0.318
UF + SF	0.363	**0.280**	0.331
UF + PF	0.365	**0.282**	0.343
UF + SF + PF	0.362	**0.280**	0.331
LUF + SF	**0.244**	0249	0.278
LUF + PF	**0.245**	0.251	0.285
LUF + SF + PF	**0.243**	0.249	0.278
EF + SF	**0.243**	0.249	0.246
EF + PF	**0.247**	0.253	0.250
EF + SF + PF	**0.242**	0.248	0.246
UF + SF + PF + EF	**0.357**	0.274	0.294
LUF + SF + PF + EF	**0.238**	0.241	0.249

The first group in Table 4 is about combinations of UF, SF and PF, we use them as feature combination baselines.

The second group shows the performance about LUF with other features. Compared to combinations about Unigram features, this group makes notable improvements (about 13 %).

The third group shows the performance about word embedding features with other features. From the results, combinations with EF show better performance than UF combinations with UF and LUF. This can verify the efficiency of EF features.

The last group in Table 4 shows combinations with all the features. From the results, combinations with LUF, SF, PF, and EF perform better than UF, SF, PF and EF. The results show than LUF can improve the performance again. In addition, this combination shows the best performance among all the combinations.

Table 5. Model performance

	LR	LSVR	SVR
Single feature	4	1	0
Feature combination	8	4	0

Regression Model Performance. Furthermore, we try to find the relationship between features and the underlying model of helpfulness prediction. For the result of each feature in Table 3 and each feature combination in Table 4, we count the best performance of three regression models. The statistical results are shown in Table 5. Linear regression gets the best in both single feature and feature combination results. Linear SVR also performs better than SVR. It shows that linear relation exists between these features and helpfulness of reviews.

6   Conclusions and Future Work

Until now, the helpfulness of reviews has been well studied with kinds of features, including Unigram features, text structural features, part-of-speech features, semantic features etc. However, features used in previous research so far produce results that are too unreliable to become a basis of a discourse-level prediction. We assert that the helpfulness of an online review should be predicted with its hidden structural information and lexical information. In this paper, we first give the definition of review helpfulness prediction, and then introduce word embedding features to predict the helpfulness score. Our experiments show that the word embedding features can lead to a substantial improvement over previous features. In addition, we test the LSA technology on Unigram features and the results show that LSA can lead to a substantial improvement over Unigram features. As a result of different features combinations, we try to analyze the hidden relationship between features and helpfulness of a review.

In the future, we will test the prediction performance on different corpus and try to do prediction with deep learning [12]. Convolutional neural network (CNN) has been proved to be efficient in modeling sentences [8], text categorization [7,23] and machine reasoning [2]. Further, we will investigate how to bring CNN into this research and predict the helpfulness of reviews.

Acknowledgments. This work is supported by the National Natural Science Foundation of China (No. 61370137, 61272361) and the 111 Project of Beijing Institute of Technology.

References

1. Agarwal, D., Chen, B.C., Pang, B.: Personalized recommendation of user comments via factor models. In: Proceedings of the Conference on Empirical Methods in Natural Language Processing, pp. 571–582. Association for Computational Linguistics (2011)

2. Bottou, L.: From machine learning to machine reasoning. Mach. Learn. **94**(2), 133–149 (2014)
3. Chen, C.C., Tseng, Y.D.: Quality evaluation of product reviews using an information quality framework. Decis. Support Syst. **50**(4), 755–768 (2011)
4. Duan, W., Gu, B., Whinston, A.B.: The dynamics of online word-of-mouth and-product salesan empirical investigation of the movie industry. J. Retail. **84**(2), 233–242 (2008)
5. Hong, Y., Lu, J., Yao, J., Zhu, Q., Zhou, G.: What reviews are satisfactory: novel features for automatic helpfulness voting. In: Proceedings of the 35th International ACM SIGIR Conference on Research and Development in Information Retrieval, SIGIR 2012, pp. 495–504. ACM, New York (2012)
6. Jindal, N., Liu, B.: Opinion spam and analysis. In: International Conference on Web Search and Data Mining, pp. 219–230 (2008)
7. Johnson, R., Zhang, T.: Effective use of word order for text categorization with convolutional neural networks. arXiv preprint arXiv:1412.1058 (2014)
8. Kalchbrenner, N., Grefenstette, E., Blunsom, P.: A convolutional neural network for modelling sentences. arXiv preprint arXiv:1404.2188 (2014)
9. Kim, S.M., Pantel, P., Chklovski, T., Pennacchiotti, M.: Automatically assessing review helpfulness. In: Proceedings of the 2006 Conference on Empirical Methods in Natural Language Processing, EMNLP 2006, pp. 423–430. Association for Computational Linguistics, Stroudsburg (2006)
10. Krishnamoorthy, S.: Linguistic features for review helpfulness prediction. Expert Syst. Appl. **42**(7), 3751–3759 (2015)
11. Landauer, T.K.: An introduction to latent semantic analysis. Discourse Process. **25**(2), 259–284 (1998)
12. Le, Q.V., Mikolov, T.: Distributed representations of sentences and documents. arXiv preprint arXiv:1405.4053 (2014)
13. Lee, S., Choeh, J.Y.: Predicting the helpfulness of online reviews using multilayer perceptron neural networks. Expert Syst. Appl. **41**(6), 3041–3046 (2014)
14. Liu, J., Cao, Y., Lin, C.Y., Huang, Y., Zhou, M.: Low-quality product review detection in opinion summarization. In: EMNLP-CoNLL, pp. 334–342 (2007)
15. Liu, Y., Jin, J., Ji, P., Harding, J.A., Fung, R.Y.K.: Identifying helpful online reviews: a product designer's perspective. Comput. Aided Des. **45**(2), 180–194 (2013)
16. Mikolov, T., Chen, K., Corrado, G., Dean, J.: Efficient estimation of word representations in vector space. arXiv preprint arXiv:1301.3781 (2013)
17. Momeni, E., Tao, K., Haslhofer, B., Houben, G.J.: Identification of useful user comments in social media: a case study on flickr commons. In: Proceedings of the 13th ACM/IEEE-CS Joint Conference on Digital Libraries, JCDL 2013, pp. 1–10. ACM, New York (2013)
18. Otterbacher, J.: helpfulnessín online communities: a measure of message quality. In: Proceedings of the SIGCHI Conference on Human Factors in Computing Systems, pp. 955–964. ACM, New York (2009)
19. Pennington, J., Socher, R., Manning, C.D.: Glove: global vectors for word representation. In: EMNLP, vol. 14, pp. 1532–1543 (2014)
20. Siersdorfer, S., Chelaru, S., Nejdl, W., San Pedro, J.: How useful are your comments? Analyzing and predicting youtube comments and comment ratings. In: Proceedings of the 19th International Conference on World Wide Web, WWW 2010, pp. 891–900. ACM, New York (2010)
21. Smola, A.J., Scholkopf, B.: A tutorial on support vector regression. Stat. Comput. **14**(3), 199–222 (2004)

22. Tsur, O., Rappoport, A.: RevRank: a fully unsupervised algorithm for selecting the most helpful book reviews. In: AAAI Conference on Weblogs and Social Media - ICWSM 2009 (2009)
23. Wang, P., Xu, J., Xu, B., Liu, C.L., Zhang, H., Wang, F., Hao, H.: Semantic clustering and convolutional neural network for short text categorization. In: Proceedings of the 53rd Annual Meeting of the Association for Computational Linguistics and the 7th International Joint Conference on Natural Language Processing, vol. 2, pp. 352–357 (2015)
24. Xiong, W., Litman, D.: Automatically predicting peer-review helpfulness. In: Proceedings of the 49th Annual Meeting of the Association for Computational Linguistics: Human Language Technologies: Short Papers, HLT 2011, vol. 2, pp. 502–507. Association for Computational Linguistics, Stroudsburg (2011)
25. Yang, Y., Yan, Y., Qiu, M., Bao, F.: Semantic analysis and helpfulness prediction of text for online product reviews. In: Proceedings of the 53rd Annual Meeting of the Association for Computational Linguistics and the 7th International Joint Conference on Natural Language Processing: Short Papers, vol. 2, pp. 38–44. Association for Computational Linguistics, Beijing (2015)
26. Zhang, R., Gao, Y., Yu, W., Chao, P., Yang, X., Gao, M., Zhou, A.: Review Comment Analysis for Predicting Ratings. In: Dong, X.L., Yu, X., Li, J., Sun, Y. (eds.) WAIM 2015. LNCS, vol. 9098, pp. 247–259. Springer, Heidelberg (2015). doi:10.1007/978-3-319-21042-1_20
27. Zhang, Z., Varadarajan, B.: Utility scoring of product reviews. In: Proceedings of the 15th ACM International Conference on Information and Knowledge Management, CIKM 2006, pp. 51–57. ACM, New York (2006)

Facial Texture Analysis for Recognition of Human Gender

Zahra Noor[1](✉), Muhammad Usman Akram[1], Mahmood Akhtar[1], and Muhammad Saad[2]

[1] NUST, College of Electrical and Mechanical Engineering,
Rawalpindi, Pakistan
engr.zahranoor@gmail.com, usmakram@gmail.com,
mahmood@ieee.org
[2] Pakistan Institute of Engineering and Technology, Multan, Pakistan
saad@piet.edu.pk

Abstract. The study of similarities and dissimilarities between human faces has been very active research topic for decades. The human face is composed of several textural representations, carrying discriminating information from one another mainly due to variations in age, gender, and facial expressions. Based on study and analysis of different facial features, the development of an accurate and efficient gender recognition system is an ultimate requirement for different useful applications such as surveillance, design of entrance and exit protocols at shopping malls, browsing of gender specific advertisement material on internet, etc. The gender recognition is essentially a binary classification process in which an input image is reported either true or false gender of the person in question. In this paper, we propose a knowledge based gender recognition system using histogram of oriented gradients (HOG) and local binary descriptor (namely LBP, Brisk and FREAK) based features, which can be used in robotics etc. To the best of our knowledge, it is the first time that Brisk and FREAK based features are used for the gender recognition problem. To evaluate our proposed system, we use standard IMM gender database (of 240 images) to extracted features and train (and test) the K nearest neighbors (KNN) classifier. Our results show that the proposed gender recognition system outperforms the existing methods when tested on above mentioned database, by giving accuracy as high as 90 % for FREAK features, and 94.16 % for HOG feature, improved by 6.5 %.

Keywords: Histogram of oriented gradients · Local binary pattern · Binary Robust Invariant Scalable Keypoints · Fast retinal keypoints descriptor

1 Introduction

The human face depicts discriminatory information about the gender that person. This makes gender estimation from face a proficient and resourceful process [1]. The gender estimation has extreme connectivity with the two dimensional (2D) facial information and three dimensional (3D) shape of the face. Most of the gender estimation studies are based on processing of 2D information than retrieval and processing of their 3D counterpart mainly due to increased computationally cost and complexity involved in the latter.

© Springer International Publishing AG 2016
F. Lehner and N. Fteimi (Eds.): KSEM 2016, LNAI 9983, pp. 134–145, 2016.
DOI: 10.1007/978-3-319-47650-6_11

The development of an accurate and efficient gender recognition system is an ultimate requirement for a number of useful applications such as surveillance, design of entrance and exit protocols at shopping malls, browsing of gender specific advertisement material on internet, etc. [2]. The computation time and complexity of gender recognition system can be reduced by excluding data for the gender we are not dealing at any time slot e.g. counting males in any video clip. Many researchers have previously reported their work on gender estimation using facial image based features. Mozaffari et al. [2] have proposed the use of combined features based on appearance and geometry of facial images. These features were relying on physiological difference between men and women faces to perform gender classification using Euclidean distance classifier.

Ko et al. [3] have used a combination of local binary pattern (LBP) and minimum distance classifier to extract the true age label for input human face images. The authors have divided images into two groups (1) below 50 years, and (2) above 50 years. The extracted features from local binary patterns and active appearance model were then combined and used to classify the gender of test images by using Euclidean distance classifier.

Bekhouche et al. [4] have proposed a method for estimating gender from facial images. Without applying any preprocessing, the raw images were used for extraction of features through multi-level local phase quantization (ML-LPQ) method to produce texture based feature. These features when passed on to SVM classifier, resulted into an accuracy of 79.1 % on group images database.

Yaoyu Tao [5] has proposed the use of Fisher face for gender estimation using the linear discriminate analysis. The mean faces and unified mean faces of both genders were generated for gender classification using linear discriminant analysis.

In this paper, we propose the use of HOG, Brisk, Freak and LBP features. İt is a knowledge based gender recognition system that works by estimating gender labels by preprocessing input images, extracting features, and classifying through KNN. The preprocessing of an input image involves detection of facial region and cropping of detected face. The cropped image is then converted into grayscale and contrast enhancement is achieved by histogram equalization. We have evaluated our proposed system on standard IMM dataset. Results and comparisons to the other estimation systems are presented in Sect. 4.

2 Existing Methods

2.1 Feature Selection

Descriptors have wide range of applications e.g. stitching two images requires proper pattern matching, reconstructing images and object recognition. Binary descriptors work with three basic components; (1) sampling points (2) pattern for sampling (3) orientation compensation. It works by applying a Gaussian kernel for smoothing the image and then random point pairs are compared [12]. These descriptors work with Hamming distance rather than Euclidean distance, which make these descriptors efficient in terms of time.

Histogram of Oriented Gradients (HOG). Histogram of oriented gradients is a feature descriptor that is widely used in image processing and computer vision. It counts occurrences of gradient orientation in an image or within a detection window [16] (Fig. 1).

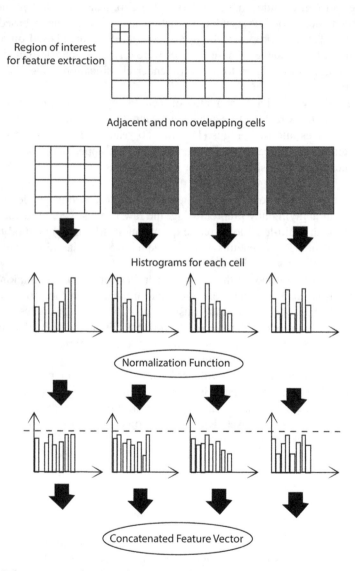

Fig. 1. HOG feature extraction by splitting area of interest (ROI) in adjacent, non overlapping cells

HOG works as follows;

- Divide an input image into cells (adjacent but non-overlapping) of size N × N [17].
- Calculate gradient orientation for each pixel in a cell.
- Group cells into lock of sizes [2 × 2] (overlapping blocks).
- Compute histogram of gradient directions for each cell of the input image.
- Normalized histograms of cells are actually block histogram.
- Concatenated normalized histograms provide descriptor.

Local Binary Pattern (LBP). LBP is a textural descriptor. It is used extensively for the detection and classification of textures [5]. To classify a pattern as a member of a class we first need to convert bitmap of textural image into grayscale [14]. Divide image into windows of size n × n. The pixel at location [1, 1] inside a window is labeled as central pixel as shown in Fig. 2. Compare intensity value of each neighboring pixel to its central pixel. If gray level value of neighboring pixel is greater than central pixel, replace the pixel value with one, zero otherwise. This will lead us to LBP for pixel [1, 1], which is converted to decimal value for ease of calculation. A histogram is constructed from the LBP values of the entire image and it provides a feature vector that portrays textural properties of that image.

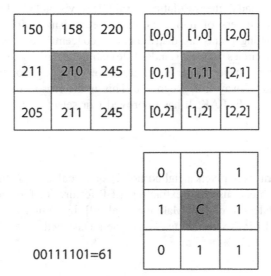

Fig. 2. Calculating LBP for central pixel located at window position [1, 1] (decimal equivalent in lower left section)

LBP has major role in facial recognition due to the textural variations between human faces. Similar is the case with textural variations across genders, which makes LBP applicable for gender recognition. A simple equation describing the working of LBP is;

$$LBP = \sum\nolimits_{p=0}^{p-1} s(gp - gc)2^p, \quad \text{while } s(x) = \begin{cases} 1, x \geq 0 \\ 0, x < 0 \end{cases} \tag{1}$$

In Eq. 1 p is the number of neighboring pixels distributed at equal distance at a circle of radius R around the center pixel. The grey level intensities of center and neighboring pixels are denoted as gc and gp respectively.

Binary Robust Invariant Scalable Keypoints (Brisk). Brisk is a binary descriptor that works to take full advantage of the descriptiveness by cautious selection of comparison of brightness [6]. It takes a small region around a sampling point and smoothes it through Gaussian smoothing. Sizes of Gaussian filters follow a standard deviation of σ for each sampling point. Brisk descriptor takes smoothed intensity for each short pair and compares smoothed intensity of first point in a pair to a second. If intensity of first is larger than the second one then it a 1 to the resultant bit of the descriptor, otherwise a zero. It works on Hamming Distance rather than Euclidean. It is fast in terms of computational power [7].

Fast Retinal Keypoints (FREAK). Fast Retina Keypoints (FREAK) follows a fixed sampling pattern instead of working with randomly distributed pairs. Processing of images by human eye is the basic motivation behind FREAK. Lower resolution is captured at periphery and higher resolution at middle of vision [15]. Light spreads from inside to the outer corners of retina. Similar Mechanism is followed in FREAK descriptor [8]. Sampling points are distributed on concentric circles, centered across keypoint. Radius of circles follows standard deviation according to the layer. Circles closer to the keypoint have denser distribution of sampling points. Concatenation decreases as we move away from keypoint. This is the reason behind the increased description accuracy of FREAK. It has improved pair selection mechanism.

2.2   LBP+SVM

Local binary pattern uses pixels neighborhood for calculation of binary values. An image divided into block, used for extracting LBP feature for four neighborhood and taking radius equivalent to one by Makinen et al. [9]. Input images were also filtered with 8 neighbors LBP. Extracted features when classified with SVM provided a recognition rate of 79.17 %.

2.3   Neural Network

Active appearance model is used for shape fitting. Superimposition of AAM model on input images helps in selecting feature points. Selected features while used with neural network after histogram equalization provided an accuracy of 81.74 % [9]. Back propagation method was used.

2.4   SVM

Histogram equalized features were also classified with SVM classifier in the literature [9]. This method provided an accuracy of 83.38 %.

2.5   Threshold Adaboost

Researchers combined discrete adaboost with haar-like features [9]. Used Haar-like features secured an accuracy of 82.60 %.

2.6   Neuro-Fuzzy System

A neuro fuzzy system with two inputs and single output was proposed by Mousavi et al. [10]. Output shows the probability of being a male face. Images with a probability greater than 75 % were labeled as male faces and those with less probability were recognized as female faces.

3   Proposed Methodology

Proposed work extracts features for input test images after preprocessing. Preprocessing involves detection of face from an input test image, cropping facial region from upper left corner of forehead to the upper right corner and deep down to the lower edge of chin, resulting a rectangular image. It converts image from RGB to grayscale and after histogram equalization images are resized at [180 × 160]. Resized images are used to extract features and classified using KNN classifier. Section below describes each preprocessing stage, in detail. Figure 3 provides a generalized block diagram of proposed estimator.

Preprocessing: Pre-processing is performed to make images properly aligned and resized. This would eliminate errors that can cause due to misalignment and un-even dimensions.

Face Detection is primary step that involves detection of area of interest from an image. Detection provides a boundary around the face that helps in cropping the images. Cropped images are converted to grayscale images and then it performs histogram equalization for contrast enhancement. Detected and cropped facial image is converted into grayscale and after histogram equalization features are extracted from input image.

Feature Extraction: The histogram equalized image is used for feature extraction. Our proposed work extracts a number of features from input images after splitting them into cells.

Histogram of oriented gradients (HOG) descriptor counts occurrences of gradient orientation in an image or from a specific segment of image, basically called region of interest. HOG feature extraction works by splitting an image into cell and calculates gradient orientation at each image pixel with in a cell. At second step, histogram of

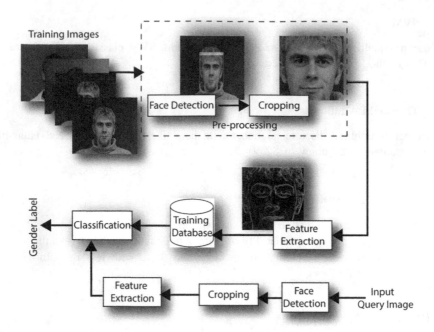

Fig. 3. Block diagram of proposed gender estimator to for input test image depicting three major steps involved in recognition (pre-processing, feature extraction and classification)

each orientation with in a cell in computed. At final step, histograms from each cell are concatenated in a single feature vector. Proposed estimation system extracts HOG features for frontal cart images (cropped and resized image after face detection).

Similarly local binary descriptors are used for feature extraction. LBP features are extracted for eight neighborhood at a radius equivalent to one. A concatenated LBP feature vector is obtained for complete image. Brisk and FREAK features are extracted for regions containing more texture based information. Resulting feature vectors are used for classification of human gender. Recognition rates of above discussed features are given in results section.

4 Classification

After extracting desired features, the KNN classifiers trained and tested for human gender recognition problem. The KNN classifier takes K number of closest training samples around an input test image and most probable class of training samples is assigned to the test image. In our case, we are working with three nearest neighbors. When the extracted features from input image are passed on to KNN, the distance is calculated from training samples on the basis of extracted features, and three closest neighbors are computed Class with more probability among three training samples is assigned to input test image.

5 Results

The proposed gender recognition system was tested using standard IMM database [11]. The database contains 240 images (198 male and 42 female faces, featuring 7 females and 33 males) [13]. These are JPEG images of size 640 × 480. The images were split through random shuffling of the database to provide randomness in image selection for testing and training purposes. After performing the above mentioned shuffling, it is very likely that images featuring same person will go to different datasets (training and testing). The shuffled images were divided into two datasets, training and testing datasets. During the training and testing phases, features were extracted for each image and stored in their respective feature vectors (column vectors) with accurate age labels before passing them to the classifier.

Proposed work suggests use of Histogram of oriented (HOG), Local binary pattern (LBP), brisk and FREAK features for recognition of gender. More than one implementations of each feature descriptor are provided.

To the best of my knowledge, it is the very first time that Brisk and FREAK features are used for gender recognition. Proposed recognition system extracts Brisk and FREAK features for complete preprocesses image (frontal cart). Out of the extracted feature points, it selects ten strongest feature points to extracts feature. Reason behind selecting strongest features is minimizing computation time, yet maintaining accuracy. Strongest selected regions include area around eye, on forehead and cheeks. These are regions that carry most of the textural information helpful in gender recognition. FREAK features results an accuracy of 89.2 % (Fig. 4).

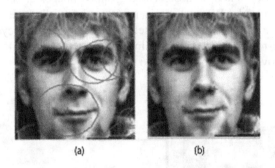

(a) (b)

Fig. 4. Selected Brisk (a) and Freak points (b) for an input image

Proposed work extracted LBP and HOG features for whole frontal cart image, and classification through KNN provided an accuracy of 85.8 % and 94 %, respectively. Similar features are also extracted by dividing an input image into a grid of 4 × 4 and calculating LBP and HOG features for each separate image from the grid of 16 images (Shown in Fig. 5). Extracted features from each grid section are concatenated in a feature vector and classified through KNN.

LBP feature extraction is also implemented by taking three distinct regions on human face. For region selection, first it takes a cropped image and locates eye and nose in the certain image. Left and right eye pixel coordinates are used for calibrating

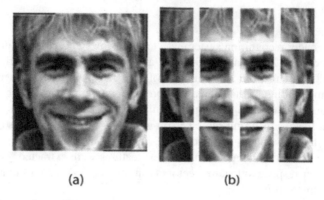

(a)   (b)

Fig. 5. Splitting an input image (a) into grid of 4 × 4 = 16 images each of size [45 × 40] (b)

an image. Saying that coordinates for left and right eyes are (xl, yl) and (xr, yr), respectively. Then β gives the angle of rotation for calibrating input image (given in Eq. 2).

$$\beta = \tan^{-1} \frac{(yl - yr)}{(xl - xr)} \tag{2}$$

Calibration makes sure that eyes are perfectly aligned lies at equal distance from the boundary of an image. Rotated image is used for selecting regions, including regions below left and right eyes and on nose. Three cropped images are created from above mentioned selection and LBP analysis is performed for all three images. Mean is calculated from each LBP vector, separately (i.e. right eye, left eye and nose) and concatenated in a row vector for each image (shown in Fig. 6). This module of implementation provides an accuracy of 86 %.

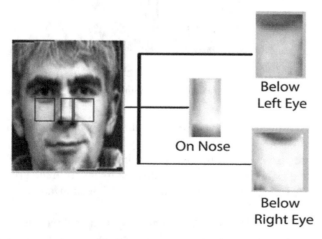

Fig. 6. Three separate images are generated from one facial image for extraction of LBP features

Results shown in Table 1 are computed for two folds cross validation. Experiments are also conducted for leave one out and ten folds cross validation. Results are shown in Table 2. Best results are achieved with two folds. All experiments are conducted for three nearest neighbors.

Table 1. Accuracies obtained though proposed work for two folds cross validation

Methods	Classification accuracy (for two folds cross validation)
Brisk	85.8 %
FREAK	90 %
LBP (for frontal cart)	84.83 %
LBP (for three selected regions)	86.5 %
LBP (for Grid 4 × 4)	83.75 %
HOG (for frontal cart)	94.16 %
HOG (for Grid 4 × 4)	92.5 %

Table 2. Accuracies obtained though proposed work for ten folds cross validation

Methods	Accuracy (10 fold)	Accuracy (leave one out)
Brisk	83.79 %	84.25 %
FREAK	85.14 %	87.5 %
LBP (for frontal cart)	82.47 %	85.83 %
LBP (for three selected regions)	81.15 %	82.5 %
LBP (for Grid 4 × 4)	82.75 %	83.44 %
HOG (for frontal cart)	86.55 %	92.53 %
HOG (for Grid 4 × 4)	85.64 %	90.83 %

Proposed gender recognition system outperforms state of the art gender estimation techniques by providing a recognition rate of 94.16 %. Comparison of our proposed work in terms of accuracies to gender recognition techniques from literature is provided in Table 3.

Table 3. Recognition rates for of state of the art algorithms

Methods	Classification accuracy
LBP+SVM [9]	83.79 %
Neural network	85.14 %
SVM	82.47 %
Threshold Adaboost	81.15 %
Neuro-fuzzy [10]	82.75 %
Proposed method (HOG+KNN)	94.16 %

IMM dataset provides an un-equal class level distribution. Performance of proposed system is tested over FG-Net database to overcome the said limitation. It is standard dataset of 1002 images, featuring 82 individuals. Accuracy of proposed system for FG-Net dataset is given in Table 4 along with the accuracies of state of the art gender recognition algorithms.

Table 4. Recognition rates for FG-Net database with proposed system in comparison to the state of the art algorithms

Methods	Classification accuracy
LDA [5]	81.3 %
AAM+SVM [18]	97.6 %
Proposed method (HOG+KNN)	65.32 %

Both of the gender recognition techniques used subsets of the database for experimental evaluation, while the proposed method is trained and tested using complete database of 1002 images. That is why the accuracy of proposed method is less than the two referenced methods.

6 Conclusion

In this paper, we have proposed a gender recognition system that exploits local binary descriptors, Brisk, FREAK, and HOG features. To measure the accuracy and compare different results, we trained and tested our classifier (KNN) on publically available standard IMM dataset. The database images were divided into two subsets (training and testing subsets) by using two fold, ten fold and leave one out cross validations. It has been shown that the proposed gender recognition system can achieve a maximum recognition accuracy of 94.16 % using HOG features as compared to accuracy of 87.5 % as achieved by one of the best existing approaches.

References

1. Hu, Y., Yan, J., Shi, P.: A fusion-based method for 3D facial gender classification. In: The 2nd International Conference on Computer and Automation Engineering (ICCAE), vol. 5, pp. 369–372. IEEE Press, Singapore (2010)
2. Mozaffari, S., Behravan, H., Akbari, R.: Gender classification using single frontal image per person: combination of appearance and geometric based features. In: 20th International Conference on Pattern Recognition (ICPR), pp. 1192–1195. IEEE Press, Istanbul (2010)
3. Ko, J.B., Lee, W., Choi, S.E., Kim, J.: A gender classification method using age information. In: International Conference on Electronics, Information and Communication (ICEIC), pp. 1–2. IEEE Press, Kota Kinabalu (2014)
4. Bekhouche, S.E., Ouafi, A., Benlamoudi, A., Taleb-Ahmed, A., Hadid, A.: Facial age estimation and gender classification using multi level local phase quantization. In: 3rd International Conference on Control, Engineering and Information Technology (CEIT), pp. 1–4. IEEE Press, Tlemcen (2015)

5. Tao, Y.: Automated estimation of human age, gender and expression. EE368 Digital Image Processing Final Project Report, p. 4 (2014)
6. Choi, S., Han, S.: New binary descriptors based on BRISK sampling pattern for image retrieval. In: International Conference on Information and Communication Technology Convergence (ICTC), pp. 575–576. IEEE, Busan (2014)
7. Schaeffer, C.: A Comparison of Keypoint Descriptors in the Context of Pedestrian Detection: FREAK vs. SURF vs. BRISK (2013)
8. Wu, Y., Cheng, Z., Jing, W., Nan, W.: Image registration method based on SURF and FREAK. In: International Conference on Signal Processing, Communications and Computing (ICSPCC), pp. 1–4. IEEE Press, Ningbo (2015)
9. Mäkinen, E., Raisamo, R.: Evaluation of gender classification methods with automatically detected and aligned faces. IEEE Trans. Pattern Anal. Mach. Intell. **30**(3), 541–547 (2008). IEEE Press
10. Mousavi, B.S., Hirad, A.: Automatic gender classification using neuro fuzzy system. Indian J. Sci. Technol. **4**(10), 1198–1201 (2011)
11. Data sets for Statistical Models of Shape. http://www.imm.dtu.dk/ ∼ aam/datasets/datasets. html
12. Lee, P., Timmaraju, A.S.: Learning binary descriptors from images
13. Nordstrøm, M.M., Larsen, M., Sierakowski, J., Stegmann, M.B.: The IMM face database-an annotated dataset of 240 face images. Technical University of Denmark, DTU Informatics, Building 321 (2004)
14. Leibstein, J., Findt, A., Nel, A.: Efficient texture classification using local binary patterns on graphics processing unit. In: Proceedings of the Twenty-First Annual Symposium of the Pattern Recognition Association of South Africa, pp. 147–152 (2010)
15. Spang, V., Henry, A.: Object Tracking Using Local Binary Descriptors (2014)
16. Dalal, N., Triggs, B.: Histograms of oriented gradients for human detection. In: IEEE Computer Society Conference on Computer Vision and Pattern Recognition (CVPR 2005), vol. 1, pp. 886–893. IEEE Press, San Diego (2005)
17. Watanabe, T., Ito, S., Yokoi, K.: Co-occurrence histograms of oriented gradients for pedestrian detection. In: Wada, T., Huang, F., Lin, S. (eds.): PSIVT 2009. LNCS, vol. 5414. Springer, Heidelberg (2009). doi:10.1007/978-3-540-92957-4_4
18. Saatci, Y., Town, C.: Cascaded classification of gender and facial expression using active appearance models. In: 7th International Conference on Automatic Face and Gesture Recognition, pp. 393–398. IEEE Press, Washington, DC (2006)

Enterprise Knowledge

Quantitative Analysis Academic Evaluation Based on Attenuation-Mechanism

Fan Li, WenLi Yu, JinJing Zhang, and Li Li[(✉)]

School of Computer and Information Science, Southwest University,
Chongqing 400715, China
lifan19930501@163.com, lily@swu.edu.cn

Abstract. The citation-based measure is known unpredictable. How-ever, it is used quite often in the cases when quantitatively evaluat-ing the academic impact is required. With the development of social networks, it is natural to ask the question: is there any trustworthy model which is able to provide quantitatively analysis of the academic impact with a huge amount of relevant information instead of peer-review only before the prevalence of social media? Many efforts have been devoted to provide the standard academic evaluation indicators, but they are either inadequate to be fully qualified or unable to become the universal applicable measure. In this paper, we propose a systematic approach, named Attenuation Mechanism, to quantitatively analysis the academic evaluation based on four estimated factors. It would bring new insights into how the academic impact takes place and the influence it has (either short term or long term). The extensive experiments on real academic search datasets show that the proposed model can perform significantly better than the baseline models in different areas and different disci-plines.

Keywords: Attenuation-Mechanism · Quantitative analysis · Academic evaluation

1 Introduction

Academic evaluation is a difficult subject which makes everyone convinced and has the general applicability, but this is a work which must be carried out. As the government, research institutions, funding agency and individual scholars need the academic evaluation indicators to make a fair evaluation to personal or academic institutions. Academic evaluation was almost entirely depended on Peer-review in the past, because there were no other trustworthy indicators. Currently, this situation has been changed, academic evaluation is increasingly depending on the quantitative indicators. There are many tangible indicators of academic evaluation, one indicator stands out in its frequency of use: *citations* [1–10], from the *Hirsch-index* [4] to the *g-index* [11], from *impact factors* [2] to *eigen factors* [12]. Those quantitative indicators lead to the following two problems that may hinder its popular use.

© Springer International Publishing AG 2016
F. Lehner and N. Fteimi (Eds.): KSEM 2016, LNAI 9983, pp. 149–158, 2016.
DOI: 10.1007/978-3-319-47650-6_12

First, the impact factor (IF) [2], conferring a journal's historical impact to a paper, in fact, it's value reflects the ratio of the several published papers and the citations (frequency) during that two years. In normal journals, reference frequency data will not be ups and downs. In addition, there are some good reviewed papers in certain areas in that two years. Otherwise there are some non-academic factors, such as excessive self-citation, and those indicators can not conduct a comprehensive evaluation to individual academic standards, because the paper's impact factor (IF) [2] is not mean the influence of the paper itself, but the impact factor of journal which the paper published in.

Second, Papers published in the same journal citations a decade later acquire widely different number of citations, from one to thousands (Fig. 1) [1,2,14] and the number of citations [1–10] collected by a paper strongly depends on the paper's age [1–4], and the number of citations from the early seminal paper is not high. It was reported by *Egghe Leo*. He studied the 75 Nobel prize winners' papers and founded that more than 10 % of papers have never been invoked. At the same time, although some articles published in the journals lower impact factors, but it was widely cited. Hence, the early impact [11–14] have no much value [1,2,14]. Frequently lost in this debate is the fact that comparison of different papers is confounded by incompatible publication, citation, and/or acknowledgment traditions of different disciplines and journals. Hence, we have to recognize the fact that the impact factor (IF) to foresee lasting impact on the papers has well-known limitations.

For better understand the academic researcher's academic ability, the *Hirsch-index* was proposed by *JorgeHirsch* (University of California, San Diego physicist) in 2005, the *Hirsch-index* takes into account the researcher's citations and the number of articles, but by the essence of *Hirsch-index*'s calculation process, the *Hirsch-index*'s evaluation capacity is still uncertain. There were, hence, a number of crucial flaws in the *Hirsch-index*, one of those flaws is the *Hirsch-index* based on different search databases, such as *WebofScience*, *Scopus*, *GoogleScholar*, often don't provide the same *Hirsch-index*, For this

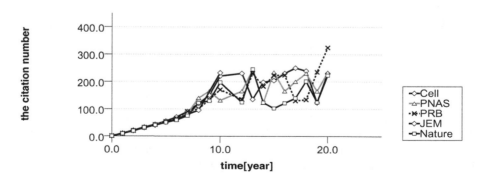

Fig. 1. Citation history of papers shown in Fig. 3 that acquired the same citations 10 years after publication, illustrating the different long-term impact despite impact factors.

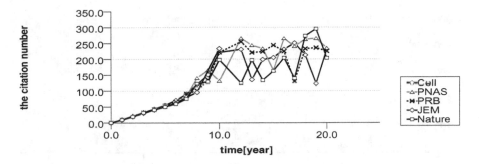

Fig. 2. Hirsch-index of papers shown in Fig. 3 that acquired the same Hirsch-index 10 years after publication, illustrating the different long-term impact despite Hirsch-index.

situation is caused by the database's coverage and the different disciplines and the retrieval's degree is widely vary from different databases. Often, the molecular biology and medicine are much higher than other disciplines and journals. Another concern flaw about the *Hirsch-index* is that there has no distinction between the actual contribution of the authors (Fig. 2). For example, author *A* and author *B* are the close collaborators, the two authors *A* and *B* have co-published the papers, co-authored 20 articles those citation has over 20 times. At this time both the *Hirsch-index* is 20, on the surface seems quite level of scientific research. However, in this 20 articles, the author *A* is the first author of 15 articles, and the author *B* is the first author of only 5 articles. Clearly, the author *A* should make a greater contribution than author *B*. So, *Hirsch-index* tend to the large research team advantageous, and the multiple partners share the higher *Hirsch-index* (large experimental discipline's *Hirsch-index* cooperation on combat is higher than ordinary man or a small group of academic disciplines), this situation confusing the actual contribution of different authors. So we propose the Attenuation-Mechanism to solve above problems.

2   Related Work

In this section, we briefly review some representative works related to quantitative analysis academic evaluation. There are some works on quantitative analysis on academic evaluation. For example, *Barabasi* and *Albert* proposed preferential attachment that captures the *well-documented* fact that highly cited papers are more visible and are more likely to be cited again than *less-cited* contributions [15,16] in a science paper in 1998. In 2011, *Medo, Cimini* and *Gualdi*'s paper published in *PRL* (*Physical Review Letters*) proposed a *Aging model* that captures the fact that new ideas are integrated in subsequent work; hence, each paper's novelty fades eventually [17,18]. And the *Fitness model* that proposed by *Bianconi* and *Barabasi* in a *EPL* (*Europhysics Letters*) paper in 2001 shows that the inherent differences between papers, accounting for the perceived novelty and importance of discovery [19,20]. The Fitness model bypass the end

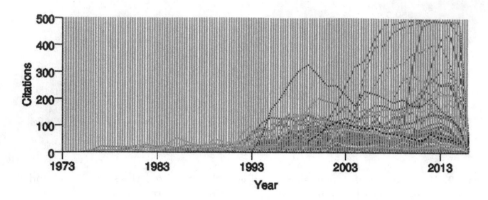

Fig. 3. Yearly citation c_i^t for 200 randomly selected papers published between 1973 and 2016 in the Cell corpus.

to evaluate a paper's intrinsic value and view fitness are a collective measure capturing the community's response to a work [12–14]. Because these three elements, preferential attachment, aging, fitness were released in the empirical, and these three elements was proved that independent to each other, so these three elements were multiplied together by *Barabasi* in 2013, The team's results based on the analytical analysis pointed out the ultimate influence of the paper (referenced throughout the life cycle to obtain the sum) was only related to fitness, and the basic relationship is proportional to $exp(lambda)$ [1–6,14]. The inherent difficulty in addressing this problem is well illustrated by the citation history of papers extracted from the *Cell corpus*, consisting of 20035 papers published between 1973 and 2016 and spanning all areas of *Cell* (Fig. 3). The fat-tailed nature of the citation distribution 43 years after publication indicates that, although most papers are hardly cited, a few do have exceptional impact (Fig. 3) [9–13]. This impact heterogeneity, coupled with widely different citation histories (Fig. 1), suggests a lack of order and hence lack of predictability in citation patterns.

3   Attenuation-Mechanism

We start by identifying four fundamental factors that drive the quantitative analysis academic evaluation of individual papers:

The temporal relaxation function $P_i(t)$ captures the fact that new ideas are integrated in subsequent work [10,11,13,14]. Hence, we can capturing the decay rate of paper i by $P_i(t)$ which can be measured directly from real data. Given that a paper i's citation is driven by four independent forces, that are difficult to separate from each other, by needing to control the influence of these factors, isolating the temporal decay, we finding that the relaxation function is best approximated by a *lognormal function*:

$$P_i(t) = \frac{1}{\sqrt{2\pi}\sigma_i t} \exp[-\frac{(\ln t - \mu_i)^2}{2\sigma_i^2}] \tag{1}$$

where t is *time*; μ indicates immediacy, governing the time for paper i to reach its citation peak; and σ is *longevity*, capturing the decay rate. This attenuation function reflects the process that papers citations from the post to achieve the ultimate influence. However, the *lognormal function* does not reflect to the complexity that the paper i to achieve its ultimate impact of intermediate, in other words, the attenuation model for predicting the ultimate impact of the paper i does not have the versatility and stability. For example, a group of papers that within a 5-year span collect the same publish time and the same number of citations are founded to have widely different ultimate impacts (Figs. 1 and 2).

Therefore, we propose the *breakpoint-time function* δ of a paper i to control a paper i's attenuation stability. For example, paper A's δ is 0.2, and paper B's δ is 0.4, then, even if the two papers have the same reference numbers or the paper B's final quote numbers is higher than the paper A's quote numbers, we also believe that the paper A's stability is higher than the paper B, because the paper B owns smaller δ, and we use the following simple model to characterize δ, and to make modifications to some papers by the complexities of a paper i:

$$b_i(t) = \delta = \frac{t_m - t_n}{t - t_0} \tag{2}$$

where

$$c_{tm}(i) = c_{tn}(i) = 0 \tag{3}$$

is the cumulative citations distribution, and we use $c_t(i)$ to description paper i's probability to be cited. The proposed model offers a journal-free methodology to evaluate the complexity of the papers. To illustrate this, we selected three journals with widely different disciplines: *Physical Review B* (*PRB*) (*in* 1992), *Proceedings of the National Academy of Sciences USA* (*PNAS*), *Nature, J Environ Monit* (*JEM*) and *Cell*. We measured for each paper published by δ, obtaining their distinct journal-specific *breakpoint-time* distribution. Then we selected all papers with $\delta \approx 0.2$ and followed their citation histories. As expected, they follow the same distribution.

Then, we consider the level of *journals impact factor* influence on the papers attenuation, although the journals' level do not reflect the real ultimate influence of a paper i, even sometimes lead to the incorrect ultimate influence to paper i, but in our paper, we refer to the existing journal impact factors, and found that the real academic literature data (*KddCup*2016) and the long-term influence meet an exponential relationship between the degree of attenuation and these papers whose journal level is β.

Finally, we propose the *energy loss function* during a paper i's attenuation according to different journals, called inelastic attenuation, to help us to understand the mechanisms that influence the evolution of the attenuation of a paper i.

In this paper, the quality factor q (Fig. 4), is presented to represent the energy loss mechanism, which is defined as a hysteresis curve on a circulating energy loss and the ratio of the total energy of the reciprocal:

$$q = 2\pi \frac{sum[1 - b_i(t)]}{T} \tag{4}$$

where T is the paper's total time. The larger the quality factor q, the energy loss is smaller, so the papers decay process of inelastic attenuation characterized by

$$z_i(t) = A \exp(-\frac{\mu_i \beta}{2q\theta}) \tag{5}$$

where θ is the inelastic attenuation factor which can be fitted by using the least square fit method.

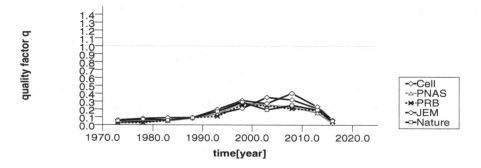

Fig. 4. The different quality factors in different journals

Combining these above four factors, we can get the probability that a paper i is cited at the time t after publication by attenuation mechanism as

$$c_i^t = m[e^{\frac{\beta \lambda_i}{A}\Phi(\frac{\ln t - \mu_i}{\sigma_i})\delta} - 1] \tag{6}$$

where

$$\Phi(x) \equiv (2\pi)^{-1/2} \int_{-\infty}^{x} e^{-y^2/2} dy \tag{7}$$

is the cumulative normal distribution, m is a global parameter measures the average number of *cell* paper's references each new paper contains (Fig. I1) [14], and A is a normalization constant. Ultimate impact (c^∞) [14,15] represents the total number of citations a paper i acquires during its lifetime [14–16]. By taking the $t \to \infty$ limit in Eq. 6, we obtain the papers attenuation process

$$c^\infty = m(e^{\frac{\beta \lambda_i \delta}{A}} - 1) \tag{8}$$

Using the Eq. 6 for each paper, the model is expected to generate a citation history that resemble the real citation of. We show in (Fig. 3) an example of randomly selected 200 papers published between 1973 and 2016, finding excellent agreement between the model and the empirical data.

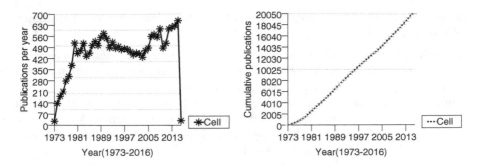

Fig. 5. The references of Cell corpus per year and the cumulative citations of Cell corpus

4   Experiments

4.1   Datasets

The dataset used in the paper was released by $KDD\ CUP$ 2016. The data set is massive after unzipped. PostgreSQL database is used to extract the particular Cell dataset to be used in our experiment. The detailed dataset can be found at: https://kddcup2016.azurewebsites.net/DataAbstract, and the *Cell* dataset consists of all papers published in the *Cell* journals from 1973 to 2016 (Fig. 3). We can count some important properties of *Cell* dataset, for example, the references of *Cell* dataset per year (Fig. 5) and the cumulative citations of *Cell* dataset (Fig. 5). The dataset is comprised of over 20,000 papers with citations. The data is unique in its longitudinal nature, spanning over 45 years. Therefore, it is ideal for understanding the paper's attenuation process.

4.2   Baseline Algorithms

To show the high performance of Attenuation-Mechanism, we compare our method against the following models:

– **Logistic Model:** The logistic function is widely used to model population growth and product adoptions, with applications in many fields. In the context of citations one could view a paper as a new product, whose adoption leads to an increase in citations. Each paper is characterized by a different increase rate r and a total number of citations c^{∞} that captures the differences in impact [11–14]. In the rate equation formalism this can be described as

$$\frac{\mathrm{d}c_i^t}{\mathrm{d}t} = r_i c_i^t (1 - \frac{c_i^t}{c_i^{\infty}}) \tag{9}$$

yielding

$$c_i^t = \frac{c_i^{\infty}}{1 + e^{-r_i(t-\tau_i)}} \tag{10}$$

where c_i^∞, r_i and τ_i correspond to ultimate citation, longevity, and immediacy of paper i.

- **Bass Model:** One of the most famous models in marketing and management sciences is the Bass model [21]. The Bass model assumes the adopters of a product are influenced by two aspects: mass media and word of mouth. The innovators correspond to people who cite the paper spontaneously, little in influenced by how many people have already cited the paper. At the same time, a papers citations are driven by word-of-mouth diffusion (the imitators).
- Mathematically, this can be expressed as

$$\frac{\mathrm{d}c_i^t}{\mathrm{d}t} = (p + qc_i^t/c^\infty)(c^\infty - c_i^t) \tag{11}$$

where p characterizes innovators, reflecting an influence that is independent of current citations (c_i^t), and q reflects the imitation part of the model. Solving (11) yields

$$c_i^t = c^\infty \frac{1 - e^{-(p+q)t}}{1 + \frac{q}{p}e^{-(p+q)t}} \tag{12}$$

- **Gompertz Model:** The Gompertz model [22], named after a Anthropologist called *Benjamin Gompertz*, was first proposed to model human mortality. In this context, early citations pave the way for new citations and drive the citation dynamics, hence the rate of research develop increases at an exponential rate. This can be formulated as

$$\frac{\mathrm{d}c_i^t}{\mathrm{d}t} = qc_i^t \ln(c^\infty/c_i^t) \tag{13}$$

yielding,

$$c_i^t = c^\infty e^{-e^{-(a+qt)}} \tag{14}$$

In (14), a sets the displacement in c_i^t, while q characterizes the growth rate of citations.

4.3   Results on Cell Data Set

In this part, we randomly selected 40 papers from *Cell* as our validation data set, and we compare our Attenuation-Mechanism with other models, Bass, Gompertz, Logistic. Figure 6 show the results find that the Attenuation-Mechanism for the selected papers published between 1973 and 1988 is consistent with three models. The Y-axis of Fig. 6 represents the number of citations (short for "c") of a particular paper. The Attenuation-Mechanism which fits the real tendency make a sharp distinction to other models in 1988 called differentiator point, and the Attenuation-Mechanism has the largest number of salient point that means the Attenuation-Mechanism has better stability than the baseline models between 1990 and 2003.

From Fig. 3, we can get the conclusion that our Attenuation-Mechanism allowing us to collapse the long-term impact of papers from *Cell* into a single curve. As shown in Fig. 7, Attenuation-Mechanism the selected papers published between 1973 and 2016.

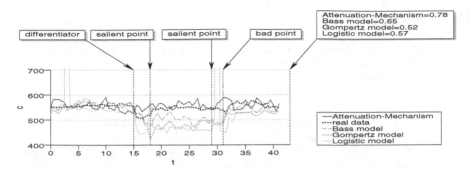

Fig. 6. The goodness of fit for selected papers published between 1973 and 2016 within the Cell corpus by using (A) Attenuation-Mechanism (Eq. 8), (B) Bass, (C) Gompertz, and (D) Logistic models.

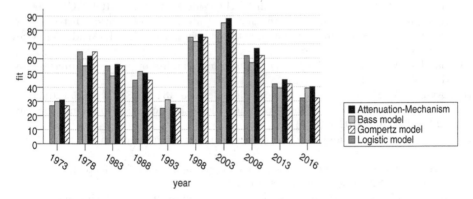

Fig. 7. Fitting the selected papers published in different years in Fig. 3 by using (A) Attenuation-Mechanism (Eq. 8), (B) Bass, (C) Gompertz, and (D) Logistic models.

5   Conclusion

In this paper, we introduced a mechanism called Attenuation-Mechanism which can be used to quantify academic papers long-term impact efficiently. The experimental results demonstrated that the proposed model can outperform the baseline models significantly in terms of accuracy. In this paper, we only considered the research papers with their citations represented by the IF. We found the experimental results on some particular journals such as Science, Cancer Letters are lower than expected. Currently we do not know the underlying reasons. We will consider different document type of each paper indexed by the Web of Science to extend our experiment. Parameter tuning is worth considering in order to enhance the whole performance.

Acknowledgement. This work was supported by National Natural Science Foundations of China (No. 61170192).

References

1. Garfield, E.: The history and meaning of the journal impact factor. JAMA J. Am. Med. Assoc. **295**(1), 90–93 (2006)
2. Price, D.J.: Networks of scientific papers. Science **149**(3683), 510–515 (1965)
3. Redner, S.: Citation statistics from 110 years of physical review. Phys. Today **58**(6), 49–54 (2005)
4. Hirsch, J.E.: An index to quantify an individual's scientific research output. Proc. Natl. Acad. Sci. USA **102**(46), 16569–16572 (2005)
5. Jones, B.F., Wuchty, S., Uzzi, B.: Multi-university research teams: shifting impact, geography, and stratification in science. Science **322**(5905), 1259–1262 (2008)
6. Lehmann, S., Jackson, A.D., Lautrup, B.E.: Measures for measures. Nature **444**(7122), 1003–1004 (2006)
7. Phan, X.H., Nguyen, L.M., Horiguchi, S.: Learning to classify short and sparse text & web with hidden topics from large-scale data collections. In: Proceedings of the 17th international conference on World Wide Web, pp. 91–100. ACM (2008)
8. Radicchi, F., Fortunato, S., Castellano, C.: Universality of citation distributions: toward an objective measure of scientific impact. Proc. Natl. Acad. Sci. USA **105**(45), 17268–17272 (2008)
9. Barabsi, A.L., Song, C., Wang, D.: Publishing: handful of papers dominates citation. Nature **491**(7422) (2012)
10. Evans, J.A., Foster, J.G.: Metaknowledge. Science **331**(6018), 721–725 (2011)
11. Evans, J.A., Reimer, J.: Open access and global participation in science. Science **323**(5917), 1025–1025 (2009)
12. Egghe, L.: Theory and practice of the g-index. Scientometrics **69**(1), 131–152 (2006)
13. Fersht, A.: The most influential journals: impact factor and eigenfactor. Proc. Natl. Acad. Sci. USA **106**(17), 6883–6884 (2009)
14. Wang, D., Song, C., Barabsi, A.L.: Quantifying long-term scientific impact. Science **342**(6154), 127–132 (2013)
15. Barabasi, A.L., Albert, R.: Emergence of scaling in random networks. Science **286**(5439), 509–512 (1999)
16. Book, G.J.: Book review: scale-free networks by G. Caldarelli. Int. J. Microstruct. Mater. Prop. **4**(4), 520–521 (2009)
17. Medo, M., Cimini, G., Gualdi, S.: Temporal effects in the growth of networks. Phys. Rev. Lett. **107**(23), 1261–1267 (2011)
18. Barabsi, A.: Evolution of networks: from biological nets to the Internet and WWW. Eur. J. Phys. **25**(5), 697 (2010)
19. Bianconi, G., Barabsi, A.L.: Competition and multiscaling in evolving networks. Physics **30**(1), 37–43 (2000)
20. Caldarelli, G., Capocci, A., De, L.R.P., et al.: Scale-free networks from varying vertex intrinsic fitness. Phys. Rev. Lett. **89**(25), 148–168 (2002)
21. Bass, F.M.: A new product growth for model consumer durables. Manag. Sci. **50**(12 Supplement), 215–227 (2010)
22. Gompertz, B.: On the nature of the function expressive of the law of human mortality and on a new mode of determining life contingencies. Philos. Trans. Roy. Soc. Lond. **115**, 513–585 (2013)

When IT Leveraging Competence Meets Uncertainty and Complexity with Social Capital in New Product Development

Shiuann-Shuoh Chen[1], Pei-Yi Chen[2(✉)], and Min Yu[1]

[1] Department of Business Administration, National Central University, No. 300, Jung-da Rd., Zhongli District, Taoyuan 320, Taiwan, Republic of China
kenchen@cc.ncu.edu.tw, m2121374@gmail.com
[2] Department of International Business, Hsin Sheng College of Medical Care and Management, No. 418, Gaoping Sec., Zhongfeng Rd., Longtan District, Taoyuan 320, Taiwan, Republic of China
peiyi01@ms47.hinet.net

Abstract. We examine how the aspects of IT leveraging competence [i.e., the effective uses of project and resource management systems (PRMS), organizational memory systems (OMS), and cooperative work systems (CWS)] and the social capital (SOCI) influence the performance [i.e., product effectiveness (PDT) and process efficiency (PCS)] by the coordination capability (COOR) and absorptive capacity (ACAP) under the uncertainty and complexity in the new product development (NPD). We find the IT leveraging competence positively affects COOR and ACAP, the links of SOCI-COOR, SOCI-ACAP, COOR-PCS, and ACAP-PDT are positive, neither uncertainty nor complexity has the moderating effect on the COOR-PCS link, the uncertainty negatively moderates the ACAP-PDT link, but the complexity has no moderating effect on this link. Our findings reveal why the NPD teams may have difficulty achieving high levels of performance and why these teams may vary in their ability to create the value from their COOR and ACAP.

Keywords: IT leveraging competence · Social capital · Project characteristics · Uncertainty · Complexity · New product development

1 Introduction

The new product development (NPD) is a strategic process wherein the firms integrate disparate inputs from the R&D scientists, engineers, and marketers to jointly develop and launch the new products [1]. The firms are in a position where NPD is no longer a strategic option but a necessity [2]. Under the rapidly changing technologies and customer needs, the firms must continuously introduce the new products to maintain pace with the changes [3].

Recent literature has found the influences of IT capability (a firm's ability to effectively acquire, deploy, and leverage its IT resources.), social capital (the actual and potential resources embedded in the relationships among actors, SOCI), coordination

© Springer International Publishing AG 2016
F. Lehner and N. Fteimi (Eds.): KSEM 2016, LNAI 9983, pp. 159–171, 2016.
DOI: 10.1007/978-3-319-47650-6_13

capability (a firm's ability to manage the dependencies among its various resources and activities, COOR), and absorptive capacity (a firm's ability to identify, assimilate, transform, and apply the valuable external knowledge, ACAP) on the NPD prospect.

Although much is known regarding the effects of IT capability, social capital, coordination, and absorptive capability on NPD performance, far less is known regarding their interactions to impact NPD performance [i.e., product effectiveness (the extent to which the new product is successful by some external criteria like quality and innovativeness, PDT) and process efficiency (the extent to which the NPD project adheres to budgets and schedules, PCS)]. Although isolated organizational capabilities are valuable, they may not be effective as single assets, particularly for complex activities such as NPD [4]. Thus, we regard IT capability and social capital as the effective organizational complements to a firm's processes for coordinating organizational activities and absorbing external knowledge. In addition, an extensive literature distinguishes between uncertainty (the newness of technologies employed in the product development effort to the development organization) and complexity (the nature, quantity, and magnitude of organizational subtasks and subtask interactions posed by the project) dimensions of NPD project characteristics that affect performance [5]. When faced with the burning debates regarding the strategic potential of IT, our study is an attempt to address this issue and therefore refine and extend comprehension of the link among IT capability, social capital, coordination capability, absorptive capacity, project characteristics, and performance in NPD.

We organize this paper as follows: the next section presents a review of theory and hypotheses. The following section shows our methodology. The final section discusses the implications, limitations and suggestions for future research, and conclusion of our work.

2   Theory and Hypotheses'

2.1   IT Leveraging Competence in NPD

In accordance with Pavlou and El Sawy [7], IT leveraging competence in NPD denotes the ability of NPD teams to effectively use IT functionalities to support IT-enabled NPD activities. The NPD teams should know what IT functionalities offer, understand when to use them, and do so effectively by utilizing their specific IT functionalities.

The IT tools that NPD teams commonly use include project and resource management systems (PRMS), organizational memory systems (OMS), and cooperative work systems (CWS) [7]. PRMS are designed for scheduling management, resource allocation, and task assignment [8]. OMS are IT tools for knowledge coding, sharing, directories, and retrieval [9]. CWS provide IT functionalities designed for real-time communication and group collaboration across time and space, such as conveyance, presentation, and convergence systems [10]. Therefore, IT leveraging competence in NPD is a three-dimensional construct that reflects how effectively these three IT tools are leveraged [11].

2.2 Coordination Capability

Coordination capability (COOR) signifies a firm's ability to manage the dependencies among its various resources and activities [12]. Okhuysen and Bechky suggest that coordinating mechanisms emerge through the accomplishment of three conceptually discrete but practically intertwined characteristics of interdependent organizational activity: accountability, predictability, and common understanding [13]. Accountability emerges from people's efforts to identify who is responsible for what task within the organizational output. Predictability emerges as actors anticipate the elements of an output and know when they are likely to occur within a pattern or sequence of tasks. Common understanding is accomplished when actors develop shared perspective on the goals and outputs of work.

Regarding PRMS [7], first, it provides an effective way to identify available resources and access real-time project data, thereby enabling better resources allocation. Second, its scheduling and time management functionality helps NPD managers effectively appoint NPD workers to relevant tasks and monitor the performance of NPD workers. Third, it provides real-time information on project status and enables aggregate project portfolios, thereby contributing to better synergies identification and synchronically collective activities.

Regarding OMS [7], first, it provides the functionality for the creation of knowledge directories, thereby enabling easy access to project information and best practices from prior projects. Second, its knowledge networking functionality enables communication forums and knowledge communities that help NPD teams discuss new product ideas. Third, it also helps NPD teams locate relevant expertise through visualization IT technologies.

Regarding CWS [7], first, its conveyance functionality enables data-based collaboration, content management, and sharing ideas. Second, its presentation functionality fosters NPD teams to transform their tacit ideas into graphic images. Third, its convergence functionality can clarify assumptions, elicit tacit knowledge, and construct product histories by enabling NPD teams to work together and review product designs in real time. This functionality supports NPD teams' brainstorming, converging ideas, finding solutions for new products, and reaching group consensus.

Consequently, IT leveraging competence in NPD can foster accountability, predictability, and common understanding, which underlying a team's COOR. Hence:

H1: IT leveraging competence is positively related to coordination capability in NPD.

2.3 Absorptive Capacity

Absorptive capacity (ACAP) denotes the dynamic capacity existing as two subsets of potential and realized ACAP (PACAP and RACAP) [14]. PACAP, which includes knowledge acquisition and assimilation, captures efforts expended in valuing, acquiring and assimilating new external knowledge. RACAP, which contains knowledge transformation and application, encompasses deriving new insights and consequences from the combination of existing and newly acquired knowledge and incorporating transformed knowledge into operations.

It leveraging competence is proposed to influence PASAP. **PRMS** improve the competence of NPD teams in knowing the true availability of people, skill, and resources to enable appropriate task assignment, and in analyzing and measuring work, tasks, and resources by task assignment and resource management, thus enabling knowledge acquisition and assimilation. **OMS** makes NPD teams more competent in acquiring product-related knowledge by storing, archiving, retrieving, sharing, and reusing project information and best practices. It also enhances the competence of NPD teams in articulating, interpreting, and synthesizing new and stored knowledge by facilitating easy access to stored knowledge, thus enabling knowledge assimilation [7].

IT leveraging competence in NPD is also proposed to influence RACAP. **OMS** help retrieve knowledge that was previously created and internalized for use, thus enabling knowledge exploitation [7, 15]. **CWS** can enhance the problem solving capability of NPD work units and the units' ability to generate new thinking [15], thereby enabling knowledge transformation. It can also enhance the ability of NPD work units to pursue new product initiatives and find new solutions [16], thus enabling superior knowledge exploitation. Therefore, it is hypothesized that:

H2:   IT leveraging competence is positively related to absorptive capacity in NPD.

2.4   Social Capability

Social capability (SOCI) is the sum of the actual and potential resources embedded within, available through, and derived from the network of relations possessed by an individual or social unit [17]. Nahapiet and Ghoshal identified three distinct dimensions of SOCI as structural, relational, and cognitive [17].

The structural dimension of SOCI refers to the overall pattern of connection between actors [17]. The close social interactions permit people to know one another, to share vital information, to create a common understanding related to task issues or goals and to gain access to others' resources. As Sparrowe et al. theorized [18], information sharing and exchange can enhance cooperation and mutual accountability.

The relational aspect of SOCI represents the type of personal relationships people have developed with one another through a history of interactions [17]. Among SOCI's key attributes is the level of trust among actors [17]. Trusting relations facilitate collaborative behaviours and collective action in the absence of explicit mechanisms to foster and reinforce those behaviours [19].

The cognitive dimension of SOCI refers to those resources providing shared representations, interpretations, and systems of meaning among parties [20]. SOCI's cognitive dimension represents the fact that, as individuals interact with one another as part of a collective, they are better able to develop a common set of goals, and a shared vision for the organization [17]. When a shared goal is present in the network, project team members have similar perceptions regarding how they should interact with one another.

According to these arguments, SOCI can enable accountability, predictability, and common understanding among participants of interdependent organizational activity, which leads to the emergence of coordinating mechanisms [13]. Therefore, we propose the following hypothesis.

H3: Social capital is positively related to coordination capability in NPD.

As Zahra and George noted [14], social integration leads to knowledge assimilation, occurring either informally or formally. Informal mechanisms are advantageous for exchanging ideas, but formal mechanisms tend to be more systematic. Formal social integration fosters information distribution as well as interpretation collection and trend identification. Research has shown that organizational structures promote interaction, encouraging problem solving and creative action [21]. Firms that build such connectedness by social integration mechanisms tend to make their employees aware of the types of data that constitute their PACAP.

Connectedness develops trust and cooperation and fosters commonality of knowledge [22]; it encourages communication and improves the efficiency of knowledge exchange through units [23]. Thus, connectedness allows units to transform and exploit new external knowledge [14]. Moreover, connectedness reduces the likelihood of conflict regarding goals and implementation [24]. Thus, connectedness facilitates the transformation and exploitation of newly acquired knowledge and develops a unit's RACAP. Therefore,

H4: Social capital is positively related to absorptive capacity in NPD.

2.5   New Product Development Performance

New product development (NPD) intrinsically regards integrating multiple functional departments to launch a new product through idea generation, product design, manufacturing ramp-up and marketing deployment [3].

In the broader capabilities' view of resource-based theory, performance entails a firm's ability to achieve a competitive advantage that ultimately is measured by superior financial returns but that, in the shorter run, is gauged in terms of improved efficiency, market share or position, or breaking into new arenas [25]. This is similar to NPD, where performance is a combination of product effectiveness (PDT) and process efficiency (PCS) [11, 27]. PDT is the extent to which the new product is successful by some external criteria, such as quality and the level of innovativeness [26]. PCS measures the extent to which the NPD project adheres to budgets and schedules [26]. Several studies suggested that harmony between quality and cost of product is a key aspect of NPD team performance [28].

Insufficient coordination between the teams tends to cause rework on certain work products [27]. Such rework can become problematic, particularly in later development phases, and often entails delays and additional development costs. Accordingly:

H5: Coordination capability is positively related to process efficiency.

In accordance with Zahra and George [14], firms with well-developed PACAP tend to be more proficient at continually improving their knowledge stock by recognizing trends in their external environment and internalizing this knowledge, thereby overcoming certain of the competence traps. Zahra and George distinguish between the timing and cost dimensions of being proficient [14]. The timing dimension denotes that a developed PACAP improves the effectiveness of changes track in the industries and facilitates the deployment of production and technological competencies. The cost dimension signifies that a developed PACAP reduces the investments in changes of resource positions and operational routines. In addition, Zahra and George noted that RACAP tends to influence performance through product and process innovation [14]. RACAP's transformation capabilities foster the development of new perceptual schema or changes to existing processes through the process of bisociation. RACAP's exploitation capabilities take this a step further and convert knowledge into new products [28]. Therefore, we conclude:

H6: Absorptive capacity is positively related to product effectiveness.

2.6 Project Characteristics

An extensive literature distinguishes between uncertainty and complexity dimensions of NPD project characteristics that affect performance [5]. The project uncertainty, measured by product newness, market newness, technology newness, and process technology newness, denotes the newness of technologies employed in the product development effort to the development organization [6]. The project complexity, measured by technology interdependency, object novelty, and project difficulty, signifies the nature, quantity, and magnitude of organizational subtasks and subtask interactions posed by the project [6].

Following Grote [31], the greater the uncertainty and complexity, the greater the information quantity that must be processed during project execution in order to achieve high levels of performance. Hence, NPD teams face more difficulties and spend more time when developing products with a higher level of novelty [30]. As Sheremata noted [21], uncertain projects increase the coordination need that results in higher costs. NPD cycle times and costs increase with product newness due to greater uncertainty and complexity [32]. Besides, higher project uncertainty and complexity imply high variability in and unpredictability of exact means to accomplish the project, in turn resulting in poorer project outcomes. Thus, a higher level of uncertainty and complexity is expected to have a negative effect on the quality of project outcome [29]. Accordingly, we have the following hypotheses:

H7: The level of project uncertainty negatively moderates the relationship between coordination capability and process efficiency.

H8: The level of project uncertainty negatively moderates the relationship between absorptive capacity and product effectiveness.

H9: The level of project complexity negatively moderates the relationship between coordination capability and process efficiency.

H10: The level of project complexity negatively moderates the relationship between absorptive capacity and product effectiveness.

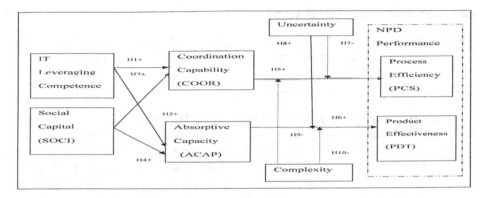

Fig. 1. Research framework

2.7   Research Framework

As shown in Fig. 1, the model proposes that there are two key areas for which the process of establishing NPD performance can be understood i.e., PDT, and PCS. Within these two key areas, constructs such as IT leveraging competence, SOCI, COOR, ACAP, and project characteristics, emerge. What is said above is a description of these constructs and a discussion on their interrelationships.

3   Methodology

3.1   Data Collection Procedure

In this survey, all of the items were measured with 7-point Likert scale (1 = strongly disagree, 7 = strongly agree). The measurements used in this study were primarily derived from the previous studies appears as in Appendix A, available if needed.

After pre-test, two rounds of survey were conducted by distributing the survey instrument in the form of questionnaire to the production managers of 770 electrical manufacturing firms in Taiwan from Jan 1 to March 31, 2012. These firms were listed in the directories of the 2012 top 2000 firms in Chinese Credit (Taiwan's leading credit company). The exclusion of 3 invalid questionnaires resulted in a total of 121 complete and effective responses for data analysis. The total response rate is 16 %. Each respondent was asked to provide their opinions about project A and B because we hope to obtain the responses from projects with different performance and thereby we got two samples from single questionnaire, consequently, there are 242 samples in this study. To examine the possibility of nonresponsive bias, a Chi-square test was conducted to compare early and late respondents on the research variables. Responses from the first mailing were 65 questionnaires. The late respondents were 56 questionnaires after a follow-up mailing. The results revealed no significant differences (p > .05) between the early and late respondents suggesting that non-response bias is not a problem in this study [33]. Appendix B shows the non-response analysis results, available if needed.

3.2 Analysis of Measurement Model

Our main data analyzing tool is the partial least squares (PLS), a component-based technique for structural equation modeling, because this study includes both formative and reflective constructs. The formative and reflective indicators require different approaches and criteria for validating reliability, convergent validity, and discriminant validity [34].

Formative constructs: We modeled the indicators of IT leveraging competence (2nd-order) and absorptive capacity (2nd-order) as formative measures since these indicators are not expected to have covariation within the same latent construct and they are causes of, rather than caused by, their latent construct [34]. Diamantopoulos and Winklhofer (2001) propose that the formative items should correlate with a "global item that summarizes the essence of the construct" [35]. PLS item weights, which indicate the impact of individual formative items [36], can be multiplied by the item values and summed, as noted by Bagozzi and Fornell [42]. In effect, our results in a modified multitrait, multimethod (MTMM) matrix of item-to-construct and item-to-item correlations similar to that analyzed by Bagozzi and Fornell as well as Loch et al. [37]. The resulting matrix, showing item-to-construct correlations as grayed out cells, appears as in Appendix C, available if needed.

Loch et al. suggest that the convergent validity is demonstrated if the items of the same construct correlate significantly with their corresponding composite construct value (item-to-construct correlation) [37]. This condition has been met in our study, as all items correlated significantly ($p < 0.01$) with their respective construct composite value. The results, therefore, indicate an acceptable level of convergent validity.

The discriminant validity can be established if the item-to-construct correlations are higher with each other than with other construct measures and their composite value [37]. This condition is also met in our study. In a sense, very high reliability can be undesirable for the formative constructs because the excessive multicollinearity among the formative indicators can destabilize the model [34]. To ensure that the multicollinearity is not a significant issue, we assessed the VIF (variance inflator factor) statistic. If the VIF statistic is greater than 3.3, the conflicting item should be removed as long as the overall content validity of the construct measures is not compromised [38]. For our formative measures, we find the VIF values of both IT leveraging competence and absorptive capacity to be 1.000 and 1.591. In summary, the results suggest that all indicators have VIF statistics lower than 3.3.

Reflective constructs: The convergent validity is demonstrated if (1) the item loadings are in excess of 0.70 on their respective factors and (2) the average variance extracted (AVE) for each construct is above 0.50 [39]. These conditions have been met in our study. Gefen and Straub also contend that the discriminant validity is demonstrated if (1) the square root of each construct's AVE is greater than the inter construct correlations, and (2) the item loadings on their respective constructs are greater than their loadings on other constructs [39], available in Appendix D, available if needed. These conditions have also been met, thereby demonstrating that the independent construct indicators discriminate well. The composite reliability scores equal to or greater than 0.70 are regarded as acceptable [39]. So the composite reliability scores of these

Table 1. Structure model

Structure model

	Baseline model Process E R2=0.348 Product E R2=0.291		Model with interaction Process E R2=0.428 f2=0.086		Model with interaction Product E R2=0.308 f2=0.024		Model with interaction Process E R2=0.391 f2=0.070		Model with interaction Product E R2=0.297 f2=0.008		Support
	Parameter	t-statistic	Parameter	t-statistic	Parameter	t-statistic	Parameter	t-statistic	Parameter	t-statistic	
H1	0.49	7.084	0.522	8.173	0.522	7.791	0.522	8.295	0.522	7.484	YES
H2	0.461	7.876	0.438	7.386	0.438	7.949	0.438	7.935	0.438	7.957	YES
H3	0.273	3.510	0.311	4.319	0.311	4.243	0.311	4.427	0.311	4.223	YES
H4	0.328	5.102	0.353	6.301	0.353	6.791	0.353	6.782	0.353	6.847	YES
H5	0.599	12.927	0.389	4.864	0.608	12.711	0.539	8.692	0.608	12.489	YES
H6	0.539	10.488	0.498	7.287	0.339	4.335	0.498	7.039	0.396	6.324	YES
H7			0.002	0.038							NO
H8					-0.169	2.287					YES
H9							-0.006	0.084			NO
H10									-0.003	0.024	NO

reflective variables are acceptable, available in Appendix E, available if needed. Our validation results suggest that all reflective measures demonstrated satisfactory reliability and construct validity and all formative measures demonstrated satisfactory construct validity and no significant multicollinearity. Therefore, all of the measures were valid and reliable.

3.3 Assessment of Structural Model

The structural model aims to examine the relationship among a set of dependent and independent constructs. In this section, we tested the amount of variance explained and the significance of the relationships. Additionally, a bootstrap re-sampling approach is suggested in order to estimate the precision of the PLS estimates [41]. Following this suggestion, a bootstrap analysis with 500 bootstrap samples [42] and the original 242 cases were performed to examine the significance of the path coefficients. The result of our structural model analysis is presented in Table 1.

Following Henseler and Fassott [40], we have used the product-indicator technique to test the moderating relationship included in our research model (H7, 8, 9, 10). As in regression analysis, the predictor and the moderator variables are multiplied to obtain the interaction term. Chin et al. recommend the standardization of the product indicators [41]. In our study, the coefficient of Task uncertainty × ACAP → Production E (−0.169) is statistically significant. The R-square for this interaction model is compared to the R-square for the baseline model, which excludes the interaction term [42]. The difference in R-square assesses the overall effect size f^2 for the interaction effect. The effect size f^2 can be calculated as $f^2 = (R^2 \text{ included} - R^2 \text{excluded})/1 - R^2 \text{ included}$. Values of 0.02, 0.15 and 0.35 indicate that the interaction term has a low, medium, or large effect on the criterion variable. In our case, the H8 of interaction term achieves a f^2 value of 0.024. Therefore, hypothesis 8 is supported.

4 Discussion

4.1 Implications

We did not find the uncertainty to negatively moderate the relationship between coordination capability and process efficiency. In addition, we failed to demonstrate the moderating effects of complexity on the link between coordination capability and process efficiency, as well as the link between absorptive capability and product effectiveness. Although managers are often tempted to undertake simple NPD projects, this study shows that projects with a higher level of uncertainty and/or complexity do not necessarily lead to poor performance. Managers should resist the temptation to fall back on me-too products. Managers should be encouraged to undertake breakthrough projects even if such projects increase the level of uncertainty and/or complexity. The competitive advantage is gained often by doing difficult tasks better than the competition.

In addition, investing in the creation of SOCI inside the NPD team eventually creates performance. NPD performance depends on the employees' complementary capabilities and the ability to manage the social interactions to achieve common goals. To effectively leverage investments in human resources, it may be imperative for NPD teams to invest in the development of SOCI to provide the necessary conduits for their participants to network and share their expertise. NPD teams that neglect the social side of individual skills and inputs and do not create synergies between their human and SOCI are unlikely to realize the potential of their members to realize superior performance. Thus, a team's efforts at hiring, training, work design, and other human resource management activities may need to focus on not only strengthening their members' specific technological skills/expertise but also developing their abilities to network, collaborate, and share information and knowledge.

4.2   Limitations and Suggestions for Future Research

First, our survey research was conducted at the production managers of 770 electrical manufacturing firms in Taiwan. Such a focus helped to account for corporate-, industry-, country-, and cultural-specific differences that might have otherwise masked significant effects. Empirical studies in a wider variety of organizations within different industries and countries are necessary to further generalize the findings.

Second, our construct measures were perceptual, based on key informants. We relied on perceptual measures because strategic capabilities are difficult to capture with self-reported survey responses. The measures of IT leveraging competence, SOCI, COOR, ACAP, uncertainty, and complexity in NPD may not be perfectly captured with primary data. Future research could use objective third-party assessments for these capabilities.

Third, it is possible, however, to measure IT leveraging competence beyond NPD or other specific processes. In this study, the IT leveraging competence construct is based on IT functionalities specifically used for NPD. Future research could develop a generalizable measure of IT leveraging competence that is not dependent on context-specific tools.

4.3   Conclusions

Our study provides an empirically grounded framework simultaneously linking various aspects of IT leveraging competence, SOCI, COOR, ACAP, uncertainty, and complexity to performance in NPD. This framework shows how NPD project teams need to combine their IT leveraging competence and SOCI to improve COOR and

ACAP under uncertainty and complexity for superior performance. This framework also provides a structure for future research, probing through more specific questions regarding the capabilities-performance link.

References

1. Clark, K.B., Fujimoto, T.: Product Development Performance. Harvard Business School Press, Boston (1991)
2. Craig, A., Hart, S.: Where to now in new product development research? Eur. J. Mark. **26**, 2–29 (1992)
3. Chen, S.-S.: A contingency perspective of R&D cross-functional communication in new product development. Unpublished Ph.D. Dissertation, Graduate School of Vanderbilt University (2000)
4. Moorman, C., Slotegraaf, R.J.: The contingency value of complementary capabilities in product development. J. Mark. Res. **36**(2), 239–257 (1999)
5. Ahmad, S., Mallick, D.N., Schroeder, R.G.: New product development: impact of project characteristics and development practices on performance. J. Product Innov. **30**(2), 331–348 (2013)
6. Tatikonda, M.V., Rosenthal, S.R.: Technology novelty, project complexity, and product development project execution success: a deeper look at task uncertainty in product innovation. IEEE Trans. Eng. Manage. **47**(1), 74–87 (2000)
7. Pavlou, P.A., El Sawy, O.A.: From IT leveraging competence to competitive advantage in turbulent environments: the case of new product development. Inf. Syst. Res. **17**(3), 198–227 (2006)
8. Rangaswamy, A., Lilien, G.L.: Software tools for new product development. J. Mark. Res. **34**(1), 177–184 (1997)
9. Stein, E.W., Zwass, V.: Actualizing: organizational memory with information systems. Inf. Syst. Res. **6**(2), 82–117 (1995)
10. Wheeler, B.C., Dennis, A.R., Press, L.I.: Groupware comes to the internet: charting a new world. Data Base **30**(3/4), 8–21 (1999)
11. Pavlou, P.A., El Sawy, O.A.: The "Third Hand": IT-enabled competitive advantage in turbulence through improvisational capabilities. Inf. Syst. Res. **21**(3), 443–471 (2010)
12. Malone, T.W., Crowston, K.: The interdisciplinary study of coordination. ACM Comput. Surv. **26**(1), 87–119 (1994)
13. Okhuysen, G.A., Bechky, B.A.: Coordination in organizations: an integrative perspective. Acad. Manage. Ann. **3**(1), 463–502 (2009)
14. Zahra, S.A., George, G.: Absorptive capacity: a review, reconceptualization, and extension. Acad. Manage. Rev. **27**(2), 185–203 (2002)
15. Tippins, M.J., Sohi, R.S.: IT competency and firm performance: is organizational learning a missing link? Strateg. Manage. J. **24**(6), 745–761 (2003)
16. McGrath, M., Iansiti, M.: Envisioning IT-enabled innovation. Insight (Magazine) **1**(1), 2–10 (1998)
17. Nahapiet, J., Ghoshal, S.: Social capital, intellectual capital, and the organizational advantage. Acad. Manage. Rev. **23**, 242–266 (1998)
18. Sparrowe, R., Liden, R., Wayne, S., Kramer, M.: Social networks and the performance of individuals and groups. Acad. Manage. J. **44**, 316–325 (2001)

19. Onyx, J., Bullen, P.: Measuring social capital in five communities. J. Appl. Behav. Sci. **36**(1), 23–42 (2000)
20. Cicourel, A.V.: Cognitive Sociology. Penguin Books, Harmondsworth (1973)
21. Sheremata, W.A.: Centrifugal and centripetal forces in radical new product development under time pressure. Acad. Manage. Rev. **25**, 389–408 (2000)
22. Rowley, T., Behrens, D., Krackhardt, D.: Redundant governance structures: an analysis of structural and relational embeddedness in the steel and semiconductor industries. Strateg. Manage. J. **21**, 369–386 (2000)
23. Galunic, D.C., Rodan, S.: Resource recombinations in the firm: knowledge structures and the potential for schumpeterian innovation. Strateg. Manage. J. **19**, 1193–1201 (1998)
24. Rindfleisch, A., Moorman, C.: The acquisition and utilization of information in new product alliances: a strength-of-ties perspective. J. Mark. **65**, 1–18 (2001)
25. Hunt, S.D.: Resource-based theory: an evolutionary theory of competitive firm behavior. J. Econ. Issues **31**(1), 59–77 (1997)
26. Sivasubramaniam, N., Liebowitz, S.J., Lackman, C.L.: Determinants of new product development team performance: a meta-analytic review. J. Prod. Innov. Manage **29**(5), 803–820 (2012)
27. Dutoit, A.H., Bruegge, B.: Communication metrics for software development. IEEE Trans. Softw. Eng. **24**(8), 615–628 (1998)
28. Kogut, B., Zander, U.: What do firms do? Coordination, identity, and learning. Organ. Sci. **7**, 502–518 (1996)
29. Tatikonda, M.V., Rosenthal, S.R.: Successful execution of development projects: balancing firmness and flexibility in the innovation process. J. Oper. Manage. **18**(4), 401–425 (2000)
30. Khurana, A., Rosenthal, S.R.: Integrating the fuzzy front end of new product development. Sloan Manage. Rev. **38**(2), 103–121 (1997)
31. Grote, G.: Management of Uncertainty. Springer, London (2009)
32. Griffin, A.: Product development cycle time for business-to-business products. Ind. Mark. Manage. **31**(4), 291–304 (2002)
33. Armstrong, J.S., Overton, T.S.: Estimating nonresponse bias in mail surveys. J. Mark. Res. **14**(3), 396–402 (1977)
34. Petter, S., Straub, D., Rai, A.: Specifying formative constructs in information systems research. MIS Q. **31**(4), 623–656 (2007)
35. Diamantopoulos, A., Winklhofer, H.M.: Index construction with formative indicators: an alternative to scale development. J. Mark. Res. **38**(2), 269–277 (2001)
36. Bollen, K., Lennox, R.: Conventional wisdom on measurement: a structural equation perspective. Psychol. Bull. **110**(2), 305–314 (1991)
37. Loch, K.D., Straub, D.W., Kamel, S.: Diffusing the internet in the arab world: the role of social norms and technological culturation. IEEE Trans. Eng. Manage. **50**(1), 45–63 (2003)
38. Diamantopoulos, A., Siguaw, J.A.: Formative versus reflective indicators in organizational measure development: a comparison and empirical illustration. Brit. J. Manage. **17**(4), 263–282 (2006)
39. Gefen, D., Straub, D.: A practical guide to factorial validity using PLS-graph: tutorial and annotated example. Commun. Assoc. Inf. Syst. **16**(25), 91–109 (2005)
40. Henseler, J., Fassott, G.: Testing moderating effects in PLS path models: an illustration of available procedures. In: Vinzi, V. E., Chin, W.W., Henseler, J.,Wang, H. (eds.) Handbook of Partial Least Squares: Concepts, Methods and Applications, pp. 713–736. Springer, Berlin (2010). Henseler, J., Ringle, C.M., Sinkovics

41. Chin, W.W.: How to write up and report PLS analyses. In: Vinzi, V.E., Chin, W.W., Henseler, J. (eds.) Handbook of Partial Least Squares. Springer Handbooks of Computational Statistics, pp. 655–690. Springer, Heidelberg (2010)
42. Chin, W.W.: The partial least squares approach for structural equation modeling. In: Marcoulides, G.A. (ed.) Methodology for Business and Management, pp. 295–336. Lawrence Erlbaum Associates, New Jersey (1998)

Adoption Factors for Crowdsourcing Based Medical Information Platforms

Till Blesik[(✉)] and Markus Bick

ESCP Europe Business School, Business Information Systems, Berlin, Germany
{tblesik,mbick}@escpeurope.eu

Abstract. Platforms increasingly utilize crowdsourcing to offer knowledge intense services such as medical diagnostics while retaining low costs. The increased interest elevated the adoption of established theories for user acceptance, risk avoidance and motivational influencing factors. We propose a combination of elements from existing concepts, extended by newly identified factors, to postulate a novel theoretical model. Our analysis of a survey with 349 respondents reveals new constructs based on users' perception of risks and features. We found that risks and features are significantly interrelated and both influence perceived usefulness, the technology acceptance model's most important construct. Usefulness is diminished by perceived risks while it is increased by crowdsourcing features. Additionally, external motivation yields important influence factors. The revealed interrelations are discussed and should be accounted for in future research and implementations.

Keywords: Information adoption · Medical platform · Crowdsourcing

1 Introduction

Crowdsourcing leverages the increased availability of people over the internet and is defined as the "act of taking a job traditionally performed by a designated agent and outsourcing it to an undefined, generally large group of people" [1]. Recently, the concept of crowdsourcing has generated significant interest in both practitioner and academic communities. Building on user participation, crowdsourcing benefits highly from the diffusion of digital mobile technology [2] and crowdsourcing based platforms gain popularity in most areas [3], including health [4, 5]. Crowdsourcing tools have proven to be able to produce results of high quality for comparatively low costs [6].

In healthcare, the utilization of crowdsourcing is still lower than in other sectors of today's digital society [7]. There are positive examples such as *Fold.it*; a game which uses crowdsourcing to predict protein structures. Besides similar applications that are designed for a specific task, crowdsourcing is still primarily used for data collection and is accordingly specified as method to enable public involvement by the United States National Library of Medicine [8]. Crowdsourcing is seen as an important approach in the vision of "Health 2050" to realize improved healthcare such as personalized medicine while decreasing costs [9]. Identifying factors that hinder or assist the approaches applicability could be an essential step to reach those goals.

© Springer International Publishing AG 2016
F. Lehner and N. Fteimi (Eds.): KSEM 2016, LNAI 9983, pp. 172–184, 2016.
DOI: 10.1007/978-3-319-47650-6_14

Following this, the aim of this paper is to develop a theoretical model of perceived risk and feature richness and analyze their impact on perceived usefulness of medical diagnosis platforms. The following research questions are the basis of this paper:

1. What factors influence the perceived usefulness of crowdsourcing based platforms for medical diagnosis?
2. How are corresponding influence factors interrelated?

The presented study was conducted by combining a literature review on user adoption of information with a survey. Hence, the paper is structured in 3 sections: (Sect. 2) Theory; we conceptualized a theory based model, (Sect. 3) Analysis; we created a fitting survey and performed an exploratory analysis to identify useable constructs which we used to designed a final structured equation model to test interrelations and confirm our hypotheses, (Sect. 4) Conclusion; we reflect on our findings and their implications.

2   Theory

2.1   Perceived (Information) Usefulness

Perceived usefulness is the most relevant construct of models for user acceptance of technology. User acceptance of technology or respectively the adoption of information is of high importance for the successful utilization of a technology. The most widely recognized and exhaustively validated model of behavioral intention is the Technology Acceptance Model (TAM) [10], where acceptance is explained by the user's attitude and the perceived usefulness. The perceived ease of use influences both the attitude and usefulness positively. In later models like TAM2 or Venkatesh's Unified Theory of Acceptance and Use of Technology (UTAUT), attitude was omitted as other constructs were included. Both TAM and UTAUT were successfully tested in healthcare settings, as shown by Yarrough and Smith, who conducted an extensive review of technology acceptance among physicians [11], and Holden and Karsh, who analyzed the models past and future in health care [12].

As a common pattern among all studies, perceived usefulness is the only construct whose paths towards dependent constructs are always significant [12]. Furthermore, perceived usefulness is included in TAM and all of its successor models. The central TAM and UTAUT constructs as well as their paths and the amount of studies which found those paths to be significant are summarized in Fig. 1.

In our research, we primarily analyze how crowdsourcing based medical information are adopted. In a similar endeavor, Sussman and Siegal analyzed how knowledge workers are influenced to adopt advice [13]. They showed that the perceived (information) usefulness is an important moderator of information adoption. Since we focus on perceived usefulness, we also include perceived ease of use as it is the main construct influencing perceived usefulness in the introduced acceptance models. We hypothesize that the relation holds true for medical platforms, too:

Hypothesis 1:   Perceived Ease of Use influences Perceived Usefulness.

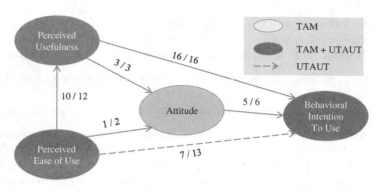

Fig. 1. TAM and UTAUT constructs with count of significant paths based on Holden and Karsh [12].

2.2 Risks

For more than a decade already, patients and physicians regularly consult the internet for health advice. While the Internet is eligible to disseminate information, problems like misinformation or confidentiality arise and are even more relevant in the context of health information [14]. Perceived risks have been shown to be a determinant of peoples' trust towards anonymous online health information and are therefore likely to be important adoption factors [15]. Implementing crowdsourcing in healthcare also transfers risks of social interaction into an already delicate environment.

Sims et al. developed DocCHIRP (Crowdsourcing Health Information Retrieval Protocol for Doctors), which uses a system of push notifications to help solve medical problems by crowdsourcing [16]. After the system's trial run of about 244 days, they analyzed log files and conducted a short online survey, where they identified potential risks that arise on crowdsourcing based medical platforms. In the first place, "distracted doctoring" was repeatedly mentioned as potential risk. If physicians use medical platforms, especially if they access it on a mobile device, it is likely that they spread their attention. Hence, neither the digital problem nor the local, on-site problems receives the required rigor. The "unwillingness to expose personal knowledge gaps" describes the worry about personal drawbacks of asking questions on public platforms. Lastly, "responsibility issues" result from the different roles between patients and physicians on the platform and between physicians on the platform and physicians in contact with their patients, who are actually liable. Consequently, users are motivated by different incentives and have to consider different repercussions.

A common drawback of electronic health records and a barrier to the use of health platforms in general are "privacy concerns" [17]. While the handling of personal health information already was an issue for patients at the beginning of telemedicine in the early 1990s [18], technological progress and social methods such as crowdsourcing amplified them to being a core concern of patients nowadays [19].

In addition to health specific risks, we analyzed characteristics and effects of crowdsourcing in general. Surowiecki named two dominant crowdsourcing risks [20]. "Groupthinking" can occur which is a dysfunctional dynamic that leads users towards supporting an already popular opinion instead of forming their own decision. The second risk is "biased results" and originates from the characteristics of the contributing crowd. Due to self-selection processes, a group might consists of people with a specific mindset, which counteracts the benefits of heterogeneous groups.

The mentioned risks relate to different risk facets. We will try to extract risk constructs and describe common characteristics. We formulate the following hypothesis:

Hypothesis 2:   Risk influences Perceived Usefulness.

2.3   Features

Opposing to risks, features provide benefit to platform users. Accordingly, the analysis of features is important to effectively implement health information platforms. In different situations, patients' feature preferences vary, depending on their actual needs [17]. Multiple studies have shown that the feature richness is linked to the perceived usefulness of related medical platforms and should be considered as a starting point during design processes [21, 22]. In addition to specific features of medical platforms, implicit features of crowdsourcing based platforms have to be considered.

Estellés and González did an extensive review on crowdsourcing literature and extracted three main elements which are further divided into a total of eight characteristics [1]. Looking for similarities among those characteristics, we derived two features: "large user base" and "monetary rewards". A core characteristic of crowdsourcing is the composition and size of the crowd; a "large user base" increases the likeliness that members of the crowd have the required knowledge or capability to solve a given problem. Furthermore, Sims et al. found that crowd size is directly related to each user's sustained use and engagement [16]. For both the participating crowd and the initiator, the benefit gained through the platform is an important characteristic. On platforms where the crowd consists of amateurs and professionals, a "monetary reward" for contributions is the most frequent feature [1].

While some risks are due to difficult resolvable psychological processes [20], privacy concerns are either technology induced or due to human ruthlessness. Technology induced risk can be handled comparatively straightforward, by implementing security features. "Anonymity" is an important feature to avoid bad personal consequences for contributing thereby exposing knowledge gaps or being ashamed of personal problems. Personal health information is highly sensitive and there are strict laws enforcing certain levels of data protection [23]. Designing a platform with features for "privacy protection" is therefore not only advantageous but legally required. Furthermore, anonymity and privacy are regularly asked for by users of medical platforms [16, 23]. Given the direct relation between those features and risks, we formulated the following hypothesis:

Hypothesis 3a: Features influence Perceived Risk.

Existing systems are the third source we used to extract possible features from. Looking at *SERMO* and *CrowdMed*, one of the big differences is their restriction system. A platform can be open for everyone or restrict certain functions for specific groups, i.e.

SERMO allows only certified physicians to contribute. The two features "posting medical cases restricted to physicians" and respectively "answering medical cases restricted to physicians" could help to depict the users desired restriction system. Involving the not risk related features, we hypothesize:

Hypothesis 3b: Features influence Perceived Usefulness.

2.4   Motivation

Medical platforms, like all crowdsourcing platforms, rely on user participation [24]. Motivation is considered a key element that has to be understood to implement successful crowdsourcing platforms [25]. Yang and Lai integrated constructs of motivation factors to examine knowledge sharing behavior [26].

"Intrinsic motivation" describes the internal satisfaction that an individual receives by a certain behavior. To intrinsically motivated individuals, enjoyment is more important than monetary gains or extrinsic rewards. On the opposite, "extrinsic motivation" is goal-oriented. Activities are performed to obtain a specific return, primarily money or reputation. While intrinsic motivation is generally positively related to knowledge sharing intentions, an extrinsically motivated person might not share his knowledge, if he perceives it to be valuable. In addition to those two traditional measurements, they also integrate self-concept-based motivation factors. A person's inherent standard is the main reason to pursue an activity, if that person's motivation is based on the "internal self-concept". A behavior has to align with their internal standards, to trigger a positive feedback. For persons whose motivation is based on "external self-concept", an activity should match with expectations of a reference group and the resulting feedback triggers a feeling of affiliation. Internal self-concept motivation can lead to very high participation because positive feedback increases the confidence in their knowledge and thereby amplify the related internal standard, which can result in a self-enhancement loop. For external self-concept based users, the platforms overall success is important and that they find a way to connect to its community.

The different types of motivations that drive users could on the other side influence how they perceive a platform. Therefore, we hypothesize

Hypothesis 4:   Motivational Factors influence Perceived Usefulness.

3   Analysis

3.1   Survey

All measurement items are either taken from validated constructs or extracted from former studies in the area of medical platforms. The resulting items were discussed with a private physician and a software developer, who are in charge of developing a hospital's medical information system and can therefore be seen as practical domain experts. Initially, they were asked to come up with possible risks on their own. Both lists overlapped respectively with three or five risks of those we identified, only "Distracted doctoring" was not proposed by neither of the two experts. Afterwards, our items were revealed, discussed and refined. In the end, the experts agreed on all items which were accordingly integrated in the survey as Likert-type scales.

A first version of the survey was created and pilot tested with six persons, two of each occupation. We found no major issues and only did some minor adjustments to wordings in the instructions and explanations. After being double-checked, we implemented the final survey as online survey, using the Google Forms platform. Since crowdsourcing is almost exclusively done online [1], an online survey is a suitable instrument to reach our target audience.

We used hospital and university mailing lists, social networks and personal referrals to distribute the survey link among practicing physicians, medical students and people without any medical education. To the last group, people without medical education, we refer as patients. The assignment of participants to a group is based on a direct question in the survey. Participants were animated to forward the survey. In total, we gained 349 completed and useable answers to our survey, which is a large sample for comparative studies. As shown in Table 1, all three groups are of similar size and there are slightly more female participants in our sample.

Table 1. Demographic overview of the sample

Total	349	
Physicians	95	27,2 %
Medical students	144	31,5 %
Patients	110	41,3 %
Female	213	61,0 %
Male	136	39,0 %
Age	7 groups, most 20–30	

3.2 Data Screening

To ensure that the data is useable, reliable and valid for testing causal relationships, data screening was performed prior to analysis. First, we checked for missing values as those could cause problems with factor analysis or represent a bias.

In a next step, we controlled for outliers. Since our survey only uses Likert-scales, univariate extremes cannot occur. Therefore we only checked for unengaged respondents, using patterns such as "1, 1, 1, 1" or "1, 2, 3, 4" to answer all questions; neither was found. Excluding multivariate outliers would not be reasonable since it is not possible to justify the removal of data points that do not fit a not yet validated theory.

The assumption of normal distribution is a requirement of most statistical methods. To ensure the applicability of those methods later on, we checked our data for skewness and kurtosis. All items were within the thresholds of ± 1.5 for skewness and ± 3 for kurtosis, except item 1 of internal self-concept which scored marginally above 3.

3.3 Constructs

After we ensured the reliability, we look for patterns in the data to determine correlations and derive constructs. An exploratory factor analysis (EFA) is performed to prepare variables for a clean structural equation modeling (SEM) [27]. We use principal axis factoring (PAF) as factor extraction method, which accounts for co-variation and therefore extracted factors fit the statistical approach of SEM. Furthermore, PAF generally outperforms other methods and is especially well suited for relatively simple factor patterns and to recover weak factors [28]. To assess the suitability of the data for factor analysis, we calculate the Kaier-Meyer-Olkin (KMO) measure of sampling adequacy and Bartlett's Test of Sphericity [29]. As shown in Table 2, the KMO of 0.777 is close to meritorious and clearly above the threshold of 0, 5, and the Bartlett's test is significant (<.05). In conclusion, there are relations among the variables and a meaningful factor analysis can be performed.

Table 2. KMO and Bartlett's test

		Initial EFA	Final EFA
Kaiser-Meyer-Olkin measure of sampling adequacy		.777	.785
Bartlett's test of sphericity	Approx. Chi-Square	3833,862	3619.589
	df	435	325
	Sig.	.000	.000
Cumulative variance explained		52.9 %	55.4 %

Our factor analysis initially results in a total of 10 extracted factors which explain about 53 % of the variance contained in the model. There are no strong thresholds regarding the required amount of explained variance but, as a rule of thumb, values above 50 % are considered sufficient [30]. In the resulting pattern matrix, the constructs which we adapted from existing theories, such as perceived usefulness or motivational factors, occur as expected. Only internal self-concept appears to have cross loadings on other motivational factors. After the kurtosis issue, those cross loadings are the second warning sing regarding the internal self-concept construct.

For our newly introduced risk and feature items, two clean constructs already emerged. R1 – R4 load onto one factor and F2 and F3 load onto another factor. Few cross loadings exist between risk and feature items and items we adopted from theory. For example, F1 and R5 slightly load on a common factor with perceived usefulness. Since we do not want to expand existing constructs, we exclude them from the factor analysis. The first analysis of all risk and feature items reveals a third new factor whereupon F1, F4 and F5 load. Furthermore, a fourth factor absorbs small loadings from multiple risk and feature items. To reach a clean pattern matrix and therefore distinctive constructs, we iteratively remove items with low communalities. The communality describes the extent to which an item correlates with all other items and low communality therefore indicates candidates for removal [27]. After successively removing R6, F6 and R5, only F5 loads on two factors and is therefore removed.

In the end, we extracted three new constructs: A large user base (F1) and anonymity (F4) are described in theory as being essential to successfully use crowdsourcing for medical purposes and the analysis shows that users value the importance of those features. Consequently, F1 and F4 build the construct "Importance of Crowdsourcing Features". "Importance of Restrictive features" consists of answer restriction (F2) and post restriction (F3) and the constructs importance is supported by implementations on existing platforms. Looking at the differences between excluded and included risks, we identified intention as possible separator. While R5 and R6 are only problematic if a person acts intentionally malevolence, R1 - R4 are based on psychological processes and organizational issues and therefore form the construct of "Perceived Unintentional Risk". Combining our newly identified constructs with the adapted ones, we end up with a total of eight constructs, as shown in Table 3.

Table 3. Overview of items, related constructs and their theoretical source(s)

Item	Construct	Source
(F1) LargeUserBase	*Importance of Medical Crowdsourcing Features (IoCF)*	Features
(F4) Anonymity		*Estellés-Arolas and González-Ladrón-de-Guevara (2012)* [1]
(F5) PrivacyProtection		
(F6) MonetaryReward		
(F2) AnswerRestriction	*Importance of Restrictive Features (IoRF)*	
(F3) PostRestriction		
(R1) Distraction	*Perceived Unintentional Risk (PUR)*	Risks
(R2) PopularityEffect		*Sims et al. (2014)* [16]
(R3) Responsibility		*Frost and Massagli (2008)* [20, 31]
(R4) BiasedResults		
(R5) HideKnowledgeGap		
(R6) UploadConcerns		
E1 – E3	*Perceived Ease of Use (PEoU)*	TAM
U1 – U3	*Perceived Usefulness (PU)*	*Davis (1989)* [32]
Int1 – Int3	*Intrinsic Motivation (Int)*	Motivational Factors
Ext1 – Ext3	*Extrinsic Motivation (Ext)*	*Yang and Lai (2010)* [26]
Esc1 – Esc3	*External Self Concept (ESC)*	
Isc1 – Isc3	*Internal Self Concept (ISC)*	

To further validate our measurement model, a confirmatory factor analysis was performed. Cronbach's alpha (Cr. α) yields results around 0.7, with the exception of IoCF which has a very low value. While Cronbachs alpha is one of the most commonly used criterions, it tends to be underestimated in PLS models and therefore researchers should rely on the average variance extracted (AVE) and composite reliability (CR) instead [33]. For AVE all constructs yield values above the recommended threshold of 0.5. Similar, CR constructs yield results very close to or above 0.7. The combination of all criteria, which are listed in Table 4, indicates the construct reliability and validity of our measurement model.

Table 4. Construct Reliability and Validity

Construct		AVE	CR	Cr. α
Int	Internal Motivation	0.84	0.94	0.90
Ext	External Motivation	0.79	0.92	0.87
ISC	Internal Self Concept	0.57	0.80	0.64
ESC	External Self Concept	0.70	0.87	0.78
IoCF	Importance of Crowdsourcing Features	0.54	0.67	0.22
IoRF	Importance of Restrictive Features	0.74	0.85	0.64
PEoU	Perceived Ease of Use	0.74	0.89	0.82
PUR	Perceived Unintentional Risk	0.52	0.81	0.67

3.4 Structured Equation Model

Using the previously identified constructs, we created a structured equation model to check our initially formulated hypothesizes. Besides the established relation between PEoU and PU, we analyzed the influence of IoCF, IoRF and PUR on PU as well as relations among risk and feature constructs. Furthermore, we calculated multiple models including all combinations of constructs for motivational factors. Of all calculated models, only the relation between Ext and PU was significant.

In our analysis, we found that risks and features are not only directly influencing PU but PUR and IoRF load significantly onto a second order construct, which we called "Perceived Risk". The underlying assumption is that the perception of risk is a combination of theoretical worst-case scenarios and implemented counter measurements. The more users are aware of risks, the more they value restrictive features that help to prevent or at least mitigate those risks. This finding is in line with the theory of avoidance of information technology threats (TTAT) [34], where perceived threat and perceived avoidability form users' behavior.

To assess the fit of our structural model, we calculated Stone-Geisser's Q^2 and the variance inflation factors (VIF). As shown in Table 5, all VIF values are below 5, which indicates the absence of multicollinearity and the Q^2 values of dependent constructs are above 0. This means that every endogenous construct can be predicted by its preceding factors, indicating a relatively high explanatory power of the model [35].

Table 5. Explanatory power of the model

	PUR	IoRF	IoCF	PEoU	Ext	PUR	PR	PU
Q^2						0.01	0.36	0.18
VIF	1.02	1.02 (PR) 1.00 (PUR)	1.02	1.01	1.02			

While the R^2 (0.26) and respectively the Q^2 (0.18) of PU seems to be low at first sight, this value has to be relativized as we deliberatively focused on extending existing models which already reveal other influence factors. Therefore, we excluded already validated influence factors such as social influence or effort expectancy. It can be assumed that adding established factors would drastically increase those values.

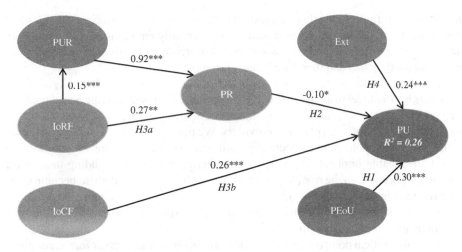

Fig. 2. Effects and Hypothesizes in the structured equation model

Our results show significant correlations for all hypothesized relationships between the latent constructs. As shown in Fig. 2, five paths are highly significant at $p < 0.001$, one path is significant at $p < 0.01$ and one path is moderately significant at $p < 0.05$. Respectively, t-values are ranging from 2.03 to 20.5. While the influence of PUR on PR ($\gamma = 0.92$) is stronger than the influence of IoRF ($\gamma = 0.27$) both are significant. The influence of PEoU ($\gamma = 0.30$) on PU is highly significant as predicted by technology acceptance theory. More interestingly, we found new significant relations between PU and Ext ($\gamma = 0.24$), IoCF ($\gamma = 0.26$), and PR ($\gamma = -0.10$).

4   Conclusion

We started our research by identifying influence factors for perceived usefulness which are therefore important drivers of adoption. Based on literature and existing systems, we hypothesized a model combined of new influence factors and factors adopted from existing adoption theories. All hypotheses are at least partially confirmed and the derived relationships in our model are significant at different levels.

The newly found relations have not yet been described in literature. The correlation between PR and PU is the only negative one in our model which is adequate, given the negative characteristic of risk. To increase the likeliness that crowdsourcing based information are adopted by users, the perceived risk related to those information has to be reduced, for example by providing quality indicators such as displaying contributors' experience level. A platform that implements crowdsourcing methods should not only implement features to restrict contribution but also make those features visible to their users. While previous studies showed the importance of various motivation factors to trigger contribution, we showed that extrinsic motivation influences the perceived usefulness for medical platforms. This relation might differ among various types of platforms and further research would be needed. For example, to analyze the proportion of career related influences in comparison of expert advice leading users towards such platforms. Our findings further suggest that crowdsourcing platforms should offer anonymity, if the context could be sensitive or potentially embarrassing. Furthermore, users' are aware that the effectiveness of crowdsourcing generally increases with the amount of participants and therefore value a large user base.

This study looks at validated models and integrates them with newly derived constructs into a model to analyze perceived usefulness of crowdsourcing based medical diagnoses platforms. Our stakeholders include scholars, both types of potential users, physicians and patients, and platform providers. We pursued our study with similar sized groups of all stakeholders in respectively suiting contexts to facilitate a representative result with useable implications. Because knowledge platforms, including those with medical purpose, are on the rise [36], the findings of this study are likely to become even more relevant in the future. For future research, it could be interesting to shed light at the specific risk perceptions of each stakeholder and derive implications for customized implementation recommendations.

From a theoretical point of view, the understanding of impact factors for the adoption of crowdsourcing based information is extended. As our dependent construct – perceived usefulness – is one of the most relevant constructs for various adoption models, this study spawns into a broad field of adoption and acceptance research, not only for knowledge platforms but for all kinds of crowdsourcing based applications and tools. Besides the newly created constructs (PUR, IoRF, IoCF and PR), the identified significant relations are the central scientific contribution of this research. In TTAT, the authors also argue that adoption and avoidance of threats are qualitatively different phenomena and using only one approach is not sufficient. Integrating both elements in one model can therefore help to improve interrelations of those models and help to lead towards a more holistic understanding.

References

1. Estellés-Arolas, E., González-Ladrón-de-Guevara, F.: Towards an integrated crowdsourcing definition. J Inf. Sci. **38**, 189–200 (2012)
2. Kauffman, R.J., Techatassanasoontorn, A.A.: International diffusion of digital mobile technology: a coupled-hazard state-based approach. Inf. Technol. Manage. **6**, 253–292 (2005)
3. Mourelatos, E., Tzagarakis, M., Dimara, E., et al.: A review of online crowdsourcing platforms. South-East. Eur. J. Econ. **14**, 59–73 (2016)
4. Ranard, B.L., Ha, Y.P., Meisel, Z.F., Asch, D.A., Hill, S.S., Becker, L.B., Seymour, A.K., Merchant, R.M.: Crowdsourcing - harnessing the masses to advance health and medicine, a systematic review. J. Gen. Intern. Med. **29**, 187–203 (2014)
5. Vogt, F., Seidl, F., Santarpino, G., van Griensven, M., Emmert, M., et al.: Healthcare IT utilization and penetration among physicians novel IT solutions in healthcare-use and acceptance in hospitals and private practice. J Health Med. Inform. **7**, 2 (2016)
6. Allahbakhsh, M., Benatallah, B., Ignjatovic, A., Motahari-Nezhad, H.R., Bertino, E., Dustdar, S.: Quality control in crowdsourcing systems: issues and directions. IEEE Internet Comput. **17**, 76–81 (2013)
7. Brabham, D.C., Ribisl, K.M., Kirchner, T.R., Bernhardt, J.M.: Crowdsourcing applications for public health. Am. J. Prev. Med. **46**, 179–187 (2014)

8. Information, N.C. for B.: Crowdsourcing – MeSH. http://www.ncbi.nlm.nih.gov/mesh?term=%22crowdsourcing%22
9. Swan, M.: Health 2050: the realization of personalized medicine through crowdsourcing, the quantified self, and the participatory biocitizen. J. Personalized Med. **2**, 93–118 (2012)
10. King, W.R., He, J.: A meta-analysis of the technology acceptance model. Inform. Manage. **43**, 740–755 (2006)
11. Yarbrough, A.K., Smith, T.B.: Technology acceptance among physicians: a new take on TAM. Med. Care Res. Rev. **64**(6), 650–672 (2007)
12. Holden, R.J., Karsh, B.-T.: The technology acceptance model: its past and its future in health care. J. Biomed. Inform. **43**, 159–172 (2010)
13. Sussman, S.W., Siegal, W.S.: Informational influence in organizations: an integrated approach to knowledge adoption. Inf. Syst. Res. **14**, 47–65 (2003)
14. Baker, L., Wagner, T.H., Singer, S., Bundorf, M.K.: Use of the internet and e-mail for health care information: results from a national survey. JAMA **289**, 2400–2406 (2003)
15. Mun, Y.Y., Yoon, J.J., Davis, J.M., Lee, T.: Untangling the antecedents of initial trust in web-based health information: the roles of argument quality, source expertise, and user perceptions of information quality and risk. Decis. Support Syst. **55**, 284–295 (2013)
16. Sims, M.H., Bigham, J., Kautz, H., Halterman, M.W.: Crowdsourcing medical expertise in near real time. J Hosp. Med. **9**, 451–456 (2014)
17. Wang, J., Lam, R.W., Ho, K., Attridge, M., Lashewicz, B.M., Patten, S.B., Marchand, A., Aiken, A., Schmitz, N., Gundu, S., et al.: Preferred features of E-mental health programs for prevention of major depression in male workers: results from a Canadian National survey. J. Med. Internet Res. **18**, e132 (2016)
18. Norton, S., Lindborg, C., Delaplain, C.: Consent and privacy in telemedicine. Hawaii Med. J. **52**, 340–341 (1993)
19. Menachemi, N., Collum, T.H.: Benefits and drawbacks of electronic health record systems. Risk Manage. Healthc. Policy **4**, 47–55 (2011)
20. Surowiecki, J.: The Wisdom of Crowds. Anchor Books, New York (2005)
21. Ozok, A.A., Wu, H., Garrido, M., Pronovost, P.J., Gurses, A.P.: Usability and perceived usefulness of personal health records for preventive health care: a case study focusing on patients' and primary care providers' perspectives. Appl. Ergon. **45**, 613–628 (2014)
22. Ruland, C.M., Maffei, R.M., Børøsund, E., Krahn, A., Andersen, T., Grimsbø, G.H.: Evaluation of different features of an eHealth application for personalized illness management support: cancer patients' use and appraisal of usefulness. Int. J. Med. Inform. **82**, 593–603 (2013)
23. Fernández-Alemán, J.L., Señor, I.C., Lozoya, P.Á.O., Toval, A.: Security and privacy in electronic health records: a systematic literature review. J. Biomed. Inform. **46**, 541–562 (2013)
24. Hossain, M.: Users' motivation to participate in online crowdsourcing platforms. In: International Conference on Innovation Management and Technology Research (ICIMTR), pp. 310–315 (2012)
25. Kietzmann, J.H., Silvestre, B.S., McCarthy, I.P., Pitt, L.F.: Unpacking the social media phenomenon: towards a research agenda. J. Public Affairs **12**, 109–119 (2012)
26. Yang, H.-L., Lai, C.-Y.: Motivations of Wikipedia content contributors. Comput. Hum. Behav. **26**, 1377–1383 (2010)
27. Gaskin, J.: Exploratory Factor Analysis. http://statwiki.kolobkreations.com/index.php?title=Exploratory_Factor_Analysis

28. De Winter, J.C., Dodou, D.: Factor recovery by principal axis factoring and maximum likelihood factor analysis as a function of factor pattern and sample size. J. Appl. Stat. **39**, 695–710 (2012)
29. Williams, B., Onsman, A., Brown, T.: Exploratory factor analysis: a five-step guide for novices. Australas. J. Paramed. **8**, 1–13 (2010)
30. Cangelosi, R., Goriely, A.: Component retention in principal component analysis with application to cDNA microarray data. Biol. Dir. **2**, 1 (2007)
31. Frost, J., Massagli, M.: Social uses of personal health information within PatientsLikeMe, an online patient community: what can happen when patients have access to one another's data. J. Med. Internet Res. **10**, e15 (2008)
32. Davis, F.D., Bagozzi, R.P., Warshaw, P.R.: User acceptance of computer technology: a comparison of two theoretical models. Manage. Sci. **35**, 982–1003 (1989)
33. Hair, J.F., Hult, G.T.M., Ringle, C.M., Sarstedt, M.: A Primer on Partial Least Squares Structural Equation Modeling (PLS-SEM), 2nd edn. Sage, Thousand Oaks (2017)
34. Liang, H., Xue, Y.: Avoidance of information technology threats: a theoretical perspective. MIS Q. **33**(1), 71–90 (2009)
35. Götz, O., Liehr-Gobbers, K., Krafft, M.: Evaluation of structural equation models using the partial least squares (PLS) approach. In: Vinzi, V.E., Chin, W.W., Henseler, J., Wang, H. (eds.) Handbook of Partial Least Squares, pp. 691–711. Springer, Heidelberg (2010)
36. Hatcliff, J., King, A., Lee, I., Macdonald, A., Fernando, A., Robkin, M., Vasserman, E., Weininger, S., Goldman, J.M.: Rationale and architecture principles for medical application platforms. In: Third International Conference on Cyber-Physical Systems (ICCPS), pp. 3–12 (2012)

Crowd Label Aggregation Under a Belief Function Framework

Lina Abassi[(✉)] and Imen Boukhris

LARODEC, Institut Supérieur de Gestion de Tunis,
Université de Tunis, Tunis, Tunisia
lina.abassi@gmail.com, imen.boukhris@hotmail.com

Abstract. Crowdsourcing emerged as an efficient human-powered concept to tackle the problem of labeling complex tasks that computer programs still cannot solve. Amazon's Mechanical Turk is one of the most popular platforms that allows to gather labels from human workers. These labels are then aggregated in order to estimate the true labels. Considering that not all labelers are experts, their answers may be imperfect and consequently unreliable. In this paper, we propose a novel label aggregation method based on the belief function theory. The proposed method grants a strong framework that does not only allow to reliably aggregate imperfect labels but also to integrate labelers expertise for more accurate results. To demonstrate the effectiveness of the proposed method, experiments are conducted on real datasets. The results show that our method is a promising solution in the crowd labeling domain.

Keywords: Crowdsourcing · Uncertainty · Label aggregation · Belief function theory · Reliability · Combination rules

1 Introduction

Crowdsourcing is today one of the most powerful concepts that allows solving tasks that cannot be computerized, such as image categorization, document transcription and data collection, through the use of human intelligence. This concept was introduced by [1] as the act of outsourcing to a large number of people in the form of an open call. It was defined later [2] as "The act of taking a job traditionally performed by a designated agent and outsourcing it to an undefined generally large group of people in the form of an open call". One of the most popular crowdsourcing platforms today is the Amazon's Mechanical Turk (AMT). It enables employers (requesters) to post batches of small jobs known as Human Intelligence Tasks (HITs). Labelers (Turkers) can then browse among existing jobs and complete them for a monetary reward set by the employer. However, a major issue in such crowdsourcing systems is the relatively low result quality since they combine the efforts of numerous anonymous workers with different levels of reliability. Because of the tedious tasks and the low payment, errors can occur even with serious workers. In some cases, workers turn out to be "spammers" as they submit arbitrary answers in the aim to

© Springer International Publishing AG 2016
F. Lehner and N. Fteimi (Eds.): KSEM 2016, LNAI 9983, pp. 185–196, 2016.
DOI: 10.1007/978-3-319-47650-6_15

gather the maximum rewards. As a solution, AMT give employers the possibility to require qualification tests to choose the most reliable workers and block spammers. However this method increases the cost and do not guarantee that the chosen workers will provide correct answers. Therefore, employers must resort to certain strategies to increase labels confidence such as redundancy that consists in assigning each task many times and aggregating labels in one of the existent aggregation methods. Several methods have been proposed [5, 11, 12] that try to effectively aggregate labels. However, none of them is able to model uncertainty in labelers answers.

In this context, we propose a novel method, the belief label aggregation (BLA), based on the belief function theory. This latter is known as a powerful framework that deals with uncertain and imprecise information and that enables, thanks to its related operations, to take into account the reliability degree of each source of information (i.e. labeler). Indeed, this theory has shown its efficiency for classifying workers and identifying experts [19]. In this work, we investigate labels aggregation besides of considering reliability degrees of crowd providing them. In fact, our method consists in a first step, in estimating the reliability of each labeler by comparing his answers to answers generated by majority voting then, with the discounting operation of the belief function theory, the reliability of each labeler is computed. In the final step answers are aggregated using the combination with adapted conflict rule (CWAC) [7] that is based on both the conjunctive [15] and Dempster combination rules [3] within the belief function framework. The CWAC has the ability to better manage the conflict induced when combining different sources of information.

The remainder of this work is structured as follows: Sect. 2, surveys the related work of label aggregation methods. In Sect. 3, belief function theory notions are presented. Later in Sect. 4, our proposed method is detailed and experimentation results are illustrated in Sect. 5. Finally, Sect. 6 concludes this paper and states future work.

2   Label Aggregation Methods

One of the most important technical challenges of crowdsourcing is label aggregation [13]. Label aggregation techniques take as input a set of all labelers' answers on a given set of questions. Every question has a true label that each aggregation method attempts to accurately estimate. Methods can either be non-iterative as they compute the final aggregated value of each question separately or iterative aggregating many questions at the same time in a sequence of computational rounds [9].

The family of non-iterative methods includes the majority voting approach [10], in which the label with highest votes is selected as the final aggregated value. Based essentially on majority voting, the Honeypot (HP) method [11] suggests that unreliable workers are removed in a preprocessing step. For that, HP includes randomly some gold standards (i.e. questions whose true label is already known) into original questions. Workers who fail to answer a certain number of these questions are neglected and removed. Then, the final label is estimated by majority voting with the remaining workers. Another non-iterative technique is proposed in [5] that offers a framework called Expert Label Injected Crowd Estimation (ELICE). It also uses gold standards, but to estimate the reliability degree of each worker by calculating the ratio of his answers which are similar to true labels of gold standards. Then, the difficulty level of each question is estimated by the number of workers who correctly answer a given number of the gold questions. The final label is obtained by the aggregation of all labels, each weighted by the worker expertise and the question difficulty.

The Expectation Maximization (EM) technique [14] figures among the iterative methods. It measures label probabilities in two steps. The first is the expectation (E) step that consists in estimating label probabilities weighting the answers of workers according to the current estimation of their reliability degree. The second is the maximization (M) step in which EM re-estimates the reliability degree of workers based on the current probability of each label. This iteration reaches an end when all label probabilities are unvaried. There is also the Iterative Learning [12] which is a belief propagation based method for annotation aggregation that can be even used to estimate question difficulty and the labeler reliability degree.

Despite their advantages, the major drawback of iterative methods is the high running time since reaching convergence requires going through a lot of steps [9].

Besides, to deal with the imperfect information occurring in crowd labeling, it is important to model the answers uncertainty to obtain more accurate results in the aggregation step. For this purpose, we propose to adopt the belief function theory.

In the next section, we present the main concepts of belief function theory that we employed in our proposed method.

3   Belief Function Theory

The belief function theory (also known as the Dempster-Shafer theory or evidence theory) was introduced by Dempster [3] in the context of statistical inference then it was formalized by Shafer [4] as a model to represent beliefs, popularized and developed by Smets under the name of Transferable Belief Model [8]. The belief function theory is often presented as a generalization of the probability and the possiblity theories and has the ability to distinguish between ignorance and equiprobability.

3.1   Basic Belief Assignments

Let $\Omega = \{\omega_1, \ldots, \omega_k\}$ be the frame of discernment of exhaustive and finite set of mutually exclusive events associated to a given problem. 2^{Ω} is the power set of Θ, it includes all the possible subsets and formed unions of events, and the empty set which matches the conflict. The piece of evidence provided by a source of information is represented by a basic belief assignment (*bba*) also called belief mass, defined as a function from 2^{Ω} to [0,1] verifying:

$$\sum_{A \subseteq \Omega} m^{\Omega}(A) = 1. \tag{1}$$

$m(A)$ is the piece of belief supporting exactly the subset A of the frame of discernment. A is considered as a focal element when $m(A) > 0$.

Since the belief function theory models several types of imperfection, special *bbas* were defined. In particular, we have:

– certain *bba*: the case where there is exactly one focal element that is a singleton. $m(\omega_i) = 1$ for one particular element of Ω.
– vacuous *bba*: the case where *bba* models the state of the total ignorance. Ω is the unique focal element: $m(\Omega) = 1$.
– categorical *bba*: the case where the *bba* has a unique focal element A.
– simple support function: the case where the evidence only supports a subset A of Ω, i.e., focal elements are $\{A, \Omega\}$.
– normalized *bba*: the case where $m(\emptyset) = 0$.

3.2   Discounting Operation

Sometimes, the reliability of information formalized with a *bba* provided by a source of information cannot be fully trusted. The discounting operation introduced by Shafer [4] enables to weaken a belief mass by a constant called the discount rate $\alpha \in [0,1]$ such as $(1 - \alpha)$ is the degree of reliability of a source. Accordingly, the belief mass becomes:

$$\begin{cases} m^{\alpha}(A) = (1 - \alpha)m(A) & \forall A \subset \Omega, \\ m^{\alpha}(\Omega) = \alpha + (1 - \alpha)m(\Omega). \end{cases} \tag{2}$$

3.3   Combination Rules

The conjunctive rule of combination (CRC) [15]. allows to combine two *bbas* induced by distinct and reliable sources of information. It is defined as the orthogonal sum of two *bbas* m_1 and m_2, whose focal elements are all the possible intersections between pairs of focal elements of m_1 and m_2 respectively. It is defined as:

$$m_1 \copyright m_2(A) = \sum_{B \cap C = A} m_1(B)m_2(C) \tag{3}$$

The mass assigned to the empty set quantifies the degree of conflict between the two *bbas*.

Dempster's rule of combination [3]. is the normalized alternative to the conjuctive rule as it does not support the existence of a mass on the empty set.

Indeed, the mass of the empty set must be reallocated over all focal elements in the case where $m_1 \bigcirc\!\!\!\!\wedge m_2(\emptyset) \neq 0$. It is defined as:

$$m_1 \oplus m_2(C) = \begin{cases} \dfrac{m_1 \bigcirc\!\!\!\!\wedge m_2(C)}{1 - m_1 \bigcirc\!\!\!\!\wedge m_2(\emptyset)} & if\, A \neq \emptyset, \forall C \subseteq \Omega \\ 0 & otherwise. \end{cases} \tag{4}$$

The combination with adapted conflict rule (CWAC) [7]. is an adaptive weighting between the two previous combination rules. It operates in the same way as the conjunctive rule when the *bbas* are contradictory (it keeps the conflict) and as the Dempster's rule when the *bbas* are similar (redistributes the conflict on all focal elements). It is formulated as follows:

$$m_{\ominus}(A) = (m_1 \ominus m_2)(A) = d(m_1, m_2)m_{\bigcirc\!\!\wedge}(A) + (1 - d(m_1, m_2))m_{\oplus}(A), \tag{5}$$

where $d(m_1, m_2)$ is a distance measure introduced by Jousselme [6] expressing the degree of dissimilarity between two sources of information:

$$d(m_1, m_2) = \sqrt{\frac{1}{2}(m_1 - m_2)^t \mathrm{D}(m_1 - m_2)}, \tag{6}$$

where D is the Jaccard index defined by

$$\mathrm{D}(A, B) = \begin{cases} 0 & if\, A = B = \emptyset, \\ \dfrac{|A \cap B|}{|A \cup B|} & \forall A, B \in 2^{\Omega}. \end{cases} \tag{7}$$

In the general case, when dealing with more than two *bbas*, a synthesis between all distances have to be made. For this, we use the maximal value of the set of distances and thus the value of D may be defined by:

$$D = max[d(m_i, m_j)], \tag{8}$$

with i $\in$ [1, N] and j $\in$ [1, N]. The rule is generalised as follows:

$$m_{\ominus}(A) = (\ominus\, m_i)(A) = Dm_{\bigcirc\!\!\wedge}(A) + (1 - D)m_{\oplus}(A) \tag{9}$$

3.4 Decision Making

Decision making's purpose is to select the suitable hypothesis for a given problem. The Transferable Belief Model (TBM) is composed by two main levels: The credal level where beliefs are represented by *bbas* and the pignistic level where *bbas* are transformed into probabilities which are used in the decision making process. This pignistic transformation is denoted by *BetP* and defined as follows:

$$BetP(w_i) = \sum_{A \subseteq \Omega, w_i \in A} \frac{1}{|A|} \cdot \frac{m(A)}{(1 - m(\emptyset))} \tag{10}$$

where $|A|$ is the number of elements of Ω in A.

4 Belief Crowd Label Aggregation

Our proposed method (BLA) is a solution for the problem of aggregating imperfect labels using the belief function tools. Consider a dataset of N questions that are labeled as either positive (+1) or negative (0). True label L_i of each question is unknown. There are M workers and the label of the jth worker for the ith question is l_{ij}, $l_{ij} \in \{-1, 0, 1\}$ where (-1) represents a vacuous label (labeler did not label it).

We propose to model the reliability degree of the jth labeler by $(1-\alpha_j)$ where α_j is its error rate. It can have value in [0, 1] where in the case $\alpha_j = 0$, it means that the labeler is totally reliable and if $\alpha_j = 1$, the labeler is unreliable and his labels must not be taken in consideration.

Our method follows four main steps namely labeler reliability degree estimation, mass function modelling, label discounting and finally label aggregation and decision making. These steps are illustrated in Fig. 1 and reviewed in-depth in what follows. Note that for simplicity's sake, we focused our work on binary labeling but our method can as well handle multi-class labeling.

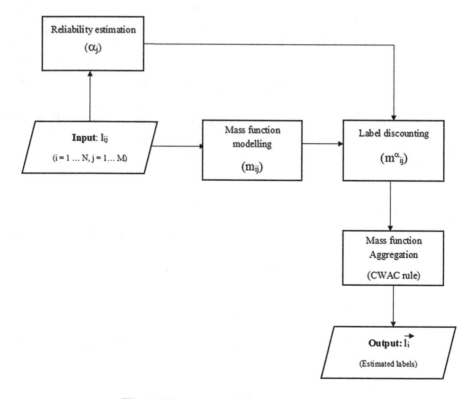

Fig. 1. Illustration of the proposed method

4.1   Labeler Reliability Degree Estimation

The idea consists in generating some ground truth and then comparing it with labelers answers in order to measure the accuracies of the M labelers. For that, we aggregate labels with a naive but relatively accurate method, namely the majority voting [10], then compare the estimated labels with labelers answers. We finally calculate α_j for each worker as follows:

$$\alpha_j = \frac{Number\_of\_incorrect\_labels}{Number\_of\_labeled\_questions} \tag{11}$$

Example 1. Let us consider the case of three distinct workers giving answers to three questions. We calculate their error rate α_j using (Eq. 11). Results are reported in Table 1.

Table 1. Example of α_j estimation for three workers

Worker(j)	Answer(l_{1j})	Answer(l_{2j})	Answer(l_{3j})	α_j
1	0	1	1	0
2	0	0	1	0.3
3	1	1	0	0.6

4.2   Mass Function Modelling

As we adopt the belief function theory in order to model conveniently imperfections in labelers answers, each label l_{ij} will be transformed into a mass function (i.e. *bba*) m_{ij}^{Ω} with $\Omega = \{\omega_1, \ldots, \omega_n\}$. Note that in the case of binary labeling, $\Omega = \{0, 1\}$.

So, if the jth worker labels 1 as the right answer on question i, the corresponding *bba* will be a certain *bba*:

$$m_{ij}(\{1\}) = 1.$$

We add that in the case of a vacuous label (-1), the bba is represented by a vacuous *bba* as follows:

$$m_{ij}(\{0, 1\}) = 1,$$

expressing total ignorance of the label.

Example 2. To illustrate this step, we consider Example 1 and we transform the workers' answers to question 1 into mass functions. Results are shown in Table 2.

Table 2. Example of mass function modelling

Worker(j)	Answer(l_{1j})	(bbam$_{1j}$)
1	0	$m_{11}(\{0\}) = 1$
2	0	$m_{12}(\{0\}) = 1$
3	1	$m_{13}(\{1\}) = 1$

4.3   Label Discounting

In this step, for each worker labels transformed into mass functions are weakened by his relative reliability degree $(1 - \alpha_j)$ using the discounting operation (Eq. 2) and consequently changed into simple support functions. Thus, worker expertise will be taken into consideration.

Example 3. Let us continue with Example 1. We consider the bba m_{12} discounted using the discount rate of the second worker $\alpha_2 = 0.3$. As a result, the certain *bba* $m_{12}(\{0\}) = 1$, after the discounting operation is transformed into a simple support function as follows:

$$m_{12}^{\alpha}(\{0\}) = (1 - 0.3) * 1 = 0.7,$$
$$m_{12}^{\alpha}(\{0, 1\}) = 0.3 + (1 - 0.3) * 0 = 0.3$$

4.4   Label Aggregation and Decision Making

During this final step, we choose the combination with adapted conflict (CWAC) rule to aggregate discounted labels because these labels can often be conflictual so CWAC is the most appropriate rule that will reveal and cope with that conflict. Moreover, CWAC showed in a previous research work [16] its accuracy superiority to the conjuntive and Dempster's rules for the classifier fusion problem [16].

When applied, we will obtain a final *bba* for each question and the decision about the final label is made with the pignistic probability (*BetP*) (Eq. 10): the possible label that has the greater pignistic probability is selected.

Example 4. For question 1 in Example 1, the *bba* obtained after applying the combination rule (Eq. 9) to combine the three workers' discounted *bbas* is as follows:

$$m_1(\emptyset) = 0.3, m_1(\{0\}) = 0.7$$

The corresponding pignistic probability will be as follows:

$$BetP(\{0\}) = 1 * (0.7/(1 - 0.3)) = 1,$$
$$BetP(\{1\}) = 0$$

and 0 is the final estimated label to question 1 as it has the greater BetP.

5   Experimentation and Results

In order to validate our proposed method, we conducted experiments on three real datasets on the AMT.

1. The duchenne dataset [17] includes images of smiling faces and smiles are either duchenne (a real smile) or non duchenne (a fake smile).
2. The event temporal ordering (Temp) dataset [18] has a pair of events and the label 1 means the first one happened before the second one.
3. The recognizing textual entailement (Rte) dataset [18] is composed of pair of phrases and label 1 means the second one can be deduced from the first one.

Table 3 shows details of these datasets.

Table 3. Description of datasets

Dataset	Labelers (M)	Questions (N)	Number of labels	Proportion of labels
Duchenne	7	159	1221	0.45
Temp	76	462	4620	0.13
Rte	164	800	8000	0.06

Our method (BLA) is compared to the majority voting through accuracy defined as the ratio of correctly labeled questions to total questions:

$$Accuracy = \frac{Number\_of\_correctly\_labeled\_questions}{Number\_of\_total\_questions} \tag{12}$$

Table 4 presents for the three datasets the average accuracies. 50 runs were executed of the majority voting, and our belief based method (BLA). It demonstrates that our method outperformed majority voting in estimation quality especially with the Rte and Temp datasets hitting values higher than 94 %.

Table 4. Average accuracies (MV vs. BLA)

Dataset	MV accuracy	BLA accuracy
Duchenne	0.72	0.74
Temp	0.93	0.95
Rte	0.89	0.94

In Fig. 2, accuracies (ratio of true estimated labels to total questions) as functions of labelers number are illustrated. Labelers are sampled randomly 50 times. Clearly, when the number of workers increases, our method (BLA) achieves higher accuracies than simple majority voting with all databases.

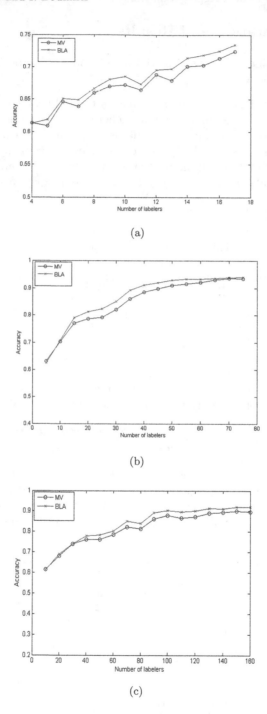

Fig. 2. Accuracies as function of the labelers' number for duchenne (a), temp (b) and rte (c) datasets

This is due to the increasing amount of bad workers that the majority voting strategy does not take into consideration. We also notice that the duchenne accuracies superiority is more obvious since it has the higher proportion of labels different from (-1).

6 Conclusion and Future Work

We propose a new method that demonstrated that aggregating labels within the belief function theory with considering labelers reliabilities improves the estimation of the true labels. We calculated reliabilities of each worker by comparing their labels to ones estimated by majority voting, then discounted labels' mass functions by these reliabilities to finally aggregate all the labels and obtain a final one for each question. We intend as future work to reconsider the method in which reliability degrees are measured as majority voting cannot always reflects the truth. The CWAC combination rule that we applied to aggregate labels mass functions also needs to be more exploited in estimating labelers accuracies or questions difficulty regarding the conflictual rate that it generates.

References

1. Howe, J.: The rise of crowdsourcing. Wired Mag. **14**(6), 1–4 (2006)
2. Howe, J.: Crowdsourcing: How the Power of the Crowd is Driving the Future of Business. Random House, New York (2008)
3. Dempster, A.P.: Upper and lower probabilities induced by a multivalued mapping. Ann. Math. Stat. **38**, 325–339 (1967)
4. Shafer, G.: A Mathematical Theory of Evidence, vol. 1. Princeton University Press, Princeton (1976)
5. Khattak, F.K., et al.: Quality control of crowd labeling through expert evaluation. In: Proceedings of the Neural Information Processing Systems 2nd Workshop on Computational Social Science and the Wisdom of Crowds (2011)
6. Jousselme, A.-L., et al.: A new distance between two bodies of evidence. Inf. Fusion **2**(2), 91–101 (2001)
7. Lefèvre, E., Elouedi, Z.: How to preserve the confict as an alarm in the combination of belief functions? Decis. Support Syst. **56**, 326–333 (2013)
8. Smets, P.: The combination of evidence in the transferable belief model. IEEE Trans. Pattern Anal. Mach. Intell. **12**(5), 447–458 (1990)
9. Quoc Viet Hung, N., Tam, N.T., Tran, L.N., Aberer, K.: An evaluation of aggregation techniques in crowdsourcing. In: Lin, X., Manolopoulos, Y., Srivastava, D., Huang, G. (eds.) WISE 2013. LNCS, vol. 8181, pp. 1–15. Springer, Heidelberg (2013). doi:10.1007/978-3-642-41154-0_1
10. Kuncheva, L., et al.: Limits on the majority vote accuracy in classifier fusion. Pattern Anal. Appl. **6**, 22–31 (2003)
11. Lee, K., et al.: The social honeypot project: protecting online communities from spammers. In: International World Wide Web Conference, pp. 1139–1140 (2010)
12. Karger, D., et al.: Iterative learning for reliable crowdsourcing systems. In: Neural Information Processing Systems, pp. 1953–1961 (2011)

13. Quinn, A.J., et al.: Human computation: a survey and taxonomy of a growing field. In: Conference on Human Factors in Computing Systems, pp. 1403–1412 (2011)
14. Dawid, A.P., Skene, A.M.: Maximum likelihood estimation of observer error-rates using the EM algorithm. Appl. Stat. **28**, 20–28 (2010)
15. Smets, P.: The transferable belief model for quantified belief representation. In: Smets, P. (ed.) Quantified Representation of Uncertainty and Imprecision, pp. 267–301. Springer, Dordrecht (1998)
16. Trabelsi, A., Lefèvre, E., Elouedi, Z.: Belief function combination: comparative study within the classifier fusion framework. In: Gaber, T., Hassanien, A.E., El-Bendary, N., Dey, N. (eds.) The 1st International Conference on Advanced Intelligent System and Informatics (AISI 2015), November 28-30, 2015, Beni Suef, Egypt. Advances in Intelligent Systems and Computing, pp. 425–435. Springer, Cham (2016)
17. Whitehill, J., et al.: Whose vote should count more: optimal integration of labels from labelers of unknown expertise. In: Neural Information Processing Systems, pp. 2035–2043 (2009)
18. Snow, R., et al.: Cheap and fast but is it good? Evaluation non-expert annotations for natural language tasks. In: The Conference on Empirical Methods in Natural Languages Processing, pp. 254–263 (2008)
19. Ben Rjab, A., Kharoune, M., Miklos, Z., Martin, A., Ben Yaghlane, B.: Characterization of experts in crowdsourcing platforms. In: 24ème Conférence sur la Logique Floue et ses Applications, Poitiers, France, November 2015

Formal Semantics and Fuzzy Logic

Weighted Node Importance Contribution Correlation Matrix for Identifying China's Core Metro Technologies with Patent Network Analysis

Mei Long[(✉)] and Tieju Ma

Business School, East China University of Science and Technology,
Shanghai, China
meilongjy@126.com, tjma@ecust.edu.cn

Abstract. The purpose of this study is to identify the core technologies in the metro domain by analyzing its patent network, which is beneficial for grasping technological trends and advancing the metro domain in China. Metro patent data (1986–2016) published in China were collected from the State Intellectual Property Office of the People's Republic of China. Then, we built a patent network with co-occurrence of information from the International Patent Classification, and improved the node importance contribution correlation matrix method to a weighted version in order to calculate the importance of each node. Nodes with high importance scores play more crucial roles in efficiency and stability of the network, and are viewed as the core metro technologies. The results can be useful for companies' technology R&D planning and government policymaking.

Keywords: Metro · Patent network · Node importance · Core technology

1 Introduction

The metro, which is mostly built underground, has become a significant transportation mode to alleviate traffic jams and energy shortages in metropoles. Statistically, the transportation capacity of metros per ton of energy consumption is 10 times higher than that of buses. Moreover, the Chinese government has always paid significant attention to building metros, and it is likely that China's total operational mileage will reach 4,500 km in 2050. Therefore, it has become increasingly important and meaningful to identify core technologies to grasp technological trends and to advance the metro industry by analyzing the overall structure of technologies and the interaction among of them [1]. This is the purpose of this study.

Meanwhile, technology forecasting and planning based on patent analysis has been a crucial method of technology management, and the most commonly used method for identifying core technologies is analysis of the patent network developed using patent citation information [2–4]. However, this method also has some shortcomings, as reported in the literature, such as an average 10-year time lag between citing–cited patents [5]. Furthermore, it is difficult to grasp technological relationships and

© Springer International Publishing AG 2016
F. Lehner and N. Fteimi (Eds.): KSEM 2016, LNAI 9983, pp. 199–208, 2016.
DOI: 10.1007/978-3-319-47650-6_16

characteristics with patent citation analysis, because the citing–cited relatedness is based on the perspective of individual patents [6].

On the contrary, patent analysis with co-occurrence information of the International Patent Classification (IPC) has some advantages over citing–cited patents [7]. This is because a patent is coded with a main IPC and several secondary IPCs, with an IPC indicating a technology field. For example, patent A is coded G03B25/00 as the main IPC, and B61B13/10 and G09F19/00 are secondary IPCs. Therefore, this study builds patent networks with co-occurrence information of IPC by considering the overall interrelationships among technologies.

In the past, centrality measures indicated the position importance of a node in a network, and are often used to identify important nodes [8]. Thus, they are helpful for discovering core technologies by identifying important nodes in the patent network. Researchers have attempted to propose and improve many measures to identify key nodes, including degree centrality, closeness centrality, betweenness centrality, and structural holes [8, 9].

However, those measures have a different emphasis to node importance appraisal for specific issues, and the node importance is affected by a variety of factors. A multi-attribute decision-making method based on the purpose of decreasing evaluation disparities was proposed to evaluate the node importance in network [10]. Meanwhile, considering the time consumption duration measure, a semi-local centrality measure was proposed that could realize a tradeoff between the low-relevant degree centrality and other time-consuming measures [11].

Although these methods solved the multi-factor problem, they neglected the importance of the contribution relationship among nodes. The node importance contribution matrix (NICM) method was proposed to reflect the node dependence relationship, in which node i contributes its original importance, denoted by betweenness, to its adjacent node j with contribution coefficient $1/D_j$, where D_j is the degree of node j [12]. While, betweenness has some limitations to reflect the node's original importance, because a node's betweenness depends on the shortest path between other nodes, there is a significant proportion of nodes that do not lie on any shortest path, and thus, these nodes obtain the same betweenness score of 0 [13]. The node importance evaluation matrix (NIEM) method redefined the node's original importance with node efficiency as follows. For node i, its efficiency, denoted as $I_i = \frac{1}{n-1}\sum_{j,j\neq i}^{n} \frac{1}{d_{ij}}$, indicates node i's original importance, where I_i is the efficiency of node i, n is the number of nodes of the network, and d_{ij} is the shortest distance from node i to node j [14].

Nevertheless, the NICM and NIEM methods take into account only the interaction between adjacent nodes. Generally, the node importance depends largely on its contribution to both adjacent nodes and non-adjacent nodes. Therefore, the node importance contribution correlation matrix (NICCM) method expanded the node contribution scope to non-adjacent nodes, which evaluated the node importance based on the sum of node importance contribution among different layers in the network [15], in which the adjacent nodes of node i were defined as the first layer nodes, and its next adjacent nodes were the second layer nodes etc. [8].

Although the NICCM method was proposed for undirected and unweighted networks, it cannot work in weighted patent networks. Most researchers have indicated that the transfer and diffusion of technology requires strong ties [16]. Therefore, it is necessary to improve the NICCM method to apply it in weighted network. In order to establish the relative importance between the number of ties and tie weights, a weighted shortest path making a trade-off between them in a weighted network was proposed [13, 17]. Thus, we introduce the weighted shortest distance into the NICCM method, and we name this novel method the weighted node importance contribution correlation matrix (WNICCM).

The purpose of this study is to identify the core technologies of metros in China using the patent network analysis method, which is significant for companies' technology R&D planning and government policymaking. First, we build a patent network with the co-occurrence information of IPC in China's metro domain. Next, we propose a novel node importance assessment method called WNICCM to indentify core nodes in patent network, and this novel method can be also used in other undirected and weighted network. Finally, in order to prove the reliability of our WNICCM method, we use the coefficient of variation (CV) to compare the performance of WNICCM with other traditional centrality measuring methods (degree centrality, closeness centrality and betweenness centrality).

The remainder of this paper is organized as follows. Section 2 introduces the methods for building patent networks and identifying the core technologies of metros. The experimental results are presented and explained in Sect. 3. Finally, Sect. 4 provides the conclusion and discussion.

2   Methodology

2.1   Building a Patent Network

As Sect. 1 shows, we can use the co-occurrence information of the IPC to build a patent network, which reveals overall interrelationships among metro technologies. Here, we introduce the process of building the patent network, and give a brief outline of each step.

2.1.1   Data Collection and Pre-processing

The patents data (1986–2016) published in China are collected from the State Intellectual Property Office of the People's Republic of China and refer to relevant search conditions, such as title = (metro), or abstract = (metro), or title = (subway), or abstract = (subway), or title = (underground), or abstract = (underground). Finally, we collect 7,267 patents of metros, including invention patents, utility-model patents, and industrial design patents. Generally, the invention patents are commonly considered as the most novel, practical and creative patents among them, and thus, most researchers choose invention patents for their studies. Therefore, we also use invention patents of metros in this study for about 2,745 patents. Then, we can obtain the IPC information for each invention patent.

2.1.2 Patent IPC Co-occurrence Matrix

Most patent IPCs are coded using five hierarchies, including section, category, sub-category, group, and subgroup, but some patents are coded only at subcategory level. In order to reduce this disparity, we intercept all metro invention patents' IPCs at the subcategory level in this study, and we obtain 244 intercepted IPCs from 2,745 invention patents, which means there are 244 metro technologies or technology fields we analyze. Two arbitrary IPCs may occur in different patents, and thus, the number of co-occurrences can be figured out between the two arbitrary IPCs. Therefore, the metro patent IPC co-occurrence matrix, denoted as $T = [t_{i,j}]$, is constructed and exemplified partly in Table 1. In this table, there are 244 IPCs, t_i indicates the i-th IPC, and the number of patent IPC co-occurrences is arranged by decimal value: $t_{i,j} = a$, for example, $t_{1,2} = 63$ means that the two IPCs of t_1 and t_2 occur together in 63 patents, while 0 means the two IPCs do not occur together.

Table 1. Example of patent IPC co-occurrence matrix

Patent IPC	t_1	t_2	...	t_{243}	t_{244}
t_1	0	63	...	0	0
t_2	63	0	...	0	0
...	...	...	0	...	...
t_{243}	0	0	...	0	0
t_{244}	0	0	...	0	0

2.1.3 Patent Network

The patent network is built with the relevant relationships shown in the patent IPC co-occurrence matrix. Accordingly, the network contains 244 nodes, and if $t_{i,j}$ is not equal to 0, there is an undirected edge between node i and j; otherwise the two nodes cannot be connected. That is, as $t_{i,j} \neq 0$ if there is an edge between node i and j, $t_{i,j} = 0$ otherwise. Moreover, when $t_{i,j}$ does not equal 0, the value of $t_{i,j}$ is the tie weight between node i and j. Therefore, we build the patent network with the patent IPC co-occurrence matrix, and use the principal components layout method to reprofile the patent network shown as Fig. 1. In Fig. 1, nodes denote IPCs, and the width of the edges between two nodes reflects their co-occurrence frequency; the higher the value is, the wider the edge is. The size of the nodes reflects their degree, which is the number of nodes that a local node is connected to, and the node with a higher degree is endowed with larger size. Clearly, the nodes are sorted by degree in descending order from left to right in Fig. 1.

2.2 WNICCM to Identify Core Technologies

In the patent network, the core nodes indicate the core technologies or technology fields of metros, and thus, we need to calculate the node importance to identify the core nodes. The node importance depends on its original importance and importance contribution to other nodes; for instance, node i contributes its initial importance to node j,

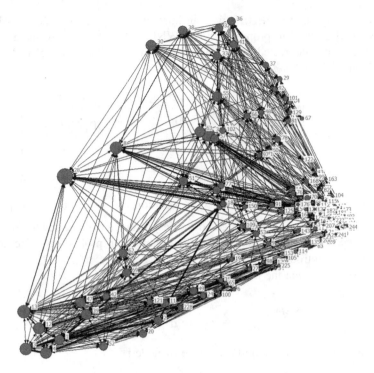

Fig. 1. The patent network

and simultaneously accepts the contribution from node j. Then, how to describe the complex contribution relationship becomes crucial. Next, we introduce the WNICCM method-building process.

2.2.1 Weighted Node Importance Contribution Correlation Matrix

In an undirected and unweighted network with n nodes, node i proportionately contributes its initial importance to node j; the contribution proportion depends on the contribution probability from node i to node j and the dependence strength from node j to node i [15]. Therefore, the NICCM method defines the NICCM denoted by H_{NICCM} as follows:

$$H_{NICCM} = \begin{bmatrix} 0 & \pi_{1,2}\delta_{1,2}I_1 & \cdots & \pi_{1,n}\delta_{1,n}I_1 \\ \pi_{2,1}\delta_{2,1}I_2 & 0 & \cdots & \pi_{2,n}\delta_{2,n}I_2 \\ \vdots & \vdots & \ddots & \vdots \\ \pi_{n,1}\delta_{n,1}I_n & \pi_{n,2}\delta_{n,2}I_n & \cdots & 0 \end{bmatrix}, \pi_{i,j} = \begin{cases} \frac{1}{d} & if\ j \in V_i^d \\ 0 & or\ else \end{cases} \quad (1)$$

where I_i is the initial efficiency of node i, which contributes $\pi_{i,j}\delta_{i,j}I_i$ to node j (adjacent node or non-adjacent node); $\delta_{i,j}$ is the contribution probability from node i to node j, which is explained in detail in Subsect. 2.2.4; $\pi_{i,j}$ is the dependence strength coefficient from node j to node i, and V_i^d is the set of nodes in layer d, which is also the shortest binary distance from node i to node j is d.

Meanwhile, as stated in Sect. 1, we can introduce the weighted shortest distance into the NICCM method to propose a novel WNICCM method to evaluate the node importance for an undirected and weighted network. Thus, we can define the WNICCM denoted by H_{WNICCM} as follows:

$$H_{WNICCM} = \begin{bmatrix} 0 & \frac{1}{d^w(1,2)}\mathcal{P}_{1,2}E_1 & \cdots & \frac{1}{d^w(1,n)}\mathcal{P}_{1,n}E_1 \\ \frac{1}{d^w(2,1)}\mathcal{P}_{2,1}E_2 & 0 & \cdots & \frac{1}{d^w(2,n)}\mathcal{P}_{2,n}E_2 \\ \vdots & \vdots & \ddots & \vdots \\ \frac{1}{d^w(n,1)}\mathcal{P}_{n,1}E_n & \frac{1}{d^w(n,2)}\mathcal{P}_{n,2}E_n & \cdots & 0 \end{bmatrix}, \qquad (2)$$

which is similar with the H_{NICCM} matrix; in addition, $\frac{1}{d^w(i,j)}$ denotes the dependence strength coefficient, $d^w(i,j)$ is the weighted shortest distance from node i to node j, E_i is the initial efficiency of node i, and $\mathcal{P}_{i,j}$ is the contribution probability. The new calculation method is introduced in Subsects. 2.2.2–2.2.4.

Finally, we can define the node importance of node j as

$$H_j^w = E_j * \sum_{i=1,i\neq j}^{n} \left(\frac{1}{d^w(i,j)}\mathcal{P}_{i,j}E_i\right), \qquad (3)$$

where H_j^w denotes the node importance score of node j.

2.2.2 Node Dependence Strength Coefficient in Weighted Network

Researchers have attempted to identify the shortest path in a weighted network, and most of these attempts have focused on defining the role of the transmission tie weights as cost or strength. The weights represent cost in global positioning system devices or the time to route Internet traffic [18], while the weights in a patent network are the mediums for information transmission, which are the tie strength, not cost. In so doing, the tie weights could be inverted to find the shortest paths [17, 19]. Thus, the shortest distance in a weighted network can be defined as follows:

$$d^w(i,j) = \min\left(\frac{1}{w_{i,h}} + \cdots + \frac{1}{w_{h,j}}\right), \qquad (4)$$

where $d^w(i,j)$ represents the weighted shortest distance from node i to node j, and $w_{i,h}$, the tie weight of node i and node h (the intermediary nodes on paths between node i and j), is equal to $t_{i,h}$. Thus, we can obtain the node dependence strength coefficient in a weighted network as $\frac{1}{d^w(i,j)} = \frac{1}{\min\left(\frac{1}{w_{i,h}} + \cdots + \frac{1}{w_{h,j}}\right)}$.

2.2.3 Node Efficiency in Weighted Network

In a patent network, node plays a very important role in information transmission, which reveals the contribution to network efficiency [13]. Thus, node efficiency can be defined as follows:

$$E_i = \frac{1}{n-1} \sum\nolimits_{i=1,j\neq i}^{n} \frac{1}{d^w(i,j)},$$ (5)

where E_i is the efficiency of node i, $d^w(i,j)$ is the weighted shortest distance between node i and node j, and n is the number of nodes in the patent network.

2.2.4 Node Contribution Probability in Weighted Network

As is known, all shortest paths in a network change when a random node is added or removed, and then bring about changes of the contribution probability between nodes. Therefore, in the NICCM method, the node contribution probability $\delta_{i,j}$ is calculated by using the node removal-based method, as explained in the following steps.

Step 1. For node i, calculate the efficiency of nodes that are located in the same layer or the same shortest distance to node j, which forms $I_i = [I_a, I_b, \cdots, I_j, \cdots I_l]$.

Step 2. Remove node i, and recalculate the efficiency for these nodes. We can build a set of new node efficiency as $I_i' = [I_a', I_b', \cdots, I_j', \cdots I_l']$.

Step 3. Compute the changes of these nodes' efficiency after removing node i, yielding $\Delta I = [\Delta I_a, \Delta I_b, \cdots, \Delta I_j \cdots \Delta I_l]$, where $\Delta I_j = |I_j - I_j'|$.

Step 4. Calculate the contribution probability $\delta_{i,j}$ as follows:

$$\delta_{i,j} = \frac{\Delta I_j}{\Delta I_a + \Delta I_b + \cdots + \Delta I_j + \cdots + \Delta I_l}.$$ (6)

Meanwhile, it seems improper that the node contribution probability is related to only these nodes located in the same shortest distance in the weighted network. Thus, we can improve $\delta_{i,j}$ from the perspective of a global network.

Following the same steps, we calculate the changes of node efficiency for all nodes after removing node i, which forms $\Delta E = [\Delta E_1, \Delta E_2, \cdots, \Delta E_j \cdots \Delta E_n]$, where $\Delta E_j = |E_j - E_j'|$, E_j is the original efficiency of node j, and E_j' is the updated efficiency after removing node i. Thus, we can define the node contribution probability $\mathcal{P}_{i,j}$ in the weighted network as follows:

$$\mathcal{P}_{i,j} = \frac{\Delta E_j}{\sum_{j=1,j\neq i}^{n} \Delta E_j}.$$ (7)

3 Results

The node importance in the patent network can be calculated as outlined in Sect. 2, and we can obtain the core nodes with the node importance ranking information. Meanwhile, the core technologies of metros can be identified. To illustrate the validity and feasibility of the WNICCM method in this study, we compare our method with other

Table 2. Ranking results of node importance in patent network

Rank	Degree		Closeness		Betweenness		WNICCM	
	Node ID	Values	Node ID	Values	Node ID	Values	Node ID	Values
1	C04B(2)	164.0000	E21F(28)	0.0254	E21F(28)	13.8870	C04B(2)	36.2375
2	E04B(1)	163.0000	B61B(8)	0.0253	B61D(31)	11.5850	E04B(1)	21.4872
3	C08L(3)	153.0000	E21D(13)	0.0252	B61B(8)	7.9390	C08L(3)	20.0919
4	C08 K(4)	135.0000	B61D(31)	0.0252	H02 J(63)	6.5120	C08 K(4)	14.8644
5	B61B(8)	119.0000	C08L(3)	0.0252	E21D(13)	6.3320	E01F(5)	13.9181
6	E21D(13)	95.0000	B60L(51)	0.0252	B32B(39)	6.3250	C04B(6)	8.9970
7	E01F(5)	91.0000	H02 J(63)	0.0251	C22C(45)	6.2600	E06B(7)	6.9539
8	E06B(7)	89.0000	E02D(12)	0.0251	H01B(53)	5.7450	B61B(8)	6.4400
9	E21F(28)	82.0000	C08 K(4)	0.0251	E02D(12)	5.1280	E21D(13)	5.4125
10	E02D(12)	80.0000	B32B(39)	0.0251	C04B(2)	5.0160	E02D(12)	4.9153
11	B29C(11)	68.0000	G06F(30)	0.0251	E06B(7)	4.4190	B29C(11)	4.7124
12	B61D(31)	66.0000	B61L(38)	0.0251	G09F(55)	4.2430	E05F(18)	3.6552
13	E05F(18)	58.0000	B61C(57)	0.0250	B01D(70)	4.0600	E21F(28)	3.4380
14	H04L(36)	53.0000	E06B(7)	0.0250	B60L(51)	4.0150	B28B(21)	3.2343
15	G06F(30)	53.0000	E01C(47)	0.0250	F24F(84)	3.9650	E04F(49)	2.9073
16	B32B(39)	51.0000	E04H(69)	0.0250	G06F(30)	3.9370	B61L(38)	2.8585
17	C04B(6)	51.0000	E05F(18)	0.0249	C08L(3)	3.7030	B61D(31)	2.5465
18	F03G(9)	50.0000	G05B(102)	0.0249	C08 K(4)	3.5400	C08 J(27)	2.5117
19	C08G(20)	50.0000	E04B(1)	0.0249	H02 K(50)	3.5220	E05D(23)	2.4913
20	F21S(14)	50.0000	G01 N(77)	0.0249	G02B(91)	3.0960	E01C(47)	2.2005

traditional centrality measuring methods, including degree centrality, closeness centrality, and betweenness centrality in Table 2, which ranks the top 20 nodes for each evaluating method. In addition, Fig. 2 plots the ranking results for node centrality, and it is easy to find that the results of the WNICCM method are similar to those of degree centrality.

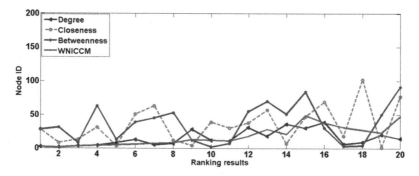

Fig. 2. Ranking results for node centrality

Considering the number of the same core nodes in Table 2, we observe that the percentage rates for the first three methods compared with the node importance method in the top 20 nodes are as follows: degree centrality 70; closeness centrality 60; betweenness centrality 45. This provides some evidence of the validity and feasibility of the node importance method.

Meanwhile, we cannot prove completely that this method performs better than other methods. Note that we can use the coefficient of variation (CV) to compare the performance of these methods, where CV is a normalized measure to quantify the spread of the data and is defined as the ratio of the standard deviation to the mean [20]. A larger CV means that it is easier to determine patent importance levels. Table 3 shows the CV for different node importance measures, and we can observe from Table 3 that the WNICCM method has the largest CV, which means that this method performs better than traditional node centrality measures in distinguishing patent ranking.

Table 3. CV for importance scores of top 20 ranked nodes

Importance measure	Average	Std dev.	CV
Degree	86.0500	39.8569	0.4632
Closeness	0.0251	0.0001	0.0052
Betweenness	5.6615	2.7606	0.4876
WNICCM	8.4937	8.8046	**1.0366**

4 Conclusion and Discussion

This study proposed a novel node importance evaluation method to identify metro core technologies in a patent network. First, we built the patent network with patents' IPC co-occurrence information. Second, we proposed a novel WNICCM method and applied it to ranking nodes based on their importance scores in weighted network. Since the WNICCM method introduces tie weights and evaluates node contributions probability from the perspective of global network structure, this method should perform better in terms of feasibility and validity than other traditional measures. This has been confirmed by the results of the CV. Finally, we can identify metro core technologies based on these core nodes in the patent network. Then, we can forecast their development tendency by utilizing technology life-cycle theory.

This research contributes to identifying China's core metro technologies with WNICCM method, and we found that these core metro technologies are focused on three key domains, which are related to transportation (section B), chemistry (section C) and building (section E). Therefore Chinese metro is short of overall development strategy, and the results can have reference significance to companies' technology R&D planning and government policymaking.

In future work, we will research the features of the underlying technological structure of metros by comparing them with developed countries. This would be beneficial for in-depth understanding of the new development direction of metros in China.

Acknowledgments. This research was sponsored by NSFC (No.71125002).

References

1. Lee, H., Kim, C., Cho, H., Park, Y.: An ANP-based technology network for identification of core technologies: a case of telecommunication technologies. J. Expert Syst. Appl. **36**, 894–908 (2009)
2. Newman, M.E.: Modularity and community structure in networks. J. Proc. Natl. Acad. Sci. **103**, 8577–8582 (2006)
3. Duch-Brown, N., Costa-Campi, M.T.: The diffusion of patented oil and gas technology with environmental uses: a forward patent citation analysis. J. Energy Policy. **83**, 267–276 (2015)
4. Rodriguez, A., Kim, B., Lee, J.M., Coh, B.Y., Jeong, M.K.: Graph kernel based measure for evaluating the influence of patents in a patent citation network. J. Expert Syst. Appl. **42**, 1479–1486 (2015)
5. Kim, C., Seol, H.: On a patent analysis method for identifying core technologies. In: Watada, J., Watanabe, T., Phillips-Wren, G., Howlett, R.J., Jain, L.C. (eds.) Intelligent Decision Technologies. Springer, Heidelberg (2012)
6. Yoon, B., Park, Y.: A text-mining-based patent network: analytical tool for high-technology trend. J. High Technol. Manage. Res. **15**, 37–50 (2004)
7. Leydesdorff, L.: Patent classifications as indicators of intellectual organization. J. Phys. **59**, 1582–1597 (2009)
8. Burt, R.S.: Structural holes and good ideas. Am. J. Sociol. **110**, 349–399 (2004)
9. Xiao-Hang, Z., Zhu, J., Wang, Q., Zhao, H.: Identifying influential nodes in complex networks with community structure. J. Knowl. Based Syst. **42**, 74–84 (2013)
10. Hui, Y., Zun, L., Yong-Jun, L.: Key nodes in complex networks identified by multi-attribute decision-making method (in Chinese). J. Acta Phys. Sin. **02**, 54–62 (2013)
11. Duan-Bing, C., Lin-Yuan, L., Ming-Sheng, S., Yi-Cheng, Z., Tao, Z.: Identifying influential nodes in complex networks. J. Fuel Energy Abstr. **391**, 1777–1787 (2012)
12. Yi-Huan, Z., Zu-Lin, W., Jing-Guo, Z., Jing, X.: Finding most vital node by node importance contribution matrix in communication networks (in Chinese). J. Beijing Univ. Aeronaut. Astronaut. **35**, 1076–1079 (2009)
13. Opsahl, T., Agneessens, F., Skvoretz, J.: Node centrality in weighted networks: generalizing degree and shortest paths. J. Soc. Netw. **32**, 245–251 (2010)
14. Xuan, Z., Feng-Ming, Z., Ke-Wu, L., Xiao-Bin, H., Hu-Sheng, W.: Finding vital node by node importance evaluation matrix in complex networks (in Chinese). J. Acta Phys. Sin. **61**, 201–207 (2012)
15. Ping, H., Wen-Li, F., Sheng-Wei, M.: Identifying node importance in complex networks. J. Phys. A Stat. Mech. Appl. **429**, 169–176 (2015)
16. Tsai, W.: Knowledge transfer in intraorganizational networks: effects of network position and absorptive capacity on business unit innovation and performance. Acad. Manage. J. **44**, 996–1004 (2001)
17. Newman, M.E.: Scientific collaboration networks. II: shortest paths, weighted networks, and centrality. J Phys. Rev. E **64**, 132–158 (2001)
18. Dijkstra, E.W.: A note on two problems in connexion with graphs. J. Numer. Math. **01**, 269–271 (1959)
19. Brandes, U.: A faster algorithm for betweenness centrality. J. Math. Sociol. **25**, 163–177 (2001)
20. Kim, D., Lee, B., Lee, H.J., Sang, P.L.: Automated detection of influential patents using singular values. IEEE Trans. Autom. Sci. Eng. **9**, 723–733 (2012)

Situation-Aware Rating Prediction Using Fuzzy Rules

Rim Dridi[1], Saloua Zammali[1]([✉]), and Khedija Arour[2]

[1] Faculty of Science of Tunis, University Tunis El-Manar, 2092 Tunis, Tunisia
dridi.rime@gmail.com, saloua.zammali@fst.utm.tn
[2] National Institute of Applied Science and Technology of Tunis, 1080 Tunis, Tunisia
Khedija.arour@issatm.rnu.tn

Abstract. Context-Aware Recommendation Systems (CARS) extend traditional recommendation systems by adapting their output to users' specific contextual situations. Rating prediction in CARS has been tackled by researchers attempting to recommend appropriate items to users. However, in rating prediction, three thriving challenges are still to tackle: *(i)* the weight of each context dimension; *(ii)* the correlation between context dimensions; and *(iii)* situation inference. A major shortcoming of the classical methods is that there is no defined way to study dependencies and interactions existing among context dimensions. Context-aware algorithms made a strong assumption that context dimensions weights are the same or initialized with random values. To address these issues, we propose a novel approach for weighting context dimensions, studying the correlation between them to infer the current situation then predict the rating based on the inferred situation. Through detailed experimental evaluation we demonstrate that the proposed approach is helpful to improve the prediction accuracy.

Keywords: Recommendation · Context weighting · Context dimension correlation · Situation · Rating prediction

1 Introduction

The huge amount of available data and services online make it difficult for users to find their desired content (movies, music, books, etc.) in a reasonable time. One way of solving this problem is using recommender systems. Researchers in the domain of recommender systems started to realize that it is useful to take contexts into account when making recommendations. Context-Aware Recommender Systems (CARS) have been introduced to provide more relevant results by generating recommendations that match the current interests of each user [10].

Context information are known as context dimensions, it presents the contextual factors, such as time, location, weather, mood and the company of other people. Each context dimension has different context conditions. For instance, the context conditions for weather dimension are: clear sky, sunny, cloudy, rainy,

© Springer International Publishing AG 2016
F. Lehner and N. Fteimi (Eds.): KSEM 2016, LNAI 9983, pp. 209–221, 2016.
DOI: 10.1007/978-3-319-47650-6_17

thunderstorm, snowing, etc. Context dimensions do not have the same degree of importance. However, in most existing approaches like [10], contextual dimensions have the same weight or initialized with random weight, limiting the accuracy of the results. Weighting contextual dimensions is rarely discussed in the previous literature. Our reflection on the recommendation is based on assigning a level of importance for each dimension and for the set of dimensions to detect interactions and dependencies existing among context dimensions.

For inference, fuzzy logic is a suitable tool that deals with such problems [3], it contains an inference engine based on a set of formulated rules, which enables to infer user's current situation. We used fuzzy logic to infer user's situation at a given moment, including weather, mood, activities that the user is currently pursuing.

Our goal is to propose a rating prediction approach by using Choquet integral to assign weight to each dimension and to each dimensions combination, then to study the correlation between context dimensions. Afterwards, inferring user's current situation by employing fuzzy logic.

The remainder of the paper is organized as follows: Section 2 describes related works. Section 3 investigates the proposed approach in detail. Experimental evaluation for the proposed approach is given in Sect. 4. Finally, we conclude and give some directions for future work in Sect. 5.

2   Related Works

CARS is a hot research branch of recommender system. It incorporates context into traditional recommender process to make more accurate predicts. In the following, we pay attention to related works focusing on context-aware rating prediction algorithms and fuzzy inference in recommender systems.

2.1   Context-Aware Rating Prediction Approaches

Rating prediction algorithms calculate a predicted rating that a user is expected to assign to an item according to his current context [11,12].

Some approaches suppose that context dimensions equally contribute to make prediction. In [8] collaborative filtering is combined with demographic information and item profile. A threshold based k-Nearest Neighbors (KNN) algorithm is used to find the K neighbors for collaborative filtering. The predicted ratings for the item will be the average rating of the top N neighbors for that item. Otherwise, several predicting items rating approaches highlight the importance of each contextual dimension. In this context, Zammali et al. [11] propose a prediction model that starts from a current situation and a user history to predict the rating of the current item. It consists on calculating the weight of each context feature based on an optimization algorithm called Particle Swarm Optimisation (PSO) [12]. The predicted rating is calculated by using the rating function presented in [11]. Differential Context Weighting (DCW) [12] is another context-aware recommender system which introduces contextual weighting in rating prediction process. A weighted similarity measure was proposed where contextual weights were assigned to each dimension for each algorithm component. The idea behind it is the assumption that the more similar the contexts given by two ratings are, the more they are influential in making predictions.

In addition to the contextual dimensions weighting, using contextual correlation for recommendation has been concerned by researchers. Zheng et al. [14], highlight the importance of contextual correlations and propose a correlation based context-aware Matrix Factorization algorithm. The underlying assumption behind the notion contextual correlation is that, more correlated two contexts are, the two recommendation lists for a same user for those two contextual situations should be similar too. When it comes to the Matrix Factorization, the prediction function can be described in [14], it contains the function "Corr" which estimates the correlation between two context situations. The contextual correlations can be learned by minimizing the rating prediction errors. In the studied works based on weighting dimensions, only [14] consider contextual correlation.

2.2 Fuzzy Inference Approaches in Recommender Systems

Fuzzy logic provides inference morphology to build a system based on a set of user supplied human language rules. In fact, fuzzy input variables $x = \{x_1, ..., x_n\}$ are transformed into one output y = Rule(x) under the rules base $R = \{Rule_1, ..., Rule_k\}$.

Thyagaraju et al. [6] propose the design and implementation of user context-aware recommendation for mobile using artificial intelligent tools like fuzzy logic and rule based reasoning. They use fuzzy linguistic variables and membership degrees to define the context situations and the rules for adopting the policies of implementing a service recommendation. In fact, the position of users can be represented by a linguistic variable $Xuser$ whose linguistic values come from the following domains (very low library, excellent admin, good classroom). Given the three locations $l1$, $l2$, $l3$, the position of the user will be associated with any one of the location based on their linguistic values for the membership functions $\mu_{excellent}$, μ_{good}, μ_{fair}, and μ_{low}. An example of the rule base for identifying the location is as follows:

IF$(\mu_{excellent}(l1) != 0$ OR $\mu_{excellent}(l2) != 0$ OR $\mu_{excellent}(l3) != 0$ AND $\mu_{excellent}(l1) > \mu_{excellent}(l2)$ AND $\mu_{excellent}(l1) > \mu_{excellent}(l3))$ THEN the location of the user is $l1$.

An approach of providing high quality recommendation [10] can be achieved by utilizing different contextual information such as time of the day, location and users mood. Then, these information are passed through a cascade of fuzzy logic models to infer the user's context, which is used to recommend music from an online music streaming service. The proposed system is able to identify some general places for users. These areas are recognized implicitly. The system monitors significant location updates and marks any visited locations as possible candidates for any of the general places in a two-step process. First, using the

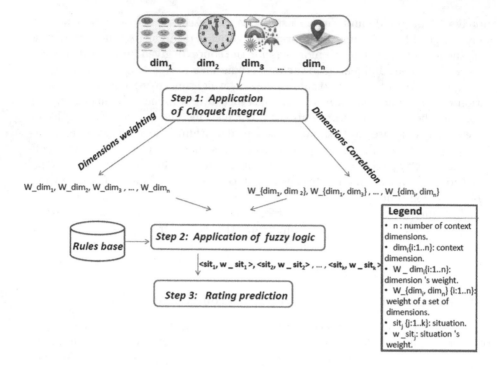

Fig. 1. The proposed FWR approach

Foursquare Venues API, the authors reverse-geocoded the location's coordinates to one of the general places. If no results are returned, the visit information is then passed through an internal fuzzy model to determine the general place categories based on fuzzy rules such as the following: IF Activity IS Stationary AND DayOfWeek IS Weekday AND TimeOfDay IS Afternoon AND Indoor/Outdoor IS Indoor AND Place IS Oce THEN Context IS Working or Studying.

3   FWR: Fuzzy Weighting Recommender Approach

In this section, we introduce our approach FWR based on context dimensions by applying Choquet integal and fuzzy logic. Our approach operates through three steps: *(i)* Application of Choquet integal; *(ii)* Application of fuzzy logic and *(iii)* Rating prediction, as shown in Fig. 1.

Step 1: Application of Choquet integral
Choquet integral has been widely applied as an aggregation operator in multiple criteria decision making. However, it has been used much less in the recommendation so far. The reason behind our choice of Choquet integral is that it allows to define a weight, not only for each criterion, but also for each subset of criteria. This operator model the importance of criteria as well as

the positive and negative synergies between them. Our approach has adapted Choquet integral to obtain the weight of importance of individual dimension and of subsets of dimensions, which enables avoidance of the overestimation and underestimation caused by possible dependencies between some dimensions. Thanks to Choquet, it is possible to define interactions between dimensions which lead to results that are human interpretable. To apply Choquet integral, initially we attribute all possible combinations of weights to the existing dimensions. Afterwards, we calculate evaluation metrics that will be described in Sect. 4.2. The best combination is the one that minimizes the rating prediction errors. Next, we have applied the least squares method in the KAPPALAB package based on the R language [4]. Then, we recalculate evaluation metrics using the combination returned by the least squares method to guarantee the best combination. We obtain finally the weight of each dimension and the weight of subsets of dimensions. Choquet has the advantage of facilitating the task of interpreting the interactions between the relevant dimensions with readily available interpretations via the interaction index.

Step 2: Application of Fuzzy Logic

Step 1's outputs are required as the inputs of the step 2. In fact, we used the weighted correlated dimensions in the fuzzy logic to infer the current situation of the user as well as its weight by exploiting fuzzy inference rules. The fuzzy inference layer handles the vagueness of some conditions of these rules and outputs a degree for each situation determined by the semantic layer. Based on these degrees, a situation is recognized. This situation allows the identification of specific tasks of a current user. Using fuzzy logic for context inference makes the system extremely flexible and easy to understand, it provides a light weight and efficient technique. More precisely, to build a fuzzy inference system, we consider only correlated dimensions to create the necessary rules for situation inference. The proposed system accepts the required contextual dimensions and calculates the weight of the current situation by using correlated dimensions' weights. More formally, let consider

- $<dim_1, w\_dim_1><dim_2, w\_dim_2>, ..., <dim_n, w\_dim_n>$: a set of weighted dimensions which represent the input variables.
- $\{L_{1dim_i}, L_{2dim_i}, ..., L_{kdim_i}\}$: a set of linguistic terms of a dimension dim_i. *Example 1.* The linguistic terms for the time dimension are: $\{morning, afternoon, evening, night\}$
- $(x_0, y_0)(x_1, y_1), ..., (x_k, y_k)$: a membership function can represent linguistic term. It is represented as piece-wise linear functions using a series of points.
  Example 2. In our case, we use a triangular membership function. For the *morning* linguistic term, the function is defined as follows: $(6,0)(10,1)(12,0)$.
- $<sit_1, w\_sit_1><sit_2, w\_sit_2>, ..., <sit_j, w\_sit_j>$: a set of weighted situations which represent the output variables.
- IF $<dim_1, w\_dim_1>$ IS L_{1dim_1} AND $<dim_2, w\_dim_2>$ IS L_{1dim_2} TNEN situation IS $<sit_1, w\_sit_1>$: a fuzzy inference rule.

Example 3. IF $<time, w\_time>$ IS *morning* AND $<mood, w\_time>$ IS *active* THEN situation IS $<work, w\_work>$

In the next step, we detail the rating prediction process which integrates the weighted inferred situation.

Step 3: Rating prediction

The central task in recommender systems is to predict the taste of a user. In this paper, we deal with the rating prediction problem that tries to predict how much a user likes a particular item by considering his actual situation. The estimated ratings can be used to recommend items to the user, e.g. what movies he might want to watch. We calculate the predicted rating by incorporating the weight of the inferred current situation in the rating prediction function $P_{a,i,\sigma}$ [12], for the item i. The key parameters in this equation are the four σ vectors, one for each component, that weight the contribution of each contextual situation in that component, and the four ϵ values that set the threshold of situation similarity $SitSim$ in each component. Formally, rating prediction function is defined as follows:

$$P_{a,i,\sigma} = \bar{\rho}(a, \sigma_3, \epsilon_3) + \frac{\sum\limits_{u \in N_{a,\sigma_1,\epsilon_1}} (\rho(u, i, \sigma_2, \epsilon_2) - \bar{\rho}(u, \sigma_2, \epsilon_2)) \times sim_w(a, u, \sigma_4, \epsilon_4)}{\sum\limits_{u \in N_{a,\sigma_1,\epsilon_1}} sim_w(a, u, \sigma_4, \epsilon_4)}$$

(1)

Equation 1 has the following components:

- $\bar{\rho}(a, \sigma_3, \epsilon_3)$: The overall average of the target user's ratings of items in similar situations.
- $N_{a,\sigma_1,\epsilon_1}$: $\{u : max_{r_{u,i,sit_n}}(SitSim(sit_a, sit_n) > \epsilon_1\}$ The set of neighbors of a user a obtained by comparing their situations sit_n for rating item i with the target situation sit_a and allowing them as neighbors if $SitSim$ is greater than a threshold ϵ_1.
- $\rho(u, i, \sigma_2, \epsilon_2)$: The weighted average of a neighbor u for an item i of all ratings issued in similar situations.
- $\bar{\rho}(u, \sigma_2, \epsilon_2)$: The overall average across all items rated by u in similar situations.
- $sim_w(a, u, \sigma_4, \epsilon_4)$: The weighted similarity between two users a and u.

4   Experimental Study

In this section, we start by describing the used datasets. Also, we introduce an open-source java based context aware recommendation engine named CARSKit which implements the evaluation metrics and baseline algorithms. Finally, we present and discuss the obtained results.

4.1   Dataset and CARSKit Java Recommendation Engine

The number of context-aware datasets is quite limited because ratings in multiple contexts are difficult to collect and user privacy is often a concern.

For the lack of dataset with context information, there are three kinds of dataset: real-world dataset, semi-simulated dataset and all-simulated dataset. The all-simulated data simulates all context information on the initial dataset and semi-simulated dataset makes initial context information richer. Of course, the real-world dataset is most credible. We take three real world datasets from the domain of context-aware recommender systems with different numbers of contextual dimensions and conditions.

Food Data is the "AIST context-aware food preference dataset" used and distributed by Hideki Asoh et al. [9]. It is based on a survey of users ratings on a menu in the context of different degrees of hunger.

DePaulMovie is a context-aware movie dataset collected from surveys [15]. Students were asked to rate movies in different contexts.

InCarMusic dataset is a context-aware music dataset collected from InCar-Music which is a mobile application offering music recommendations to the passengers of a car [1]. The characteristics of these datasets are summarized in Table 1.

Table 1. Description of used datasets

Dataset	# of items	# of ratings	# of contextual dimensions	# of contextual conditions
Food Data	20	6360	6	8
DePaul Movie	79	5035	3	12
InCarMusic	139	3940	8	34

To compare our approach to the state of the art recommendation algorithms, we have used CARSKit tool [15]. It is an open-source java-based context-aware recommendation engine, which provides a standard platform for evaluations and practical use.

4.2   Evaluation Metrics and Baseline Methods

There are three popular classes of evaluation metrics in the domain of recommender systems: prediction errors (Mean Absolute Error (MAE) and Root Mean Squared Error (RMSE)), IR metrics (precision, recall) and ranking metrics (Normalized Discounted Cumulative Gain (NDCG) and Mean Average Precision (MAP)). In this paper, we chose the first two classes of metrics as they are the most commonly employed for evaluating the performance of predictions [7,13], and they are defined as below:

$$MAE = \frac{1}{n} \sum_{i=1}^{n} |f_i - y_i|$$

$$RMSE = \sqrt{\frac{1}{n} \sum_{i=1}^{n} (f_i - y_i)^2}$$

$$Precision = \frac{N_r}{N_t}$$

$$Recall = \frac{N_r}{N_{tr}}$$

where:

- n: Total number of predictions.
- f_i: Predicted rating of an item i.
- y_i: Real rating of an item i known from offline experiments.
- N_r: Number of returned relevant items for each user.
- N_t: Number of total returned items.
- N_{tr}: Number of total relevant items.

For comparison purposes, we chose the well known Context-Aware Matrix Factorization algorithm (CAMF) proposed by Baltrunas et al. [1], it has been well recognized as the standard baselines in this domain. It usually incorporate a contextual rating deviation term which is used to estimate users rating deviations associated with specific contexts. CAMF algorithm assumes that the contextual rating deviation is dependent on items. Therefore, this approach is named as CAMF_CI. Alternatively, this deviation can also be viewed as being dependent on users resulting in the CAMF_CU variant. In addition, CAMF algorithm assumes that the contextual rating deviation is independent of users and items, this approach is named as CAMF_C. We tried all three variants: CAMF_CI, CAMF_CU and CAMF_C.

4.3   Evaluation Protocol

A crucial preliminary step to the calculation of dimensions weights by the Choquet operator is the identification of fuzzy measures. In this respect, we split the dataset into two parts: a training set (TR.S) and a testing set (TE.S). In TR.S, for each combination we calculate the MAE and RMSE using combinations of dimensions weights. Then, we select the combination which gave the best result in terms of MAE and RMSE throughout TR.S.

For reasons of lack of sufficient space, we present an application of Choquet integral only on Music dataset, we did the same procedure for other datasets.

In our approach, we have tested 63 combinations of dimensions weights. The combinations are obtained with a pitch of 0.1 and such that the sum of four partial individual weights of the four context dimensions: mood (M), Weather (W), Time (T) and Activity (A) is equal to 1. The weight of a subset of dimensions is the sum of the weights of the individual dimensions contained in this subset.

For reasons of lack of sufficient space, we present, in Table 2, seven combinations with a correspondent weight μ^i among the 63 combinations. For each combination, we apply Choquet integral on the four context dimensions using the

Table 2. Choquet-based context dimensions weights of Music dataset

Combination	μ_W	μ_M	μ_T	μ_A	$\mu_{\{W,M\}}$	$\mu_{\{W,T\}}$	$\mu_{\{W,A\}}$	$\mu_{\{M,T\}}$	$\mu_{\{M,A\}}$	$\mu_{\{T,A\}}$	$\mu_{\{W,M,T\}}$	$\mu_{\{W,M,A\}}$	$\mu_{\{W,T,A\}}$	$\mu_{\{M,T,A\}}$	MAE	RMSE
μ^1	0.1	0.1	0.1	0.7	0.2	0.2	0.8	0.2	0.8	0.8	0.3	0.9	0.9	0.9	1.104	1.369
μ^2	0.4	0.3	0.2	0.1	0.7	0.6	0.5	0.5	0.4	0.3	0.9	0.8	0.7	0.6	1.012	1.160
μ^3	0.1	0.7	0.1	0.1	0.8	0.2	0.2	0.8	0.8	0.2	0.9	0.9	0.3	0.9	1.076	1.338
μ^4	0.1	0.1	0.7	0.1	0.2	0.8	0.2	0.8	0.2	0.8f	0.9	0.3	0.9	0.9	1.024	1.208
μ^5	0.7	0.1	0.1	0.1	0.8	0.8	0.8	0.2	0.2	0.2	0.9	0.9	0.9	0.3	1.066	1.248
μ^6	0.2	0.3	0.3	0.2	0.5	0.5	0.4	0.6	0.5	0.5	0.8	0.7	0.7	0.8	1.064	1.240
μ^0	0.1	0.4	0.4	0.1	0.5	0.5	0.2	0.8	0.5	0.5	0.9	0.6	0.6	0.9	**0.989**	**1.173**
μ^*	0.12	0.36	0.27	0.26	0.10	0.09	0.12	−0.10	−0.09	−0.02	0.78	0.80	0.75	0.78	**0.966**	1.171

Table 3. Comparison between CAMF and FWR for the three datasets

	Food				Movie				Music			
	CAMF_C	CAMF_CI	CAMF_CU	FWR	CAMF_C	CAMF_CI	CAMF_CU	FWR	CAMF_C	CAMF_CI	CAMF_CU	FWR
MAE	1.001	2.361	2.368	**1.167**	1.396	1.811	1.815	**1.247**	1.275	2.540	2.548	**0.966**
RMSE	1.236	2.647	2.654	**1.114**	1.571	2.223	2.193	**1.244**	1.465	2.914	2.923	**1.171**
Precision	0.045	0.063	0.043	**0.066**	0.037	0.033	0.027	**0.055**	0.008	0.018	0.011	**0.076**
Recall	0.045	0.063	0.043	**0.064**	0.030	0.009	0.101	**0.044**	0.004	0.010	0.006	**0.041**

training set and we compute the two metrics. According to the obtained results, μ^0 presents the best combination which minimize MAE and RMSE obtained across TR.S. Next, in TE.S, we have applied the least squares method on the combination of μ^0 to obtain the final combination μ^*. It is situated in the last row of Table 2. The least squares method based on quadratic programming uses the best combination returned by the training step μ^0 to provide an optimal solution which represents the combination of μ^*. We obtain finally the weight of each dimension and the weight of subsets of dimensions through μ^*.

We notice that mood dimension has the highest individual weight (equals to 0.36). For example, after a long tiring day, Alexander prefers to listen classical music rather than disco music. The Choquet integral responded well to this preference by assigning an important weight to (M) in recommending music. Although, mood has an important individual weight, we observe that the obtained weights of the subsets {M,T} and {M,A} are negative. Here we speak of negative synergy: the union of dimensions does not bring anything, and the importance of the pair is almost the same as the importance of the single dimension. However, the combination of {W,M}, {W,T} and {W,A} have positive interaction degree. It is explained by a positive interaction between these dimensions. We conclude that our approach gives more importance to these correlated dimensions {W,M}, {W,T} and {W,A}.

For the triplet combinations, we notice that all the obtained results by the least squares method are positive and important. Therefore, the correlation between three dimensions lead to a good interaction between them. We analyze the dependence between dimensions by the computation of interaction index [5].

Given a dimension dim_i its interaction with a dimension dim_j is defined by $I_{dim_i dim_j}$.

We obtain the following values: $I_{W,M} = 0.05$, $I_{W,T} = 0.07$, $I_{W,A} = 0.08$, $I_{T,M} = -0.07$, $I_{A,M} = -0.04$, $I_{T,A} = -0.4$. Positive values demonstrate positive interaction between relevance dimensions considered when combined together. We conclude that only the subsets of {W,M}, {W,T} and {W,A} present correlated dimensions.

After obtaining dimensions weights and correlated dimesions by Choquet from the first step, we apply the fuzzy logic to infer user situation.

To determine the situation, we used an open source Java library called jFuzzyLogic [2]. The main goal of jFuzzyLogic is to bring the benefits of open source software and reduce programming work. It offers a fully functional and complete implementation of a Fuzzy Control Language (FCL) format. FCL uses exclusively a Fuzzy Inference System (FIS). All fuzzy language definitions should be within a FIS, it includes: input and output variables, membership functions and fuzzy inference rules. The output and the input variables are fuzzy variables and expressed as linguistic variables. Each linguistic variable is characterized by a variety of terms called linguistic terms. Each term is composed by a name and a membership function. For example:

- **Linguistic variable:** time
- **Linguistic terms:**
    - morning: if time $\in$ [6 am, 12 pm]

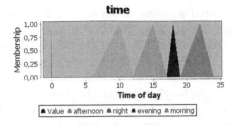

Fig. 2. Fuzzy membership function for time

- afternoon: if time $\in [12\,\text{pm}, 5\,\text{pm}]$
- evening: if time $\in [5\,\text{pm}, 7\,\text{pm}]$
- night: if time $\in [7\,\text{pm}, 12\,\text{am}]$

The membership function for the input variable time is shown in Fig. 2. In this example the membership function is triangular. We chose some common situations in which participants listen to music: waking up, commuting, working/studying, exercising, relaxing, housework and sleeping. To determine the situation, we define some IF-THEN rules. Within each rule, the antecedent (the IF part) is composed of only correlated context dimensions connected by AND operators. In our case the antecedent for IF-THEN rules is composed of {W,M} or {W,T} or {W,A} dimensions. We have already proven previously through Choquet and the interaction index that these dimensions are correlated. Now we define some fuzzy rules:

RULE 1: IF weather IS cloudy AND time IS morning THEN situation IS sleepy;

RULE 2: IF weather IS sunny AND mood IS active THEN situation IS working;

RULE 3: IF activity IS relax AND weather IS rainy THEN situation IS relaxing;

After specifying the weights of input variables, JFuzzyLogic will calculate and will show the output variable, in our case the situation as well as the degree of support for each rule. In the last step, we predict the rating according to the obtained weighted situation.

4.4 Results and Discussion

The study on the effectiveness of our approach was conducted based on all test combination and in comparison with contextual baselines (CAMF_C, CAMF_CI, CAMF_CU). Table 3 illustrates all the experiment results measured by MAE, RMSE, precision and recall on three datasets. This table shows that FWR achieves the best results consistently which indicates the accuracy of our approach and demonstrates the effectiveness on using weighted and correleted contextual dimensions in the prediction process. In fact, our approach is able to beat CAMF in the presented datasets across all four metrics. It is able to obtain comparable results with CAMF-C in the Food data. Comparing with the best performance of other algorithms, FWR improves the RMSE values by 11 %, 26 % and 25 % on the Food, Movie and Music datasets respectively. In terms of the MAE on Movie and Music datasets respectively, FWR gives an improvement of 12 % and 31 % over the best performance of the mentioned baselines. However, the MAE of the proposed approach is a little lower on the Food dataset, this may be because the limited number of contextual conditions in this dataset. Let us note that our approach has the highest relative progression on precision with a 5 %, 49 % and 332 % improvement in the Food, Movie and Music datasets respectively, compared with the best performing baseline CARS algorithm. It is obvious that FWR improves greatly on the Music dataset, which proves FWR to be particularly helpful for the dataset with rich contextual dimensions.

In short, those experimental results demonstrate that the importance of contextual dimensions and their interaction directly influences the inference of the current situation thus recommendations accuracy. Imagine for instance, it is raining, the user is on lunch break, and he likes to eat in an outdoor restaurant, the importance of mood should be little. However, the weather is very important, also the time of day dimension should be combined to the weather dimension to distinguish the appropriate meal for the appropriate time (breakfast, lunch or dinner). We conclude that assigning weight to each contextual dimension and to each dimensions combination using the proposed weighting technique and then applying it for situation inference can substantially improve the rating prediction accuracy than the conventional and the other methods.

5   Conclusion

In this paper, we presented an overview of works concerning prediction rating and fuzzy inference in context-aware recommender systems. We have proposed a new approach for rating prediction according to the current situation. Based on the Choquet integral, the proposed approach is able to weight contextual dimensions then model the interactions that may exist between them. The study focuses on fuzzy rules which takes the correlated context dimensions into account to infer user's current situation. The evaluation of our approach, in rating prediction task is based on three real-world datasets, it showed its superiority over standard prediction methods. This approach is helpful to improve rating prediction quality, consequently to improve recommendation performance. There are many ways in which this work can be applied and/or extended. First of all, we plan to integrate our approach in other recommendation algorithms, using additional metrics, such as ranking-oriented metrics (MAP, NDCG). Additionally, it is needed to evaluate our approach using more larger contextual datasets, this task is currently under achievement by our team.

References

1. Baltrunas, L., Ludwig, B., Ricci, F.: Matrix factorization techniques for context aware recommendation. In: Conference on Recommender Systems, pp. 301–304. ACM, New York (2011)
2. Cingolani, P., Alcalá-Fdez, J.: jFuzzyLogic: a robust andflexible fuzzy-logic inference system language implementation. In: IEEE International Conference on Fuzzy Systems, Australia, pp. 1–8. IEEE (2012)
3. Ephzibah, E.P.: Advances in computing and information technology. In: First International Conference, India (2011)
4. Grabisch, M., Kojadinovic, I., Meyer, P.: A review of methods for capacity identification in Choquet integral based multi-attribute utility theory. Eur. J. Oper. Res. **186**, 766–785 (2008)
5. Grabisch, M., Labreuche, C.: A decade of application of the choquet and sugeno integrals in multi-criteria decision aid. Ann. Oper. Res. **175**, 247–286 (2009)
6. Gs, T., Kulkarni, U.P.: Design and implementation of user context aware recommendation engine for mobile using Bayesian network, fuzzy logic and rule base. Int. J. Comput. Appl. U.K. **40**, 47–63 (2012)
7. Liu, X., Aberer, K.: SoCo: a social network aided context-awarerecommender system. In: International Conference on World Wide Web, pp. 781–802. ACM, New York (2013)
8. Mehta, S.J, Javia, J.: Threshold based knn for fast and more accurate recommendations. In: 2nd IEEE International Conference on Recent Trends in Information Systems, India, pp. 109–113. IEEE (2015)
9. Ono, C., Takishima, Y., Motomura, Y., Asoh, H.: Context-aware preference model based on a study of difference between real and supposed situation data. In: Houben, G.-J., McCalla, G., Pianesi, F., Zancanaro, M. (eds.) UMAP 2009. LNCS, vol. 5535, pp. 102–113. Springer, Heidelberg (2009). doi:10.1007/ 978-3-642-02247-0_12
10. Sen, A., Larson, M.: From sensors to songs: a learning-free novelmusic recommendation system using contextual sensor data. In: Location-Aware Recommendations, co-located with the Conference on Recommender Systems, Austria, pp. 40–43. CEUR-WS.org (2015)
11. Zammali, S., Arour, K. Bouzeghoub, A.: A context features selecting and weighting methods for context-aware recommendation. In: 39th IEEE Computer Software and Applications Conference, Italy, pp. 575–584. IEEE Computer Society (2015)
12. Zheng, Y., Burke, R., Mobasher, B.: Recommendation with differential context weighting, user modeling, adaptation, and personalization. In: 21th International Conference, Italy (2013)
13. Zheng, Y., Burke, R., Mobasher, B.: The role of emotions in context-aware recommendation. In: International Workshop on Human Decision Making in Recommender Systems, China, pp. 21–28. CEUR-WS.org (2013)
14. Zheng, Y., Mobasher, B., Burke, R.: Incorporating context correlation into context-aware matrix factorization. In: Proceedings of International Conference on Constraints and Preferences for Configuration and Recommendation and Intelligent Techniques for Web Personalization, Germany, pp. 21–27. CEUR-WS.org (2015)
15. Zheng, Y., Mobasher, B., Burke, R.D.: CARSKit: A Java-based context-aware recommendation engine. In: IEEE International Conference on Data Mining Workshop, Atlantic City, NJ, USA, 14–17 November 2015, pp. 1668–1671. IEEE Computer Society, Los Alamitos (2015)

Adaptive Conceding Strategies for Negotiating Agents Based on Interval Type-2 Fuzzy Logic

Jieyu Zhan and Xudong Luo$^{(\boxtimes)}$

Department of Philosophy, Institute of Logic and Cognition, Sun Yat-sen University,
135 Xingang West Road, Guangzhou 510275, People's Republic of China
luoxd3@mail.sysu.edu.cn

Abstract. In human-agent automated negotiations, one of crucial problems is how a negotiating agent updates conceding strategies in the light of the new information during the course of a negotiation. To this end, this paper proposes a novel model of a seller negotiating agent, which can be used in human-agent negotiations. More specifically, it can dynamically change its conceding strategies according to the remaining time and opponents' cooperative degree. We use type-2 fuzzy rules to determine such changes because the rules of this kind can well reflect uncertain information in human-computer negotiations. Finally, our agent is evaluated by both agent-agent and human-agent experiments.

Keywords: Automated negotiation · Fuzzy logic · Conceding strategy · Human-computer negotiation · Type-2 fuzzy set

1 Introduction

Due to the popularity of electronic commerce, it has become more and more necessary to employ software agents [9] negotiating on behalf of human users in the online trading, so that they can save the time and other costs, reduce logical errors caused by fatigue when dealing with too much information during the course of a negotiation [4,9,10]. So, huge numbers of automated negotiation systems (e.g., [1,7,10,21]) have been proposed. In an automated negotiation system, the main components are the negotiation protocol and strategy. The former one determines how the negotiation is carried out, and the latter determines how to generate offers during a negotiation, which is critical for good negotiation results. Therefore, lots negotiation strategies have been designed [2,3,14,17], but most of them are designed for negotiating with another agent, rather than a real human, which is more sophisticated than a software agent.

However, with the rapid growth of B2C markets, human-computer negotiation (*i.e.*, a software agent acting on behalf of one party negotiates with a human counterpart) will be more applicable [1,8,22]. Different from complete rational agents, humans are often boundedly rational, easily affected by mood, relationships or other social factors, and use heuristics to make decision, sometimes making mistakes and so on. Since the assumptions such as complete information

F. Lehner and N. Fteimi (Eds.): KSEM 2016, LNAI 9983, pp. 222–235, 2016.
DOI: 10.1007/978-3-319-47650-6_18

or rationality may not hold any more for many such examples in daily negotiation [8], more powerful negotiation strategies are required for a software agent to negotiate properly with different types of humans.

Thus, in this paper, we will propose a new negotiating agent of seller, equipped with an adaptive strategy that can be used to negotiate with both agents and humans well about one issue (price). Our main contributions are: (i) propose a method for evaluating the cooperative degree of opponent based on the opponent's offers and the seller's reservation price; (ii) identify a group of type-2 fuzzy rules for the inference from the time left and the cooperative degree of opponent to the concession rate and the reservation price; and (iii) give the method for determining the offer of a negotiating agent according to the updated concession rate and reservation price. The effectiveness of our model is confirmed by both agent-agent negotiation experiments and agent-human experiment.

The rest of this paper is organised as follows. Section 2 recaps the basic concepts and notations of interval type-2 fuzzy logic. Section 3 proposes the model of our negotiating agent. Section 4 does our experimental analysis. Section 5 discusses the related work. Finally, Sect. 6 concludes the paper with future work.

2   Preliminaries

In this section, we recap some necessary concepts and notations in interval type-2 fuzzy sets (IT2FSs) [13], and interval type-2 fuzzy logic systems (IT2FLSs) [11].

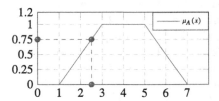

Fig. 1. An example of T1FS

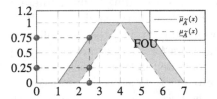

Fig. 2. An example of IT2FS

We start with the definitions of interval type-2 fuzzy sets (IT2FSs).

Definition 1. *An interval type-2 fuzzy set $\widetilde{A}$ on a nonempty set X is represented as*

$$\widetilde{A} = \int_{x \in X} \frac{[\underline{\mu}_{\widetilde{A}}(x), \overline{\mu}_{\widetilde{A}}(x)]}{x}, \tag{1}$$

where $\underline{\mu}_{\widetilde{A}}$ and $\overline{\mu}_{\widetilde{A}}$ are two mapping from X to $[0,1]$, called the upper membership function (UMF) and lower membership function (LMF) of interval type-2 fuzzy set $\widetilde{A}$, respectively, and $\forall x \in X$, $0 \leq \underline{\mu}_{\widetilde{A}}(x) \leq \overline{\mu}_{\widetilde{A}}(x) \leq 1$.

In the above definition, if $\forall x \in X$, $\underline{\mu}_{\widetilde{A}}(x) = \overline{\mu}_{\widetilde{A}}(x)$, then $\widetilde{A}$ is a type-1 fuzzy set. The main difference between type-1 fuzzy sets (T1FSs) and interval type-2 fuzzy sets (IT2FSs) is their membership functions. In T1FSs, the membership degrees are crisp values, while in IT2FSs, they are intervals. Hence, IT2FSs are more useful than T1FSs when it is hard to estimate the precise membership functions. For example, the linguistic term: *about four* can be regarded as a T1FS (denoted as A) and an IT2FS (denoted as $\widetilde{A}$) in Figs. 1 and 2, respectively. In Fig. 1, $\mu_A(2.5) = 0.75$ can be interpreted as the degree to which 2.5 is about four is 0.75; in Fig. 2, $\underline{\mu}_{\widetilde{A}}(2.5) = 0.25$ and $\overline{\mu}_{\widetilde{A}}(2.5) = 0.75$, which can be interpreted as the degree to which 2.5 is about four is between 0.25 and 0.75.

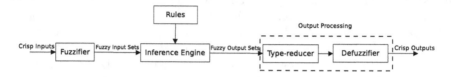

Fig. 3. The block diagram of an IT2FLS

Now we brief the framework of interval type-2 fuzzy logic system. As shown in Fig. 3, the IT2FLS consists of four parts: the fuzzifier, which translates crisp inputs into fuzzy input sets according to the membership functions of linguistic variables of inputs; the inference engine, which maps fuzzy input sets into fuzzy output sets according to fuzzy inference based on fuzzy rules; the type-reducer, which performs a type reduction (TR) procedure to map a type-2 fuzzy set into a type-1 fuzzy set; and the defuzzifier computes the final crisp outputs. The type-reducer and defuzzifier are the output processing of IT2FLS. There are many methods used in type reduction, and in this paper, we employ the center-of-sets type-reducer, which is the most commonly used one.

The fuzzy rules are expressed as the IF-THEN forms as follows:

R^n: IF x_1 is $\widetilde{X}_1^n$ and $\cdots$ and x_I is $\widetilde{X}_I^n$, THEN y is Y^n, n=1, 2, ..., N

where $\widetilde{X}_i^n$ ($i = 1, \ldots, I$) are the antecedent IT2-FSs and $Y^n = [\underline{y}^n, \overline{y}^n]$ is an interval. The inference from crisp inputs, $\mathbf{x}' = (x_1', x_2', \ldots, x_I')$, to crisp outputs is shown as follows:

(i) Calculate both the upper and lower membership degrees of x_i' on each $\widetilde{X}_i^n$, $[\overline{\mu}_{\widetilde{X}_i^n}(x_i'), \underline{\mu}_{\widetilde{X}_i^n}(x_i')]$, $i = 1, \ldots, I$, $n = 1, 2, \ldots, N$.

(ii) Calculate the firing interval of the nth rule as follows:

$$F^n(\mathbf{x}') \equiv \left[\underline{f}^n, \overline{f}^n \right], \tag{2}$$

where $\underline{f}^n = \underline{\mu}_{\widetilde{X}_1^n}(x_1') \times \cdots \times \underline{\mu}_{\widetilde{X}_I^n}(x_I')$ and $\overline{f}^n = \overline{\mu}_{\widetilde{X}_1^n}(x_1') \times \cdots \times \overline{\mu}_{\widetilde{X}_I^n}(x_I')$.

(iii) Perform type-reduction to combine $F^n(\mathbf{x}')$ and the corresponding rule consequents. Here we just discuss the center-of-sets type-reducer as follows:

$$Y_{cos}(\mathbf{x}') = \bigcup_{\substack{f^n \in F^n(\mathbf{x}') \\ y^n \in Y^n}} \frac{\sum_{n=1}^{N} f^n y^n}{\sum_{n=1}^{N} y^n} = [y_l, y_r], \tag{3}$$

where

$$y_l = \frac{\sum_{n=1}^{L} \overline{f}^n y^n + \sum_{n=L+1}^{N} \underline{f}^n \overline{y}^n}{\sum_{n=1}^{L} \overline{f}^n + \sum_{n=L+1}^{N} \underline{f}^n}, \qquad y_r = \frac{\sum_{n=1}^{R} \underline{f}^n \overline{y}^n + \sum_{n=R+1}^{N} \overline{f}^n \underline{y}^n}{\sum_{n=1}^{R} \underline{f}^n + \sum_{n=R+1}^{N} \overline{f}^n},$$

where L and R are calculated by using Karnik-Mendel (KM) algorithms.[1]
(iv) Calculate the crisp output by $y = \frac{y_l + y_r}{2}$.

3   A Negotiating Agent

This section proposes a seller agent to negotiate adaptively with human or agent buyers.

The negotiation scenarios we consider in this paper is a single issue (e.g., price) human-agent negotiation. Our purpose is to build a seller agent that is able to negotiate with other agents or humans automatically and adaptively changes its negotiation strategies responding to the opponent's actions during the course of a negotiation. In this paper, we use the most common negotiation protocol: an alternating offers protocol [20]. That is, a buyer and a seller give their offers to each other and generate counter-offers according to their negotiation strategies alternately. Both the seller and the buyer have a range of acceptable prices, which are presented as $[\underline{P}_s, \overline{P}_s]$ and $[\underline{P}_b, \overline{P}_b]$, respectively, where $\overline{P}_s$ and $\underline{P}_b$ can be regarded as the opening prices offered by the seller and the buyer, and $\underline{P}_s$ and $\overline{P}_b$ are their reservation prices, respectively. Then interval $[\underline{P}_s, \overline{P}_b]$ is the zone of agreement, in which a mutually acceptable outcome they could reach is.

Firstly, we present the model of our negotiating seller agent as follows:

Definition 2. *A seller agent is 7-tuple* $(\overline{P}_s, \underline{P}_s, D_s, U_s, A_s, S_s, IT2FLS)$,[2] *where:*

(i) $\overline{P}_s$ *is the opening price of the seller.*
(ii) $\underline{P}_s$ *is the reservation price of the seller. That is, any offer lower than the price is not acceptable to the seller.*
(iii) D_s *is the seller agent's negotiation deadline of time.*

---

[1] It takes about two pages to display the whole algorithm. So, for the sake of space, we do not show it here but the reader can check out its details in [12].
[2] The methodology of the seller agent model can also be applied to a buyer agent. But in this paper, we will pay more attention in the situation where the seller is an agent with adaptive strategies, while the buyer is a human, or an agent with fixed strategies.

(iv) Let $P = [\underline{P}_b, \overline{P}_b]$ be the set of all possible offers of a buyer and $T = \{1, 2, 3, \dots\}$ be the set of negotiation rounds, the seller's utility function is:

$$\forall p \in P, \forall t \in T, U_s(p, t) = \begin{cases} p - \underline{P}_s & \text{if } t \leq D_s, \\ 0 & \text{if } t > D_s. \end{cases} \tag{4}$$

(v) Let $p_{b \to s}^t$ denote the price offered by the buyer to the seller at time t and $p_{s \to b}^{t+1}$ be the price offered by the seller to the buyer at time $t + 1$. Then the action that the seller agent can take is represented as follows:

$$A_s(p_{b \to s}^t, t) = \begin{cases} Quit & \text{if } t > D_s, \\ Accept & \text{if } U_s(p_{b \to s}^t, t) \geq U_s(p_{s \to b}^{t+1}, t+1), \\ Propose\ counter\text{-}offer\ p_{s \to b}^{t+1} & \text{otherwise.} \end{cases} \tag{5}$$

(vi) S_s represents the seller agent's negotiation strategy, denoted as 2-tuple $(\underline{P}_s, \psi_s)$, where ψ_s depicts the concession rate of the seller agent.

(vii) IT2FLS is an interval type-2 fuzzy system that generates negotiation strategies adaptively according to the information of negotiation time and opponent's offers.

Negotiation strategies refer to the way in which how the agent generates counter-offer during the course of a negotiation. One of the most significant strategies is the kind of time-dependent strategies [2]. Offers that are generated by this kind of negotiation strategies can be represented as follows:

$$p_{s \to b}^t = \underline{P}_s + (1 - \phi_s(t))(\overline{P}_s - \underline{P}_s), \tag{6}$$

$$p_{b \to s}^t = \underline{P}_b + \phi_b(t)(\overline{P}_b - \underline{P}_b), \tag{7}$$

where $\phi_a(t) = \left(\frac{t}{D_a}\right)^{\psi_a}$, in which D_a is the deadline of agent $a \in \{seller, buyer\}$, and ψ_a is the concession rate parameter, which reflects different ways for changing the speed of conceding as time goes on. Fatima et al. [3] distinguish three time-dependent strategies as follows: (1) Boulware [16]: when $\psi_a > 1$, the speed of concession increases with the time; (2) Conceder [15]: when $\psi_a < 1$, the speed of concession decreases with the time; and (3) Linear: when $\psi_a = 1$, the speed of concession is constant.

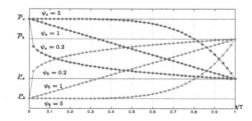

Fig. 4. An example of three time-dependent concession strategies of the seller and the buyer

Figure 4 illustrates the three time-dependent concession strategies that sellers and buyers can use in a negotiation, respectively. In this example, we suppose the deadlines of both a seller and a buyer are the same, denoted as T. The horizontal axis shows the negotiation time and the vertical axis represents the concession the agents could make. The curves with star markers describe the tendency of generating offers by the seller with the three concession rate parameters of $\psi_s = 5$, $\psi_s = 0.2$ and $\psi_s = 1$, respectively. Similarly, the curves with circle markers show the strategies of the buyer. From this figure, we can see that the different combinations of strategies of the seller and buyer will lead to different outcomes and cost different negotiation times. For example, if both agents use *Boulware* strategies, it will take a lot of time to achieve a fair agreement. However, if the seller agent uses a *Conceder* strategy but the buyer agent uses a *Boulware* strategy, the buyer agent will get a very low price.

However, it is difficult to use such fixed strategies to handle dynamical situations in human-agent negotiation [1,8], because they will lead to enormous reduction of utility and time consumption in some cases (e.g., using a Conceder strategy to deal with a Boulware strategy). Especially, in the human-agent negotiation, humans' strategies change very differently and so are hard to be predicted. Hence, in this paper, we construct an adaptive negotiation strategy, i.e., the agent can change the current strategy in the light of the new information during the course of a negotiation. By formula (6), a strategy is determined by the concession rate parameter and the reservation price. So, we will establish an interval type-2 fuzzy system (IT2FLS) based on fuzzy rules to adaptively adjust these two parameters in the process of negotiation, which can facilitate the efficiency of negotiation and improve the utility of the seller agent. The reason why we apply interval type-2 fuzzy system is as follows. Firstly, it can well reflect reasoning based on uncertain information in human-computer negotiations. Secondly, interval type-2 fuzzy sets are flexible enough to describe the membership degree of linguistic terms given by humans in a range and make the reasoning of high uncertainty possible, because it is hard to represent the precise membership functions in terms of type-1 fuzzy set. Finally, this approach provides the user with a flexible update method by adjusting the rules of reasoning in the face of different scenarios.

There are two inputs and two outputs in our interval type-2 fuzzy system. The inputs are Remaining Time (RT) and Cooperation Degree (CD), and the outputs are Concession Rate (CR) and Reservation Price (RP), respectively. Specifically, the system will calculate the corresponding CR and RP for the seller agent in the next round according to the remaining time and the cooperation degree to which the buyer shows its cooperation in the last round in the negotiation. The percentage of remaining time, which is based on negotiation time t, can be simply calculated by $RT(t) = 1 - \frac{t}{D}$, while the calculation of cooperation degree will be detailed later on in this section. The reasoning process of the fuzzy system is based on our fuzzy rules. We will introduce the fuzzy linguistic terms and fuzzy rules in this section.

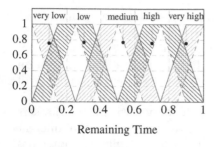

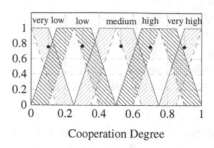

Fig. 5. Membership functions of RT   **Fig. 6.** Membership functions of CR

In our fuzzy system, two fuzzy input variables (RT and CD) take linguistic terms, including *very low, low, medium, high* and *very high*, respectively, which is an effective way to reflect people's perception [24]. The meanings of such five parameters' linguistic terms are intuitionistic. The *very high* remaining time indicates that there is a lot of time left for the agent to negotiate before the deadline, while *very low* remaining time means that the time left for the negotiation agent is very limited before the deadline. Similarly, we can understand the other linguistic terms (*i.e., low, medium, high*) of both two parameters (*i.e.,* remaining time and cooperation degree).

In an interval type-2 fuzzy logic system, these linguistic terms can be modelled by the interval between the upper membership function (UMF) and the lower membership function (LMF). In this paper, we employ the trapezoidal type of upper membership functions and trigonal type of lower membership functions, respectively, as follows:

$$
\overline{\mu}_{\widetilde{A}}(x) = \begin{cases} 0 & \text{if } x \leq a, \\ \frac{x-a}{b-a} & \text{if } a < x < b, \\ 1 & \text{if } b \leq x \leq c, \\ \frac{d-x}{d-c} & \text{if } c \leq x \leq d, \\ 0 & \text{if } x \geq d. \end{cases}
\qquad
\underline{\mu}_{\widetilde{A}}(x) = \begin{cases} 0 & \text{if } x \leq a, \\ \frac{x-a}{b-a} & \text{if } a \leq x \leq b, \\ \frac{c-x}{c-b} & \text{if } b \leq x \leq c, \\ 0 & \text{if } x \geq c. \end{cases}
$$

The membership functions of the five linguistic terms of remaining time and cooperation degree are shown in Figs. 5 and 6, respectively.

The outputs of the interval fuzzy system are calculated according to fuzzy rules, which are summarised in Table 1. Every rule consists of the antecedent and the consequent. The former includes the information about remaining time before the negotiation deadline and the cooperation degree of the opponent shown in the last negotiation round. The latter indicates how the agent responds to the opponent's offer through adjusting both the concession rate parameter and reservation price. For example, rule 1 means that if there is a plenty of negotiation time left and the opponent adopts a more cooperative posture, then the agent concedes more quickly (ψ_s is in an interval between 0.1 and 0.2) and the reservation price remains constant (the increase rate of $\underline{P}_s$ is 0). Similarly,

Table 1. Fuzzy rules

1	If RT is *very high* and CD is *very high*, then CR is $[0.1\ 0.2]$ and RP is $[0\ 0]$
2	If RT is *very high* and CD is *high*, then CR is $[0.2\ 0.5]$ and RP is $[0\ 0]$
3	If RT is *very high* and CD is *medium*, then CR is $[0.5\ 1]$ and RP is $[0\ 0]$
4	If RT is *very high* and CD is *low*, then CR is $[0.5\ 1]$ and RP is $[0\ 0.1]$
5	If RT is *very high* and CD is *very low*, then CR is $[1\ 2]$ and RP is $[0.1\ 0.2]$
6	If RT is *high* and CD is *very high*, then CR is $[0.1\ 0.2]$ and RP is $[0\ 0]$
7	If RT is *high* and CD is *high*, then CR is $[0.2\ 0.5]$ and RP is $[0\ 0]$
8	If RT is *high* and CD is *medium*, then CR is $[1\ 1]$ and RP is $[0\ 0]$
9	If RT is *high* and CD is *low*, then CR is $[1\ 2]$ and RP is $[0\ 0.1]$
10	If RT is *high* and CD is *very low*, then CR is $[2\ 3]$ and RP is $[0.1\ 0.2]$
11	If RT is *medium* and CD is *very high*, then CR is $[0.5\ 0.8]$ and RP is $[0\ 0]$
12	If RT is *medium* and CD is *high*, then CR is $[0.8\ 1]$ and RP is $[0\ 0]$
13	If RT is *medium* and CD is *medium*, then CR is $[1\ 2]$ and RP is $[0\ 0]$
14	If RT is *medium* and CD is *low*, then CR is $[2\ 3]$ and RP is $[0\ 0.1]$
15	If RT is *medium* and CD is *very low*, then CR is $[3\ 4]$ and RP is $[0\ 0.2]$
16	If RT is *low* and CD is *very high*, then CR is $[0.8\ 1]$ and RP is $[0\ 0]$
17	If RT is *low* and CD is *high*, then CR is $[1\ 2]$ and RP is $[0\ 0]$
18	If RT is *low* and CD is *medium*, then CR is $[2\ 3]$ and RP is $[0.1\ 0.2]$
19	If RT is *low* and CD is *low*, then CR is $[2\ 3]$ and RP is $[0.1\ 0.2]$
20	If RT is *low* and CD is *very low*, then CR is $[3\ 4]$ and RP is $[0.2\ 0.3]$
21	If RT is *very low* and CD is *very high*, then CR is $[0.5\ 0.8]$ and RP is $[0\ 0]$
22	If RT is *very low* and CD is *high*, then CR is $[0.8\ 1]$ and RP is $[0\ 0]$
23	If RT is *very low* and CD is *medium*, then CR is $[2\ 3]$ and RP is $[0.1\ 0.2]$
24	If RT is *very low* and CD is *low*, then CR is $[3\ 4]$ and RP is $[0.1\ 0.2]$
25	If RT is *very low* and CD is *very low*, then CR is $[4\ 5]$ and RP is $[0.2\ 0.3]$

we can understand other rules. It is important to note that the interval of the outputs in the fuzzy rules are based on both experience and experiments. We set these parameters in order to improve the capability of negotiating agents.

How to calculate the cooperation degree to which the opponent shows cooperation personality is one of the most significant issues when designing our negotiating agent. In general, the most effective way is to measure the degree through the opponent's conceding speed or the offer of opponent [18]. Nevertheless, when doing experiments with humans, we find that the humans' speed of making concession changes somehow randomly, so from which it is difficult to see their cooperation personality. Hence, we assess the opponent's cooperation degree according to the opponent's offers. In addition, in real life, people have different attitudes toward the equivalent concessions with different offers. For example, if the buyer's offer in the last round is very low and he gives the same

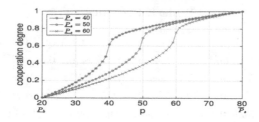

Fig. 7. An example of cooperation degree function

concession with another buyer whose offer is in the last round is close to the sell's offer, it is obvious that the latter buyer is more cooperative than the former one. According to prospect theory [5], one of the most important characteristics when humans evaluate outcomes or offers is loss aversion (i.e., people tend to strongly prefer avoiding losses to acquiring gains). That is, consumers firstly evaluate the potential change by distinguishing it as a gain or a loss according to a reference point. The changes that are regarded as losses will have more impact than that regarded as gains. In our system, we apply the S-shape value function in prospect theory to evaluate cooperation degree from the opponent's offer, where the reference point is the seller's reservation price, because an offer of buyer that higher than the price will result in a profit to the seller, while the one that lower than the price will lead to a loss to the seller. Formally, we have:

Definition 3. *Let* $P = [\underline{P}_b, \overline{P}_b]$ *be the set of all possible offers of a buyer. A cooperation degree function of a buyer is a mapping* $CD : P \to [0, 1]$, *given by:*

$$
CD(p) = \begin{cases} \dfrac{(p-\underline{P}_s)^\alpha - \underline{CD}}{\overline{CD} - \underline{CD}} & \text{if } p \geq \underline{P}_s, \\[2mm] \dfrac{-\lambda(\underline{P}_s - p)^\beta - \underline{CD}}{\overline{CD} - \underline{CD}} & \text{if } p < \underline{P}_s, \end{cases} \tag{8}
$$

where $\overline{CD} = (\overline{P}_s - \underline{P}_s)^\alpha$ *and* $\underline{CD} = -\lambda(\underline{P}_s - \underline{P}_b)^\beta$.

In the above definition, α and β are diminishing sensitivity parameters, which reflect how the power of gains and losses change, and λ is the loss aversion parameter, which depicts the characteristic of loss aversion. In our system, we set $\alpha = \beta = 0.7$ and $\lambda = 2.25$. Consider the following simple example where the opening prices of a buyer and a seller are 20 and 80, respectively. Then we can see how the cooperation degree changes with the price offered by the buyer in three situations where the seller's reservation prices are 40, 50 and 60, respectively. From Fig. 7, we can see that the speed of the increase of cooperation degree rises first and then falls slowly with the increase of the buyer's offer. The major demographic changes appear near the reference points. It reflects that our modelling of cooperation degree concerns not only the buyer's offer, but also the seller's reservation price.

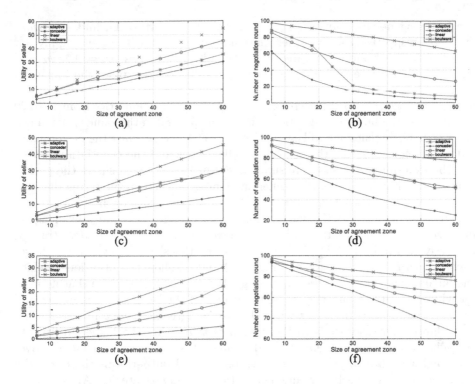

Fig. 8. Outcomes via four strategies when countering different types of buyers

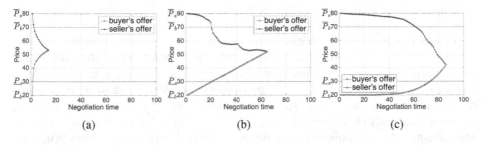

Fig. 9. Seller with adaptive strategy to buyers with different fixed strategies

4 Experiment

This section will analyse our negotiating agent via two kinds of experiments.

The first experimental is about agent-agent, which setup is as follows. A buyer agent wants to buy a mobile hard drive from a seller agent. The acceptable price intervals of the seller and the buyer are $[\underline{P}_s, \overline{P}_s]$ and $[\underline{P}_b, \overline{P}_b]$, respectively, where $\overline{P}_s = 80$ and $\underline{P}_b = 20$, but $\underline{P}_s$ and $\overline{P}_b$ will be set differently in different experiments. The negotiation deadline for both agents are 100. The agents do not know the reservation prices and the deadlines of each other. In order to make the negotiation results more comparable, we set three negotiation scenarios where the buyer agents use *Conceder* ($\psi_b = 0.2$), *Linear* ($\psi_b = 1$), and *Boulware* ($\psi_b = 5$), respectively. In each negotiation scenario, the seller agents apply our adaptive strategy and other three time-dependent strategies ($\psi_s = 0.2$, $\psi_s = 1$, $\psi_s = 5$) to negotiate with the buyer agents. In order to explore the impact of the size of agreement zone to the negotiation outcomes, we gradually expand the agreement zone $[\underline{P}_s, \overline{P}_b]$ by decreasing the seller's reservation price from 47 to 20 and increase the buyer's reservation price from 53 to 80 in the same speed. Then the size of agreement zone will change from 6 to 60. We evaluate the performance of the strategies through the utilities of seller agent and the negotiation times.

The experimental results are summarised in Fig. 8, which shows the performances of our seller agents with different strategies when encountering different types of buyer agents. More specifically, Figs. 8(a) and (b) show how the utilities of our seller agent and the negotiation times change with the size of agreement zone in the negotiations where the buyer agent uses a *Conceder* strategy, respectively. Similarly, Figs. 8(c) and (d) show the situation where the buyer agent uses a *Linear* strategy; and Figs. 8(e) and (f) show the situation in which the buyer agent uses a *Boulware* strategy. From Fig. 8, we can see that no matter what type of the buyer agent is, the utility of the seller agent increases and the negotiation time decreases when the size of agreement zone increases. And our adaptive strategy leads to the outcomes that bring the higher utilities to the seller agent than the *Linear* and *Conceder* strategies when facing buyer equipped with *Linear* or *Boulware* strategies, and outperforms the *Conceder* strategies when facing buyer equipped with *Conceder* strategies. Although the *Boulware* strategies always get the highest utilities for our seller agents when encountering different types of buyer agent, it spends the highest negotiation time compared to the other strategies.

Compared to other strategies, our adaptive strategy has achieved a good balance between utility and time consumption. Specifically, Fig. 9 illustrates how our adaptive strategies respond to other strategies in a situation in which $\underline{P}_s = 30$ and $\overline{P}_b = 70$. We can see that the proposed strategy will adjust its concession rate to different types of buyer agents according to fuzzy reasoning based on the updated information. For example, in Fig. 9(a), the seller agent makes a great concession when the buyer is compromised in the beginning. In Figs. 9(b) and (c), the seller adopts totally different styles of strategies to face the buyers with *Linear* and *Boulware* strategies, respectively.

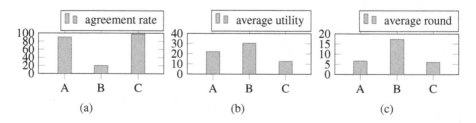

Fig. 10. Results of agent-human negotiation

Now we turn to the second experiment about agent-human negotiation, which purpose is to evaluate the performance of our negotiating agent with real human opponent. In this experiment, 40 human subjects are asked to play role of buyer and are told that their reservation price is 70. They can see and respond to the seller agent's offer on a computer platform we developed. And they are assigned to negotiate with three types of sellers equipped with adaptive strategies (*A*), *Boulware* strategy (*B*) and *Conceder* strategy (*C*), respectively. The reservation price of the seller agent is 30 and its deadline is 20. Both sides have no ideas about the reservation prices and deadlines of each other.

Figure 10 summarises the agent-human experimental results, including the percentage of agreement rate (Fig. 10(a)), the average utility of seller agent in successful negotiation (Fig. 10(b)), and the average negotiation round needed in successful negotiation (Fig. 10(c)). We can see that, although the *Boulware* strategy achieves the highest average utility when there is an agreement between the seller and the buyer, the agreement rate is very low (20 %) and the average round is very high (about 17 rounds). It means that if the seller agent uses *Boulware* strategy to respond to a real human, it is likely to lead to a breakdown in negotiations. This may be because the strategy is so competitive that the opponent loses patience, and does not think that there is an opportunity for cooperation with each other. However, the adaptive strategy gets a high agreement rate with human opponent and spends little time in negotiation, which is similar to the *Conceder* strategy, while the average utility is higher than that of *Conceder* strategy. Hence, the proposed strategy is suitable for our negotiating agent in agent-human negotiations.

5   Related Work

Ren et al. [19] study the adaptive change of agents' strategies in open, dynamic markets, and design market-driven negotiation strategies. Actually, in this paper we extend their strategies via changing the concession rate in the light of negotiation information. However ours can negotiate with human, which theirs is not concerned with. Ours can also adjust the reservation price to prevent being excessively exploited by human opponents.

Shyur and Shih [21] propose a purchase negotiation support system, in which preferences of negotiators are represented by the S-shape value functions [5]. Their experiments confirm that value function can well reflect behaviours of negotiators and their attitudes towards risk. However, we use a value function differently, i.e., to use it to evaluate the cooperative attitude of its opponent, thus better determine how to adjust its strategy in the new round during the course of a negotiation.

Kolomvatsos et al. [6] develop a negotiating agent with an adaptive fuzzy logic system, and their fuzzy rules and membership functions can be updated. Their fuzzy logic system is of type-1, while ours is of type-2; and their negotiation is just for agent to agent, while ours can also for computer to human. And our experiments show our agent performs well in negotiation scenarios with real humans.

Some researchers design intelligent agents that concern both economic outcomes and socio-psychological outcomes in human-computer negotiation. For example, in [23], what developed is a win-win seeking negotiation agent using strategies of *simultaneous equivalent offers* and *delayed acceptance*. Our negotiating agent also takes the opponent's socio-psychological outcomes into consideration via the construction of fuzzy rules. For example, our agent can adjust strategies to achieve an agreement quickly to prevent opponents' losing his patience.

6 Conclusion

This paper designs a seller agent that can adaptively changes its conceding strategies during a negotiation with other agents or humans. Specifically, the model of our seller negotiating agent is equipped with an interval type-2 fuzzy system (a good tool for handling uncertain reasoning). Moreover, we developed a method to evaluate opponent's cooperative attitude, based on which our agent can adjust its strategies to achieve an agreement more quickly. Finally, two experiments shown that compared with the fixed strategies, our negotiating agent that takes adaptive strategy performs well in both agent-agent negotiation and human-agent negotiation scenarios.

Acknowledgments. This research is supported by the Bairen Plan of Sun Yatsen University, the Natural Science Foundation of Guangdong Province, China (No.2016A030313231) and the National Fund of Social Science (No. 13BZX066).

References

1. Cao, M., Luo, X., Luo, X.R., Dai, X.: Automated negotiation for e-commerce decision making: a goal deliberated agent architecture for multi-strategy selection. Decis. Support Syst. **73**, 1–14 (2015)
2. Faratin, P., Sierra, C., Jennings, N.R.: Negotiation decision functions for autonomous agents. Robot. Auton. Syst. **24**(3), 159–182 (1998)
3. Fatima, S.S., Wooldridge, M., Jennings, N.R.: An agenda-based framework for multi-issue negotiation. Artif. Intell. **152**(1), 1–45 (2004)
4. Jennings, N.R., Faratin, P., Lomuscio, A.R., Parsons, S., Wooldridge, M.J., Sierra, C.: Automated negotiation: prospects, methods and challenges. Group Decis. Negot. **10**(2), 199–215 (2001)
5. Kahneman, D., Tversky, A.: Prospect theory: an analysis of decision under risk. econometrica J. Econometric Soc. **47**(2), 263–292 (1979)
6. Kolomvatsos, K., Trivizakis, D., Hadjiefthymiades, S.: An adaptive fuzzy logic system for automated negotiations. Fuzzy Sets Syst. **269**, 135–152 (2015)
7. Lang, F., Fink, A., Brandt, T.: Design of automated negotiation mechanisms for decentralized heterogeneous machine scheduling. Eur. J. Oper. Res. **248**(1), 192–203 (2016)
8. Lin, R., Kraus, S.: Can automated agents proficiently negotiate with humans? Commun. ACM **53**(1), 78–88 (2010)

9. Luo, X., Jennings, N.R., Shadbolt, N., Leung, H.F., Lee, J.H.: A fuzzy constraint based model for bilateral, multi-issue negotiations in semi-competitive environments. Artif. Intell. **148**(1), 53–102 (2003)
10. Luo, X., Miao, C., Jennings, N.R., He, M., Shen, Z., Zhang, M.: KEMAND: a knowledge engineering methodology for negotiating agent development. Comput. Intell. **28**(1), 51–105 (2012)
11. Mendel, J.M., John, R.I., Liu, F.: Interval type-2 fuzzy logic systems made simple. IEEE Trans. Fuzzy Syst. **14**(6), 808–821 (2006)
12. Mendel, J.M., Liu, X.: Simplified interval type-2 fuzzy logic systems. IEEE Trans. Fuzzy Syst. **21**(6), 1056–1069 (2013)
13. Mendel, J.M., Wu, D.: Interval type-2 fuzzy sets, pp. 35–63. Perceptual Computing (2010)
14. Pan, L., Luo, X., Meng, X., Miao, C., He, M., Guo, X.: A two-stage win-win multiattribute negotiation model: optimization and then concession. Comput. Intell. **29**(4), 577–626 (2013)
15. Pruitt, D.G.: Negotiation Behavior. Academic Press, New York (2013)
16. Raiffa, H.: The Art and Science of Negotiation. Harvard University Press, Cambridge (1982)
17. Ren, F., Zhang, M.: Bilateral single-issue negotiation model considering nonlinear utility and time constraint. Decis. Support Syst. **60**, 29–38 (2014)
18. Ren, F., Zhang, M., Bai, Q.: A dynamic, optimal approach for multi-issue negotiation under time constraints. In: Marsa-Maestre, I., Lopez-Carmona, M.A., Ito, T., Zhang, M., Bai, Q., Fujita, K. (eds.) Novel Insights in Agent-based Complex Automated Negotiation. SCI, vol. 535, pp. 85–108. Springer, Heidelberg (2014). doi:10.1007/978-4-431-54758-7_5
19. Ren, F., Zhang, M., Sim, K.M.: Adaptive conceding strategies for automated trading agents indynamic, open markets. Decis. Support Syst. **46**(3), 704–716 (2009)
20. Rubinstein, A.: Perfect equilibrium in a bargaining model. Econometrica J. Econometric Soc. **50**(1), 97–109 (1982)
21. Shyur, H.-J., Shih, H.-S.: Designing a multi-issues negotiation support system based on prospect theory. Inf. Sci. **322**, 161–173 (2015)
22. Yang, Y., Singhal, S.: Designing an intelligent agent that negotiates tactfully with human counterparts: a conceptual analysis and modeling framework. In: Proceedings of the 42nd Hawaii International Conference on System Sciences, pp. 1–10 (2009)
23. Yang, Y., Singhal, S., Xu, Y.: Alternate strategies for a win-win seeking agent in agent-human negotiations. J. Manag. Inf. Syst. **29**(3), 223–256 (2012)
24. Zadeh, L.A.: The concept of a linguistic variable and its application to approximate reasoning - I. Inf. Sci. **8**(3), 199–249 (1975)

Semiotic Rules Generation and Inferences Reasoning for Movie Documents

Manel Fourati[✉], Anis Jedidi, and Faiez Gargouri

Laboratory MIR@CL, University of Sfax, Sfax, Tunisia
Manel.Fourati@fsegs.rnu.tn, {anis.jedidi,faiez.gargouri}@isimsf.rnu.tn

Abstract. With the rapid but still incomplete maturation of the information retrieval research related to multimedia document, the progress of a new solution requires the extraction of semantic information from the content. It should be however, not only extracted from the content but it should also present the different semiotics meaning conveyed in the content. In our work, we concentrate our efforts on the movie documents. In fact, the knowledge extracted separately can conceal the global vision on the sequence of events or analysis of history conveyed in a film. In this context, we are interested in this paper to generate relationships either between sub-parts of the same movie or between movies. Consequently, we propose an inference reasoning to build these relationships in order to reveal the hidden knowledge semantics of resources. These relationships are basically based on the semiotic description that poses a major challenge. A case study where we substantiate and prove the accurate performance of our proposed process is highlighted.

Keywords: Semiotic · Movie · Rule · Inference

1 Introduction

The information retrieval in the multimedia document is increasingly reaching a certain maturity. In fact, it remains a very useful phenomenon despite its young discipline. Indeed, the audiovisual documents are resources of high descriptive capacity but are miss used in the querying process because of the lack of extracting descriptive metadata. Several search engines based on the manual annotation such as VideoAnnEx or semi-automatic annotation like Advene, ANVIL of the audiovisual document, which seems to be the first solution. However, if this approach seems natural and brings about an audiovisual search problem, it still remains an unsatisfactory way. Indeed, the disadvantages of this type of description are not limited only to the multiple meanings of the document but also appear on the lack of the description related to the content. Consequently, an automation of audiovisual documents description process is required. Indeed, the core aspect related to a pertinent metadata extraction process is the extraction of the descriptions that are related to the content. Therefore, we propose, in this paper, not only the analysis of the audiovisual content to extract

© Springer International Publishing AG 2016
F. Lehner and N. Fteimi (Eds.): KSEM 2016, LNAI 9983, pp. 236–248, 2016.
DOI: 10.1007/978-3-319-47650-6_19

semantic description but also the extraction of semantic relationships between the extracted description in order to achieve the highest semantic level. So, the extraction of descriptions requires the passage from one semantic level to a higher one. In this regard, the extraction of the semantic relationship should be able to develop the structures of semantic and contextual descriptions. In fact, a key advantage of this solution is that it ensures consistency description of interrogation by the interoperability between the different resources. To explain our proposed process, the following organization is proposed: after presenting the existing problems and some possible solutions in this section, we will discuss in the second section a brief overview in the first place on the related work based on the extraction of semantic relationships and in the second place on semiotic relationships. We will be interested thereafter in Sect. 3 to describe the different steps of our proposed process to generate the different relationships. Section 4 is a case study in which our proposed solution has been substantiated and tested. Finally, Sect. 5 contains conclusions and some reflections on possible openings and on our future work.

2   Related Works

The annotation and the extraction description of the multimedia document represents the main point of information retrieval. However, many applications require a finer granularity of information such as indexing and searching from the content. In this context, several works in the literature propose different methods in order to extract description. These methods relay essentially either on low-level or high-level descriptions. However, even these high-level descriptions used for indexing of video content, are not sufficient to provide information semantically conveyed in the content. For this reason, we propose in this paper a process that defines not only the semiotic description presenting the meaning from the movie document but also a set of semiotic relations defined semantically in order to extract the hidden semantics. In this section, we present an overview of the most relevant works proposed in the literature related to the extraction of semantic relationships and we define thereafter the different semiotic relationships.

2.1   Semantic Relationships

Semantic relationship extraction is a bio-inspired processing of the human mental system. This process, as its name suggests, allows identifying the semantic from a set of information. In the literature, a few studies have been devoted to extract semantic relationships of multimedia resources which aim to facilitate the annotation and the retrieval of these documents. In [1], the authors accomplished in their work a classification approach based on the semantic rules generation for image annotation. A formal writing for these rules is proposed via the logic of predicates. The classification techniques are based on a thematic grouping by using the fuzzy c-means algorithm. Recently, in [7], the authors have proposed a method for interlinking multimedia resources. This method is based on the extraction of the semantic through some defined rules. It defines a set of relationships such as Talk, Talk About, Speak, Speak About,... Indeed, the objective behind these relationships is to simplify and automate discovering the hidden semantics between multimedia resources. Other research works have focused on extracting the semantic relations of multimedia document by using the external textual description related to the document. In order to extract the semantic relationships between concepts in the image, the authors in [6] propose a cross-modal method which is based on the tagged images on Flickr. It considers the features extracted as a latent space to calculate a representative distribution of latent variables for each concept. In the same context, several works are interested in extracting the semantic relations of the textual document. In order to detect the semantic relations between Wikipedia entities, Xu, M in [10] used a learning model to identify features from the entity in Wikipedia infoboxes. This process facilitates the creation of RDF links between DBpedias instances. Again by using the textual document, [5] proposed a method based on the entity linking in a web text. The suggested method determines a collective entity linking based on a graph which can model the global interdependence between entities. Then, an inference algorithm is used. Despite the efficiency proven by these studies, some lacks are highlighted. Indeed, these works extract methods which have identified the relations either between the annotated concepts or between the predefined relationships. In fact, the annotated concepts do not consider the semantic conveyed in the content. According to the feature or description used in the process of generation relationship, we notice a lack of some predefined extracted descriptions and other relations are inaccurate. Consequently, the more the description used is semantically and related to the content, the more the relationships generated are pertinent. In this work, we focus on the cinematic audiovisual document by using a set of semiotic description in order to extract semiotic relationships between resources. The goal behind this work is to ameliorate the movie retrieval. In the following section, we will define the semiotic description and presents the different semiotic audiovisual relationship proposed in the literature.

2.2 Semiotic Relationships

The semiotics is defined as a process which allows to study the meanings of the content. In fact, the semiotic audiovisual is the process of the significance of audiovisual document in order to have a bridge with different semantics and meaning conveyed which is contained in the audiovisual document. In this context, [8] P.Stockinger was interested in his work by extracting the semiotic descriptions of audiovisual documents. Indeed, he defines different thematic descriptions. The semiotic description that we have used in our work are: 'Genre'(represents the genres related to the audiovisual segmentsuch as drama, Action, adventure'...), 'Dominant Theme'(represents the topic related in each audiovisual segment), 'theme of discourse'(represents all extracted themes in each segment), 'taxeme'(represents the pertinent 'theme of discourse') and 'specified theme'(represents the keywords related to the segment. A combination process between the keywords extracted form the synopsis [2] and the adapted LDA model [3]). Besides, P. Stockinger defines in his study a set of semiotic relationships such as: 'Taxonomic'(A taxonomic relationships allow giving an explicit description of the relative similarity between filmic situations. This relationship can be classified into two types such as: *Generalization/Specification* between filmic situations or events and *similarity/contrast* between them), 'Meryonimic'(A meryonimic relationship explains the various components or constituents that link filmic situations. We distinguish different types of relationships such as: composition, dependence, membership and location) and 'Narrative'(A narrative relationship allows defining the different thematic links between filmic situations such as: Chronology (eg. Situation1 follows Situation1), Scalability of events or action, rhetoric, cooperative, polemic).

3   Process of Semiotic Relationship Extraction

Faced with the growing mass of audiovisual document, and the user's dissatisfaction with information retrieval in the movie document content, we find it essential to overcome these difficulties. Indeed, we focus in our work on solutions to properly operate these huge quantities produced. In fact, the description extraction of an audiovisual movie from the content is becoming an optimal solution accordingly. In this context, such an approach takes as input semiotic descriptions extracted from the content via an automatic determination of semiotic relationships between films and segments of films. Hence, we propose a set of semantic rules that allow identifying the possible links in order to eventually consider other processes of structuring and search for information. The main goal of this analysis is designed to bridge the gap between, on one hand, the segment in the same movie and on the other hand between movies. Figure 1 illustrates the general principle of our proposed process. It consists of three phases: (i) semiotic description extraction, (ii) semiotic rules extraction and (iii) inference reasoning. This process follows the different steps to follow in order to define our grammar (Fig. 1).

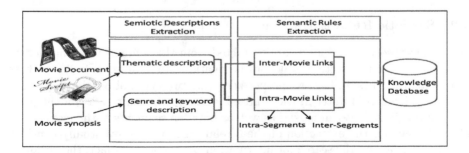

Fig. 1. The steps of the semiotic relationships process

3.1   Semiotic Description Extraction

As our work focuses on the movie audiovisual documents and in order to obtain semiotic relationships that improve the retrieval process, we have been inspired by our description proposed by [2,3]. In [2], we proposed a method based on the extraction of the genre and keyword from the content of the movie. This method based mainly on the pre-production document such as synopsis of the movie by incorporating both semantic and semiotic analysis. From the synopsis, a statistical process emphasizes the adaptation of the TF*IDF that allows identifying the pertinence of each term in the synopsis. This process is proceeded by a step of removing non-significant terms. So, a semantic process is highlighted. This process based on the use of the lexical and semantic database wordnet and the combination of a set of semantic measures. By combining the pre-production documents namely the scenario of the movie and the post-production documents such as the superposed text in the image [4], we carried out in [3] a thematic process in order to extract the topic and the theme related to the movie content. The principle of this process consists in extracting a semiotic description proposed by [9] such as: **genre**(the genre of the movie), **D_TH** (the Dominant Theme of the movie), **TH_DISC**(the theme of discourse of the movie), **TXM**(the taxeme of the movie) and **SPEC_TH** (the specified theme of the movie). For this reason, a manual segmentation process is set up. The segmentation provides a set of audiovisual segments following criterion thematic areas. The results of these phases were taken as a starting point to our proposed method to extract the semantic relationships between the segments of a movie (or between the films). The structure of semiotic descriptions obtained represents basic support to create a grammar. As defined, a grammar is composed of a set of terminal(T), non-terminal (NT) symbols and a set of production rules. In our work, we define that: D_TH $\supset$ {th_disc1, th_disc2,....}, TXM $\in$ {th_disc1, th_disc2,....}, {TH_DISC} $\Rightarrow$ {spec_th1, spec_th2, spec_th3.....}

From these definition, we can define the following elements of our grammar: T={ {th_disc_i},{spec_th_j}, {txm_k}, {genre_l} }}, NT = {D_TH, TH_DISC, TXM, SPEC_TH, GENRE}. Based on these elements, the following section presents the different production rules.

3.2   Semiotic Rules Extraction

Once the semiotic descriptions are extracted from the content of the movie, we are interested in this section to define the relationship in order to build the first level semantic rules. In fact, we propose defining two types of relationships such as 'Intra-Movie' and 'Inter-Movie'. Indeed, these relationships allow identifying the possible links between movies as well as between segments of the movie. The objective behind the extraction of these relationships is to eventually consider the structuring and the search for information. As a matter of fact, the obtained segments described separately are not always possible to have a global vision on the sequence of events or analysis of history conveyed in a film. In this respect, the consideration of relations between segments will allow to solve this problem. Starting from the semiotic relations proposed by Peter Stockinger [8], we present in this section, an overview of our semiotic relations giving an overview of our solution. Mainly, three types of relations were taken such as Taxonomic, Mcryonimic and Narrative. Based on the different semiotic description (Desc) extracted Di, several semantic rules can be formulated and extract relationships. Consequently, each semantic rule can be expressed through the following rules:

$\forall S \in \{Segment\}, \forall D \in \{Desc\}Di\ Relation\ Dj \implies semantic\ relation(Si, Sj)$

$\forall M \in \{Movie\}, \forall D \in \{Desc\}Di\ Relation\ Dj \implies semantic\ relation(Mi, Mj).$

Intra-movie Relationship Generation. As mentioned previously the result of our semiotic description process is a set of segments and each segment is represented by a 'Dominant Theme', a set of 'theme of discourse', a 'taxeme' and a set of 'specified theme'. In this regard, based on the obtained structure, we propose extracting two types of relations such as an intra-segment relation and an inter-segment relation.

– The intra-segment relation: In order to ensure the generic overview of all filmic situations in a movie segment, we propose extracting five types of semantic relations in the same segment. Indeed, from the description mentioned previously, we propose defining a set of semantics through the concept of predicate logics. A result of the thematic segmentation, each segment S presents a Dominant Theme and a set of keywords called a specified theme. We notice some highlighted relationships such as the Dominant Theme which is composed by a set of themes of discourse. Since a taxeme represents the pertinent theme of discourse, we notice Membership relations between them. In addition to these result, we consider a dependence relation between the taxeme and the specified theme. In the following, we explain the different proposed rules (Ri) production such as: $\forall S \in \{Segment\}$
  R1: $D\_TH(S) \wedge TH\_DISC(S) \implies Composition(D\_TH(S), TH\_DISC(S))$
  $\equiv \{TH\_DISC(S)\}\ Talk\_about\ D\_TH(S)$
  R2: $TXM(S) \wedge TH\_DISC(S) \implies Membership(TXM(S), TH\_DISC(S))$
  $\equiv TXM(S)\ Belong\_to\ \{TH\_DISC(S)\}$
  R3: $TH\_DISC(S) \wedge SPEC\_TH(S) \implies Composition(TH\_DISC(S),$
  $SPEC\_TH(S)) \equiv TH\_DISC(S)\ Related\_to\ \{SPEC\_TH(S)\}$
  R4: $SPEC\_TH(S) \wedge TXM(S) \implies Dependence(SPEC\_TH(S), TXM(S))$
  Following these extracted rules, we propose production rules of our defined grammar in the intra-segments relations.
  P = { S ⇒ D_TH | TH_DISC | TXM
  D_TH ⇒ TH_DISC | SPEC_TH
  TXM ⇒ TH_DISC
  TH_DISC ⇒ SPEC_TH }
– The inter-segments relation:
  With the growing of the multimedia data volume, the challenges in information retrieval is related to the discrepancy between the user's query and the

information contained in the multimedia document especially the movie document. In this context, based on a set of varied semiotic descriptions extracted, we aim to define a semantic relationship between movie segments. Indeed, the objective behind the extraction of these relationships is probably a better retrieval of the semantic information. In fact, the structure of the description obtained allows the creation of different links between audiovisual resources. Consequently, we propose, in this section, defining a semantic rule between the different thematic segments extracted. Furthermore, the description of separate segments is not always possible to have a global vision on the sequence of events or analysis of history conveyed in a film. In the following, we explain the different proposed rules such as:

R5: $\forall Si, Sj \in \{Seg\}\ \exists (D\_TH(Si) \equiv D\_TH(Sj)) \wedge (Location(Si) < Location(Sj)) \implies Chronology(Si, Sj) \equiv Sj\ follows\ Si$

R6: $\forall Si, Sj, Sk \in \{Seg\}\ \exists (D\_TH(Si) \equiv D\_TH(Sj)) \wedge (D\_TH(Sj) \equiv D\_TH(Sk)) \wedge (Location(Si) < Location(Sj)) \wedge (Location(Sj) < Location(Sk))$
$\implies Scalability(Si, Sj, Sk) \equiv Si\ begin\ Sj\ medium\ Sk\ end$

R7: $\forall Si, Sj \in \{Seg\}\ \exists (D\_TH(Si) \equiv D\_TH(Sj)) \implies Similarity(Si, Sj) \equiv Si\ similarto\ Sj$

R8: $\forall Si, Sj \in \{Seg\}\ \exists (TH\_DISC(Si) \cap TH\_DISC(Sj)) \implies Dependence(Si, Sj) \equiv Si\ correlatedto\ Sj$

R9: $\forall Si, Sj \in \{Seg\}\ \exists (TXM(Si) \equiv TXM(Sj) \implies Membership(Si, Sj) \equiv Si\ associatedto\ Sj$

R10: $\forall Si, Sj \in \{Seg\}\ \exists (SPEC\_TH(Si) \cap SPEC\_TH(Sj)) \implies Composition(Si, Sj) \equiv Si\ concordto\ Sj$

Following these extracted rules, we propose the production rule of our proposed grammar in the inter-segments relation.

P = {S ⇒ Si follows Sj | Si similarto Sj | Si correlatedto Sj | Scalability(Si, Sj, Sk) | Si associatedto Sj | Si concordto Sj

follows ⇒ E1 ∧ E2	| E3 ⇒ Desc2 ≡ Desc5	|Desc6 ⇒ TXM (Si))
similarto ⇒ E1	| E4 ⇒ Locat2 < Locat3	| Desc7 ⇒ TXM (Sj)
correlatedto ⇒ Desc3 ∩Desc4	| E5 ⇒ Desc6 ≡ Desc7	| Desc8 ⇒ SPEC_TH(Si)
Scalability ⇒ E1 ∧ E3 ∧ E2∧ E4	| Desc1 ⇒ D_TH (Si)	| Desc9 ⇒ SPEC_TH(Sj)
associatedto ⇒ E5	| Desc2 ⇒ D_TH (Sj)	| Locat1 ⇒ time(Si)
concordto ⇒ Desc8 ∩Desc9	| Desc3 ⇒ TH_DISC(Si)	| Locat2 ⇒ time(Sj)
E1 ⇒ Desc1 ≡ Desc2	| Desc4 ⇒ TH_DISC(Sj)	| Locat3 ⇒ time(Sk)}
E2 ⇒ Locat1 < Locat2	| Desc5 ⇒ D_TH (Sk) }	

Inter-movie Relationship Generation: In order to assist the user in finding his individual preferences according to the movie content, we propose extracting different relationships between them. In this regard, we focus on the use not only of the genre of the movie but also the different semiotic and thematic descriptions used in the previous section. As a result for the rule defined, we notice that a relationship exists between these rules and the rule proposed in the intra-movie.

R11 $\forall Mi, Mj \in \{Movie\}\ \exists (Genre(Mi) \cap Genre(Mj)) \wedge \exists (SPEC\_TH(Mi) \cap SPEC\_TH(Mj)) \implies Dependence(Mi, Mj) \equiv Mi\ homonymto\ Mj$

R12 $\forall Mi, Mj \in \{Movie\}\ \exists(Genre(Mi) \cap Genre(Mj)) \wedge \exists(D\_TH(Mi) \cap D\_TH(Mj)) \Longrightarrow Membership(Mi, Mj) \equiv Mi\ fellows\ Mj$

R13 $\forall Mi, Mj \in \{Movie\}\ \exists(Genre(Mi) \cap Genre(Mj)) \wedge \exists(TH\_DISC(Mi) \cap TH\_DISC(Mj)) \Longrightarrow Composition(Mi, Mj) \equiv Mi\ correlatedto\ Mj$

Following these extracted rules, we propose the production rule of our proposed grammar in the inter-movies relation.

P = { S $\Rightarrow$ Mi homonymto Mj | Mi fellows Mj | Mi correlatedto Mj

    homonymto $\Rightarrow$ E6 $\wedge$ E7 | E8 $\Rightarrow$ Desc12 $\cap$ Desc13 | Desc13 $\Rightarrow$ D_TH (Mj)

    fellows $\Rightarrow$ E6 $\wedge$ E8 | E9 $\Rightarrow$ Desc14 $\cap$ Desc15 | Desc14 $\Rightarrow$ TH_DISC(Mi)

    correlatedto $\Rightarrow$ E6 $\wedge$ E9 | Desc10 $\Rightarrow$ Genre(Mi) | Desc15 $\Rightarrow$ TH_DISC (Mj)}

    E6 $\Rightarrow$ Desc10 $\cap$ Desc11 | Desc11 $\Rightarrow$ Genre(Mj)

    E7 $\Rightarrow$ Desc8 $\cap$ Desc9 | Desc12 $\Rightarrow$ D_TH(Mi)

Faced with the obtained production rule, we propose in the following section an inference reasoning. The main goal is, roughly, to underpin the process of deduction and to prove the hidden semantic.

3.3 Inference Reasoning

Once the preliminary relations referring to the movie estimated relationship are obtained, an inference process will be held. The purpose of this inference process is to move from one semantic level to another until the most generic level. In fact, the objective behind the semantic level is to bring the semantic gap between the user's preferences and the semantic extracted from the movie. Therefore, we relied on the previous set of rules to generate new semantics under a logical formalism. Our process of inference comprises the following inferences rules deducted from the proposed rule of intra-segment and inter-segment. In fact, this process follows the cascading effect to obtain the inference rules.

Intra-segment Inference: Based on the semantic relationship extracted from R1, R2 and 3, we propose two inferences rules such as R'1 and R'2.

Rule'1 $\forall D\_TH, TH\_DISCandSPEC\_TH \in Description\ \forall R1, R3 \in Realtion$
$\exists R'1: D\_TH(S) \wedge SPEC\_TH(S) \Longrightarrow Composition\ D\_TH(S), (SPEC\_TH(S))$
$\equiv D\_TH(S) relatedtoSPEC\_TH(S)$

Rule'2 $\forall D\_TH, TH\_DISCandTXM \in Description\ \forall R1, R2 \in Realtion$
$\exists R'2: D\_TH(S)) \wedge TXM(S) \Longrightarrow Membership\ (D\_TH(S), SPEC\_TH(S))$

The identification of these rules allows obtaining different semantics between the theme related to each segment. In fact, this semantic level is the first step to the creation of a knowledge base for a movie document. In the following section we will present an example which illustrates this.

Inter-segment Inference: Based on the semantic relationship extracted from R1, R7, R5, R6, R8, R9, R'1, R'2, R'3 and R'8, we propose five inferences rules such as:

Rule'3 $\forall D\_THandTH\_DISC \in Description \forall R7, R1 \in Realtion$
$\forall Si, Sj \in Seg \exists R'3: Composition(D\_TH(Si), (TH\_DISC(si) \bigcup TH\_DISC$

(sj))) $\equiv D\_TH(Si)talkabout(TH\_DISC(si)\bigcup TH\_DISC(sj))$

Rule′4 $\forall D\_THandSPEC\_TH \in Description\forall R'3, R'1 \in Realtion$

$\forall Si, Sj \in Seg\exists R'4 : Composition(D\_TH(Si), (SPEC\_TH(si)\bigcup SPEC\_TH$

(sj))) $\equiv D\_TH(Si) \; relatedto(SPEC\_TH(si)\bigcup SPEC\_TH(sj))$

Rule′5 $\forall D\_TH, TH\_DISCandTXM \in Description\forall R'2, R'3 \in Realtion$

$\forall Si, Sj \in Seg\exists R'5 : Membership(D\_TH(Si), (TXM(si)\bigcup TXM(sj)) \equiv$

$D\_TH(Si)belongto(TXM(si)\bigcup TXM(sj))$

Rule′6 $\forall D\_TH, TH\_DISCandTXM \in Description\forall R1, R5 \in Realtion$

$\forall Si, Sj \in Seg\exists R'6 : TH\_DISC(Si)followsTH\_DISC(Sj))$

Rule′7 $\forall D\_TH, TH\_DISCandTXM \in Description\forall R1, R6 \in Realtion$

$\forall Si, Sj \; andSk \in Seg\exists R'7 : TH\_DISC(Si)begin, TH\_DISC(Sj)medium,$

$TH\_DISC(Sk)end$

4 Case Study

We will devote this section to presenting a case study results in order to substantiate our proposed process. Such a process is based on the generation of the semiotic relationship of the movie audiovisual document. The objective behind the case study is to demonstrate the importance of our proposed rule to extract the hidden semantics between not only the thematic segment in the movie but also between two movies. In order to carry out our experiments, we have used the representative test database movies from the online database "Imdb". This base is a collection of movies with different cinematic genres. In this context, we propose studying firstly our semiotic relationship extraction process and therafter we discuss our obtained results. For this reason, two series of studies are performed. The first series of study is designed to study the semiotic relationship generated. Finally, in the second series we will concentrate our effort to show the different inference reasoning rules.

4.1 Semiotic Rule Study

In this section, there are two cases study where we present successively the different relationships extracted from a movie and between two movies. By taking "Toy Story" as an example, we notice different links between segments in the movie such as "Follow", "Similar to", "concord to", "correlated to" and "associated with". The following Fig. 2 shows these extracted semiotic relations. We notice that many relationships have as type "concord to" and this means that there is a composition relation between segments. We find that the obtained results are logical because there is a high probability to find common specified themes between segments. This first analysis level allows facilitating the user's query. In fact, in order to find the segment that has the same dominant theme for example, we just need to see the relationships that are connected to this topic as "follow" and "similar to". The semiotic relationships extracted from the intra-movie focus on extracting semantics that follow a bottom-up gait for the construction of a knowledge base related to the movie domain. We present in the Fig. 3 an extract of the results of the movie test "Toy story". Once the relationships intra and inter segments are obtained, an inference reasoning must be set up in order to extract the hidden semantics between resources. Based on the different inference rules we obtain these new relationships:

R1́: '*Social_issue*' related to {toy, andy, play, favorite, woody, prepare, birthday, party}

'*initiative*' related to {toy, woody, declare, andy, prepare, birthday, party, get, decide, soldier, mission, new, present}

'*Caring*' related to {toy, soldier, start, act, declare, friend, room, new, andy}

'*freshstart*' related to {buzz, woody, prepare, set, order, toy, realize, fall, chance}

As is shown in the Fig. 3 and based on the inference rule, we can observe the cascading effect to extract hidden semantic. For the example used in this case study we obtain the following relationship:

R3́ : '*Social_issue*' talk about {Attack, Chosen, Dance, Doctor, Entitlement, Indulgence, Legend, Memorial, Remembrance, Teasing} ∪ {Dance, Identity, Memorial, Remembrance}

R4́ : '*Social_issue*' related to {toy, andy, play favorite, woody prepare, birthday, party} ∪ {toy, woody, declare, andy, prepare, birthday, party, get, decide, soldier, mission, new, present}

R5́ : '*Social_issue*' belong to {Attack, dance}

R6́ : {Attack, Chosen, Dance, Doctor, Entitlement, Indulgence, Legend, Memorial, Remembrance, Teasing, belong} follows {Dance, Identity, Memorial, Remembrance}. In order to study the relationship between movies, we have selected two movies that have the same genre as "Toy story" and "The Lion King": {'Animation', 'Comedy'} and a common specified theme: {fight, rivalry, child, ...}. For this reason, a confusion matrix is used. This matrix presents the number of similar specified themes between segments of movies. Based on the relationship R11, we notice that "Toy story" is homonym to "The Lion King".

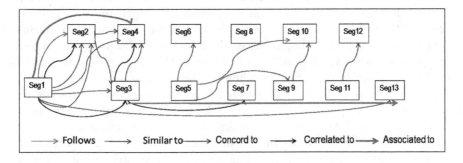

Fig. 2. Result of semiotic relationships between segments in the movie

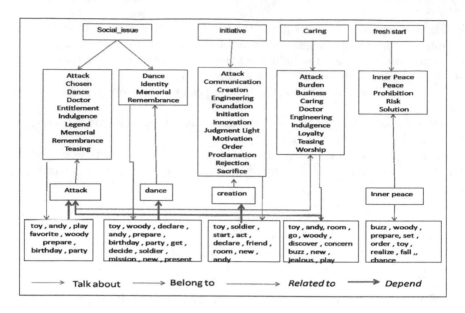

Fig. 3. Result of semiotic relationships intra segments in the movie

4.2 Discussion

In this section, the above-mentioned study proves the relevance of our proposed rules. Indeed, our process allows obtaining a knowledge set (see Fig. 3) existing between the pertinent concepts of the movie (theme of discourse, dominant theme,...). The knowledge is identified from the extraction of the relationships between segments as well as from the movie (for instance "Toy story Movie") presented in the Fig. 2. Otherwise, the set of knowledge is extracted from both the initial proposed rules and from the obtained rules through the inference reasoning. Relying on the proposed knowledge extraction allows to effectively extract the hidden semantic. In the previously subsection (Case study), we presented that the movies "Toy story" and "The Lion King" are homonym. Initially, the homonymy is defined in linguistics as the relationship between several forms having a same indicating and a different meanings. In this regard, we inspired from this definition in our work to define the homonymy between two movies. In fact, two movies are homonym if they have the same genre and a common keyword set. For this purpose, we identified the fact that the movies "Toy story" and "The Lion King" are considered as dependent. An important issue emerges in this context: what is the usefulness of this dependency relationship? This interesting question answers the user requirement in the interrogation process.

5   Conclusion and Future Works

This paper proposes a semantic rule-based approach using a semiotic description to represent movie audiovisual semantics. The proposed approach performs semiotic descriptions extraction, semantic rules identification and inference rules generation simultaneously. Indeed, these steps help to increase the precision to the querying process and enhances the semantic gap between the user's query and the movie audiovisual content. In fact, we focus on the first phase of the extraction of semiotic descriptions from the content across the different types of analysis. Then, from the obtained level of structure between these descriptions, we have built a set of semantic rules both intra and inter movie. Thereafter, once the sematic relationship between description and even between segments from the movie obtained, we applied an inference reasoning in order to generate the hidden semantics. In this context, the fundamental issue that arises consists in the way to achieve the automation in order to learn these rules. To respond this issue, we propose the pertinent use of the supervised classification process based on the thematic segments and especially on the semiotic relationships defined. As a first perspective, we consider the validation and automation of the generation of the proposed semiotic rules. As a second perspective, we consider the creation of RDF triples to build a network of semantic link between the different relationships extracted. By referring to the network of semantic link, a semantic retrieval process of search for information and interrogation of film audiovisual resources should be highlighted.

References

1. Ayadi, Y., Amous, I., Gargouri, M., Gargouri, F.: Semantic rules classification for images annotation. In: 2011 11th International Conference on Hybrid Intelligent Systems (HIS), pp. 72–77. IEEE (2011)
2. Fourati, M., Jedidi, A., Gargouri, F.: Automatic audiovisual documents genre description. In: 6th International Joint Conference on Knowledge Discovery and Information Retrieval (KDIR 2014), Rome, Italy, pp. 21–24 (2014)
3. Fourati, M., Jedidi, A., Gargouri, F.: Topic and thematic description for movies documents. In: Arik, S., Huang, T., Lai, W.K., Liu, Q. (eds.) ICONIP 2015. LNCS, vol. 9492, pp. 453–462. Springer, Heidelberg (2015). doi:10.1007/978-3-319-26561-2_54
4. Fourati, M., Jedidi, A., Hassin, H.B., Gargouri, F.: Towards fusion of textual and visual modalities for describing audiovisual documents. Int. J. Multimedia Data Eng. Manag. (IJMDEM) 6(2), 52–70 (2015)
5. Han, X., Sun, L., Zhao, J.: Collective entity linking in web text: a graph-based method. In: Proceedings of the 34th International ACM SIGIR Conference on Research and Development in Information Retrieval, pp. 765–774. ACM (2011)
6. Katsurai, M., Ogawa, T., Haseyama, M.: A cross-modal approach for extracting semantic relationships between concepts using tagged images. IEEE Trans. Multimedia 16(4), 1059–1074 (2014)
7. Kharrat, M., Jedidi, A., Gargouri, F.: Defining semantic relationships to capitalize content of multimedia resources. In: Amine, A., Bellatreche, L., Elberrichi, Z., Neuhold, E.J., Wrembel, R. (eds.) CIIA 2015. IAICT, vol. 456, pp. 367–378. Springer, Heidelberg (2015). doi:10.1007/978-3-319-19578-0_30
8. Stockinger, P.: Les archives audiovisuelles: description, indexation et publication. Lavoisier (2011)

9. Stockinger, P.: Audiovisual Archives: Digital Text and Discourse Analysis. Wiley, New York (2013)
10. Xu, M., Wang, Z., Bie, R., Li, J., Zheng, C., Ke, W., Zhou, M.: Discovering missing semantic relations between entities in Wikipedia. In: Alani, H., et al. (eds.) ISWC 2013. LNCS, vol. 8218, pp. 673–686. Springer, Heidelberg (2013). doi:10.1007/978-3-642-41335-3_42

Knowledge Engineering

Generic Model for Adaptable Caching in the Knowledge-Oriented Web Engineering

Jiří Štěpánek and Vladimír Bureš[✉]

Faculty of Informatics and Management, University of Hradec Králové,
Rokitanského 62, Hradec Králové, Czech Republic
{jiri.stepanek,vladimir.bures}@uhk.cz

Abstract. Web applications represent a commonly used type of technology in many domains. They are applied for various knowledge-intensive purposes ranging from the external marketing activities to the internal content presentation. Therefore, the speed in which they can be utilised represents quite significant variable of their evaluation. Based on the domain knowledge expressed in a form of generic model this paper presents a possibility of the adaptable cache utilisation. The model is based on existing technologies and its aim is to significantly reduce data load and time demands associated with the use of web pages. The model is implemented in practice with various fragments of codes provided in text. This implementation is consequently tested by means of the three most common types of web pages. Acquired results prove significant reduction of data load in comparison with the traditional request-response model. The paper also outlines directions for further research in this area.

Keywords: Knowledge-oriented web engineering · Adaptability · Model · Asynchronous download · Cache · Local storage

1 Introduction

Knowledge-based software engineering is currently associated with various research directions. Since web applications belong to one of the wide spread information technologies commonly implemented in various fields [1, 2], the attention to knowledge-oriented activities has recently spread to this domain as well. Very often, architecture of web applications enables them to be used quite smoothly in the business settings [3], for any web application can be used almost immediately, without necessity to install drivers, middleware, or any specific tools. The application communicates with a user by means of HTML (hyper-text mark-up language) pages that represents its GUI (Graphical User Interface). Communication between the thin client and the application is based on dispatching of requests, waiting for and processing of corresponding answer. One web page usually comprises several clients, which causes its quite significant overload from time to time, especially when several users run demanding operations at the same moment [4]. Any help that would help to reduce this overload is of high significance. Thus, this manuscript deals with quite specific technical research problem from knowledge-oriented web engineering domain. The paper has the structure as follows. Next

© Springer International Publishing AG 2016
F. Lehner and N. Fteimi (Eds.): KSEM 2016, LNAI 9983, pp. 251–261, 2016.
DOI: 10.1007/978-3-319-47650-6_20

section briefly formulates the problem and depicts applied methodology. It also introduces concept of adaptable cache. The third section deals with implementation of the proposed solution. The fourth section focuses on testing and the fifth section suggests further research steps. The last section concludes the paper.

2   Model Development

This section focuses on selected aspects of model development. Firstly, the main problem is depicted. Then, the applied methodology is described.

2.1   Problem Formulation

In the traditional conception the operation request-response is represented by the request processing, execution of the application logic and creation of the final output (very often the HTML page together with active implementation of templates and similar supportive mechanisms). All page related data are transferred while many page elements remain unchanged. Therefore, AJAX operations are considered to be beneficial for reduction of the data load [5]. They are able to reduce data load and computational complexity to reasonable values. When AJAX operations are applied, requests are sent asynchronously. Hence, only necessary data are transferred and completed by already downloaded data from the past with the help of the client script. However, AJAX-based approach does not cope with the issue associated with requesting the same data by the same client in relatively short period of time. Similar situations are usually solved by application of a certain type of cache [6]. This architecture is connected with solutions located at the server level. Thus, existing response to URL (Uniform Resource Locator) is provided to any client directly from cache. However, this concept does not avoid server to process repeating requests. Therefore, the main research question is formulated as: Is there any way in which existing issues with traditional request-response processing operation can be overcome by knowledge-oriented development of web applications? The main objective is to develop and present a generic model, in which relevant knowledge associated with a potential to improve (or better to say, accelerate) interaction between a user and the application under certain circumstances is embedded. Several studies dealing with the aforementioned issue have already been published. The concept of adaptable cache is frequently used in this context. Early in the mid of 2000's, D'Orazio et al. [7] suggested Adaptable Cache Service, a framework which allows building adaptable cache services. The framework presents a generic cache definition and provides a description of models implemented. Furthermore, Sato et al. [8] deal with way-adaptable caches, which adjust the number of cache ways available to a running application based on assessment of its working set size. Xu et al. [9] apply cache-related concepts in practice and design a real-time mass data exchange system for Internet of Things. This system is among others based on deployment of asynchronous socket to allow connection from mass TCP clients, or three-level data storage model which handle concurrent access and modification to the cache space by multiple threads.

2.2   Research Methodology

The answer to the aforementioned research question is based on solution that takes advantage of knowledge-oriented engineering of web applications and the existing technology Local Storage which is available for local caching. The main idea is that it is possible to develop generic model with embedded knowledge associated with possibility to avoid multiple processing of the same requests by server, if responses are already available. The proposed model is implemented in the ASP.NET MVC application. Its behaviour is consequently tested. Therefore, specific use case simulating real run of the web application is developed. This simulation is used for model verification and measuring of selected behavioural attributes.

The model of adaptable cache is based on the WebStorage technology, which represents an integral part of the HTML5 standard. This technology enables web pages to store data in a local storage and access them later on demand [10]. In connection with this technology, web browser usually provide a repository in size of 5 MB. This technology can be used as a cache available for the web application. It is not necessary to limit it merely to storage of previously displayed pages (data). It can be also successfully used for prediction, which data will be very likely used by the users during the next steps and retrieve them in advance.

Developed model is grounded in the knowledge of which pages (generally data) should be cached for the downloaded page. For instance, it seems reasonable to cache the list with news or company contacts in relation to download of the first introductory page. Significant is that this type of knowledge (search patterns, association) can be empirically determined [11], estimated or derived from user profiles. In case of the latest, it is necessary to carry on the identification procedure, which is however pretty usual in case of business web applications [12].

The adaptable cache model runs from the first request of the browser. The web application not only processes the request, but it also finds out (based on preferences that can created on various principles) which other pages (data) are suitable for caching as well. This information is sent together with the original page data. Downloaded page needs to decide whether caching is suitable or not.

3   Model Implementation

This model needs to be implemented in both the server side, and the client side of the application. The reference application displaying fictive contracts signed in a given day was created. This application should demonstrate model-related benefits and behaviour based on simulation of real use case. The application applies standard type of layout, where the content is encompassed by controls.

The client side of the model has the following tasks:

• Decision whether it is suitable to download further content on the background.
• Execution of the asynchronous content download.
• Saving of the downloaded content in the local storage.

- Capture of user's requests and their processing with the help of the local repository instead of execution of the traditional request processing.
- Decision whether data in cache are up-to-date.

Main tasks and their implementations are outlined further in this section.

3.1 Decision to Use Adaptable Cache

The speed of connection between the user and the web application represents the main decision criterion. However, this cannot be measured precisely. The rough estimation of the connection speed (i.e. time required for the page download) is the only alternative that can be used in this case. It can be performed by subtracting two time stamps. While the first one is created at the beginning of the download of the HTML document, the second one is created at the time of event which is produced as soon as the whole document content is downloaded. Principally, one time stamp is created and one variable consequently used as an indicator, that the speed is measured, is set. Speed measurement can be performed only once due to potential download of data from cache, not original URL. Once the event connected with the document download is created, it is necessary to perform two fundamental operations. First, it is important to find out time need for the page download. Based on the acquired result the decision whether cache should be used might be made. The following code fragment determines the time needed for the page download.

```
var loadTime;
    if (pageLoadAlreadyMeasured) {
        loadTime = Date.now() - loadingStart;
        console.log("Page load took " + loadTime + "ms");
        pageLoadAlreadyMeasured = true;
    }
```

Second, it is necessary to decide which data will be cached and with which pages. This decision has to be made at the web application level. Specific implementation techniques may vary and represent another knowledge-related aspect of the model implementation. For instance, it is possible to download a list of other pages (data) suitable for caching simultaneously with the every page. Alternatively, it is possible to mark links on the page, which should be incorporated in the mechanism of adaptable caching. Data attributes available in the HTML5 standard can be used for this purpose (see the code fragment below). Similar marking of links enables quite universal method for incorporation of HTML components into the process of adaptable caching.

```
<a href="@Url.Action("Index", "Home", new {year = 2016,
week = WeekNum - 1})" data-
adaptabilecache="1">Previous</a>
```

3.2 Asynchronous Download and Storage of Data

The code presented below executes a process of each hypertext link marked by particular attribute (data-adaptabilecache). This process finds out if the source with specific URL already exists in the cache. If it does not or if it is not actual, asynchronous request for this sources is performed. Response data together with the time stamp are stored in the local repository (Session Storage) as soon as the response data arrive. API available for manipulation with the repository enables to store chains in a form of key-value only. In order to store the whole object with several attributes, it is necessary to use any of the existing serial techniques, such as the JSON (JavaScript Object Notation) format. It is possible to get the serial chain for the JSON object with the help of its method stringify. It can be consequently stored by means of standard API. URL source is used as a key in this case, which represents the simplest and the most straightforward solution.

```javascript
// store url for each adaptabile link
$("[data-adaptabilecache='1']").each(function (index,
element) {
            var href = element.href;
            var startCacheTime = Date.now();
            // look if this page already exist in storage
            if (sessionStorage.getItem(href) == null ||
Date.now() -
JSON.parse(sessionStorage.getItem(href)).dateSaved >
ac_data_age ) {
                // get data asynchronously
                $.get(href, function (data, success) {
                    console.log("cached url: " + this.url
+ " / " + data.length + " bytes" + " / in " + (Date.now()
- startCacheTime)   + " ms");
                    // prepare cached data into storage
                    var savedPage = { dateSaved:
Date.now(), pageData: data }
                    var savedPageString =
JSON.stringify(savedPage);
                    // save data into session storage
                    sessionStorage.setItem(this.url,
savedPageString);
                });
            }
});
```

3.3 Capture of Requests

For the successful model implementation the mechanism for asynchronous data download in the local repository needs to be supported by other mechanisms that focuses on request processing. Sources from links that are specifically marked by particular attribute are downloaded in the presented model. In such case, it is required to modify implicit behaviour of these links, as depicted in the following code.

```
// bind click action on adaptabile cache links
        $("[data-adaptabilecache='1']").click(function
(e) {
            e.preventDefault();

            var link = e.target.href;
            // search for data in session storage
            var pageData = sessionStorage.getItem(link);
            if (pageData !== null) {
                var pageDataObject =
JSON.parse(pageData);
                if (Date.now() - pageDataObject.dateSaved
> ac_data_age) {
                    window.Location = link;
                }

$(ac_content_element_id).html(pageDataObject.pageData);
                console.log("Page " + link + " loaded
from cache");
            } else {
                window.Location = link;
            }
        });
```

There is a reaction on the click event prepared of every link. First of all, it prevents implicit behaviour, i.e. redirection to required URL. Further, it is found out whether the required URL has already been stored in cache and if data are up-to-date. Maximal acceptable data age is determined by configuration. If any of this cases takes place, the browser is redirected on the URL. Conversely, data are acquired from cache, de-serialised and located into the designated container (usually the content part of web pages).

3.4 Improvement of the Application Run

Sometimes, it is convenient to slightly modify the architecture in case of asynchronous data download. This change should ensure that data are not represented by a fragment of the HTML code, but that there is another format more convenient for data transfer used. Asynchronous transport of HTML code fragments is quite general and utilisable mechanism that does not require implementation of further supportive tools. The fragment is simply injected with the help of javascript into the page, which makes display of data certain. However, it is appropriate to avoid this general model in some instances due to reduction of the data load. Usage of javascript framework that enables development of user environment such as KnockoutJS or BackboneJS represent one of the usable and already verified methods. Each of these frameworks has its own specifics and their mutual comparison is not aimed in this manuscript. However, they mutually share the approach from the architecture point of view – the server side of the application is adjusted in a way to enable dispatching data in the structured format. Very of the JSON format, which is quite economical from this perspective, is used. Transformation of collection of objects that we want to display to the JSON format is pretty easy in majority of development platforms (see the code fragment below).

```
IList<Contract> contracts = contract-
Dao.GetContractsForDay(workingDateTime);

return Json(contracts, JsonRequestBehavior.AllowGet);
```

The sample code demonstrates how easy the collection of object can be transferred to the JSON format in case of the ASP.NET MVC application. As soon as the data acquired by the asynchronous request are received by the page, the used framework applies them correctly to the page. Apparently, the page needs to be prepared for similar operations. Modification of the classical construct of the HTML table with the help of data-* attributes enables the framework to identify repeating patterns. Moreover, it can serve as basis for settings of required behaviour. The framework automatically bind data with the displayed component and displays data. Mechanism of cache does not store fragments of the HTML code in this case, but JSON objects are used. In case of loading it has to call framework functions. However, implementation of this procedure requires individual evaluation, since it is not the most appropriate course of action in all instances.

4 Testing

The presented model of adaptable cache was tested under the following conditions. Attributes of passing through records of the whole months, where each day has its own page (URL) are investigated in the reference application. Simultaneously, non-zero load of the web application is simulated during passing through records of particular days. Standard experimental evaluation in which the only client is connected to the web application instance and every request is processed immediately would be unrealistic. Simulation of the non-zero load simply delays request processing for randomly generated time in the interval from 0 to 2000 ms. At the same time, controls for switching from the last to the next day are marked by the attribute for adaptable loading into the cache.

Data from neighbouring days are automatically downloaded during the first download of the page (i.e. the previous and the next day). In this way, data are downloaded by every change of the displayed day (if data are not already downloaded). Consequently, the application is tested with the help of another two other types of web pages. The first type with the structured content is similar to web blogs. The second type represents to contact form with fields for contact details. These web pages types can be considered as the most commonly used in the business settings. Thus, testing should verify usability of the adaptable cache mechanism and possibility to apply it outside the tested use case (passing through page records).

4.1 Data Demand

The whole pass through the month (31 pages of records) represents the overall data load over 97 kB with the application of the model. The average amount of loaded data is 3 kB for each day. Every page has 26 kB in average. This means that the passing through all pages would refer to over than 800 kB without implementation of the model (i.e. traditional requesting mechanism applied). Application of asynchronous data download can decrease the data load to one tenth of the initial value. Furthermore, this mechanism prevents redundant data downloading. When the KnockoutJS framework was used and the application for transfer of data in the JSON format was modified, the data load decreased by 28 %, from the average size of the request in the value of 2247 B in case of the HTML format to the value of 1618 B in the JSON format. Naturally, consistency or functionality of the application was not violated. The second type of web pages (blog) is represented by 37 kB of HTML code in average, while the average size of the content is 12 kB. The third type of web pages (contact form) takes 24 kB of the space, while only 2.5 kB represent the content. The same results are acquired also in case of these two types of web pages – asynchronous downloading leads to significant reduction of the data load and another reduction is associated with repeated requesting of the page that can be due to mechanism of adaptable cache downloaded from the local repository.

4.2 Time Demand

The utilisation of the application was simulated during testing, which was implemented as a random delay of the request processing set by the interval from 0 to 2000 ms. The fasted response was acquired in 44 ms during testing, while the slowest one was 2005 ms. These random times were evenly distributed in this interval. During the passing through data when the at least minimum attention is paid to displayed data, the time delay generated by the simulation of the load was completely absorbed. The reason is that during the mover to previous/next page these pages are very often already downloaded. In such case, if the data are not obsolete, there is not any communication taking place with the web application, since data are downloaded from the local storage.

Table 1. Reduction of data demand (source: authors).

Web page type	Avg. data demand tied to the traditional approach	Avg. data demand tied to the asynchronous transfer + adapt. cache	Avg. data demand reduction tied to asynchronous transfer + adapt. cache	Data demand reduction tied to repeated web page download
Records	26	3	88 %	100 %
Blog	37	12	67 %	100 %
Contact info	24	2,5	90 %	100 %

In the next step of the experimental evaluation two links to two subsequent days and two links to two previous days were added. This links were marked by particular attribute to become an integral part of the adaptable cache mechanism. This step results in down-loading of four additional records which takes place asynchronously in the background. Similarly to the previous case, data are stored in the local storage. Since all requests are executed asynchronously, times needed for processing do not have to be added up. Thus, passing through particular pages can be considered as continuous even in case of faster pace of displaying. The following Table 1 summarises results acquired during the testing.

5   Further Research Directions

Situation in which this adaptable model is not appropriate can take place when the user utilises the mobile device and is using mobile data. This type of connection does not have to be quick enough and it is very often associated with the FUP (Fair User Policy) limit [13]. In this case the used would burden his/her mobile data connection more than it would suitable, which would negatively influenced the processing speed. It is tech-nologically impossible to find out the type and speed of the connection in the web appli-cation. However, this functionality absence can be overcome and a rough estimate can be obtained. Hence, the next research agenda can focus on development of similar model and mechanism available for mobile devices.

Furthermore, knowledge embedded in the model is quite general. Thus, despite its simplicity, there are relatively wide implementation possibilities. This model can store not only page records, but also heterogeneous data types. For instance, the model can be connected with analytical information associated with the web application. In his way, the most probable patterns of passing through pages can be determined for empir-ical data. These patterns can be saved in the adaptable cache and this mechanism can be further explored. Furthermore, connection of the model with the real-time run of web application can investigated as well. In this case, data might be stored in the cache any time during the web page life cycle based on demand of the web application.

6   Conclusions

This manuscript contributes to the growth of knowledge-oriented web engineering with a proposal how to solve one specific technological issue. The idea of a model of adaptable cache for web applications is associated with extension of traditional web application engineering by knowledge-oriented principles. This is realised with the help of inclusion of knowledge available in the field into the model and its connection with asynchronous requests and HTML5 local storage technology. The whole mechanism takes advantage of combination of both technologies. It downloads additional data in the background and in the time when user is busy by dealing with the page content. Basically, there are several ways how to control data download [14]. The simplest method is presented in the paper. It is based on marking the hyper-text links with the specific attribute. The mechanism consequently detects such links and download associated URL. Download results are then stored in the cache, which can be used in case of necessity, without a need to communicate with the web application. This mechanism can be also successfully improved with the help of client framework, which enables further reduction of data load. Testing of this model proved its positive impact on the run of the web application. Quite significant reduction of data load is primarily associated with the asynchronous requesting, which responses comprise the required data only. Moreover, when repeated requests for the same URL are taking place, the reduction is absolute, since all data are acquired from the local repository. Utilisation of the model in business applications is apparent, since requests have repetitive nature in this environment. Application of the adaptable cache increases not only the processing speed, but also user comfort.

Acknowledgements. The support of the FIM UHK Specific Research Project "Solution of production, transportation and allocation problems in agent-based models" is gratefully acknowledged.

References

1. Grolinger, K., Capretz, M.A.M., Cunha, A., Tazi, S.: Integration of business process modeling and web services: a survey. SOCA **8**(2), 105–128 (2014)
2. Ndou, V., Del Vecchio, P., Passiante, G., Schina, L.: Web-based services and future business models. In: Information Resources Management Association (ed.) Economics: Concepts, Methodologies, Tools, and Applications, vol. 3, pp. 1564–1576. IGI Global, Hershey (2015)
3. Mustapha, S.M.F.D.S.: CoP sensing framework on web-based environment. Adv. Inf. Know. Proc. **46**, 333–357 (2015)
4. Tian, M., Voigt, T., Naumowicz, T., Ritter, H., Schiller, J.: Performance impact of web services on internet servers. Parallel Distrib. Comput. Syst. **15**(2), 763–768 (2003)
5. McClure, W.B.: Beginning Ajax with ASP.NET. Wiley, Indianapolis (2006)
6. Nakano, Y., Kamiyama, N., Shiomoto, K., Hasegawa, G., Murata, M., Miyahara, H.: Web performance acceleration by caching rendering results. In: 17th Asia-Pacific Network Operations and Management Symposium: Managing a Very Connected World, APNOMS 2015, Busan, China, pp. 244–249 (2015)
7. D'Orazio, L, Jouanot, F., Labbé, C., Roncancio, C.: Building adaptable cache services. In: MGC 2005. ACM International Conference Proceedings Series, 117, 1101502 (2005). doi: 10.1145/1101499.1101502
8. Sato, M., Egawa, R., Takizawa, H., Kobayashi, H.: A majority-based control scheme for way-adaptable caches. In: Keller, R., Kramer, D., Weiss, J.-P. (eds.) Facing the Multicore-Challenge. LNCS, vol. 6310, pp. 16–28. Springer, Heidelberg (2010)
9. Xu, H., Xu, X., Fan, Y., Guo, Y.: Real-time exchange of mass data in the internet of things. Inf. Technol. J. **12**(24), 8081–8087 (2013)
10. Matsumoto, S., Sakurai, K.: Acquisition of evidence of web storage in HTML5 web browsers from memory image. In: 9th Asia Joint Conference on Information Security, AsiaJCIS 2014, 7023253, Wuhan, China, pp. 148–155 (2014)
11. Gooding, P.: Exploring the information behaviour of users of Welsh Newspapers Online through web log analysis. J. Doc. **72**(2), 232–246 (2016)
12. Yu, X., Liao, Q.: User password repetitive patterns analysis and visualization. Inf. Comput. Secur. **24**(1), 93–115 (2016)

13. Chong, S., Skalka, C., Vaughan, J.A.: Self-identifying data for fair use. J. Data Inf. Qual. **5**(3), 11 (2015)
14. Bureš, V., Štěpánek, J.: Structure-oriented algorithms for comparison of web pages resemblance. Int. J. Commun. Antenna Propag. **4**(6), 221–228 (2014)

Automatic Construction of Generalization Hierarchies for Publishing Anonymized Data

Vanessa Ayala-Rivera[✉], Liam Murphy, and Christina Thorpe

Lero@UCD, School of Computer Science, University College Dublin, Dublin, Ireland
vanessa.ayala-rivera@ucdconnect.ie, {liam.murphy,christina.thorpe}@ucd.ie

Abstract. Concept hierarchies are widely used in multiple fields to carry out data analysis. In data privacy, they are known as Value Generalization Hierarchies (VGHs), and are used by generalization algorithms to dictate the data anonymization. Thus, their proper specification is critical to obtain anonymized data of good quality. The creation and evaluation of VGHs require expert knowledge and a significant amount of manual effort, making these tasks highly error-prone and time-consuming. In this paper we present AIKA, a knowledge-based framework to automatically construct and evaluate VGHs for the anonymization of categorical data. AIKA integrates ontologies to objectively create and evaluate VGHs. It also implements a multi-dimensional reward function to tailor the VGH evaluation to different use cases. Our experiments show that AIKA improved the creation of VGHs by generating VGHs of good quality in less time than when manually done. Results also showed how the reward function properly captures the desired VGH properties.

1 Introduction

Microdata (i.e., records about individuals) is a valuable resource for organizations. By exploiting it, companies acquire knowledge to improve or create new business models. For this reason, many organizations are actively collecting and publishing data. However, data must be anonymized before being shared for analysis as it may contain sensitive personal information (e.g., medical conditions) that can bring harm to the involved parties if it is disclosed (e.g., negative publicity, fines, identity theft). Privacy-Preserving Data Publishing (PPDP) offers methods for publishing data without compromising individuals' confidentiality, while trying to retain the data utility for a variety of tasks [5].

k-Anonymity is a fundamental principle to protect privacy in the release of microdata [5,21]. It requires that each record appears at least with k occurrences with respect to the quasi-identifiers (QIDs), i.e., attributes that can be linked to external information and reidentify individuals in anonymized datasets. *Generalization* is the most widely used technique to achieve k-anonymity [21]. It consists in replacing the original QIDs' values with less precise (but semantically consistent) ones, reducing the risk of reidentification (e.g., *"surgeon"* with *"doctor"*). Generalization is usually conducted using concept hierarchies, known as *Value Generalization Hierarchies* (VGHs), which indicate the transformations that an

© Springer International Publishing AG 2016
F. Lehner and N. Fteimi (Eds.): KSEM 2016, LNAI 9983, pp. 262–274, 2016.
DOI: 10.1007/978-3-319-47650-6_21

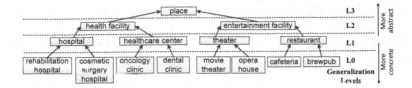

Fig. 1. A VGH for the attribute *place*

attribute can undergo. Figure 1 shows an example of a VGH. The leaves (L0) correspond to the real values of an attribute in the dataset, and the ancestors (L1 to L3) correspond to the candidate values used for generalization.

VGH design is a burdensome process for data publishers (i.e., people involved in the dissemination of data in a safe and useful manner; hereinafter referred as *users*) as one VGH needs to be created per QID, based on the input dataset. If the input values change, VGHs must be modified accordingly, which requires additional manual effort. While it is feasible to create VGHs of small size, the effort considerably increases when larger VGHs are required (e.g., open-ended surveys), or in scenarios where data constantly changes (e.g., streaming data). To tackle this issue, various approaches to generate VGHs automatically have been proposed [8]. However, most of them are designed for numerical attributes, while methods applicable to categorical data remain scarce. Numerical approaches often consist in creating intervals that fit the distribution of the input data. Thus, they are not suitable for categorical data, as its inherent semantics is ignored (a key factor to preserve its meaning). The construction of categorical VGHs presents even more challenges [12]: Disambiguation of the concepts' senses, defining meaningful labels to represent clustered lower level concepts, etc.

Traditionally, categorical VGHs are designed by users based on their own knowledge and experience, as it is commonly assumed that they are fully capable of bringing adequate domain expertise to the construction of VGHs [8]. A key problem of this practice is that the quality of VGHs is evaluated in a subjective and informal way. This issue can lead to misclassifications or inconsistencies which significantly impact the quality of the anonymized data. To mitigate this issue, knowledge engineers often participate in the evaluation process. However, the process may become expensive due to the limited availability of experts and the laborious work involved. Consequently, the design of VGHs is normally a highly error-prone and time-consuming process.

Considering these challenges, our paper has the following contributions:

1. A knowledge-based framework (AIKA) to automatically construct and evaluate categorical VGHs for anonymization, which considers users' preferences.
2. A comprehensive practical evaluation of AIKA, consisting of a prototype and a set of experiments to assess the benefits of AIKA for the creation and evaluation of VGHs for anonymization, as well as the costs of using AIKA.
3. A case-study comparing the quality and efficiency of the VGHs generated by AIKA against VGHs manually created.

2   Related Work

Several methods for creating "good" VGHs (i.e., those that yield a good utility in the data after anonymization) have been proposed in literature. However, most of them focus on numerical attributes. For instance, the authors of [8] presented an approach for creating numerical hierarchies on-the-fly based on agglomerative hierarchical clustering. In general, these approaches are unsuitable for categorical data, as semantics is ignored. Most of the existing work focusing on categorical data belongs to the field of knowledge engineering. There, various techniques exist to create concept hierarchies whose aim is usually to facilitate the understanding of documents and processes, or to enhance semantic interoperability [11, 22]. However, their direct applicability in PPDP is limited as they do not consider the particular characteristics needed by a VGH in the context of data anonymization. For example, those techniques usually validate how well the domain of interest has been covered (i.e., granularity). However, in anonymization, a trade-off exists between the granularity and the privacy vulnerability that a VGH should have. This is because, the finer the granularity, the more useful the anonymized data is, but also the more vulnerable it could be to inferences. Alternatively, some authors [10, 15] have proposed the use of ontologies (instead of VGHs) to anonymize data. However, this can bring significant restrictions to anonymization. For instance, ontologies cannot be easily tailored to diverse publishing scenarios. Also, the fine granularity of ontologies can overexpose information to an adversary. For these reasons, our work only uses ontologies as a source of knowledge for the creation and evaluation of VGHs; leveraging the fact that various large and consensus ontologies have been made available [9].

3   AIKA Framework

In this section, we provide the context of our solution and describe the methods proposed for the automatic construction and evaluation of VGHs for PPDP.

3.1   Overview

To address the need for assisting the users in the design of VGHs, we followed a typical design science research approach [17] to develop our solution. It consists of a knowledge-based framework (AIKA) for the automatic construction and evaluation of VGHs to be used in data anonymization. Our goal is to offer a mechanism that not only reduces the human effort and expertise required to design and evaluate VGHs, but also improves the quality of the generated VGHs. Figure 2 depicts the contextual view of AIKA in PPDP: (1) A trusted entity collects personal data and is required to publish it. Thus, datasets must be anonymized before being disseminated. (2) The user selects the QIDs to be generalized from the datasets. (3) For each QID, the user manually creates candidate VGHs modeling their corresponding domains. (4) Once the user is confident about the created VGHs, they are used to anonymize the data. (5) The

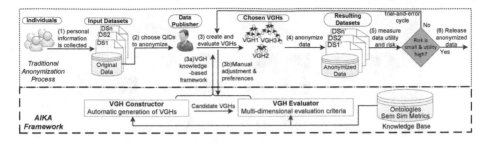

Fig. 2. Contextual view of AIKA framework in PPDP

user then evaluates the utility and disclosure risk of the anonymized data. (6) If they are acceptable, the data is released. Otherwise, a new anonymization cycle starts (Step 3). AIKA fits into Step 3, where the VGHs are designed. (3a) AIKA consists of two components: a *constructor* and an *evaluator*. The constructor (see Sect. 3.2) automatically generates various candidate VGHs for a particular domain by exploiting information from a knowledge base and the original dataset. Note that the constructor does not generate a single "optimal" VGH but a set of VGHs that can fulfill the needs of different use cases. The candidate VGHs are passed to the evaluator (see Sect. 3.3), where the VGHs are objectively assessed with quantifiable metrics from multiple perspectives. (3b) The user can inspect the VGHs and adjust (or re-evaluate) them as needed. (4) After evaluation, the best VGHs can be used to drive the data anonymization with more guarantees that those VGHs will help to retain the desired level of data usefulness and disclosure risk (hence eliminating the need of costly trial-and-error anonymization cycles).

3.2 VGH Constructor

The constructor consists of a method that automatically generates and tailors VGHs, based on the input datasets and a knowledge source that models the domain expert knowledge and human judgment. Next, we describe the elements involved in the VGH construction process (depicted in Fig. 3).

Knowledge Base (KB). AIKA exploits a KB to perform various tasks such as: semantic relationships exploitation, word-sense disambiguation, and measurement of similarity between words. The KB is encapsulated in ontologies, which act as a gold standard in which the domain expert knowledge is reflected. Since ontologies often represent the consensus opinion of a panel of experts, the risk of having partial interpretations and single judgments over the domains represented in the VGHs is mitigated. The semantic content of the ontologies is exploited by semantic similarity metrics to measure the proximity between the original values of a dataset and their possible generalizations. In this work we use WordNet as KB, and Wu and Palmer as semantic similarity metric (two of the most widely used resources in knowledge-based systems) [16]. Note that relying on a single

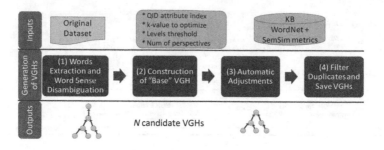

Fig. 3. AIKA - VGH constructor

ontology does not represent a limitation for AIKA, as several works support the integration of ontologies [20]. Also, the ontology used by AIKA is configurable.

(1) Words Extraction and Word Sense Disambiguation (WSD). First, the constructor identifies the leaf nodes of the VGH (by extracting the distinct values of the QID from the input dataset), and calculates their frequencies of occurrences. Next, WSD is performed, which involves defining the right sense for the words. In AIKA we use the adapted Lesk algorithm [7] which is a gloss-based method that relies on the definition of a word (using WordNet as gloss dictionary). This technique is suitable for microdata anonymization because there is no background context that can be used (e.g., documents or corpus). To mitigate the possibility of any remaining noise (i.e., incorrect senses), AIKA allows users to provide (or adjust) the senses of the individual participant words.

(2) Construction of "base" VGH. To start the generation of VGHs, AIKA extracts the minimal hierarchy that subsumes all the leaf values from the ontology. That is, for each leaf, it extracts the hypernym tree from WordNet. Then, all branches are merged into the "base" VGH. This VGH forms the basis for all other candidate VGHs, which will be later derived from it. Using the subsumption hierarchy is appropriate in our scenario as it reflects the principle of specialization/generalization used by data generalization techniques.

(3) Automatic Adjustments. This step consists in applying a series of automatic transformations to the "base" VGH with the objective of deriving multiple candidate VGHs that can be used to fulfill the requirements of different use cases. This is because the released anonymized data is intended to be used by multiple parties for different purposes. In general, such transformations vary the taxonomic structure and degree of data semantics of the "base" VGH, hence the characteristics of the derived candidate VGHs are diversified. Below, we describe the different types of adjustments performed by the VGH constructor:

a. Reduce abstractness (Fig. 4a) prunes the hierarchy at the lowest level where all the branches are connected. This adjustment naturally meets the monotonicity property [13] extensively used in anonymization: if the generalization T^* at level i preserves privacy, then every generalization of T^* at level $i + 1$ also preserves privacy. That is, all successors of an anonymous state are also anonymous.

b. Reduce outliers (shown in Fig. 4b) avoids over-generalizing the data by reducing the possible outliers in the VGH (e.g., due to data sparseness). The aim is to tailor the VGHs for a given syntactic privacy model (e.g., k-value for k-anonymity) so that the privacy condition is satisfied at the lowest possible level (where the information loss is lower). When it is possible (i.e., the frequency sum of the outliers is $\geq k$ and the semantic consistency of the VGH is respected), the outliers can be aggregated into groups so that k is satisfied. The new node (common ancestor of a group of outliers) can be one of three possibilities: one of the parents of the outliers; one of the outliers, promoted as parent; or the root node replicated (implying the full suppression of the values). All these alternatives are viable depending on the data anonymization scenario.

c. Reduce levels (shown in Fig. 4c) removes full generalization levels in the VGH based on a desired threshold of taxonomic levels to be preserved. The aim is to make the VGHs taller (fine-grained) or flatter (coarse-grained), depending on how the user wants to refine the anonymizations. This is useful to manipulate the level of safety of the anonymized data, as fine-grained usually means a higher utility but also a higher risk of disclosure. Hence, diverse profiles of data recipients (with different trustworthiness levels) can be supported such as: releasing data to an outsourced partner, or public in general.

d. Diversify perspectives (shown in Fig. 4d) applies facets to the candidate generalizations by organizing the concepts in alternative ways. The aim is to offer different perspectives about a domain. The "base" VGH is mainly created using the subsumption relationship (*is-a*) in an ontology. However other semantic relationships can be used, such as: meronym (e.g., *part-of, substance-of*) and sibling (e.g., *sister terms*). For example, *animals* can be organized in *vertebrate* and *invertebrate* but also in *ectotherm* and *homeotherm*, depending on the user's needs. The nodes to be replaced by a facet are the ancestors. The feasibility of a concept to be considered as a facet is given by a semantic similarity boundary, which determines its relevance with respect to the ancestor to be replaced.

(4) Filter Duplicates and Save VGHs. Once the automatic adjustments have been applied, the constructor cache has VGHs of diverse characteristics (potentially including repeated ones). Thus, this step consists in filtering out the duplicate VGHs so that only unique VGHs are preserved. To identify if a VGH is "equal" to another one, we use a filtering strategy based on adjacency matrix representation. Unlike techniques based on graph traversing, this approach allowed to accurately capture the equality of two VGHs for the PPDP context. That is, two VGHs qualify as equal if they have the same nodes connected in the taxonomy; even if the branches are not arranged in the same order. Finally, the unique candidate VGHs are saved (in XML format) into disk so they can be inspected by the user to be adjusted, evaluated, or used in anonymization.

3.3 VGH Evaluator

The evaluator (shown in Fig. 5) consists of a method for the multi-dimensional evaluation and ranking of VGHs. It is based on the combination of a set of

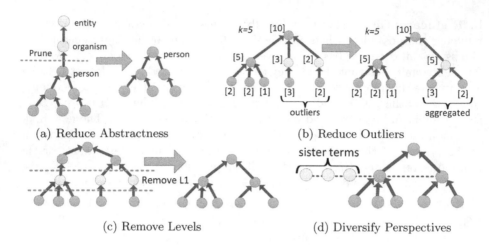

(a) Reduce Abstractness   (b) Reduce Outliers

(c) Remove Levels   (d) Diversify Perspectives

Fig. 4. Automatic adjustments in VGH constructor

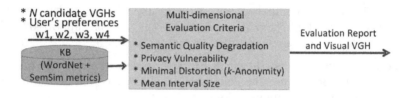

Fig. 5. AIKA - VGH evaluator

metrics that capture, in an objective and quantifiable way, the quality of VGHs from different perspectives that are relevant in anonymization. The input is a list of candidate VGHs and a set of weights that represent the user's preferences with respect to the evaluation metrics. The usage of weights allows the evaluation phase to be tailored to assess a particular use case. A KB (described in Sect. 3.2) is also used to evaluate the semantic similarity between attribute values.

Multi-dimensional Evaluation Criteria. In the following paragraph, we describe the evaluated VGH aspects and how they are measured: (1) *Semantic quality degradation* assesses the proper specification of the VGH concepts in terms of semantics. For this purpose, we apply the generalization semantic loss metric [6], which captures the quality of a VGH in terms of the semantic consistency and taxonomic organization. (2) *Privacy vulnerability* assesses the VGH susceptibility to inferences by attackers. For this purpose, we apply the semantic variance metric [19], which measures the semantic dispersion of the modeled concepts. The idea is that fine-grained taxonomies can give more information to an adversary and make the data vulnerable to attacks, thus, the more spread the concepts are, the more the privacy vulnerability. Considering this aspect helps to exclude extremely detailed VGHs. (3) *Minimal distortion* captures the ratio between the minimum level of the VGH at which the syntactic privacy condition (e.g., k-value) is satisfied, and the total height of the VGH. This metric was inspired by the minimal distortion principle discussed in [21]. (4) *Mean interval size* captures the average size of the generalization intervals (i.e., average number of children of the ancestor nodes); the more coarse the groups are, the more indistinguishable the original values are and thus, the more the information loss.

As first step, the evaluator calculates the scores of all aspects, per VGH. Then, the VGHs are ranked (per aspect) in order to obtain a reward equivalent to their ranking position in descending order. That is, VGHs with a low score will obtain a higher reward. This is because lower values are better for all evaluation metrics. Next, the user preferences (weights) are applied to the given reward, and an overall score for a VGH is calculated by the function $E(V)$ given by (1):

$$\mathcal{E}(V) = w_1 \cdot semq + w_2 \cdot priv + w_3 \cdot distrn + w_4 \cdot isize \qquad (1)$$

where $semq, priv, distrn, isize$ are the aspects evaluated in the VGH according to the multi-dimensional criteria; and w_1, w_2, w_3, w_4 are the weights assigned by the user to indicate the importance of each aspect. The best VGH is the one that maximizes $E(V)$ given the chosen weights. This is given by (2):

$$f(\mathcal{E}(V)) = \max(\mathcal{E}(V)|w_1, w_2, w_3, w_4) \qquad (2)$$

4   Experimental Evaluation

The experiments aimed three objectives: (1) to assess the benefits of using AIKA (i.e., its capability to create good quality VGHs and estimate their effectiveness in anonymization); (2) to assess the costs of using AIKA (in terms of computational resources); and (3) to compare AIKA's benefits and costs against those of manually generated VGHs. As evaluation data, we used four publicly available datasets: Adult [14] consists of census information; German Credit [14] contains credit applicants information; Chicago Homicide [1] has information about homicides filed by the Chicago police; Insurance [2] contains personal information useful for risk assessment. For each dataset, we chose the categorical attributes with the most heterogeneous values as QIDs (Table 1) to diversify the tested domains; then, we generated VGHs for them using AIKA. To assess the performance of the VGHs in anonymization, we used the commonly-used anonymization algorithm Datafly [21] (from the UTD Anonymization Toolbox [3]). We also tested a broad range of privacy levels, varying the k-values $\in [2..100]$. All experiments were done in a computer with an Intel Core i7-4702HQ CPU at 2.20 Ghz, 8 GB of RAM, Windows 8.1 64-bit, and HotSpot JVM 1.7 with a 1 GB heap. Finally, AIKA's prototype was developed in Java, internally using the WS4J library [4].

AIKA's benefits. This analysis focused on assessing the quality of the VGHs, by measuring their effectiveness for anonymizing datasets (our use case). For this purpose, we firstly evaluated the VGHs using AIKA's multi-dimensional criteria (Eq. 1). We tested the full spectrum of weights (i.e., [0..100 %]) in increments of 25 % per aspect. This strategy involved 35 sets of weights, one for each possible

Table 1. QIDs considered for VGH creation and anonymization

Dataset	Attribute	Card	Col index	Dataset	Attribute	Card.	Col index
HomicideVictims	Location	96	46	Adult	Occupation	14	7
	PHome	11	48	GermanCredit	Purpose	12	3
	POutdoor	33	56	Insurance	Occupation	60	3
	CausalFactor	47	59		Workplace	29	4
	VicRelation1	95	71		Hobby	40	5
	OffRelation1	95	72		PlaceOfHobby	32	6
	WClub	57	106				
	WKnife	25	109				

weight permutation and the four aspects (e.g., $w_1=75$, $w_2=25$, $w_3=w_4=0$). This allowed us to rank the VGHs from best to worst per weighted aspect. Next, we conducted the anonymization of the datasets using the VGHs and calculated the usefulness of the resulting datasets using four utility/risk metrics (each one associated with a desired aspect of the VGH). To measure the data utility, we used three commonly-used task-independent metrics: Semantic Sum of Squared Errors (SSE) [10], Generalized Information Loss (GenILoss) [5], and Average Equivalence Class Size (C_{AVG}) [5] which are related to the *semq*, *distrn*, and *isize* aspects, respectively. To measure the data disclosure risk (DR), we used record similarity [15], associated with the *priv* aspect. Due to space constraint, we only present the most relevant results (as this experiment involved the generation/evaluation of approximately 1.4 K VGHs and 138 K anonymized solutions).

To assess how well the properties of the VGHs were captured by AIKA's evaluator, we calculated the degree of correlation between the VGH quality scores and the quality of the anonymized datasets. For this purpose, we used the Spearman's rank order correlation (r_{Spm}), which measures the strength of a monotonic (but not necessarily linear) relationship between paired data. r_{Spm} can take values from -1 to $+1$. The closer the value is to ± 1, the stronger the relationship. The results showed that AIKA worked well (Fig. 6), as a strong level of correlation (i.e., $r_{Spm} \geq 0.60$) was achieved by all metric/aspect combinations when a high weight was used (e.g., 75 % and 100 %). Figure 6a shows the results of the *semq* aspect. There, it can also be noticed how the correlation level gradually decreases following a trend similar to the decrease in the *semq* weight. This is consequence of considering other aspects and exemplifies the trades-off that are experienced in anonymization (i.e., one sacrifices utility to enforce privacy). This behavior is also reflected in the standard deviations of the low weights, which tend to be higher than those of higher weights. Figures 6b, c, and d depict the results of the other aspects. It can be noticed how the aspects behaved similarly, as they achieved comparable levels (and trends) of correlations.

To complement this analysis, an example of the correlation plots is shown in Fig. 7. It can be noticed how the VGH quality rankings closely resemble the disclosure risk of the anonymized solutions. We also carried out a breakdown of the correlation results per dataset. No serious variations in the results were observed, showing that AIKA's rankings were accurate irrespectively of the dataset. This

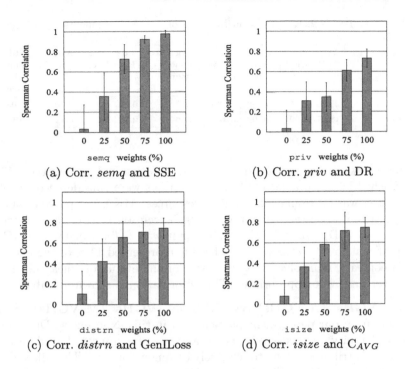

Fig. 6. Correlations between VGH evaluator criteria and data quality metrics

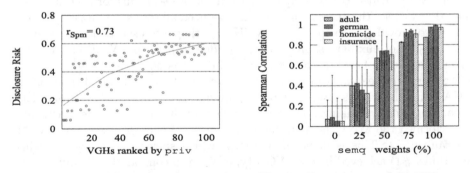

Fig. 7. *priv* (w_2=100 %) vs DR

Fig. 8. Corr. *semq* and SSE per dataset

behavior is exemplified by Fig. 8, which present the correlation breakdown for the *semq* aspect. Similar results were obtained for the other aspects.

AIKA's costs. We also studied the costs of using AIKA, in terms of computational resources: memory consumption, CPU usage, and execution time. Garbage collection (GC) was also monitored as it is a key performance concern in Java [18]. Results showed that AIKA is lightweight in terms of CPU and memory: its average CPU usage did not exceed 26 % (peak reached by the

Table 2. Resources' utilizations of AIKA components

AIKA component	Avg CPU (%)	Std CPU (%)	Avg MEM (MB)	Std MEM (MB)	Avg Exec. Time (sec)	Std Exec. Time (sec)	Avg MaGC Time (sec)	Std MaGC Time (sec)
Constructor	25.25	1.27	247.80	54.35	2.73	0.66	0.30	0.03
Evaluator	18.87	0.69	846.09	316.82	21.64	2.80	0.71	0.15

constructor), while its average memory consumption did not exceed 847 MB (peak reached by the evaluator). Both utilizations were considered tolerable as the computer was far from exhausting its resources. AIKA also proved to be efficient in terms of execution time: the average execution time of the constructor was 2.7 s (per QID), while for the evaluator it was 21.6 s. Finally, the GC was only significant for the constructor, where it represented 11 % of the execution time. In contrast, it involved less than 3 % for the evaluator (meaning that its memory settings were appropriate). This information is shown in Table 2.

AIKA's VGHs (A-VGHs) vs. Manual VGHs (M-VGHs). Sixteen researchers from our department participated in this experiment. Due to their limited availability, we focused on one dataset (i.e., Insurance). This dataset was chosen as its attributes belong to relatively common domains. This allowed us to define an improvement baseline (as the gains in more complex domains would be higher). We provided the participants with a set of leaf terms for each domain. They then defined the ancestor nodes and organized all terms, ending at the root node (also provided). To specify the VGHs, participants used their own knowledge, plus other auxiliary sources (e.g., dictionaries) except WordNet (AIKA's current knowledge base). Finally, the experiment was not time-bounded.

To compare the quality of the two VGH sets, we firstly evaluated them using AIKA (with the 35 sets of weights previously discussed) and analyzed their corresponding quality rankings. This analysis showed that the A-VGHs drastically outperformed the M-VGHs, as in more than 95 % of the 140 cases, an A-VGH was ranked #1. This is depicted in Fig. 9, which shows the number of wins (i.e., ranks #1) achieved by each VGH type. We also compared their differences in rankings and reward scores. This showed that when an A-VGH was not the best (i.e., did not win), the ranking difference was minimal (only 1 place). On the contrary, M-VGHs always lost by several places (an average of 14). The same behavior was observed in terms of reward scores. Also, in the few cases where M-VGHs won, those VGHs were created by the participants who invested the longest time designing the VGHs (meaning that they were expensive wins).

Next, we assessed the time-savings gained by AIKA. First, the time required by AIKA to create/evaluate $(C + E)$ one VGH was compared against the time reported by the participants. This comparison showed that AIKA offers significant time-savings, as its unitary cost was 99.99 % smaller. We also compared the time required to create all VGHs of each type. This also proved AIKA's usefulness, as the time-savings were also significant (an average decrease of 99.95 %).

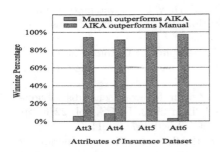

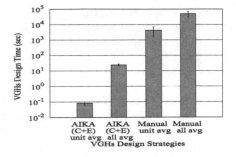

Fig. 9. Winning percentages   **Fig. 10.** Efficiency comparison

These results are depicted in Fig. 10. It is also worth noting that: (i) The manual effort only considers the intrinsic evaluation performed during the construction of the VGHs. If any extrinsic evaluation would be performed, the time-savings would be higher; (ii) AIKA created/evaluated more VGHs (an average of 100) than the participants (16), meaning that the domains were more exhaustively explored.

5   Conclusions and Future Work

This paper presents AIKA, a knowledge-based framework to automatically construct and evaluate VGHs for the anonymization of categorical data. Our experiments proved that AIKA can accurately create and determine which VGH is the most appropriate for a given scenario. AIKA also proved to be lightweight in terms of computational resources. Finally, results showed that AIKA's VGHs are not only better than manual ones, but also AIKA was significantly faster. As future work, we plan to evaluate AIKA with other ontologies, extend it to support phrases, and make it more configurable to release AIKA as a publicly-available tool. Finally, although AIKA has been tested in anonymization, its applicability can be broader. Thus, we plan to apply it to other areas where concepts are hierarchically ordered and data semantics is the main property to be preserved.

Acknowledgments. This work was supported with the financial support of the Science Foundation Ireland grants 10/CE/I1855 and 13/RC/2094.

References

1. Chicago Homicides. https://data.cityofchicago.org
2. Insurance. https://github.com/ucd-pel/Datasets/tree/master/Insurance
3. UTD ToolBox. http://cs.utdallas.edu/dspl/cgi-bin/toolbox/
4. WS4J library. https://code.google.com/p/ws4j/

5. Ayala-Rivera, V., McDonagh, P., Cerqueus, T., Murphy, L.: A systematic comparison and evaluation of k -anonymization algorithms for practitioners. Trans. Data Priv. **7**(3), 337–370 (2014)
6. Ayala-Rivera, V., McDonagh, P., Cerqueus, T., Murphy, L.: Ontology-based quality evaluation of value generalization hierarchies for data anonymization. In: PSD (2014)
7. Banerjee, S., Pedersen, T.: An adapted lesk algorithm for word sense disambiguation using WordNet. In: Gelbukh, A. (ed.) CICLing 2002. LNCS, vol. 2276, pp. 136–145. Springer, Heidelberg (2002). doi:10.1007/3-540-45715-1_11
8. Campan, A., Cooper, N., Truta, T.M.: On-the-fly generalization hierarchies for numerical attributes revisited. In: Jonker, W., Petković, M. (eds.) SDM 2011. LNCS, vol. 6933, pp. 18–32. Springer, Heidelberg (2011). doi:10.1007/978-3-642-23556-6_2
9. D'Aquin, M., Natalya, N.F.: Where to publish and find ontologies? A survey of ontology libraries. Web Semant. (online) **11**, 96–111 (2012)
10. Domingo-Ferrer, J., Sánchez, D., Rufian-Torrell, G.: Anonymization of nominal data based on semantic marginality. Inf. Sci. **242**, 35–48 (2013)
11. Kröll, M., Fukazawa, Y., Ota, J., Strohmaier, M.: Concept hierarchies of health-related human goals. In: KSEM, pp. 124–135 (2011)
12. Lee, S., Huh, S.-Y., McNiel, R.D.: Automatic generation of concept hierarchies using WordNet. Expert Syst. Appl. **35**(3), 1132–1144 (2008)
13. LeFevre, K., DeWitt, D.J., Ramakrishnan, R.: Incognito: efficient full-domain k-anonymity. In: International Conference on Management of Data, pp. 49–60 (2005)
14. Lichman, M.: UCI Machine Learning Repository (2013)
15. Martínez, S., Sánchez, D., Valls, A., Batet, M.: Privacy protection of textual attributes through a semantic-based masking method. Inf. Fusion **13**, 304–314 (2012)
16. Meng, L., Huang, R., Gu, J.: A review of semantic similarity measures in WordNet. Int. J. Hybrid Inf. Technol. **6**(1), 1–12 (2013)
17. Peffers, K., Tuunanen, T., Gengler, C.E., Rossi, M., Hui, W., Virtanen, V., Bragge, J.: The design science research process: a model for producing and presenting information systems research. DESRIST **24**, 83–106 (2006)
18. Portillo-Dominguez, A.O., Wang, M., Magoni, D., Perry, P., Murphy, J.: Load balancing of java applications by forecasting garbage collections. In: ISPDC (2014)
19. Sánchez, D., Batet, M., Martínez, S., Domingo-Ferrer, J.: Semantic variance: an intuitive measure for ontology accuracy evaluation. EAAI **39**, 89–99 (2015)
20. Solé-Ribalta, A., Sánchez, D., Batet, M., Serratosa, F.: Towards the estimation of feature-based semantic similarity using multiple ontologies. Knowl. Based Syst. **55**, 101–113 (2014)
21. Sweeney, L.: Achieving k-anonymity privacy protection using generalization and suppression. Int. J. Uncertain. Fuzziness Knowl. Based Syst. **10**(05), 571–588 (2002)
22. Wang, Y., Liu, W., Bell, D.: A concept hierarchy based ontology mapping approach. In: Bi, Y., Williams, M.-A. (eds.) KSEM 2010. LNCS (LNAI), vol. 6291, pp. 101–113. Springer, Heidelberg (2010). doi:10.1007/978-3-642-15280-1_12

Schema-Based Query Rewriting in SPARQL

Lili Jiang and Jie Luo[✉]

State Key Laboratory of Software Development Environment,
School of Computer Science and Engineering, Beihang University,
Beijing 100191, People's Republic of China
{jianglili,luojie}@nlsde.buaa.edu.cn

Abstract. SPARQL query in the semantic web has drawn consideration attention from the OWL and RDF communities. In this paper, we present SPARQL-S, a new system for SPARQL entailment regime focused on fast and efficient querying with meaningful ontology schema and large volumes of RDF data. The basic idea of SPARQL-S is different from the previous SPARQL entailment implements which are mostly archived by deriving additional facts. SPARQL-S focuses on how to rewrite the BGP in SPARQL to include more entailment BGP information absorbed from ontology schema by concept entailment and then query the rewritten SPARQL in a distribution graph computation framework such as GraphX.

Keywords: SPARQL · Ontology · Entailment · Query rewriting · GraphX

1 Introduction

SPARQL is the main query language for the Semantic Web in query answering. In its early version, SPARQL query evaluation mechanism is based on subgraph matching, also called simple entailment relation between RDF graphs. SPARQL 1.1, the new revision of SPARQL, includes several entailment regimes in order to use more elaborate entailment relations, such as those induced by RDF schema or OWL. Query answering under such entailment regimes is more complex as it may involve retrieving results that could only follow implicitly from the queried graph with special entailment regime.

There are several implementations for SPARQL entailment regimes, mainly focused on two ways: rule-based reasoning and query rewriting. Rule-based reasoning means using rules in the RDFS or OWL to extend queried graph for more RDF assertions by entailment reasoning. Jena [1] is a rule-based reasoning SPARQL entailment regime supporting RDFS entailment regime. OWL Direct Semantics entailment regime is been proposed in [2] and Pellet [3] is yet another case supporting DL entailment regime, etc. Most SPARQL entailment regimes are implemented by using reasoner interface plugged in the graph data to derive more additional RDF triples to extend the graph data size. However, these implementations are still worked in standalone mode to reduce the inference complexity.

© Springer International Publishing AG 2016
F. Lehner and N. Fteimi (Eds.): KSEM 2016, LNAI 9983, pp. 275–285, 2016.
DOI: 10.1007/978-3-319-47650-6_22

Recent researches of SPARQL query rewriting are mainly based on OWL QL. Ontop [4] gives a way for rewriting SPARQL to Datalog using TBox information and then querying ABox information in relation database such as MySQL and Oracle. The core query rewriting algorithm in Ontop is PerfectRef [5], which is similar to our rewriting algorithm as both of them try to get entailment axiom or entailment triple pattern from the schema data. However, the results of PerfectRef is a set of conjunctive queries which are the entailment queries of the origin query. While the results of our algorithm is a big union query which is also entailed by the origin query. The disadvantage of Ontop [4] is that the mapping from TBox to relational data has to be manually designed. Thus, it is not suitable for TBox which contains large amount of concepts because this kind of TBox may lead to the explosion of the number of tables in relation database.

In this paper, a SPARQL query rewriting method based on concept entailment in the ontology schema, named SPARQL-S, is proposed, which exploits both the schema reasoning to get the full meaning of ontology schema and SPARQL rewriting to match more result in the RDF data. The advantage of the proposed method is that it does not need to introduce any additional RDF assertion which is required by the previous approachs so it can reduce the query time during subgraph matching. The rewritten SPARQL query is executed distributively by using Spark GraphX to further reduce the query time and get more sound entailment results.

The rest of this paper is organized as follows. Section 2 introduce a few concepts related to SPARQL-S. In Sect. 3, the SPARQL-S query rewriting algorithm is proposed. In Sect. 4, a series of experiments are performed to evaluate the effectiveness and efficiency of SPARQL-S. We conclude the paper and address future work in Sect. 5.

2   Preliminaries

In this section, brief introductions to RDF and OWL are given, followed by the definitions of syntax, semantics, and entailment regime of SPARQL.

2.1   RDF

The Resource Description Framework (RDF) is a framework for representing and interchanging information in the Web. The language of RDF contains the following pairwise disjoint and countably infinite sets of symbols: $\mathbf{I}$ for *IRIs*, $\mathbf{L}$ for *RDF literals*, and $\mathbf{B}$ for *Blank nodes*. RDF terms are elements of the set $\mathbf{T} = \mathbf{I} \cup \mathbf{B} \cup \mathbf{L}$. An RDF graph is a collection of triples of the form *(s, p, o)*, where $s \in \mathbf{I}, p \in \mathbf{I} \cup \mathbf{B}$, and $o \in \mathbf{T}$. A triple *(s, p, o)* intuitively expresses that s and o are related by p. The element *s, p, o* in a triple is also called *subject, predicate* and *object*. In this paper, instance data are presented in RDF form.

2.2   OWL

The Web Ontology language (OWL) [6], or the latest edition OWL 2 [7], contains richer vocabulary than RDF or RDF schema and can process the content of OWL information instead of just presenting information to users. OWL facilitates greater machine interpretability of Web content than that supported by RDF and RDF Schema by providing additional vocabulary along with a formal semantics. OWL has three increasingly-expressive sublanguages: OWL Lite, OWL DL, and OWL Full. We can use the sublanguages to reason OWL file to extract more information implied in it. The reasoning time and space of these sublanguages are increasing respectively, and so is the additional assertions derived by the reasoning. In this paper, schema data is presented in OWL DL form and can be reasoned by using DL reasoners such as Fact++ [8], Pellet [3], Racer [9].

2.3   SPARQL Syntax and Semantics

SPARQL is a standard query language for RDF. For formal purposes we will use the algebraic syntax of SPARQL and graph pattern defined in the standard [10]. The SPARQL language considered in this paper has the same sets of symbols as RDF: *Blank nodes, IRIs, and literals*. In addition, it adds a countably infinite set V of variables. The SPARQL algebra is constituted by the following graph pattern operators: BGP (basic graph pattern), Join, Optional, Filter, and Union. A BGP can contain several triples and can be viewed as the result of Join between its constituent triple patterns. Algebra operators can be nested freely and BGP is the smallest inseparable unit. Each of these operators return the result of the sub-query it describes.

The basic evaluation mechanism for SPARQL query is based on simple entailment or BGP simple entailment. The SPARQL query language is currently being extended to SPARQL 1.1 entailment regimes [11]. An SPARQL entailment regime defines how queries are evaluated under more expressive semantics such as RDFS or OWL other than simple entailment.

3   SPARQL-S Query Rewriting Method

The whole process of the SPARQL-S method can be naturally divided into three procedures: data initialization, query rewriting, query evaluation, as shown in Fig. 1.

Data initialization means loading the instance data into GraphX and initializing the pellet reasoner with ontology schema before querying. **Query rewriting** means rewriting the input query based on the schema. **Query evaluation** means executing rewritten SPARQL query in a query engine.

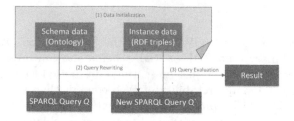

Fig. 1. The process of the SPARQL-S method

3.1 SPARQL-S Data Initialization

SPARQL-S data initialization consists of two steps in which the input schema data and instance data get analyzed and optimized so as to improve the efficiency of the rewriting and query algorithms.

Instance data loading. Loading the instance data files into property graph in GraphX. Instance data is a set of simple RDF triples of the form (s, p, o). It can be interpreted as an edge from s to o labeled with $p : (s \rightarrow p \rightarrow o)$ in GraphX. The triple sets will become edges in GraphX and s and o will become vertices in GraphX, and the edges and vertices will become triplets in GraphX.

Schema data loading and reasoning. Schema data or TBox data consists of two types of terminological axioms: class axioms and property axioms. The class axioms include all axioms about the classes and relationships between classes, such as subclass, equivalent class. The property axioms include axioms about properties and relationships between properties, such as sub-property, property role. After loading the TBox data through the OWL API [12], we use Pellet as the TBox reasoner to classify and reason the class and property axioms. The core component of Pellet is a tableaux reasoner, which repeatedly applies the tableaux expansion rules until a contradiction is detected, or until no rule is applicable. In Pellet, the tableaux reasoning has different expansion rule applying strategies which adapt to different characteristics of the TBox. The SHIN strategy is used in this paper.

3.2 SPARQL-S Query Rewriting

An input SPARQL query is first translated to a SPARQL algebra expression, and then the rewriting is performed on the SPARQL algebra expression. Before presenting the SPARQL-S query rewriting algorithm, we give some preliminary knowledge about SPARQL algebra. SPARQL algebra is constituted by a set of operators for evaluation, including BGP (basic graph pattern), Join (conjunctions), LeftJoin (optional conditions), Filter, Union, and Project. In principle, each of these operators returns the result of the sub-query it describes. In what follows, OpBGP, OPJoin, OpLeftJoin, OpFilter, OpUnion, and OpProject are used to denote BGP, Join, LeftJoin, Filter, Union, and Project operators in

SPARQL Algebra. The SPARQL-S query rewriting procedure includes two steps of rewriting: concept rewriting and ABox rewriting.

Concept rewriting. Concept question or TBox question means the variable in question can be answered by using TBox axioms along. Intuitively, concept rewriting uses TBox axioms to rewrite SPARQL algebra expressions by answering questions related to concepts. In order to do this, we need to traverse the abstract syntax tree of SPARQL algebra expressions to find and answer each concept question one by one. Since BGP is the smallest non-nested operator in SPARQL algebra, we use depth first method to traverse the abstract syntax tree of SPARQL algebra expressions to get and process each BGP operator. Each variables in concept questions in BGP operator shall be replaced by concrete values. After that, the new generated SPARQL algebra expression need to be simplified. The simplification process is performed by deleting concept question triple that has been fully answered by the TBox Information. After the concept rewriting, the new SPARQL algebra expression only contains ABox questions. The detailed of concept rewriting is listed in Algorithm 1:

Algorithm 1. Concept Rewriting

// **step 1** scanning the SPARQL algebra expression to get concepts related to variables

1: **function** GETCONCEPTANSWER($classifier$, bgp)
2: $varMap : Map < var, concept > \leftarrow \emptyset$
3: **for** each $t \in bgp$ **do**
4:   **if** $t \in ConceptAssertion$ && $canAnswerBy(t, classifer)$ **then**
5:    **for** each $var \in t$ **do**
6:     $varMap = varMap.add(var, classifer.getAnswer(t, var))$
7:    **end for**
8:    $bgp.delete(t)$
9:   **end if**
10: **end for**
11: **return** $varMap$
12: **end function**
 // **step 2** scanning the SPARQL algebra expression to substitute every var in it
13: **function** ANSWERVARCENCEPT($varMap$, bgp)
14: $bgp' : OpBGP \leftarrow \emptyset$
15: **for** each $t \in bgp$ **do**
16:   **for** each $var \in t$ **do**
17:    **if** $var \in varMap$ **then**
18:     $t' = replaceVar(t, varMap.get(var))$
19:     $bgp' = bgp'.add(t')$
20:    **else**
21:     $bgp'.add(t)$
22:    **end if**
23:   **end for**
24: **end for**
25: **return** bgp'
26: **end function**

ABox rewriting. ABox rewriting is the procedure to reformulate the new SPARQL algebra expression by unioning more equivalent or entailment SPARQL algebra expressions. This means inclusion axioms in TBox need to be taken into account. The procedure of ABox rewriting is reduced to unfold the SPARQL algebra expression to rewrite the BGP operators. The detail of rewriting and reconstructing the BGP operators are listed in Algorithm 2.

Algorithm 2. ABox Rewriting

// **function Rewrite:** rewrite the bgp to get equivalent or entailment triples
1: **function** REWRITE($classifier, bgp$)
2:     $result : Op \leftarrow \emptyset \quad preBGP : OpBGP \leftarrow \emptyset$
3:     **for** each $t \in bgp$ **do**
4:         $t' =$ EQUIVALENTOPTRIPLE($classifier, t$)
5:         **if** $t' == t$ **then**
6:             $preBGP.add(t)$
7:         **else**
8:             $result \leftarrow result \cap t'$
9:         **end if**
10:     **end for**
11:     **return** $result \cap preBGP$
12: **end function**
// **function EquivalentOpTriple:** get the equivalent triple of a triple Op
13: **function** EQUIVALENTOPTRIPLE($classifier, t$)
14:     $result : Op \leftarrow \emptyset$
15:     **if** $t \in classTypeAssertion$ (i.e. $t = (?x\ rdf : type\ A)$) **then**
16:         $equivaClasses : List < Class >= classifier.getequivalentClass(A)$
17:         **for** each $equivalence \in equivaClasses$ **do**
18:             $result = result \cup OpTriple(?xrdf : typeequivalence)$
19:         **end for**
20:         **return** $result$
21:     **else**
22:         **if** $t \in PropertyAssertion$ (i.e. $t = (?x\ p\ ?y)$) **then**
23:             $equivaProps : List < Property >=$
                                $classifier.getequivalentProperties(p)$
24:             **for** each $p' \in equivaProps$ **do**
25:                 $t' : Op = constructOP(p', t)$
26:                 $result = OpUnion(result, t')$
27:             **end for**
28:             **return** $result$
29:         **end if**
30:         **return** $result$
31:     **end if**
32: **end function**

Example 1. Consider the query rewriting problem of the following schema and query.

Schema: **subClassOf**(girl, person), **subPropertyOf**(likes, knows)
Query: **q** = select ?x, ?y where
$$\{ \text{?x rdf:type ?C. ?x knows ?y. ?C } \textbf{subClassOf} \text{ person} \}.$$
The above query **q** can be translated to the following SPARQL algebra expression

$$\textbf{op}_q = \text{OpProject}(\text{Vars}(?x, ?y),$$
$$\text{OpBGP}((?x \text{ rdf:type ?C}), (?x \text{ knows ?y}), (?C \textbf{ subClassOf } \text{person}))).$$

After applying the concept rewriting algorithm, the new generated SPARQL algebra expression $\textbf{op}'_q$ become

$$\textbf{op}'_q = \text{OpProject}(\text{Vars}(?x, ?y), \text{OpBGP}((?x \text{ rdf:type girl}), (?x \text{ knows ?y}))).$$

We can see that all variables related to concepts have been replaced by concrete values. If we further applying the ABox rewriting algorithm, then the resulting SPARQL algebra expression $\textbf{op}''_q$ become

$$\textbf{op}''_q = \text{OpProject}(\text{Vars}(?x, ?y), \text{OpJoin}((?x \text{ rdf:type girl}),$$
$$\text{OpUnion}((?x \text{ knows ?y}), (?x \text{ likes ?y})))).$$

As we can see, the most important step in the ABox rewriting is rewriting the triple pattern in BGP and reconstruct the triple pattern with its sets of entailment patterns. If a property or class has a large number of inclusion axioms in the TBox, the result of rewriting shall also contain a large number of nested union operators which increase the complexity of the result SPARQL algebra expression. Therefore, we need to limit the number of union operators during triple pattern rewriting in order to reduce the complexity of the rewritten SPARQL algebra expression. For SPARQL-S, a parameter is introduced to control whether only direct sub or full entailment shall be used for ABox rewriting. The SPARQL-S query rewriting algorithm can be obtained by sequentially apply the concept rewriting and ABox rewriting algorithms. The result SPARQL algebra expression of SPARQL-S query rewriting algorithm will be evaluated by the SPARQL-S evaluation procedure.

3.3 SPARQL-S Query Evaluation

In this procedure, since all variables of the SPARQL algebra expression related to TBox axioms in schema has been replaced in query rewriting procedure, we only need to consider the evaluation of SPARQL algebra expression on instance data (ABox axioms). The evaluation of SPARQL algebra expressions is performed on the GraphX graph-parallel computation framework of Spark. We use the S2X system [13] to process the query matching problem. First, SPARQL algebra expression will be parsed into an execution queue. The elements of the queue are SPARQL algebra operators. Each operator is a simple matching on the graph object in GraphX or the sub-results during the queue processing. In general, the graph matching problem is a graph parallel problem and the sub results matching problem is a data parallel problem. The detail of the graph and data parallel matching algorithm can be found in [13].

4   Experiment and Evaluation

In this section, we shall evaluate the effectiveness and performance of SPARQL-S and compare with the S2X system.

4.1   Dataset and Experiment Setup

The Lehigh University Benchmark LUBM [14] is used to evaluate the performance of SPARQL-S. The ontology of LUBM contains 43 classes, 32 object and data properties and 243 axioms. The benchmark also includes a set of 14 queries q_1–q_{14}. Another 7 queries with second-order variables ranging over class and property names: q_4', q_4'', q_9', q_9'', and q_2^{obg}, q_4^{obg}, q_{10}^{obg} taken from [15] are also used for the experiment. Data generated by LUBM shall be converted into a big N-triple file and stored in the HDFS file system. The experiment is performed on three generated datasets: $LUBM_1$ (103 K triples), $LUBM_9$ (1.2 M triples), and $LUBM_{100}$ (14 M triples).

We use pellet 2.3.1 for loading and reasoning schema data, Jena 3.0.0 for parsing the SPARQL queries and SPARQL algebra expressions, and S2X for evaluating SPARQL algebra expressions in GraphX framework. We conduct experiments on a cluster with 10 nodes. Each node has 2 cores, 10 GB memory. All the nodes run on Ubuntu 12.04.5 operating system with Spark 1.6.0. The query time and result count of different queries with SPARQL-S and S2X on three LUBM datasets is shown in Table 1.

In Table 1, SPARQL-S stands for querying with SPARQL-S query rewriting, where S2X stands for querying using the original query. QRT stands for the query rewriting time, T stands for the total querying time (including the query rewriting time), R stands for the total query result count. Time is measured in seconds. — means the query time exceeds the 10 minus time limit and no result is returned.

4.2   Evaluation

For all queries, we can see that the query rewriting time QRT is extremely stable with an average of about 1.5 s. The reason is that the SPARQL query rewriting algorithm is a traversal algorithm and its time complexity depends on the size of schema (TBox axioms) and query. Since the schema is the same for all queries and the queries have similar complexity, there is no remarkable difference in the rewriting time.

In Table 1, we can see that SAPRQL-S result count are significantly higher than S2X in many cases, which means that SPARQL-S can help to get more complete results through schema entailment and query rewriting with the price of slightly higher query time. For all the 21 queries in SPARQL-S, only 7 queries do not have results in $LUBM_1$ and 6 queries do not have results in $LUBM_9$ and $LUBM_{100}$. However, there are 19 queries in S2X do not have results in the three datasets. As the dataset become larger, more result counts of queries in SPARQL-S increase than that in S2X. The reason why q_{11} and q_{12} does not have

Table 1. The query results of SPARQL-S and S2X

Q	QRT	LUBM$_1$				LUBM$_9$				LUBM$_{100}$			
		SPARQL-S		S2X		SPARQL-S		S2X		SPARQL-S		S2X	
		T	R	T	R	T	R	T	R	T	R	T	R
q_1	2.1	16	4	16	4	28	4	27	4	36	4	33	4
q_2	1.6	33	0	29	0	74	**21**	61	0	105	**264**	87	0
q_3	1.7	34	6	18	6	58	6	37	6	66	6	46	6
q_4	1.6	41	**35**	22	0	103	**35**	60	0	153	**35**	94	0
q_5	1.7	33	**29**	16	0	43	**29**	24	0	48	**29**	28	0
q_6	1.7	26	**7790**	14	0	53	**87939**	24	0	60	**1048532**	27	0
q_7	1.6	41	**67**	19	0	116	**67**	51	0	167	**67**	86	0
q_8	1.5	42	**9300**	22	0	111	**9300**	59	0	141	**10170**	90	0
q_9	1.5	54	0	25	0	178	0	82	0	394	0	170	0
q_{10}	1.5	35	**4**	16	0	58	**4**	23	0	116	**4**	27	0
q_{11}	1.5	19	0	16	0	26	0	22	0	46	0	27	0
q_{12}	1.5	29	0	17	0	33	0	26	0	57	0	32	0
q_{13}	1.5	37	0	16	0	48	**8**	23	0	69	**62**	27	0
q_{14}	1.4	19	5916	17	5916	37	66676	38	66676	60	795970	66	795970
q_4'	1.5	38	**42**	27	0	98	**42**	1071	0	161	**42**	3917	0
q_4''	1.5	48	**35**	21	0	128	**35**	78	0	169	**35**	118	0
q_9'	1.6	51	0	34	0	184	0	1575	0	—	—	—	—
q_9''	1.6	60	0	31	0	170	0	872	0	—	—	—	—
q_2^{obg}	1.6	40	**547**	24	0	75	**6164**	516	0	102	**72927**	—	—
q_4^{obg}	1.4	28	**2**	19	0	40	**2**	533	0	47	**2**	—	—
q_{10}^{obg}	1.5	44	0	335	0	73	0	—	—	109	0	—	—

result is that both queries have transitive property which is not supported by SPARQL-S right now. The reason why q_9, q_9', q_9'', and q_{10}^{obg} does not have results is that only direct sub-properties are been substituted in the query rewriting algorithm to reduce query time. However, the results of these query can be obtained, if non-direct sub-properties are also allowed to be substituted during query rewriting.

For the total query time T in Table 1, we can see that the query time become longer as the size of dataset increase. The total query time also increases as the rewritten query become more complex. Compared with all the other operators in execution queues, the BGP operator is the most time-consuming operator. That is the reason for the higher total query time spent in q_9', q_9'' and q_{10}^{obg}. Too many sub-results can also cause the time of data communication in GraphX increase.

5   Conclusions

In this paper, a new query rewriting algorithm for SPARQL is proposed. A sound but not complete OWL DL entailment regime is being utilized by SPARQL-S to rewrite queries to enrich the query results. Compare to the approach that extend the instance data through rule-based reasoning, our approach changes query other than the instance data. Hence, it shall not increase the cost of subgraph matching during the query evaluation. Since the time complexity of SPARQL-S query rewriting is mostly depended on the size of schema data, it can be used for large scale knowledge graphs, such as DBpedia, Wikidata, and YAGO, which have a relatively smaller schema data (TBox axioms) compare to its instance data (ABox axioms). The effectiveness of SPARQL-S is being evaluated using an extended LUBM dataset. It can produce more query results with the cost of slightly higher query time.

For the future work, heuristic strategies can be added to the ABox query rewriting algorithm to improve the efficiency of the resulting SPARQL algebra expression and control its complexity. The S2X platform used for query evaluation also need to be optimized to improve the efficiency of BGP matching and get a better use of data parallelization.

Acknowledgements. This work is supported by National Natural Science Foundation of China (Grand No. 61502022) and State Key Laboratory of Software Development Environment (Grand No. SKLSDE-2015ZX-22).

References

1. Jena, A.: Reasoners and rule engines: Jena inference support. The Apache Software Foundation (2013)
2. Kollia, I., Glimm, B., Horrocks, I.: SPARQL query answering over OWL ontologies. In: Antoniou, G., Grobelnik, M., Simperl, E., Parsia, B., Plexousakis, D., Leenheer, P., Pan, J. (eds.) ESWC 2011. LNCS, vol. 6643, pp. 382–396. Springer, Heidelberg (2011). doi:10.1007/978-3-642-21034-1_26
3. Parsia, B., Sirin, E.: Pellet: an OWL DL reasoner. In: Third International Semantic Web Conference-Poster, vol. 18 (2004)
4. Calvanese, D., Cogrel, B., Komlaebri, S., Kontchakov, R., Lanti, D., Rezk, M., Rodriguezmuro, M., Xiao, G.: Ontop: answering SPARQL queries over relational databases (2016)
5. Calvanese, D., De Giacomo, G., Lembo, D., Lenzerini, M., Rosati, R.: Tractable reasoning and efficient query answering in description logics: the DL-Lite family. J. Autom. Reasoning **39**(3), 385–429 (2007)
6. McGuinness, D.L., van Harmelen, F., et al.: OWL web ontology language overview. W3C Recommendation **10**(10), (2004)
7. Group, W.O.W., et al.: OWL 2 web ontology language document overview (2009)
8. Tsarkov, D., Horrocks, I.: FaCT++ description logic reasoner: system description. In: Furbach, U., Shankar, N. (eds.) IJCAR 2006. LNCS (LNAI), vol. 4130, pp. 292–297. Springer, Heidelberg (2006). doi:10.1007/11814771_26
9. Haarslev, V., Möller, R.: Racer: an OWL reasoning agent for the semantic web. In: Proceedings of the International Workshop on Applications, Products and Services of Web-Based Support Systems, in Conjunction with the with 2003 IEEE/WIC International Conference on Web Intelligence, pp. 91–95 (2003)
10. Prud, E., Seaborne, A., et al.: SPARQL query language for RDF (2006)
11. Glimm, C.O.B.: SPARQL 1.1 Entailment Regimes (2012)
12. Horridge, M., Bechhofer, S.: The OWL API: a Java API for OWL ontologies. Semant. Web **2**(1), 11–21 (2011)

13. Schätzle, A., Przyjaciel-Zablocki, M., Berberich, T., Lausen, G.: S2X: graph-parallel querying of RDF with GraphX. In: Wang, F., Luo, G., Weng, C., Khan, A., Mitra, P., Yu, C. (eds.) Big-O(Q)/DMAH -2015. LNCS, vol. 9579, pp. 155–168. Springer, Heidelberg (2016). doi:10.1007/978-3-319-41576-5_12
14. Guo, Y., Pan, Z., Heflin, J.: LUBM: a benchmark for OWL knowledge base systems. Web Semant. Sci. Serv. Agents World Wide Web 3(2), 158–182 (2005)
15. Kontchakov, R., Rezk, M., Rodríguez-Muro, M., Xiao, G., Zakharyaschev, M.: Answering SPARQL queries over databases under OWL 2 QL entailment regime. In: Mika, P., et al. (eds.) ISWC 2014. LNCS, vol. 8796, pp. 552–567. Springer, Heidelberg (2014). doi:10.1007/978-3-319-11964-9_35

Knowledge Enrichment and
Visualization

Finding the Optimal Users to Mention in the Appropriate Time on Twitter

Dayong Shen, Zhaoyun Ding$^{(\boxtimes)}$, Fengcai Qiao, Jiajun Cheng, and Hui Wang

College of Information System and Management,
National University of Defense Technology,
Changsha 410073, People's Republic of China
{dayong.shen,zying,zyding,fcqiao,jiajun.chen,huiwang}@nudt.edu.cn

Abstract. Nowadays, Twitter has become an important platform to expand the diffusion of information or advertisement. Mention is a new feature on Twitter. By mentioning users in a tweet, they will receive notifications and their possible retweets may help to initiate large cascade diffusion of the tweet. In order to maximize the cascade diffusion, two important factors need to be considered: (1) The mentioned users will be interested the tweet; (2) The mentioned users should be online. The second factor was mainly studied in this paper. If we mention users when they are online, they will receive notifications immediately and their possible retweets may help to maximize the cascade diffusion as quickly as possible. In this paper, an unbalance assignment problem was proposed to ensure that we mentioned the optimal users in the appropriate time. In the assignment problem, constraints were modeled to overcome the overload problems on Twitter. Further, the unbalance assignment problem was converted to a balance assignment problem, and the Hungarian algorithm was took to solve the above problem. Experiments were conducted on a real dataset from Twitter containing about 2 thousand users and 5 million tweets in a target community, and results showed that our method was consistently better than mentioning users randomly.

Keywords: Twitter · Mention · Time · Assignment problem

1 Introduction

On Twitter, people can carry out social interactions by reading, commenting and forwarding the latest updates from others. However, due to the large amount of user-generated content, people become easily distracted during information seeking. As a result, the effect of social media marketing (SMM) (also known as marketing practices on social media platforms), such as efforts to gain attentions and encourage participation by posting attractive contents, can be very limited if the right audiences cannot be properly identified in time.

This work was supported by National Natural Science Foundation of China (No. 71331008).

© Springer International Publishing AG 2016
F. Lehner and N. Fteimi (Eds.): KSEM 2016, LNAI 9983, pp. 289–301, 2016.
DOI: 10.1007/978-3-319-47650-6_23

Fortunately, as a new feature on Twitter, Mention can help ordinary users to improve the visibility of their tweets and go beyond their immediate reach in social interactions. Mention is tagged as @username. All the users mentioned by a tweet will receive a mention notification. By using Mention, one can draw attention from specific users. Properly using mention can quickly help an ordinary user spreading his tweets.

Due to the significance of the mention feature, Mention Recommendation has attracted some studies in previous work. The representative work was proposed by Wang et al. [1]. In their work, features of users' interest and influence were considered to recommend users by mentioning. However, the temporal pattern of mentioned users was neglected in their work. If they mentioned users who were not online, these users would be not notified as quickly as possible. Users usually preferred to retweet the instant information. If users are online in next time and they discover the information is outdated, the probability of retweets would become lower. Moreover, due to the overload problems on Twitter, large number of other mentioned tweets maybe drown earlier mentioned tweets and caused these earlier mentioned tweets to be read with a lower probability.

Moreover, Fig. 1 gave the distribution of time gaps according to the real data set on Twitter. The time gaps were defined as the difference between time stamps of source tweets and time stamps of retweets. Experimental results showed that most of retweets occurred in a short time gap. So, when the information was outdated, the probability of retweets would become lower.

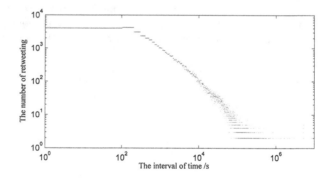

Fig. 1. The distribution of time gaps

In order to expand the diffusion of tweets by @ recommendation on Twitter, it is important to consider whether mentioned users are online. For an account on Twitter, if we want to expand the diffusion of tweets by @ recommendation by this account, the naive method is to mention as many users as possible in each hour. However, there is an overload problem on Twitter. If an account posts too many tweets in each hour, it is likely to be treated as a spam and the Twitter Service Provider will seal this account. Moreover, if the account posts tweets automatically by the API of Twitter, it will be restricted by the frequency of

Twitter API. Usually, only 60 tweets at most are allowed to publish in an hour by the API of Twitter.

The above problem was modeled as a linear programming problem in this paper. Time windows for an account in a day were divided into 24 windows. Due to the overload problem and limits of Twitter API, the mentioned users in each window were lower than a threshold. Moreover, in order to avoid the account be treated as a spam by the Twitter Service Provider, the threshold in each window should be different. The number of mentioned users in each window for an account was set up in advance. We should find such a number of users on Twitter and put them to each window in order to maximize the probability of online. Figure 2 illustrated the basic idea of our method.

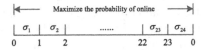

Fig. 2. The basic idea of our method

In order to get the probability of online in each window for a user on Twitter, Both the dispersion of time stamps in a window and the dispersion of this window when a user tweeted in a long time were simultaneously considered in this paper. Then, an unbalance assignment problem was proposed to maximize the probability of online, where the mentioned users were called as tasks and windows for 24 h were called as agents. The target function was to maximize the probability of online when the task was assigned the most appropriate agent. Further, the unbalance assignment problem was converted to a balance assignment problem, and the Hungarian algorithm was took to solve the above problem.

It is worthwhile to highlight the following three aspects of our scheme in this paper.

(1) Both the dispersion of time stamps in a window and the dispersion of this window when a user tweeted in a long time were simultaneously considered to get the probability of online in each window for a user on Twitter.
(2) An unbalance assignment problem was proposed to maximize the probability of online, and the target function was to maximize the probability of online when the task was assigned the most appropriate agent.
(3) We converted the unbalance assignment problem to a balance assignment problem, and the Hungarian algorithm was took to solve the above problem.

2   Related Work

In this section we discuss some work related to our study, including the recommendation on Twitter and the assignment problem.

With the development of microblogs in recent years, the recommendation on Twitter has attracted more attention by researchers, mainly including the news recommendation and the user recommendation. Diaz-Aviles et al. [2] considered collaborative filtering as an online ranking problem and presented RMFO, a method that created, in real-time, user-special rankings for a set of tweets based on individual preferences that were inferred from the user's past system interactions. Chen et al. [3] proposed a method of making tweet recommendations based on collaborative ranking to capture personal interests, and their final method considered three major elements on Twitter: tweet topic level factors, user social relation factors and explicit features such as authority of the publisher and quality of the tweet. Abel et al. [4] investigated different strategies for mining user interest profiles from microblogging activities ranging from strategies that analyzed the semantic meaning of Twitter messages to strategies that adapted to temporal patterns that could be observed in the microblogging behavior, and they evaluated the quality of the user modeling methods in the context of a personalized news recommendation system. Son et al. [5] proposed a novel news article recommendation that reflected the geographical context of the user. Phelan et al. [6] described a novel approach to news recommendation that harnessed real-time micro-blogging activity, from a service such as Twitter, as the basis for promoting news stories from a users favourite RSS feeds. Gupta et al. [7] described and evaluated a few graph recommendation algorithms implemented in Cassovary, including a novel approach based on a combination of random walks and SALSA.

Kuhn [8] first gave a Hungarian method to solve the assignment problem in 1955. Then, the assignment problem has been successfully applied to the field of data mining. Li et al. [9] proposed the Social Event Organization (SEO) problem as one of assigning a set of events for a group of users to attend. Qian et al. [10] proposed SCRAM, a sharing considered route assignment mechanism for fair taxi route recommendations in order to provide recommendation fairness for a group of competing taxi drivers, without sacrificing driving efficiency. Liu et al. [11] explored the expert retrieval problem and implemented an automatic paper-reviewer recommendation system that considered aspects of expertise, authority, and diversity. Spasojevic et al. [12] presented 'LASTA' (Large Scale Topic Assignment), a full production system used at Klout, Inc., which mined topical interests from five social networks and assigned over 10,000 topics to hundreds of millions of users on a daily basis.

3   Methods

3.1   Constructing the Benefit Matrix

In this paper, only the temporal pattern was considered as the behavior of a user. If we want to spread topics or events on Twitter, the target audience would be found according to the interest similarity. A series of keywords $\{w_1, w_2, \ldots, w_i\}$ were constructed to represent a topic or an event. Then, the target audience $\{u_1, u_2, \ldots, u_n\}$ who were interested in the topic or the event could be retrieved by the series of keywords $\{w_1, w_2, \ldots, w_i\}$. Also, if you want to give your product advertising on Twitter, the target audience $\{u_1, u_2, \ldots, u_n\}$ could be retrieved by the series of keywords $\{w_1, w_2, \ldots, w_i\}$ which represented the idea of the advertising.

Definition 1 *(Target community). A set of audience $\{u_1, u_2, \ldots, u_n\}$ with the similar interest which is represented by series of keywords $\{w_1, w_2, \ldots, w_i\}$.*

The main goal of this paper was to maximize the benefit in the constraints of the overload problem and the upper limits of Twitter API for all users in a target community.

Then, benefits for all users in a target community in 24 discrete windows need to be inferred by the temporal pattern of each user. Intuitively, the more tweets posted by a user in an hour window, the higher probability this user was on line, except for spams. So, the probability of online for an hour window could be inferred by the number of tweets and the dispersion of tweets in the hour window. For all tweets of an hour window in a period of time T posted by a user, the more tweets posted by the user in the hour window, the higher probability this user was on line; the higher dispersion in the hour window, the higher probability this user in this window was on line. The variance indicated the dispersion of a data set. So, based on the variance, the probability of online in an hour window was defined as follows.

$$p = E\{[X - E(X)]^2\} \times n = \sum_n [X - E(X)]^2 \tag{1}$$

Here, the X indicated the time interval of a user in an hour window for a period of time T. The $E(X)$ indicated the expectation of the time interval. And the n indicated the number of tweets posted by this user in an hour window for a period of time T.

In order to count the time interval of a user in an hour window for a period of time T, all time stamps $\{t_1, t_2, \ldots, t_n\}$ were got from tweets in an hour window posted by the user in a period of time T, where the t_i was converted the second base unit. The integral point in each hour window was considered as the start point of the time interval. Then, the time interval could be counted as follows.

$$\Delta t_i = (t_i - t_0)/3600 \tag{2}$$

Here, the t_0 indicated the second base unit of the integral point in each hour window, such as One o'clock, Two o'clock, Three o'clock, etc. The t_i indicated the second base unit of a time stamp for a user in an hour window.

Then, all time intervals in any window for a user could be counted as $\{\Delta t_1, \Delta t_2, \ldots, \Delta t_n\}$. So, the probability of online for a user in an hour window could be converted as follows.

$$p_h = \sum_{i=1}^{n} (\Delta t_i - \overline{\Delta t})^2 \tag{3}$$

Here, the $\overline{\Delta t}$ indicated the expectation of the time interval $\{\Delta t_1, \Delta t_2, \ldots, \Delta t_n\}$. The above formula indicated that the probability of online for a user in an hour window was related to two factors: the number of tweets and the dispersion in an hour window.

The above formula was inferred by counting all tweets posted by a user in an hour window for a period of time T. However, some users may be active in a short time period and they posted larger number of tweets in the short time period. According to the above formula, we could infer the probabilities of online for these users were higher. In fact, these users were only online in a short time, and they were not online in most of time. So, it was not enough to only count all time stamps in an hour window posted by a user in the period of time T. In order to overcome the above problem, besides for counting all time stamps in an hour window posted by a user in the period of time T, the dispersion of days when a user posted was inferred to compute the probability of online more accurately. Intuitively, if a user often posts in an hour window for every day, he would be online in the hour window with a higher probability.

So, for any a user, we counted all days $\{d_1, d_2, \ldots, d_m\}$ when the user posted in the period of time T. In order to uniform dimension with the above time stamp in an hour, the method of the normalization was defined as follows.

$$\Delta d_i = \frac{d_i - d_0}{|T|} \tag{4}$$

Here, the time stamp was measured in the day unit. The d_0 was equal to 0. The $|T|$ indicated the size for the period of time T.

Similarity, all day intervals in the period of time T for a user could be counted as $\{\Delta d_1, \Delta d_2, \ldots, \Delta d_n\}$. So, the probability of online for a user in the period of time T could be converted as follows.

$$p_y = \sum_{i=1}^{n} \left(\Delta d_i - \overline{\Delta d}\right)^2 \tag{5}$$

Here, the $\overline{\Delta d}$ indicated the expectation of the time interval $\{\Delta d_1, \Delta d_2, \ldots, \Delta d_n\}$. The above formula indicated that the probability of online for a user in the period of time T was related to two factors: the number of tweets and the dispersion of this window in the period of time T.

So, the incorporate probability of online for a user in an hour window was related to two factors: the time stamp dispersion and the whole time stamp for the hour window in the period of time T. The whole probability of online for a user in an hour window was defined as follows.

$$p = p_h \times p_y \tag{6}$$

Here, besides for counting all time stamps in an hour window posted by a user in the period of time T, the dispersion of days when a user posted was inferred to compute the probability of online. Obviously, the more dispersion of time stamps and days in an hour window, the higher probability of online was.

Then, for 24 windows in a day of each user, we could compute the probability of online in each hour for any a user with the above similar methods. So, for any a user in 24 h, we could construct a vector for the probability of online as follows.

$$\boldsymbol{p}_i = (p_1, p_2, \ldots, p_{24})^T \tag{7}$$

Here, the i indicated the i user in the target community.

Then, for all users in the target community, a benefit matrix for the probability of online could be constructed as follows.

$$P = \begin{bmatrix} p_{1,1} & p_{1,2} & \cdots & p_{1,24} \\ p_{2,1} & p_{2,2} & \cdots & p_{2,24} \\ \vdots & \vdots & & \vdots \\ p_{n,1} & p_{n,2} & \cdots & p_{n,24} \end{bmatrix} \tag{8}$$

Here, the n indicated the number of users in the target community.

3.2   Optimization Model

For an account on Twitter, if we want to expand the diffusion of tweets by @ recommendation by this account, the naive method was to mention as many users as possible in each hour. However, there was an overload problem on Twitter. If an account posts too many tweets in each hour, it would be likely to be treated as a spam and the Twitter Service Provider would seal this account. Moreover, if the account posts tweets automatically by the API of Twitter, it would be restricted by the frequency of Twitter API. Usually, only 60 tweets at most were allowed to publish in an hour by the API of Twitter. So, in order to maximize the benefit in the constraints of the number of tweets in each hour, we should mention users when they would be online. Then, more users could accept the topic or the event as quickly as possible and the topic or the event may be spread as quickly as possible.

So, for an account on Twitter, the optimization model was defined as follows.

$$\max z = \sum_{i=1}^{m} \sum_{j=1}^{n} c_{ij} x_{ij}$$

$$\begin{cases} \sum_{i=1}^{m} x_{ij} \leq 1, j = 1, 2, \ldots, n \\ \sum_{j=1}^{n} x_{ij} \leq \sigma_i, i = 1, 2, \ldots, m \\ \sum_{i=1}^{m} \sum_{j=1}^{n} x_{ij} \leq \delta \\ x_{ij} = 0 \ \ or \ \ 1 \end{cases} \tag{9}$$

Here, the x_{ij} indicated whether the j user was mentioned in the i hour window. If the j user was mentioned in the i hour window, the x_{ij} was 1, otherwise, it was 0. c_{ij} indicated the probability of online, and it was the element of the matrix P^T. The m indicated the number of windows, and the maximum number of the m was equal to 24. The n indicated the number of all users. In order to overcome the overload problem, the number of tweets in each hour was limited as σ_i; that was $\sum_{j=1}^{n} x_{ij} \leq \sigma_i$. Moreover, the total number of users mentioned was less a threshold δ, that was $\sum_{i=1}^{m} \sum_{j=1}^{n} x_{ij} \leq \delta$. In this paper, only a user was mentioned in a tweet to overcome the overload problem; that was $\sum_{i=1}^{m} x_{ij} \leq 1$. A tweet mentioning a lot of users was likely to be treated as a spam tweet, which would decrease others interest in retweeting it.

Moreover, the limits σ_i of the number of tweets in each hour was different. In general, the probability of online for a user was lower in periods from One A.M. to Six A.M. In these periods, we could mention less users. In contrast, the probability of online for a user was higher in periods from Nine P.M. to Eleven P.M. In these periods, we could mention more users. So, we counted all tweets in each hour to discover the pattern of online for all users. The pattern of online for all users was defined as follows.

$$p = \{s_1, s_2, \ldots, s_{24}\} \tag{10}$$

Here, the s_i indicated the total number of tweets in the i hour. The number of mentions in each hour was reduced by a constant c in order that the greatest number of mention was less than limits. That is, the σ_i was defined as follows.

$$\sigma_i = s_i / c \tag{11}$$

Usually, the total number of users mentioned by an account was less than the number of users in a target community. That was as follows.

$$\sum_{i=1}^{24} \sigma_i < \delta \tag{12}$$

The above optimization model was called as an unbalanced assignment problem. Here, users mentioned by the virtual account were called as the tasks. And windows for each hour were called as the agents. The target function was to maximize the benefit when each task was assigned the most suitable agent. The first constraint indicated each task was accomplished only by an agent. The second constraint indicated each agent could accomplish multiple tasks and the number of tasks was less than a threshold σ_i. The third constraint indicated the total number of tasks was less a threshold δ. Because the total number of users mentioned by a virtual account was usually less than the number of users in a target community, parts of tasks were not accomplished by agents. The fourth constraint indicated a task was accomplished or was not accomplished.

3.3   Solution

Traditional assignment problem was usually balance, where a task was only accomplished by an agent and an agent only accomplished a task. The number of tasks was equal to the number of agents. Hungarian algorithm [8] was took to solve the traditional balance assignment problem.

If the number of tasks was more than the number of agents, the assignment problem was unbalance. In order to take the Hungarian algorithm, some fictitious agents who accomplished the tasks with 0 overhead were added to the task matrix in order to ensure the number of tasks was equal to the number of agents.

However, our model was different from traditional unbalanced assignment problem. In traditional unbalanced assignment problem, an agent only accomplished a task. In our model, many users were mentioned by the virtual account in an hour window. So, an agent should accomplish multiple tasks in the assignment problem.

In order to take the Hungarian algorithm, the benefit matrix should be reformed to accommodate the traditional unbalanced assignment problem. The basic idea was that a number of the same agents were added to the benefit matrix if an agent accomplished multiple tasks. Also, the total number of users mentioned by a virtual account was usually less than the number of users in a target community and parts of tasks were not accomplished by agents. So, the rest of tasks were accomplished by the fictitious agents.

So, the new coefficient matrix H was constructed according to the above benefit matrix P^T.

$$H = \begin{bmatrix} \sigma_1 \begin{cases} M - p_{1,1} \; M - p_{2,1} \ldots M - p_{n,1} \\ M - p_{1,1} \; M - p_{2,1} \ldots M - p_{n,1} \\ \vdots \qquad \vdots \qquad \vdots \\ M - p_{1,1} \; M - p_{2,1} \ldots M - p_{n,1} \end{cases} \\ \vdots \qquad \vdots \qquad \vdots \\ M - p_{1,24} \; M - p_{2,24} \ldots M - p_{n,24} \\ 0 \qquad 0 \qquad 0 \qquad 0 \\ \vdots \qquad \vdots \qquad \vdots \qquad \vdots \end{bmatrix} \tag{13}$$

Here, the σ_i indicated the i agent accomplished σ_i tasks. So, the σ_i agent was added to the coefficient matrix. Moreover, the $\delta - \sum_{i=1}^{24} \sigma_i$ rows with 0 were added to the coefficient matrix, which indicated that fictitious agents who accomplished the rest of tasks. Because the initial optimization model without reforms was to maximize the benefit when each task was assigned the most suitable agent. An huge constant M was added to the coefficient matrix in order to convert to a minimization problem.

Based on the coefficient matrix H, the new optimization model was defined as follows.

$$\min z = \sum_i \sum_j h_{ij} x_{ij}$$

$$\begin{cases} \sum_i x_{ij} = 1, j = 1, 2, \ldots, \delta \\ \sum_j x_{ij} = 1, i = 1, 2, \ldots, \delta \\ \qquad x_{ij} = 0 \;\; or \;\; 1 \end{cases} \quad (14)$$

The above optimization model was called as the balanced assignment problem. In this paper, the Hungarian algorithm was took to solve the above problem.

4   Experiments

In order to illustrate the effectiveness of our model, we collected tweets of users from a target community in 2015. In the data set, the number of users in the target community was 2098. In our data set, the number of tweets published by these users was about 5 million. Also, we extracted time stamps of tweets.

Table 1. The value of the constant c

$c1$	$c2$	$c3$	$c4$	$c5$	$c6$
3364	4036	5045	6727	10090	20180

According to time stamps of tweets, we computed the benefit matrix containing the probability of online for 2098 users. We summed the probability of each hour for 2098 users in Fig. 3. According to Fig. 3, we could find that the probability of online for all users was higher in periods from Nine PM to Eleven PM and from Nine AM to Eleven AM. Also, the probability of online for all users was lower in periods from One AM to Six AM. So, in order to mention users reasonably by a virtual account, we set the constant c in Table 1 to get the number of mention in each hour by the virtual account.

Based on the constant c in Table 1, we could get the number of mention in each hour by a virtual account in Fig. 4. The pattern of pi in Fig. 4 was constructed by the constant ci in Table 1.

4.1   Evaluation Method

In order to evaluate the effective of our model, we divided the data set into two parts. The time stamps from 2015 January to 2015 October were collected to model. And time stamps in 2015 November were collected to evaluate the effective of our model. According to the data set from 2015 January to 2015 October, we could get mention users list in each hour $\{h_0 = \{u_{0,1}, u_{0,2}, \cdots\}, h_1 = \{u_{1,1}, u_{1,2}, \cdots\}, h_{23} = \{u_{23,1}, u_{23,2}, \cdots\}\}$. For each recommended user $u_{i,j}$ in the mention users list, the precision was computed by checking whether the user

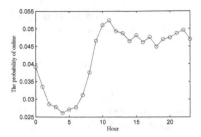
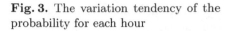

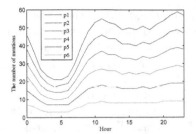

Fig. 3. The variation tendency of the probability for each hour

Fig. 4. The number of mention in each hour

was online on the i hour in every day for 2015 November. So, the precision was defined as follows.

$$p_i = \frac{\sum\limits_{j=0}^{30} f_{ij}}{30} \quad if \quad online \quad f_{ij} = 1, \quad else \quad f_{ij} = 0 \tag{15}$$

So, the average precision of all users was defined as follows.

$$p = \sum_{i=0}^{s} p_i/s \tag{16}$$

Here, the s was the total number of users in the mention users list.

We compared the mention users list by recommending users randomly. Our model was named as AP, and the randomized method was named as RM.

4.2 Experimental Results

According to the number of mentions in each hour in Fig. 4, we got the average precision of all users by taking advantage of the method of AP and RM. The experimental results were illustrated in Fig. 5.

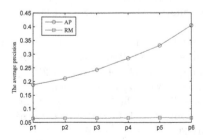

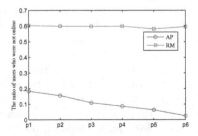

Fig. 5. The average precision of all users

Fig. 6. The ratio of users who were not online

We could find our method was more effective. The average precision of the method by recommending users randomly was lower. Most of users recommended by randomly were not even online in a whole 2015 November.

Moreover, the precision of these users who were not online in whole November was equal to 0. The more these users were mentioned, the worse results to expand the diffusion of tweets in time. Because these users may be not online in whole month. We counted the average ratio of these users who were not online in whole November. The experimental results were illustrated in Fig. 6. According to experimental results, we could find that the ratio of users who were not online for our method was lower than the method by recommending users randomly, and more valuable users were mentioned by our method.

5   Conclusions

In order to find the optimal users to mention in the appropriate time on Twitter, an unbalance assignment problem was proposed in this paper, and results showed that our method was consistently better than mentioning users randomly.

In the future work, besides for the time-series pattern of users, other factors such as interests of users, behaviors of users, et al. will be considered in order to find the better users to mention on Twitter.

References

1. Wang, B., Wang, C., Bu, J., Chen, C., Zhang, W.V., Cai, D., He, X.: Whom to mention: expand the diffusion of tweets by @ recommendation on micro-blogging systems. In: Proceedings of the the 22nd International Conference on World Wide Web (WWW 2013), Rio de Janeiro, Brazil, pp. 1331–1340, May 2013
2. Diaz-Aviles, E., Drumond, L., Gantner, Z., Schmidt-Thieme, L., Nejdl, W.: What is happening right now ... that interests me?: online topic discovery and recommendation in twitter. In: Proceedings of the 21st ACM International Conference on Information and Knowledge Management (CIKM 2012), Maui, Hawaii, pp. 1592–1596, October 2012
3. Chen, K., Chen, T., Zheng, G., Jin, O., Yao, E., Yu, Y.: Collaborative personalized tweet recommendation. In: Proceedings of the 35th International ACM SIGIR Conference on Research and Development in Information Retrieval (SIGIR 2012), Portland, Oregon, USA, pp. 661–670, August 2012
4. Abel, F., Gao, Q., Houben, G.J., Tao, K.: Twitter-based user modeling for news recommendations. In: Proceedings of the 23rd International Joint Conference on Artificial Intelligence (IJCAI 2013), Beijing, China, pp. 2962–2966, August 2013
5. Son, J.W., Kim, A.Y., Park, S.B.: A location-based news article recommendation with explicit localized semantic analysis. In: Proceedings of the 36th International ACM SIGIR Conference on Research and Development in Information Retrieval (SIGIR 2013), Dublin, Ireland, pp. 293–302, July 2013
6. Phelan, O., McCarthy, K., Smyth, B.: Using twitter to recommend real-time topical news. In: Proceedings of the 3rd ACM Conference on Recommender Systems (RecSys 2009), New York, NY, USA, pp. 385–388, October 2009

7. Gupta, P., Goel, A., Lin, J., Sharma, A., Wang, D., Zadeh, R.: WTF: the who to follow service at twitter. In: Proceedings of the 22nd International Conference on World Wide Web (WWW 2013), Rio de Janeiro, Brazil, pp. 505–514, May 2013

8. Kuhn, H.W.: The Hungarian method for the assignment problem. Nav. Res. Logist. Q. **2**(1), 83–97 (1955)

9. Li, K., Lu, W., Bhagat, S., Lakshmanan, L.V.S., Yu, C.: On social event organization. In: Proceedings of the 20th ACM SIGKDD International Conference on Knowledge Discovery and Data Mining (KDD 2014), New York, NY, USA, pp. 1206–1215, August 2014

10. Qian, S., Cao, J., Mouel, F.L., Sahel, I., Li, M.: SCRAM: a sharing considered route assignment mechanism for fair taxi route recommendations. In: Proceedings of the 21th ACM SIGKDD International Conference on Knowledge Discovery and Data Mining (KDD 2015), Sydney, Australia, pp. 955–964, August 2015

11. Liu, X., Suel, T., Memon, N.: A robust model for paper reviewer assignment. In: Proceedings of the 8th ACM Conference on Recommender Systems (RecSys 2014), Foster, Silicon Valley, USA, pp. 25–32, October 2014

12. Spasojevic, N., Yan, J., Rao, A., Bhattacharyya, P.: LASTA: large scale topic assignment on multiple social networks. In: Proceedings of the 20th ACM SIGKDD International Conference on Knowledge Discovery and Data Mining (KDD 2014), New York, NY, USA, pp. 1809–1818, August 2014

Extracting Knowledge from Web Tables Based on DOM Tree Similarity

Xiaolong Wu[1,2(✉)], Cungen Cao[1], Ya Wang[1], Jianhui Fu[1], and Shi Wang[1]

[1] Key Laboratory of Intelligent Information Processing, Institute of Computing Technology,
Chinese Academy of Sciences, Beijing, China
Wuxiaolong2016@ict.ac.cn
[2] University of Chinese Academy of Sciences, Beijing, China

Abstract. Structured (semi-structured) knowledge extraction from Web tables is an important way to obtain high quality knowledge. Unlike most extraction methods which need to understand the tables with external knowledge bases, our method uses the inherent similarities of tables to determine the semantic structure of tables. With a comprehensive analysis of table structures of various forms, we provide a novel way for calculating the DOM tree similarity between various web tables based on DTW and for clustering tables. By using 5000 Wikipedia tables which were extracted at random as the corpus, experiments show that the result of table clustering is close to the result of classification based on empirical approaches, and without the use of external knowledge bases, the quality of knowledge extracted from the tables is satisfactory.

Keywords: Knowledge extraction · Web tables · DOM tree similarity · Table clustering

1 Introduction

Web tables are a kind of concise and effective knowledge containers, which are abundant in structured and semi-structured knowledge. A Google study shows that more than 150 million high-quality relational tables exist on the Web [1]. Through a large-scale survey of tables on a large crawl of the Web, 75 % of the pages contain at least one table with an average of 9.1 tables per document [2].

Knowledge extraction from the Web tables is an important way to obtain high quality knowledge, and has many applications in knowledge acquisition, information retrieval, question answering, Web mining and so on. So it has become one of hot issues in the field of knowledge acquisition in recent years.

An important feature of Web tables is that the tables are presented in the form of a DOM tree, which is composed of various combinations of HTML tags. A HTML table usually begins with a <table> tag, and uses tags such as <tr>, <td> and <th> as its child nodes. Therefore, if we get a HTML subtree which begins with <table> and ends with </table>, we could get a table with knowledge. However, these tags are sometimes problematic, because they are usually used for decorative purposes. A group of

© Springer International Publishing AG 2016
F. Lehner and N. Fteimi (Eds.): KSEM 2016, LNAI 9983, pp. 302–313, 2016.
DOI: 10.1007/978-3-319-47650-6_24

researchers have focused on this problem about the identification of the genuine Web table [3, 4].

When we have got a genuine table, extracting knowledge from it becomes another important research question. This article aims to provide a new method of unsupervised classification, by considering the characteristics of the DOM tree, uses the tree similarity to measure the similarity between the tables, and then clusters the tables into classes. Finally, the method extracts knowledge in each table class, respectively.

The rest of this paper is organized as follows. Section 2 presents an overview of the related work. Section 3 gives a comprehensive analysis of Web tables from the perspectives of our experience and practical applications. Section 4 describes our knowledge extraction method based on DOM tree similarity. Section 5 shows the experiment results and Sect. 6 concludes the paper.

2   Related Work

Structured (semi-structured) knowledge tables have always been used as an important source of knowledge to construct a large-scale knowledge base or graph. DBpedia [5] and YAGO [6] are built on Wikipedia infoboxes and other structured data sources. Google's Knowledge Vault also uses HTML tables as a source for knowledge fusion [7].

Like knowledge extraction from unstructured data, we want to obtain terms of entities, concepts, and semantic classes, and their relationships, attributes and attribute values from the Web tables. Roughly, table knowledge extraction can be roughly divided into two tasks: one is to detect the critical elements of the table and the other is to discover the content structure of the table.

By detecting the critical elements of a table, we can infer the distribution of knowledge of the table. With the help of a universal probabilistic taxonomy called Probase, Wang [8] presented a framework based on a table header detector and entity detector to harvest knowledge from HTML tables. Nagy [9] provided a trainable critical cell location algorithm for extracting and factoring header paths and RDFs. Dalvi [10] extracted concept-instance pairs by clustering similar terms found in HTML tables and then assigned concept names to these clusters using Hearst patterns. Limaye [11] proposed a system, and it used an existing catalog and type hierarchy for annotating table columns and cells. Oz and Hogan [12] described a novel method to extract facts by using an existing Linked Data knowledge base to find existing relations between entities in Wikipedia tables, suggesting that the same relations may hold for other entities in analogous columns on different rows.

By discovering the content structure of a table, we can infer the semantic structure of the table. The content structure is mainly reflected in the schema of the critical elements of the table. Chen [13] used the similarity between cells, rows and columns based three metrics (i.e. String similarity, Named entity similarity, and Number category similarity) to identify the schema of the table and captured attribute-value relationships among table cells. Pivk and Cimiano et al. showed the TARTAR model which can normalize a table into a logical matrix [14]. In their early work [15], they distinguished between two functional types of cells, i.e. Attribute cells and Instance cells from HTML tables, by the core steps of the methodology: (1) Cleaning and Normalization, (2) Structure Detection, (3) Building of the Functional Table Model (FTM), and (4) Semantic Enriching of the FTM. Cafarella described a WebTables system which can extract relational information from HTML tables by attribute correlation statistics, and provided schema auto-complete and attribute synonym finding capability for getting more similar tables [1]. A similar work was presented in [17]. Adelfio extracted the schemas of HTML tables (attribute names, values, data types, etc.) by using a classification technique based on conditional random fields [18].

The above two ways are only directed to one specific table, and some other methods also use the information around the table to enhance the performance of the table extraction. Wang and Phillips [16] presented a table structure understanding algorithm designed by using optimization methods and improved the table detection result by optimizing the whole page segmentation probability. Govindaraju [19] used experiments to evaluate their ideas inspired by a simple human study that the use of context information around the table can be combined to extract a higher quality and more entity relationships.

Our approach focuses more on the formal structure of tables. We define the DOM tree similarity between the tables, and use it to cluster the tables which have the similar semantic structure. This makes our method domain-independent. Unlike most other methods mentioned above, our method does not need to understand the table with external knowledge bases.

3   Analysis of Web Tables

Before diving into the technical details, we present some analysis of Web tables first.

3.1   Web Tables Classification: An Experiential Analysis

Webpages contain a large number of tables, and unlike Web texts, tables are relatively structured information. Tables organize information in the concise form. A well-structured relational table can be structured by the entities and their relationships, such as a Wikipedia infobox, and extracting knowledge from it becomes much easier.

In this paper, we define a piece of knowledge as a triple of the form $<S, P, O>$, where S is the subject, P is a predicate or relation and O is its object. For example, some triples can be extracted from infobox table of Fig. 1, such as <China, Capital, Beijing> and <China, Largest city, Shanghai>.

| Capital | Beijing[a] 39° 55′ N 116° 23′ E |
| Largest city | Shanghai [1] |

Fig. 1. An example of Wikipedia infoboxes (https://en.wikipedia.org/wiki/China)

Largest cities or towns in China						
Sixth National Population Census of the People's Republic of China (2010)						
Rank	Name	Province	Pop.	Rank	Name	Province	Pop.
1	Shanghai	Shanghai	20,217,700	11	Chengdu	Sichuan	6,316,900
2	Beijing	Beijing	16,446,900	12	Nanjing	Jiangsu	6,238,200
3	Chongqing	Chongqing	11,871,200	13	Shenyang	Liaoning	5,718,200
4	Guangzhou	Guangdong	10,641,400	14	Hangzhou	Zhejiang	5,578,300
5	Shenzhen	Guangdong	10,358,400	15	Xi'an	Shaanxi	5,399,300
6	Tianjin	Tianjin	9,562,300	16	Harbin	Heilongjiang	5,178,000
7	Wuhan	Hubei	7,541,500	17	Suzhou	Jiangsu	4,083,900
8	Dongguan	Guangdong	7,271,300	18	Qingdao	Shandong	3,990,900
9	Hong Kong	Hong Kong	7,055,071	19	Dalian	Liaoning	3,902,500
10	Foshan	Guangdong	6,771,900	20	Zhengzhou	Henan	3,677,000

Fig. 2. An example of complicated Web tables

However, in many cases, the tables which we are dealing with in this paper are much more complicated than infobox tables consisting of "Attribute-Value" pairs. Figure 2 is an example of a common relational table. We can see that multiple relationships and facts can be extracted, but the way of the extraction is not like the case of infoboxes.

In a good summary, [20] provided a taxonomy of table types along with their frequencies on the Web, as shown in Fig. 3. A similar taxonomy was presented in [2]. However, this taxonomy for practical use is limited.

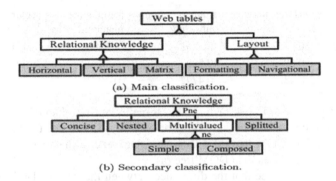

(a) Main classification.

(b) Secondary classification.

Fig. 3. A taxonomy of Web tables

Visually, a table reveals the organizational structure of the data. There is a certain degree of structural similarity between different tables, and a kind of similar structure contains all kinds of similar entity relationships. So if we know that some of the tables have the same structure, we can use the same way to extract the knowledge from them.

3.2 Formal Structure and Similarity of Web Tables

Different styles of table structure come from the "Rowspan" and "Colspan", but the structure of the DOM tree has no change and just differ in HTML tag's attribute, such as <tr colspan = "3">. More importantly, some of the tags such as <th> and <caption> would also affect the form of the semantic structure, and the distribution of these tags in the tree has great impact on the similarity of DOM trees.

Fig. 4. Some common Web tables

As shown in Fig. 4, we can find that the semantic structure in (a) and (b) is the same from the table-style perspective. But when we use the traditional methods [21–23] to measure the similarity between table (a) and (b), we get the result as shown in Table 1.

Table 1. Traditional methods to measure the similarity between three tables in Fig. 4

Methods	Edit distance	Longest common substring	Simple tree matching	Our method
Similarity(a, b)	0.5257	0.5232	0.5186	1.0
Similarity(a, c)	0.5024	0.5003	0.5854	1.0
Similarity(b, c)	0.4859	0.4792	0.4677	1.0

Actually, as shown in Fig. 4, table (b) has three more rows than table (a), and table (c) has two more columns than table (a); that is to say, DOM tree (b) has three more <td> leaf nodes than DOM tree (a). Likewise, DOM tree (c) has two more <th> and eight more <td> leaf nodes than DOM tree (a).

From Fig. 5, we can see that the difference between the various DOM trees is the number of the repeated or same substructures, such as the content marked in the red border. The methods in the literature [21–23] cannot solve the problem of the repeated substructures, and that is why the similarity results as described in Table 1 range from 0.4 to 0.6.

Through the analysis above, we can assume that the number of the same substructures does not influence the inherent global structure of a table, and it is merely a continuation of the function of some elements of the structure. Intuitively, the continuation of the function of HTML tags is like the extension of phoneme of the speech. When we repeat an utterance, although the speech rate may not be the same, what the phonemes the utterance contains is the same. In other words, when we use the features of the phonemes and time to describe different speeds of an utterance to be spoken and form different speech waveform sequences, if we are going to calculate the similarity of these sequences, the results should be that they are very similar or the same. As shown in Table 1, the result of our method to measure the similarities of between table (a) and (b), between (a) and (c), and between (b) and (c) are all 1.0, as we expect.

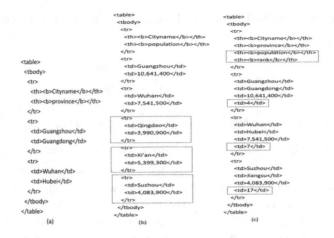

Fig. 5. DOM trees for tables (a), (b) and (c) in Fig. 4, respectively

4 Knowledge Extraction from Web Tables Based on DOM Tree Similarity

4.1 Acquiring Web Tables

A table is presented in the form of a DOM tree. DOM tags are used to represent a DOM tree. Usually, we can obtain the table by the tag <table>, no matter whether it is a genuine table or not. We think that the tables used for decorative or layout purposes may also have certain fixed structure, and it may be extracted by way of classification, without separately analyzing whether it is genuine or non-genuine. We use the HTML parser to get tables and retain their tree structure.

4.2 Defining the Similarity

Dynamic time warping (DTW) is a sequence alignment method for comparing two sequences' similarity [24]. It is widely used in speech recognition to deal with the problem of different speech rates. To some extent, the diversification of tables comes from the results of "warping". Therefore, based on our observation, we introduce the DTW method to measure the similarity between two DOM trees.

Let $Table_i$ and $Table_j$ be two Web tables, which are represented by DOM $Tree_i$ and $Tree_j$, respectively. As a tree, we let the tag <table> be the root node of the tree, and then traversing it gets a $path_i$ for a leaf node i. For example, a path can be represented as <table> ~ <tbody> ~ <tr>~<td> ~ <text>, and the <text> only represents the text content of the leaf node. The number of leaf nodes of a tree equals to the number of the paths it has. Comparing the similarity of two trees is a comprehensive comparison of the similarity between the relevant paths.

Suppose $Tree_i$ has m paths and $Tree_j$ has n paths. Let $\gamma\left(path_i, path_j\right)$ be the string edit distance [21] of two paths, let $\left|path_i\right|$ be the length of $path_i$ and $\left|path_j\right|$ be the length of

$path_j$. Let $\psi(path_i, path_j)$ be the similarity between $path_i$ and $path_j$, which we can get it by normalizing:

$$\psi(path_i, path_j) = \frac{\gamma(path_i, path_j) - Min(\gamma)}{Max(\gamma) - Min(\gamma)} \tag{1}$$

Where, $Min(\gamma) = -a \times min\left\{ |path_i|, |path_j| \right\} - b \times \left| |path_i| - |path_j| \right|$,   and

$Max(\gamma) = a \times min\left\{ |path_i|, |path_j| \right\}$. In the above formula, a denotes the weight of the string edit operation of "match", b denotes the weight of the string edit operation of "delete" or "insert". When we compare the two paths with each other, we only need to determine whether each tag of $path_i$ can match with a tag of $path_j$ or not.

After calculating the similarity of the two paths, we convert it to the distance between the two paths. By using the DTW method with the minimum cumulative distance, we can achieve the greatest match of two DOM trees which are composed of leaf paths. Formula (2) below shows how we calculate the cumulative distance between two DOM trees:

$$\Phi[i,j] = \phi(path_i, path_j) + min\{\Phi[i,j-1], \Phi[i-1,j], \Phi[i-1,j-1]\} \tag{2}$$

In the formula, $\Phi[i,j]$ denotes a $(m+1) \times (n+1)$ matrix, m and n denotes the number of paths of $Tree_i$ and $Tree_j$ respectively. $\Phi[i,0]$ and $\Phi[0,j]$ are initiated with ∞, $\Phi[0,0] = 0$, and $\phi(path_i, path_j) = 1 - \psi(path_i, path_j)$. So $\Phi[m,n]$ denotes the minimum cumulative distance i.e. the distance between two trees. Let N be the number of nodes in the warping path. In order to avoid the influence by the size of the tree and the length of the warping path, we can get the similarity between two trees by normalizing:

$$Sim(Tree_i, Tree_j) = 1 - \Phi[m,n]/N \tag{3}$$

4.3   Table Clustering

Through the establishment of two trees' similarity measure, we can obtain the similarity between two tables, which can be used to perform cluster analysis. We refer to the method described in [25], which proposed an approach based on the idea that cluster centers are characterized by a higher density than their neighbors and by a relatively large distance from points with higher densities. For each table (i.e. $Tree_i$), two quantities should be computed: its local density ρ_i and its distance δ_i from points of higher density. Both these quantities depend only on the distances d_{ij} between any two tables. The local density ρ_i of table i is defined as $\rho_i = \sum_j \chi(d_{ij} - d_c)$, where $\chi(x) = 1$ if $x < 0$ and $\chi(x) = 0$ otherwise, and d_c is a cutoff distance. The main innovation of our paper is to calculate the similarity between two tables, as shown in Formula (3), we can get the distance d_{ij} by $d_{ij} = 1 - Sim(Tree_i, Tree_j) = \Phi[m,n]/N$. Firstly we calculate the similarity between any two tables which need to be clustered and turn them into distance d_{ij}, then we refer to the clustering algorithm [25] and get the result of table clustering. Finally we use the method NCMDS[1] to represent the result.

In order to obtain a good structure of the table, we choose HTML tags <caption>, <td> or <th> as leaf nodes, which are embedded with the corresponding cell content of the table. But in practice, each cell of the table may have some other forms for dividing, such as tags <span>, <li> and <div>. If we use these tags as leaf nodes, the number of paths of one DOM tree would increase dramatically, which greatly increases the complexity ($O(mn)$) of the dynamic programming algorithm. Therefore, we implement a two-stage clustering strategy:

1. We use tags <caption>, <td> and <th> as the leaf nodes, and get the result of the coarse-grained clustering;
2. For tables of each cluster, we use tags <span>, <li>, <div> and the like as leaf nodes, increase the weight of <th> appropriately in the process of paths alignment for highlighting the headers, and get the result of the fine-grained clustering.

4.4 Knowledge Extraction

After characterizing the DOM tree's similarity between the tables to cluster the tables which have the similar semantic structure, we extract knowledge in each class, respectively.

For a table of one class of Wikipedia tables as shown in Fig. 6, horizontal relational tables are the most common tables in Wikipedia. We know that "Name" and "Position" are labeled by tag <th>, which means the header of the table. Meanwhile, all the other remaining cells are labeled by tag <td>. So we can extract knowledge (i.e. facts) such as <David Sablan, Position, National Committeeman>.

Name	Position
Michael "Mike" Benito	Chairman
Evelyn Casil	Treasurer
David Sablan	National Committeeman
Margaret McDonald Glover Metcalfe	National Committeewoman
Jose "Joe" Duenas	Executive Director

Fig. 6. Example for a horizontal relation table (https://en.wikipedia.org/wiki/Republican_Party_of_Guam)

In this paper, as shown in Fig. 7, unlike the traditional methods, our method does not rely on any resource such as Linked Data, Existing entities library, or Concept-instance base. With artificial judgment to decide which class one table belongs to, we extract knowledge from it and the remaining tables of this class by the corresponding extraction method which had been designed based on the semantic structure of the class.

---

[1] http://cn.mathworks.com/help/stats/examples/non-classical-multidimensional-scaling.html.

Figure 7 is an extraction method for the class which only contains horizontal relation tables. The algorithm has a complexity of O(*MN*).

Algorithm. Knowledge Extraction to Horizontal Relation Table

    Suppose that the table contains *M* rows and *N* columns, < *S(k)*, *P(k)*, *O(k)*> represents knowledge, the header (th) are in the first row and the *Subject* are in the first column by default (if not, some filter rules can be enabled such as the number or alphabetic order cannot be *S(k)*, *S(k)* should be an entity).

1: for each row (*i*=0 to *M*) do
2: for each column (*j*=0 to *N*) do
3:    if th[*i*=0,*j*] and td[*i*,*j*=0] accord with filter rules do
        P(k)← th[*i*=0,*j*]; *O(k)*← td[*i*,*j*]; *S(k)*← td[*i*,*j*=0];
4: end for
5: end for

Fig. 7. Algorithm for knowledge extraction

5   Experimental Results

By using 5000 Wiki tables which were extracted from Wikipedia at random as the source corpus, we obtain the clustering result of the experiments as shown in Fig. 8.

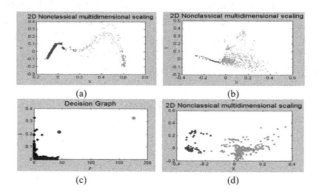

Fig. 8. Table clustering results of 5000 tables

    As Fig. 8(a) shows, 5000 Wiki tables can be classified into two clusters (divide 5000 into 4484 and 516) in the first stage, and then the tables in one cluster (the red point one) are clustered into another three clusters (divide 4484 into 2679, 1023 and 782) in the second stage, as shown in Fig. 8(b). Meanwhile, the tables in the other cluster (the green point one) in the first stage are clustered into another two clusters (divide 516 into 408 and 108) in the second stage, as shown in Fig. 8(d). As seen from the results, with artificial judgment, the 5000 Wiki tables are clustered into *relational tables* and *nested tables* in the first stage, then the *relational Tables* (4484) are classified into *horizontal relation Tables* (2679), *matrix relation Tables* (1023) and *no-header Tables* (782), and the *nested Tables* (516) are classified into *nested relational Tables* (408) and *decorated Tables* (108) in the second stage. Specially, the *vertical relation tables* cannot be well separated from the *matrix relation tables* by table clustering. We can see that most of the Wikipedia tables are relational tables, and some tables for decorative purposes can be filtered out so that we do not need to judge the genuineness of the table, as done in [3, 4]. To some extent, the clustering result is close to the empirical taxonomy like [2, 20].

Because each class of tables has a corresponding extraction method which are based on the semantic structure of this class, the quality of table clustering directly affects the quality of the extracted knowledge. We choose 100 tables which are the class of *horizontal relation tables* at random as the test corpus, then implement the preliminary extraction as described in Fig. 7, and the result shows that roughly 65.5 % of the facts are correct and 34.5 % are incorrect.

An analysis shows that incorrect facts mainly consist of three cases:

1. Case 1 (35 %): Tables do not belong to their right class due to some clustering error. The tables which have a caption (or multi-row headers) mixed in the *horizontal relation tables* which have one-row header. So the *Subject* (or Predicate) is incorrect. As Fig. 9 shows, *Predicate* (*"2011"*) and *Predicate* (*"1970"*) should be added with the captions (*"Ten largest metropolitan areas by population"* and *"Arizona Racial Breakdown of Population"*) of the two tables, respectively.

	S	P	O	Wiki Article	Case
Correct	Pat Saiki	Position	State Chairman	Hawaii Republican Party	
	Crisis Aftermath: The Spectre	ISBN	1-4012-1506-8	Spectre (comics)	
	Shopping	Compare Taiwanese Hokkien	監街（sèh-koe）	Singaporean Hokkien	
	1st	Indonesian	pertama	Indonesian language	
Incorrect	Portuguese	Number of word	131	Indonesian language	2
	Jennifer Lewis	Year	1992 (2nd team)	Duke_Blue_Devils	2
	US Billboard Adult Top 40	Peak position	33	Save Me (Remy Zero song)	2
	Vancouver	2011	2313328	British Columbia	1
	White	1970	90.60%	Arizona	1
	Pension Insurance	Yearly ceiling	West: 72,600.00 €/ East: 62,400.00 €	Taxation in Germany	3
	att tala, to speak	Indicative	talar, speak(s)	Swedish grammar	3

Fig. 9. Some facts (<*S, P, O*>) of the result of knowledge extraction

2. Case 2 (33 %): The *Subject* does not appear in the table at all, it could be the article's title of a Wikipedia page, or appears in the context of the table. As Fig. 9 shows, *Subject* (*"Portuguese"*) should be combined with the table's context (*"Indonesian Language origin"*), *Subject* (*"Jennifer Lewis"*) should be transformed by the table's context (*"Duke Women's Soccer program received NSCAA All-American honors"*); in other words, the *Subject* and the table's context mentioned before are *part-whole* relations. *Subject* (*"US Billboard Adult Top 40"*) should be added with the article's title of the Wikipedia page (*"Save Me (Remy Zero song)"*).

3. Case 3 (32 %): Due to some other reasons, such as the content of a cell having multi-valued or complex forms. As Fig. 9 shows, the *Subject* (*"att tala, to speak"*) and *object* (*"West: 72,600.00 €/East: 62,400.00 €"; "talar, speak(s)"*) represent multiple values.

6 Conclusion

With a detailed analysis of the table structure of various forms, we design a method for calculating the DOM tree similarity between different web tables based on DTW. By analyzing the characteristics of DOM trees, we use the tree's similarity to measure the similarity between the tables, and then cluster the tables into some classes, extract knowledge in each class respectively. Experimental results show that the quality of the extracted of knowledge is satisfactory.

By unsupervised clustering, we can obtain various types of tables based on the corpus of any form, and this has more practical value. Meanwhile, we do not need to spend time to ponder over the genuineness of the tables and rely on existing resources to support the knowledge extraction.

As future research direction, we will improve the performance of the clustering algorithm to adapt to the massive data and make full use of knowledge which has been extracted for mutual verification in order to achieve a higher accuracy.

Acknowledgment. This work is supported by the National Science Foundation of China (under grant Nos. 91224006 and 61173063) and the Ministry of Science and Technology (under grant No. 201303107).

References

1. Cafarella, M.J., Halevy, A., Wang, D.Z., Wu, E., Zhang, Y.: WebTables: exploring the power of tables on the web. Proc. VLDB Endow. **1**(1), 538–549 (2008)
2. Crestan, E., Pantel, P.: Web-scale table census and classification. In: Proceedings of the Fourth ACM International Conference on Web Search and Data Mining, pp. 545–554. ACM, New York (2011)
3. Wang, Y., Hu, J.: A machine learning based approach for table detection on the web. In: Proceedings of the 11th International Conference on World Wide Web, pp. 242–250. ACM, New York (2002)
4. Son, J.W., Lee, J.A., Park, S.B., Song, H.J., Lee, S.J., Park, S.Y.: Discriminating meaningful web tables from decorative tables using a composite kernel. In: IEEE/WIC/ACM International Conference on Web Intelligence and Intelligent Agent Technology, vol. 1, pp. 368–371. IEEE Computer Society (2008)
5. Auer, S., Bizer, C., Kobilarov, G., Lehmann, J., Cyganiak, R., Ives, Z.G.: DBpedia: a nucleus for a web of open data. In: Aberer, K., et al. (eds.) ASWC 2007 and ISWC 2007. LNCS, vol. 4825, pp. 722–735. Springer, Heidelberg (2007)
6. Suchanek, F.M., Kasneci, G., Weikum, G.: Yago: a core of semantic knowledge. In: International Conference on World Wide Web, vol. 272, pp. 181–221. ACM, New York (2007)
7. Dong, X., Gabrilovich, E., Heitz, G., Horn, W., Lao, N., Murphy, K., et al.: Knowledge vault: a web-scale approach to probabilistic knowledge fusion. In: Proceedings of the 20th ACM SIGKDD International Conference on Knowledge Discovery and Data Mining, pp. 601–610. ACM, New York (2014)

8. Wang, J., Wang, H., Wang, Z., Zhu, K.Q.: Understanding tables on the web. In: Atzeni, P., Cheung, D., Ram, S. (eds.) ER 2012 Main Conference 2012. LNCS, vol. 7532, pp. 141–155. Springer, Heidelberg (2012)
9. Nagy, G.: Learning the characteristics of critical cells from web tables. In: International Conference on Pattern Recognition, pp. 1554–1557. IEEE (2012)
10. Dalvi, B.B., Cohen, W.W., Callan, J.: WebSets: extracting sets of entities from the web using unsupervised information extraction. In: ACM International Conference on Web Search and Data Mining, pp. 243–252. ACM, New York (2013)
11. Limaye, G., Sarawagi, S., Chakrabarti, S.: Annotating and searching web tables using entities, types and relationships. Proc. VLDB Endow. **3**(3), 1338–1347 (2010)
12. Oz, E., Hogan, A., Mileo, A.: Using linked data to mine RDF from Wikipedia's tables. In: ACM International Conference on Web Search and Data Mining, pp. 533–542. ACM, New York (2014)
13. Chen, H.H., Tsai, S.C., Tsai, J.H.: Mining tables from large scale HTML texts. In: Conference on Computational Linguistics, pp. 166–172. ACL, Stroudsburg (2000)
14. Pivk, A., Cimiano, P., Sure, Y., Gams, M., Rajkovič, V., Studer, R.: Transforming arbitrary tables into logical form with TARTAR. Data Knowl. Eng. **60**(3), 567–595 (2007)
15. Pivk, A., Cimiano, P., Sure, Y.: From tables to frames. Web Semant. Sci. Serv. Agents World Wide Web **3**(2–3), 132–146 (2005)
16. Wang, Y., Phillips, I.T., Haralick, R.M.: Table structure understanding and its performance evaluation. Pattern Recogn. **37**(7), 1479–1497 (2004)
17. Bhagavatula, C.S., Noraset, T., Downey, D.: Methods for exploring and mining tables on Wikipedia. In: ACM SIGKDD Workshop on Interactive Data Exploration and Analytics, pp. 18–26. ACM, New York (2013)
18. Adelfio, M.D., Samet, H.: Schema extraction for tabular data on the web. Proc. VLDB Endow. **6**(6), 421–432 (2013)
19. Govindaraju, V., Zhang, C., Ré, C.: Understanding tables in context using standard NLP toolkits. In: Meeting of the Association for Computational Linguistics, vol. 2, pp. 658–664. ACL (2013)
20. Lautert, L.R., Scheidt, M.M., Dorneles, C.F.: Web table taxonomy and formalization. ACM SIGMOD Rec. **42**(3), 28–33 (2013)
21. Tai, K.C.: The tree-to-tree correction problem. J. ACM **26**(3), 422–433 (1979)
22. Bergroth, L., Hakonen, H., Raita, T.: A survey of longest common subsequence algorithms. In: Proceedings of the Seventh International Symposium on String Processing Information Retrieval, pp. 39–48. IEEE Computer Society, Washington, DC (2000)
23. Yang, W.: Identifying syntactic differences between two programs. Softw. Pract. Exp. **21**(7), 739–755 (1991)
24. Sakoe, H., Chiba, S.: Dynamic programming algorithm optimization for spoken word recognition. IEEE Trans. Acoust. Speech Sig. Process. **26**(1), 43–49 (1978)
25. Rodriguez, A., Laio, A.: Clustering by fast search and find of density peaks. Science **344**(6191), 1492–1496 (2014)

Single Image Dehazing Using Hölder Coefficient

Dehao Shang, Tingting Wang, and Faming Fang[✉]

Shanghai Key Laboratory of Multidimensional Information Processing,
East China Normal University, Shanghai 200241, China
fmfang@cs.ecnu.edu.cn

Abstract. Restoring the true scene appearance from hazy image is a challenging task and one of a most necessary part in image processing system. As we know, the clear-day image must have higher contrast compared to the hazy image. Our main idea is that the hazy image is enhanced based on this observation to achieve dehazing objective. Firstly, we use a new metric, simple but powerful Hölder coefficient, to estimate the hazy density roughly. In order to make the estimation density map more reasonable, we apply proposed energy function to refine it. Based on the refined map, we propose a new method to estimate the atmosphere light. Secondly, three new terms, which are used to enhance image, are modeled into energy function. Solving this energy function, transmission map can be obtained. Finally, we get haze-free image by using the transmission map. Experiment results demonstrate that our algorithm has similar or better performance compared to the state-of-art algorithms.

Keywords: Dehazing · Contrast enhancement · Hölder coefficient

1 Introduction

Nowadays, computer vision has become increasingly important in the modern world. Many applications of computer vision system, for example, urban transportation system, tracking system, detection system, and more high-level vision system, usually require high-quality images. However, images captured in outdoor environments often suffer from degradation, the main reason for which is there are suspended particles such as dust, mist, and fumes in the atmosphere. Since these particles absorb the lights from objects and scatter the atmosphere lights, captured images generally suffer from non-trivial degradations. Photographs of those image typically suffer having low contrast and offer a limited visibility of the scene [2]. In other words, the hazy images lose the image detail information and color fidelity.

In the area of computer vision, the monochrome atmosphere scattering model can be used to describe the physical imaging principle under weather conditions of haze or fog [12]. Based on this assumption of imaging principle, the process of restoring the true scene appearance is generally called dehazing.

Image dehazing has been drawing more and more attention for two major reasons, as stated in the following. Firstly, the free haze photography has clear

© Springer International Publishing AG 2016
F. Lehner and N. Fteimi (Eds.): KSEM 2016, LNAI 9983, pp. 314–324, 2016.
DOI: 10.1007/978-3-319-47650-6_25

results, vivid color information and abundant image details. These characters of image will make people pleasing. Secondly, most existing vision algorithms, no matter in low-level image process area or high-level object recognition area, assume that the input image is the true scene, which is unpractical. Therefore, it is necessary to preprocess the input image with dehazing algorithm.

Early in the dehazing field, haze removal methods strongly depend on additional requirements. [14] requires multiple images captured at different weather conditions. [16,17] take two images through a polarizer at different angles. Schechner develops an efficient algorithm to remove haze based on two images in the same scene. The other methods is to require additional information about the haze image scene, such as depth information or 3D model [8,13]. Additional information also can be provided in interactive mode [13]. Although these algorithms have good robustness and are proven to have excellent experiment results, these additional requirements or information restrict their application range.

Due to the limitations in a real application, there is an increasing focus on single image haze removal. Lacking of extra information about the real scenario, single image dehazing is a more challenging problem compared to other methods. However, in recent years, many excellent approaches have been proposed to remove effects of haze or fog only via a single haze image, which typically rely on prior assumptions. Tan [18] obtain data term to maximize the local contrast and smooth term to avoid color saturation respectively according to two important prior assumptions he introduced on contrast and scene depth. Then, he use Markov random fields to model these two terms, and subsequently solve all values of airlight by using the existing inference techniques, such as belief-propagation. Based on refined image formation model derived from atmosphere scattering model, Fattal [1] introduce a physics-based assumption that the transmission and surface shading function are locally uncorrelated. He propagates the transmission from thin haze area to dense haze area by modeling the transmission map using Markov random field. He et al. [6] found a simple statistical-based dark-channel prior through observing lots of haze-free outdoor images. Under this prior information, rough estimation of transmission can be obtained directly from haze image. Moreover, soft matting [10] or guided filter [5] can be used to refine the transmission. Researchers also develop some real-time methods to remove haze. Under the assumption that atmospheric veil change smoothly, Tarel [20] remove hazy by using median filter to estimate atmospheric veil. Gibson [3] develop a fast method to remove haze by using locally adaptive Wiener Filter.

Recently, some new method based on learning have also been introduced into dehazing field. Tang [19] investigates different haze-relevant features in a learning framework and comes to the conclusion that the dark-channel feature is the most importance and other haze-relevant features contribute significantly in a complementary way. Zhu [22] find a simple but powerful color attenuation prior, namely the linear correlation between depth and color attenuation. Under certain conditions, these methods can get excellent results.

In this paper, our main idea is that we eliminate the haze effects by enhancing local contrast which can be measured by Hölder coefficient.

2 Proposed Method

2.1 Optical Model

Due to aerosols present in the atmosphere in bad weather, the direction of light propagation will be defected. Meanwhile, some of the light ray from the scene objects to viewer will be absorbed. Therefore, this physical process [12] can be described as a combination of absorption and scattering as follows:

$$I(x) = J(x)t(x) + A(1 - t(x)) \tag{1}$$

where I is the input RGB image, J is the underlying scene radiance, namely the clear-day image we would like to obtain, x stands for the pixel coordinates, A is the airlight color vector. $J(x)t(x)$ is called direct attenuation and denotes that some of the light ray from scene objects is absorbed by suspended particles. $A(1 - t(x))$ is called airlight and denotes that the light reflected from objects surface is mixed with ambient light. The scalar t in $[0, 1]$ is the transmission rate indicating the fraction of the light reaching the camera without being scattered. Generally, assuming the atmosphere condition is homogenous in entire image, t can be expressed in detail as follows:

$$t(x) = e^{-\beta d(x)} \tag{2}$$

where β is the medium extinction coefficient that we assume to be global constant across entire image for the same kind of weather, $d(x)$, generally called scene depth, is the distance between the scene object and the observer at point x. This equation shows transmission rate decrease exponentially with the scene depth d.

For model (1), under the assumption that t is the same in each color channel, there are 3 constrains but 7 unknowns for every pixel. It is obvious that recovering haze-free image (solving the J) is an under-constrained problem. In order to solve this ill-posed problem, general prior constrains that may work for most weather condition are added into Eq. (1). These prior assumptions are described in the next section.

2.2 Estimate the Haze Density

In haze or foggy weather, captured image always approach white and leads to relative smooth of scene depth. It means that images captured in hazy or foggy weather always have lower contrast than the clear one in the same scene. Based on this observation, we can seek a certain metric to measure the density of the haze. The good news is the Hölder coefficient can roughly meet our needs.

As a powerful regularity, the Hölder coefficient is an effective and power metric of local contrast and feature, and it has been applied to many image processing tasks, such as image interpolation and denoising [9, 21]. Let ω_x be a certain neighbourhood of x, formally, the Hölder coefficient $H_I(x)$ of the image I in ω_x is defined as [7]:

$$H_I(x) = \max_{y \neq z \in \omega_x} \left\{ \frac{\sum\limits_{R,G,B} \|I(y) - I(z)\|}{\|y - z\|} \right\} \tag{3}$$

According to (3), the $H_I(x)$ is in proportion to the contrast of the image in ω_x, and in inverse proportion to the density of the haze. Thus the inverse Hölder coefficient can be used to measure the thickness of the haze, which means the haze density map L_I can be obtained by:

$$L_I(x) = \frac{c}{H_I(x)},$$

where c is a constant that makes the range of $L_I(x)$ into $[0, 1]$.

We show the estimated map L_I in the second row of Fig. 1, from which we can see that the estimation of hazy density is in accord with the actual situation but a little rough. Therefore, to reduce the influence of the noise and soft edges of input images, we refine the L_I as follows:

$$L_I = \min_{L_I} E(L_I) = \frac{1}{2} \left\| M \odot (L_I - L_I^0) \right\|^2 + \lambda \|\nabla L_I\| \tag{4}$$

where $L_I^0 = \frac{c}{H_I(x)}$ is a rough version of L_I, $\odot$ is the dot product, L_I is the haze density map which we would like to obtain, λ is a constant and M is defined as follows:

$$M(x) = \begin{cases} 0, & L_I^0(x) \leqslant \delta, \\ 1, & \text{else.} \end{cases} \tag{5}$$

where the threshold δ filter small image details. The refined L_I suppress the image details by adding the constraints that the haze is somewhat smooth and piecewise constant, which makes the estimation map L_I more reasonable. The refined L_I is displayed in the third row of Fig. 1. The noise is significantly less than the rough estimation map.

2.3   Recover the Haze-Free Image

As aforementioned, with the haze become denser, the image contrast become lower. So, the essence of the dehazing is to enhance the contrast of the image. Since Hölder coefficient is a good description of local features and image contrast, our model is proposed as follows:

$$\min_J E(J) = \|W \odot (J - I)\|^2 - \beta \|H_J\|_1 \quad \text{s.t. } J \in [0, 1], \tag{6}$$

where J is the haze-free image, I is the input hazy image. The weight W is defined as follows:

$$W = e^{-\mu L_I}, \tag{7}$$

Fig. 1. First row: the input hazy image, second row: the rough estimation of the haze density, third row: the refined haze density map, fourth row: the atmosphere light estimated from red pixels (Color figure online)

where μ is a constant. It can be seen that $\mathbf{W}$ is inversely related to $\mathbf{L}_I$ (the density of haze). Thus, the dehazed image $\mathbf{J}$ can be significantly enhanced in the region with very dense haze. Whereas in the region that contains little haze, the difference between the original image and dehazed one will be very small. Due to the larger weight, that's why (6) can be used to dehazing.

2.4   Algorithm

The constrained problem (6) can be converted into unconstrained one:

$$\min_{\mathbf{J}} E(\mathbf{J}) = \|\mathbf{W} \odot (\mathbf{J} - \mathbf{I})\|^2 - \beta\|\mathbf{H}_J\|_1 + P(\mathbf{J}), \tag{8}$$

where $P(J)$ is projection operator that defined as follows:

$$P(\mathbf{J}) = \begin{cases} 0, & 0 \leqslant \mathbf{J} \leqslant 1 \\ \infty, & \text{o.w.} \end{cases}. \tag{9}$$

Theoretically, optimal $\mathbf{J}$ in (8) can be obtained by minimizing the energy function using some classical iterative algorithms. However, since the energy is non-convex, and the derivation of the term $\|\mathbf{H}_J\|_1$ is hard to obtain, the classical algorithms is non-ideal. In what follows, a new approximate method for our minimization problem is presented.

Our algorithm is based on the two observations:

- As stated in many literature such as [10], the scene depth of an image generally changed smoothly. Thus, the transmission in neighboring pixels tend to be the same or have tiny difference.

– the Eq. (1) can be rewritten as follows:

$$\mathbf{J}\left(\boldsymbol{x}\right) = \frac{\mathbf{I}\left(\boldsymbol{x}\right) - \boldsymbol{A}}{\mathbf{t}} + \boldsymbol{A} \tag{10}$$

where $\boldsymbol{A}$ is assumed to be known. For each pixel $\boldsymbol{x}$, firstly, N values of transmission (namely, $t_1(\boldsymbol{x}), t_2(\boldsymbol{x}), ...t_N(\boldsymbol{x})$) are uniformly sampled from $[0, 1]$. Then these corresponding $J(\boldsymbol{x})$ (namely, $J_i(\boldsymbol{x}), i = 1, .., N$) can be obtained by using (10). Substituted $J_i(\boldsymbol{x})$ into unconstrained energy function (8), N values of E (namely, $E_i^{\boldsymbol{x}}$) are calculated. For the minimum of the energy is preferred, the $t_i(\boldsymbol{x})$ related to $E_{min} = \min\limits_{i=1,..,N} E_i^{\boldsymbol{x}}$ is selected as optimal transmission for $\boldsymbol{x}$. Using the optimal transmission, the final dehazed result can be easily obtained with (10). The formal flow of our algorithm is described in Algorithm 1.

Algorithm 1. Our Proposed Algorithm

1: Input: hazy image I and parameters μ, β and N.
2: **for** each pixel $\boldsymbol{x}$ **do**
3:     **for** $i = 1, ..., N$ **do**
4:         $J_i(\boldsymbol{x}) = \frac{I(\boldsymbol{x}) - A}{t_i(\boldsymbol{x})} + A$
5:         $E_i^{\boldsymbol{x}} = E(J_i)$
6:     **end for**
7:     $E_{min} = \min\limits_{i=1,..,N} E_i^{\boldsymbol{x}}$.
8: **end for**
9: Output: $J = \frac{I-A}{t+A}$.

2.5   Estimate the Atmosphere Light

According to Eq. (2), when scene depth approach infinity, transmission t will be equal to zeros. In other words, Eq. (1) reduces to $\mathbf{I}\left(x\right) = \mathbf{A}$. Hence, the atmosphere light can be approximately estimated from these pixels with infinite depth. In previous works, these pixels can be found in the following ways: user interactive mode [15], searching for the brightest patch [18], or looking for the brightest pixel based on certain prior, such as dark channel prior [6].

In this paper, we present a new method to estimate atmosphere light. According to the inverse relationship between the haze density and the scene depth, we can consider the region with highest haze density as infinity region. Specifically, we pick the top 0.1 % brightest pixels from the haze density map (for example, the region that marked as red in the fourth row of Fig. 1), then regarded them as the infinity region. Among these pixels, the pixels with the highest intensity in the hazy image $\mathbf{I}$ is selected as the atmospheric light.

3   Experimental Results

In this section, we compare our algorithm versus current state-of-the-art dehazing algorithms from both qualitative and quantitative aspects. In some cases, our experimental results could have a small amount of halo-effects. To prevent these halo-effects from happening, we apply the guilded filter [5] to refine the transmission map. We conduct our experiments on a 2.7 GHz Intel Core i5 CPU.

Qualitative Comparison
Figures 2 and 3 show a number of the comparisons we made against state-of-the-art methods. As we can see from (d) in Fig. 2 and (b) in Fig. 3, although Tan's method [18] can yield impressive results in various scenarios, However, it also has strong halo-effects at the region of image edge, and produce over-saturated color. Compared with (b) (c) (e) in Fig. 2, our experimental result have comparable performance except for the differences in the overall tint of colors. As is shown in Fig. 3 (c) (d) (e), some hazes are not fully removed (e.g. at centre-right region). In contrast, except for exhibiting similar visibility and local contrast, our algorithm can well remove most hazes in the image and produce more clear result with vivid color information.

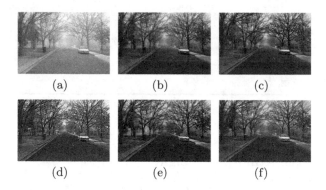

Fig. 2. (a) Input image, (b) He et al. [6], (c) Fattal [2], (d) Tan [18], (e) Nishino [15]. (f) Our method. (Color figure online)

Qualitative Evaluation
In order to further prove the effectiveness of our algorithm, the following three experiments are conducted for quantitative analysis.

Two indicators [4] are introduced to quantitatively measure experimental results: e is the rate of edges newly visible after restoration, σ is the percentage of pixels which becomes completely black or completely white after restoration. From these two indicators perspective, the aim of dehazing algorithm is to maximum the visual edge, and minimum the percentage of pixels which become white or black.

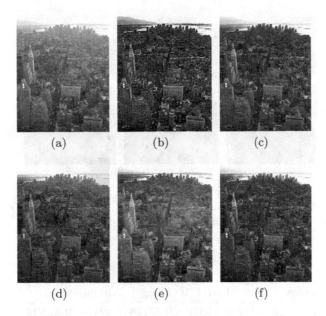

Fig. 3. (a) Input image, (b) Tan [18], (c) He et al. [6], (d) Fattal [1], (e) Kopf [8]. (f) Our method. (Color figure online)

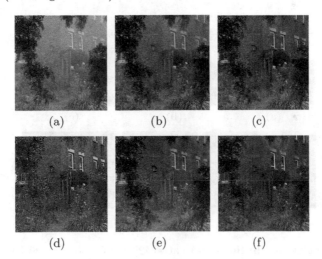

Fig. 4. (a) Input image, (b) He et al. [6], (c) Fattal [2], (d) Gibson [3], (e) Fattal [1], (f) Our method. (Color figure online)

Table 1. Red wall Fig. 4

Index	He et al.	Fattal [2]	Gibson	Fattal [1]	Our
e	0.06637	0.07998	0.03999	0.08829	**0.10817**
$\sigma\%$	3.25960	2.53840	1.14950	4.49800	**0.23687**

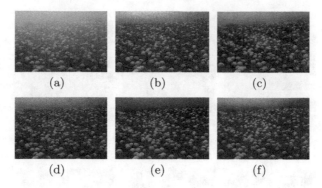

Fig. 5. (a) Input image, (b) Gibson [3], (c) Fattal [1], (d) Meng [11], (e) Nishino [15], (f) Our method. (Color figure online)

Table 2. Pumpkins Fig. 5

Index	Gibson	Fattal [1]	Meng	Nishino	Our
e	0.37162	0.14063	0.37178	0.32554	**0.41524**
$\sigma\%$	1.86170	0.07833	0.07250	0.22958	**0.03041**

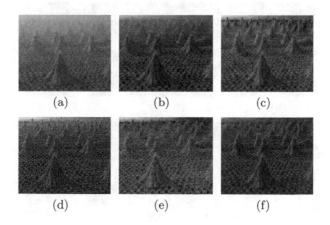

Fig. 6. (a) Input image, (b) He [6], (c) Fattal [2], (d) Gibson [3], (e) Nishino [15], (f) Our method. (Color figure online)

Table 3. Wheat Fig. 6

Index	He	Fattal [2]	Gibson	Nishino	Our
e	0.35117	0.27669	0.11340	0.16292	**0.37147**
$\sigma\%$	**0.00000**	0.01176	3.68170	0.16745	**0.00000**

Figures 4, 5, and 6 show the comparisons of our approach with He et al. [6], Fattal [2], Gibson [3], Fattal [1], Nishino [15], and Meng [11]. Setting the superiority of visual effects (i.e. the thoroughness of dehazing and the good color reproduction properties) aside, we pay attention to comparisons of the two mentioned indicator, which is shown in Tables 1, 2 and 3. It can be demonstrated that our algorithm reflect more details and improve the contrast of the image (i.e. grass leaf in Fig. 4 and textures in Fig. 6) with minimalist set of information lost.

4 Conclusion

In this paper, we make full use of general rules to develop new dehazing algorithm. Firstly, we apply a new measure, Hölder coefficient, to describe image contrast. Based on this new measure, energy function of haze density is proposed, and is solved to obtain haze density map, from which the atmosphere light can be estimated from the haze density map. Secondly, we propose a new energy function to estimate transmission. Subsequently, new method is proposed to solve this energy function. Finally, we can use the estimated transmission map and atmosphere light to obtain free image images.

Acknowledgement. This work is supported by the National Science Foundation of China (No. 61273298), and Science and Technology Commission of Shanghai Municipality under research grant No. 14DZ2260800.

References

1. Fattal, R.: Single image dehazing. ACM Trans. Graph. **27**(3), 72–72 (2008)
2. Fattal, R.: Dehazing using color-lines. ACM Trans. Graph. **34**(13) (2014)
3. Gibson, K.B., Nguyen, T.Q.: Fast single image fog removal using the adaptive wiener filter. In: IEEE International Conference On Image Processing, pp. 714–718 (2013)
4. Hautiere, N., Tarel, J.P., Aubert, D., Dumont, E., et al.: Blind contrast enhancement assessment by gradient ratioing at visible edges. Image Anal. Stereology J. **27**(2), 87–95 (2008)
5. He, K., Sun, J.: Guided image filtering. IEEE Trans. Pattern Anal. Mach. Intell. **35**(6), 1397–1409 (2013)
6. He, K., Sun, J., Tang, X.: Single image haze removal using dark channel prior. IEEE Trans. Pattern Anal. Mach. Intell. **33**(12), 2341–2353 (2011)
7. Kim, S., Eom, I., Kim, Y.: Image interpolation based on statistical relationship between wavelet subbands. In: 2007 IEEE International Conference on Multimedia and Expo (2007)
8. Kopf, J., Neubert, B., Chen, B., Cohen, M., Cohen-Or, D., Deussen, O., Uyttendaele, M., Lischinski, D.: Deep photo: model-based photograph enhancement and viewing. ACM Trans. Graph. **27**(5), 116–116 (2008)
9. Legrand, P., Vehel, J.: Local regularity based image denoising. In: Proceedings of International Conference on Image Processing, 2003, ICIP 2003. vol. 3, p. III-377-80 (2003)

10. Levin, A., Lischinski, D., Weiss, Y.: A closed-form solution to natural image matting. IEEE Trans. Pattern Anal. Mach. Intell. **30**(2), 228–242 (2008)
11. Meng, G., Wang, Y., Duan, J., Xiang, S., Pan, C.: Efficient image dehazing with boundary constraint and contextual regularization. In: IEEE International Conference on Computer Vision, pp. 617–624 (2013)
12. Narasimhan, S.G., Nayar, S.K.: Vision and the atmosphere. Int. J. Comput. Vision **48**(3), 233–254 (2002)
13. Narasimhan, S.G., Nayar, S.K.: Interactive deweathering of an image using physical models. In: IEEE Workshop on Color and Photometric Methods in Computer Vision, in Conjunction with ICCV, vol. 6, no. 1 (2003)
14. Nayar, S.K., Narasimhan, S.G.: Vision in bad weahter. In: IEEE International Conference on Computer Vision, vol. 2, no. 2, pp. 820–827 (1999)
15. Nishino, K., Kratz, L., Lombardi, S.: Bayesian defogging. Int. J. Comput. Vision **98**(3), 263–278 (2012)
16. Schechner, Y.Y., Narasimhan, S.G., Nayar, S.K.: Instant dehazing of images using polarization. In: IEEE Conference on Computer Vision and Pattern Recognition, pp. 325–332 (2001)
17. Shwartz, S., Namer, E., Schechner, Y.Y.: Blind haze separation. In: IEEE Conference on Computer Vision and Pattern Recognition, vol. 2, pp. 1984–1991 (2006)
18. Tan, R.T. Visibility in bad weather from a single image. In: IEEE Conference on Computer Vision and Pattern Recognition, pp. 1–8 (2008)
19. Tang, K., Yang, J., Wang, J.: Investigating haze-relevant features in a learning framework for image dehazing. In: IEEE Conference on Computer Vision and Pattern Recognition, pp. 2995–3002 (2014)
20. Tarel, J.P., Hautiere, N.: Fast visibility restoration from a single color or gray level image. In: IEEE International Conference on Computer Vision, pp. 2201–2208 (2009)
21. Véhel, J.: Signal enhancement based on holder regularity analysis. In: Barnsley, M.F., Saupe, D., Vrscay, E.R. (eds.) Fractals in Multimedia. The IMA Volumes in Mathematics and its Application, vol. 132, pp. 197–209. Springer, New York (2002)
22. Zhu, Q., Mai, J., Shao, L.: A fast single image haze removal algorithm using color attenuation prior. IEEE Trans. Image Process. **24**(11), 3522–3533 (2015)

Transfer Learning Based on A+ for Image Super-Resolution

Mei Su, Sheng-hua Zhong[✉], and Jian-min Jiang

Research Institute for Future Media Computing, College of Computer Science and Software
Engineering, Shenzhen University, Shenzhen 518000, People's Republic of China
sumei@email.szu.edu.cn, {csshzhong,jianmin.jiang}@szu.edu.cn

Abstract. Example learning-based super-resolution (SR) methods are effective to generate a high-resolution (HR) image from a single low-resolution (LR) input. And these SR methods have shown a great potential for many practical applications. Unfortunately, most of popular example learning-based approaches extract features from limited training images. These training images are insufficient for super resolution task. Our work is to transfer some supplemental information from other domains. Therefore, in this paper, a new algorithm Transfer Learning based on A+ (TLA) is proposed for image super-resolution task. First, we transfer supplemental information from other datasets to construct a new dictionary. Then, in sample selection, more training samples are supplemented to the basic training samples. In experiments, we seek to explore what types of images can provide more appropriate information for super-resolution task. Experimental results indicate that our approach is superior to A+ when transferring images containing similar content with original data.

Keywords: Image super resolution · Transfer learning · Example learning-based

1 Introduction

Image super resolution (SR) is a process to generate a high resolution (HR) image from input low resolution (LR) image and minimize visual artifacts. It has been widely used in many fields including computer vision, video surveillance and remote-sensing images [1]. As a low-cost post-processing technique, the SR technique can break through the limitation of both imaging equipment and the environment to produce an HR image that traditional digital cameras cannot capture from a real scene [2]. SR reconstruction has attracted extensive attention for these reasons. For the last few decades, many image super-resolution methods have been developed to display high quality images and provided a remarkable progress [3]. There is a general agreement that the existing SR methods can be categorized into three groups: interpolation-based methods, reconstruction-based methods, and example learning-based methods [4].

Example learning-based methods utilize a set of external training images to predict the missing high-frequency details in LR input. These example learning-based approaches have gained significant improvement in super-resolution task. And most example learning-based methods construct a dictionary, which contains a large number

© Springer International Publishing AG 2016
F. Lehner and N. Fteimi (Eds.): KSEM 2016, LNAI 9983, pp. 325–336, 2016.
DOI: 10.1007/978-3-319-47650-6_26

of low-resolution and high-resolution patch pairs [5]. In the learning procedure, an optimization function is proposed just like other applications [6]. An efficient method named
A+ [7] has better performance and relatively lower running time than other example
learning-based methods. In training stage, A+ method extracts features from the training
set to construct sparse dictionary. Then, A+ method employs samples taken from the
training pool to generate the neighborhood used for regression. The training set used in
A+ was proposed by Yang *et al.* [8]. This dataset contains only 91 nature images and
has been employed in many papers [7, 9]. As we know, the size of the training dataset
is less than that used in many other image processing tasks. Thus, we have such a question: are this commonly used training dataset containing sufficient information for super-
resolution task?

Transfer learning has the property to transfer information from different domains to
a specific domain. Thus, in this paper, we propose a novel algorithm named Transfer
Learning based on A+ (TLA). Our goal is to utilize some supplemental information
from other domains to achieve better performance for super-resolution task. And we
seek to explore what types of images can provide more appropriate information for us
to conduct super-resolution task.

The rest of this paper is organized as follows. In Sect. 2, we briefly review the
previous works on transfer learning. Section 3 presents the proposed Transfer Learning
based on A+ method in detail. We conduct experiments and demonstrate the performance of our method in Sect. 4. Finally, in Sect. 5, we conclude this paper.

2   Related Work

As one of the most important research directions in machine learning, transfer learning
has been well studied in various fields in recent years. Transfer learning is a new
approach to improve the performance of unknown target domain by utilizing previously
acquired knowledge learned from a source domain which may be different with the target
domain.

Approaches to transfer learning can be divided into four categories based on "what
to transfer" [10]: instance-based approaches, feature-based approaches, parameter-
based approaches and relational approaches. Instance-based transfer techniques reuse
data from the source tasks to augment the target task's training data to learn in the target
domain. Instance-based transfer learning approaches deal with the incomplete knowledge on examples and borrow data from other similar source tasks to improve the
performance of target tasks. These approaches have been widely employed in many
applications, such as image retrieval, web document classification [11].

As far as we known, there are two approaches used transfer learning for super-resolution task. The difference between these works and ours is the form of the transferred
information. Dai *et al.* [12] transferred the manifold structure from the space of HR
patches into LR patches to imitate the metrics computed in HR patches. Dong *et al.* [13]
transferred the shallow convolutional neural network learned in a relatively easier task
to initialize a deeper or harder network. Different from other methods, our method tries

to transfer useful content from other images to construct a better model for super-resolution task.

3 Proposed Method

We propose an algorithm called Transfer Learning based on A+. Our TLA method is closest related to A+, while at the same time improving over its performance. We will explain the general formulation for TLA in this section.

3.1 Feature Extraction and Representation

The number of the original training dataset used in A+ is denoted as T_o. At this stage, we transfer images from other domain to provide more information for training. We randomly selected T_t images from other datasets as the transferred images. The new training dataset is formed by the original training images and the transferred images. These images are taken as HR images.

We use the same feature extraction method as Zeyde $et\ al.$ [9] and A+ [7]. First, we obtain LR images from the HR images by down-sampling. For these given HR images, the corresponding LR images are downscaled from HR images for a given upscaling factor u. Then, the LR images is bicubically interpolated to an interpolated HR images by the same factor. Since the high-frequency details are missing, the interpolated HR images are also called LR images.

In order to extract local features that correspond to their high-frequency content, all LR images (interpolated HR images) are filtered using R high-pass filters. Thus, each interpolated HR image X_i leads to R filtered images from $f_r * X_i$, for $r = 1, 2, ..., R$ (where * stand for a convolution) [9].

Then, we extract small patches from these images. In this paper, the size of LR patches is 3×3 pixels, so we work with patches of $(3 \times 3) \times u = 3u \times 3u$ pixels. We extract low-resolution features from the filtered images $f_r * X_i$. The feature size is $3u \times 3u$ and every feature is vectorized to the length of $9u^2$. Then, every corresponding R such low-resolution features are merged into one vector of length $9u^2R$.

For all the HR images, we remove their low-frequency information. Then, the corresponding HR features y are extracted from the same locations in LR images. Thus, all LR features x and their corresponding HR features y are obtained $D = \{(x_1, y_1), (x_2, y_2), ..., (x_F, y_F)\} \in R^M \times R^N$.

3.2 Learning Phase

Since the representation of LR features is quite high-dimensional, we apply the Principal Component Analysis (PCA) dimensionality reduction to project features to a low-dimensional subspace while preserving 99.9 % of the energy. The same PCA projection matrix produced when reducing the dimension of training features is also applied during testing. Then, we use the dictionary training method proposed by Zeyde $et\ al.$ [9] to obtain a low-resolution sparse dictionary $D_L = \{d_{L,k}\}_{k=1}^K$ using the following formulation:

$$\min_{\delta} \|D_L\delta - x\|_2^2 + \lambda\|\delta\|_1 \tag{1}$$

where δ is the sparse representation, x are low resolution features and λ is a weighting factor. Then, we reconstruct the corresponding high resolution dictionary D_H by enforcing the coefficients in the HR and LR patch decompositions to be the same.

A+ employs samples taken from the training pool to generate the neighborhood used for regression. This training pool is formulated by LR features, which are extracted from the original training dataset. The number of these training samples is S_o. In order to make training samples more sufficient, we increase a fixed number of S_t samples transferred from other domain. Our new training samples are defined as S and it consists of original S_o samples and the new S_t samples. The optimization problem becomes:

$$\min_{\gamma} \|x_i - N_{L,k}\gamma\|_2^2 + \lambda\|\gamma\|_2 \tag{2}$$

where $N_{L,k}$ containing m training samples from the S training samples. And these m training samples lie closest to the dictionary atom to which the input feature x_i is matched. The closed-form solution is given in Eq. (3).

$$\gamma = \left(N_{L,k}^T N_{L,k} + \lambda I\right)^{-1} N_{L,k}^T x_i \tag{3}$$

We define the HR neighborhood corresponding to $N_{L,k}$ as $N_{H,k}$. Therefore, the super-resolution problem can be solved by calculating for each LR feature x_i its nearest neighbor atom in the dictionary, and then reconstructed HR patch $y_{H,i}$ as show in Eq. (4).

$$y_{H,i} = N_{H,k}\left(N_{L,k}^T N_{L,k} + \lambda I\right)^{-1} N_{L,k}^T x_i, k = 1, 2, \ldots, K \tag{4}$$

The projection matrix P_k can be defined as show in Eq. (5). Thus, every dictionary atom d_k has its corresponding projection matrix P_k and it can be computed offline.

$$P_k = N_{H,k}\left(N_{L,k}^T N_{L,k} + \lambda I\right)^{-1} N_{L,k}^T \tag{5}$$

3.3   Reconstruction Phase

In the reconstruction phase, for these given test images, we first extract LR features from them by the same operations as used in training. For each LR feature $x_{T,j}$, we calculate its nearest dictionary atom d_k from the LR dictionary D_L. And then, the projection matrix P_k of d_k is obtained. Therefore, the reconstructed HR patch $y_{T,j}$ for test images is computed using the following equation:

$$y_{Tj} = P_k x_{Tj} \qquad (6)$$

These reconstructed HR patches y_{Tj} are combined by the same way they decomposed and form the final reconstructed HR image.

4 Experiments

This section includes three experiments, which are conducted to investigate the performance of the proposed TLA and compare it with A+ method. The first experiment is to explore the effect when transferring similar information on TLA method. In the second experiment, we try to explore whether the improvement also exist when supplemental information comes from texture images. In the third experiment, we seek to analyze the performance of transfer learning based on different types of nature scene images.

4.1 Experiment Setting

We use the standard training set proposed by Yang et al. [8] which contains 91 images for training (original training images number T_o is 91). In each experiment, we newly increase one training dataset as transfer learning materials, including: face images, texture images and nature scene images, respectively. The standard nature scene set Urban and Natural Scene dataset [14] used in the third experiment is composed of 2688 images with eight categories and each one with an average of 336 images. Thus, we also use 336 images for face and texture images to conduct Experiment 1 and Experiment 2. For test, we employ two datasets Set14 and BD100. Set14 was proposed by Zeyde et al. [9] including 14 commonly used images for super-resolution evaluation, while BD100 contains the 100 testing images from the Berkeley Segmentation Data Set 300 (BSDS300) [15]. This Dataset is widely used for various computer vision tasks including super-resolution.

In our experiment, we compare our method with A+ [7] under the same condition, since it is an efficient dictionary-based SR method and the closest related method with our TLA. And our method is based on A+. The implementations are from the publicly available codes provided by the authors. For performance evaluation, methods are evaluated in terms of Peak Signal-to-Noise Ratio (PSNR) and structural similarity (SSIM) [16]. PSNR usually correlates well with the visual quality and SSIM is used for measuring the similarity between two images.

We conduct most of our experiments related to the internal parameters of A+ method in order to compare with A+ as fairly as possible. For the parameters used both in our method and A+, we follow the setting of them in A+. For instance, we set the upscaling factor u equals to 3 since it is the most widely used value in super resolution task, the original training samples number S_o is 5 million, the dictionary size K is 1024, the neighborhood size m is 2048 and λ is 0.1 [17]. These are the best or common parameters as reported in its respective original work. The high-pass filter number R is set to 4 including the first and second order gradients filters. We also set the transferred images number T_t equals to 100 and new samples number S_t is 1 million because a larger number of S_t does not improve significantly by experimental verification. All the experiments in this paper are repeated five times.

4.2 Experiment 1: What Is the Effect of Adding Similar Information

In this subsection, we seek to explore what is the effect on the performance when we try to transfer some information similar with the original information in training and test datasets. Faces are important in human perception and the images of face are widely used in image processing since they convey a wealth of information. We also find that face images are often selected as materials in super-resolution task. In order to make the samples of face images abundant, we selected 336 face images from the training data of ImageNet dataset [18] in our experiment. We decided to select images from ImageNet because it is large in scale and diversity in types. Some samples of these face images are shown in Fig. 1.

Fig. 1. 10 face images selected from ImageNet dataset.

For the images in the test datasets, we find that there are six human face images, and one animal face image in Set14 dataset. For dataset BD100, there are 17 human face images and eight animal face images. We regard all of these images as face images in our experiment. However, the original training dataset only has five face images, and most of them just contain some parts of human face. Therefore, we believe that sufficient face images do provide some useful information for transfer learning.

Table 1 shows two groups of results for PSNR and SSIM on two test datasets, Set14 and BD100. The first group is conducted on all images in test datasets. And the second one is the results of face images in test datasets. From these two groups of results, we find that our proposed method TLA shows performance improvement than A+ method, especially on BD100 dataset. Meanwhile, the improvement of face images is better than other images on BD100 dataset. It indicates that in super-resolution, the supplemental information could be useful. Therefore, the transfer learning based on face images is beneficial to the performance of super-resolution task.

Table 1. Average PSNR (dB) and SSIM results comparison on two test datasets and the face images in these datasets.

Benchmark		PSNR (dB)		SSIM	
		Set14	BD100	Set14	BD100
All test images	A+	29.130	28.180	0.89330	0.77970
	TLA	**29.138**	**28.192**	**0.89412**	**0.78090**
Face images	A+	29.200	28.140	0.88651	0.80174
	TLA	**29.203**	**28.158**	**0.88748**	**0.80301**

4.3 Experiment 2: What Is the Effect of Increasing Texture Images

The first experiment has proved that supplemental images have similar content with original data are beneficial to improve the super-resolution performance. But others may have such a question: whether this improvement comes from more related information or just more images? To rule out the second possibility, in Experiment 2, we use transfer learning based on texture images in our experiment.

In this part, a standard dataset Describable Textures Dataset (DTD) [19] is utilized in our experiment. As a texture database, DTD consist of 5640 images. And it is organized according to a list of 47 categories (e.g. banded, dotted, grid, stratified) and 120 images for each category. For a fair comparison with other categories in Experiment 1, we also select 336 texture images from DTD and conduct our experiment based on these texture images. Figure 2 shows some sample images selected from the dataset.

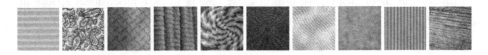

Fig. 2. 10 texture sample images selected from DTD.

The PSNR and SSIM results for texture images are shown Table 2. From Table 2, we find that transfer learning based on texture images cannot improve the performance on PSNR. We also find the performance has been improved on SSIM metric. The reason is that structural information in texture images could be useful to improve the performance of SSIM metric. However, although TLA based on texture images has a little improvement on SSIM metric, the increase is smaller than the case of transfer learning based on face images in Table 1. This indicates that if our incremental information is far different from the original data, the performance cannot be improved effectively. These results also suggest that the PSNR and SSIM of TLA with face images are better than that of TLA with texture images, and these two models are significantly different in a paired t-test ($p < 0.001$).

Table 2. Average PSNR (dB) and SSIM results comparison on Set14 and BD100.

Benchmark		PSNR		SSIM	
		Set14	BD100	Set14	BD100
Texture images	A+	**29.130**	**28.180**	0.89330	0.77970
	TLA	29.064	28.178	**0.89362**	**0.78046**

In common sense, we believe face images and texture images are different. To explore why the performance of transfer learning based on face images is better than texture images, we analyze several variables such as Discrete Cosine Transform (DCT) Coefficient Ratio (DCT-CR), Intensity Mean (IM) and Standard Deviation (SD). These variables are utilized to represent the essential nature of images in Experiment 1, Experiment 2 and the basic training and test images.

The DCT Coefficient Ratio for an image is $c = n_1/n$ where n_1 is the number of nonzero value in DCT coefficient and n is the number of all the values. DCT Coefficient Ratio is used to present the complexity of an image while Intensity Mean represents image's luminance. SD is a measure that is used to quantify the amount of variation or dispersion of each image. In our experiment, we normalize the values of three variables between 0 and 1.

Figure 3 shows the 3-D representation of the relationship between DCT Coefficient Ratio, Intensity Mean and SD for Set14, BD100, training images, face images and texture images. In Fig. 3, the x, y, z axis represents DCT Coefficient Ratio, Intensity Mean and SD, respectively. Green triangles, light blue triangles, blue points, red points and red hollow points represent Set14, BD100, training images, face images and texture images, respectively. From three variable results in Fig. 3, we find that face images may be more close to original datasets than texture images. In order to observe the detailed distributions of them clearly, we project these results in two-dimensional spaces in Fig. 4.

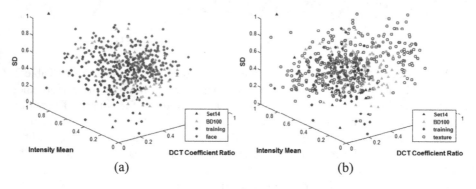

Fig. 3. 3-D representations for DCT Coefficient Ratio, Intensity Mean and SD from Set14, BD100, training images, face images and texture images. (a) Visualization on 3-D space for variables results of original datasets and face images; (b) Visualization on 3-D space for variables results of original datasets and texture images. (Color figure online)

In Fig. 4(a) and (b), x axis is DCT Coefficient Ratio and y axis is Intensity Mean. In Fig. 4(c) and (d), the x axis stays still and the y axis is SD. Green triangles, light blue triangles, blue points, red points and red hollow points represent the same objects with Fig. 3. In Fig. 4(a) and (c), we can find the points come from face images could cover the regions of the points in original training images and test images. It means they have similar visual or structural properties. On the contrary, the distribution of the variables in texture images is more disperse than face images. In Fig. 4(b) and (d), we can find some points from original datasets are isolated with the points from texture images. These results could help us understand the differences in their performance for transfer learning.

In general, we could get the conclusion that if the supplemental information has similar characters with the original data, transfer learning can improve the performance.

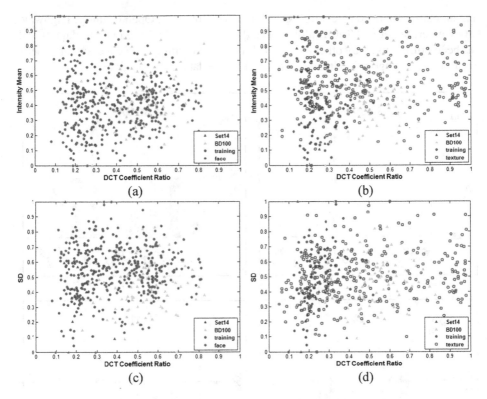

Fig. 4. 2-D representations for different variables from Set14, BD100, training images, face images and texture images. (a) Visualization on 2-D space for DCT-CR and IM of original datasets and face images; (b) Visualization on 2-D space for DCT-CR and IM of original datasets and face images; (c) Visualization on 2-D space for DCT-CR and SD of original datasets and face images; (d) Visualization on 2-D space for DCT-CR and SD of original datasets and texture images. (Color figure online)

And we can get opposite results when our supplemental information is different with the original ones.

4.4 Experiment 3: What Is the Effect of Different Categories

Experiment 1 and Experiment 2 have shown the performance of transfer learning based on face images and texture images, respectively. These experiments have proved that similar information contributes to improve the performance of super-resolution. However, we only use two types of images, the common used face images and an extreme case, texture images. Therefore, in this subsection, we seek to employ more categories of nature images to analyze whether the performance changes with the categories or not.

To statistically analyze this problem, we utilize a standard dataset called Urban and Natural Scene dataset [14]. This dataset includes 2688 authentic images with eight semantically organized categories: Coast, Forest, Mountain, Open Country, Highway, City Center, Street and Tall Building. The dataset is a widely used nature dataset and has been utilized in many other papers [20]. In this experiment, we conduct our experiment on each category in Urban and Natural Scene dataset, respectively.

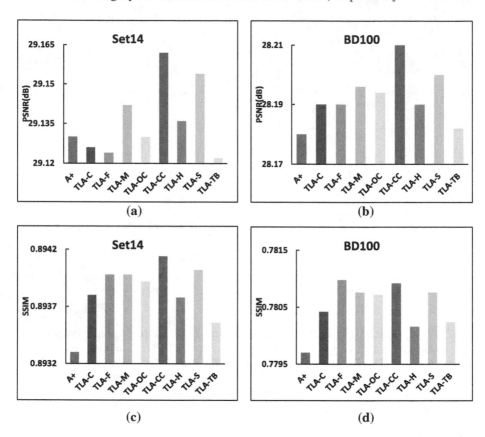

Fig. 5. Average PSNR and SSIM results comparison on test datasets for different categories. (a) PSNR on Set14 for A+ method and TLA based on eight categories. (b) PSNR on BD100 for A+ method and TLA based on eight categories. (c) SSIM on Set14 for A+ method and TLA based on eight categories. (d) SSIM on BD100 for A+ method and TLA based on eight categories.

Figure 5 shows the average PSNR and SSIM results of various categories on test datasets. Figure 5(a) and (b) are PSNR values of A+ and TLA based on the eight categories while Fig. 5(c) and (d) are the SSIM values on the same condition. In Fig. 5, the captions TLA-C, TLA-F, TLA-M, TLA-OC, TLA-CC, TLA-H, TLA-S and TLA-TB are used to represent the TLA method when supplementing Coast, Forest, Mountain, Open Country, City Center, Highway, Street and Tall Building images, respectively. The results indicate that when supplemental information is from different categories, the PSNR and SSIM results are also different. From Fig. 5(a), we can find that the images from three categories including Coast, Forest and Tall Building make the performance worse. For most of categories, the performances of TLA method improve significantly than A+ method, especially for City Center and Street images. We infer the reason is that the images from Coast, Forest and Tall Building have different characters with the original data while images from City Center and Street can provide more similar information. Therefore, transfer learning based on different categories will demonstrate different effects on the performance of super-resolution task.

5 Conclusion

In this paper, we present a novel transfer learning-based super-resolution algorithm called TLA. We transfer some useful information from other datasets to improve the performance of super-resolution task. Three experiments are conducted in our paper. The first experiment indicates that the transfer learning based on face images is beneficial to the performance of super-resolution task. In the second experiment, we can find that if our incremental information is far different from the original data, the performance cannot improve. The third one proves that transfer learning based on different categories has different effects on super-resolution task. Our experimental results indicate that images containing similar content with original data are helpful. As further research directions, transfer learning can combine with other super-resolution approaches to extract more useful information.

Acknowledgements. This work was supported by the National Natural Science Foundation of China (No. 61502311), the Natural Science Foundation of Guangdong Province (No. 2016A030310053), the Science and Technology Innovation Commission of Shenzhen under Grant (No. JCYJ20150324141711640), the Strategic Emerging Industry Development Foundation of Shenzhen (No. JCY20130326105637578), the Shenzhen University research funding (201535), Special Program for Applied Research on Super Computation of the NSFC-Guangdong Joint Fund (the second phase), and the Tencent Rhinoceros Birds Scientific Research Foundation (2015).

References

1. Deng, C., Xu, J., Zhang, K., Tao, D., Gao, X., Li, X.: Similarity constraints-based structured output regression machine: an approach to image super-resolution. IEEE Trans. Neural Netw. Learn. Syst. (2015). doi:10.1109/TNNLS.2015.2468069
2. Yu, J., Gao, X., Tao, D., Li, X., Zhang, K.: A unified learning framework for single image super-resolution. IEEE Trans. Neural Netw. Learn. Syst. 25(4), 780–792 (2014)
3. Zhu, Y., Li, K., Jiang, J.: Video super-resolution based on automatic key-frame selection and feature-guided variational optical flow. Sig. Process. Image Commun. 29(8), 875–886 (2014)
4. Zhang, K., Zhou, X., Zhang, H., Zuo, W.: Revisiting single image super-resolution under internet environment: blur kernels and reconstruction algorithms. In: Ho, Y.-S., Sang, J., Ro, Y.M., Kim, J., Wu, F. (eds.) PCM 2015. LNCS, vol. 9314, pp. 677–687. Springer, Heidelberg (2015). doi:10.1007/978-3-319-24075-6_65
5. Li, K., Zhu, Y., Yang, J., Jiang, J.: Video super-resolution using an adaptive super-pixel guided auto-regressive model. Pattern Recogn. 51, 59–71 (2016)

6. Li, J., Qiu, M.K., Ming, Z., Quan, G., Qin, X., Gu, Z.: Online optimization for scheduling preemptable tasks on IaaS cloud systems. J. Parallel Distrib. Comput. **72**(5), 666–677 (2012)
7. Timofte, R., De Smet, V., Van Gool, L.: A+: adjusted anchored neighborhood regression for fast super-resolution. In: Cremers, D., Reid, I., Saito, H., Yang, M.-H. (eds.) ACCV 2014. LNCS, vol. 9006, pp. 111–126. Springer, Heidelberg (2015). doi:10.1007/978-3-319-16817-3_8
8. Yang, J., Wright, J., Huang, T., Ma, Y.: Image super-resolution as sparse representation of raw image patches. In: CVPR, pp. 1–8. IEEE (2008)
9. Zeyde, R., Elad, M., Protter, M.: On single image scale-up using sparse-representations. In: Boissonnat, J.-D., Chenin, P., Cohen, A., Gout, C., Lyche, T., Mazure, M.-L., Schumaker, L. (eds.) Curves and Surfaces 2011. LNCS, vol. 6920, pp. 711–730. Springer, Heidelberg (2012). doi:10.1007/978-3-642-27413-8_47
10. Pan, S.J., Yang, Q.: A survey on transfer learning. IEEE Trans. Knowl. Data Eng. **22**(10), 1345–1359 (2010)
11. Zhang, D., Si, L.: Multiple instance transfer learning. In: IEEE International Conference on Data Mining Workshops, pp. 406–411. IEEE (2009)
12. Dai, D., Kroeger, T., Timofte, R., Van Gool, L.: Metric imitation by manifold transfer for efficient vision applications. In: Proceedings of the IEEE Conference on Computer Vision and Pattern Recognition, pp. 3527–3536 (2015)
13. Dong, C., Deng, Y., Change Loy, C., Tang, X.: Compression artifacts reduction by a deep convolutional network. In: Proceedings of the IEEE International Conference on Computer Vision, pp. 576–584 (2015)
14. Oliva, A., Torralba, A.: Modeling the shape of the scene: A holistic representation of the spatial envelope. Int. J. Comput. Vis. **42**(3), 145–175 (2001)
15. Martin, D., Fowlkes, C., Tal, D., Malik, J.: A database of human segmented natural images and its application to evaluating segmentation algorithms and measuring ecological statistics. In: IEEE International Conference on Computer Vision, pp. 416–423. IEEE (2001)
16. Wang, Z., Bovik, A.C., Sheikh, H.R., Simoncelli, E.P.: Image quality assessment: from error visibility to structural similarity. IEEE Trans. Image Process. **13**(4), 600–612 (2004)
17. Zhang, Y., Zhang, Y., Zhang, J., Wang, H., Dai, Q.: Single image super-resolution via iterative collaborative representation. In: Ho, Y.-S., Sang, J., Ro, Y.M., Kim, J., Wu, F. (eds.) PCM 2015. LNCS, vol. 9315, pp. 63–73. Springer, Heidelberg (2015). doi:10.1007/978-3-319-24078-7_7
18. Deng, J., Dong, W., Socher, R., Li, L., Li, K., Fei-Fei, L.: Imagenet: A large-scale hierarchical image database. In: CVPR, pp. 248–255 (2009)
19. Cimpoi, M., Maji, S., Kokkinos, I., Mohamed, S., Vedaldi, A.: Describing textures in the wild. In: CVPR, pp. 3606–3613 (2014)
20. Zhong, S., Liu, Y., Chen, Q.: Visual orientation inhomogeneity based scale invariant feature transform. Expert Syst. Appl. **42**(13), 5658–5667 (2015)

Intuitive Knowledge Connectivity: Design and Prototyping of Cross-Platform Knowledge Networks

Michael Kaufmann[1(✉)], Andreas Waldis[1], Patrick Siegfried[1], Gwendolin Wilke[1], Edy Portmann[2], and Matthias Hemmje[3]

[1] Lucerne School of Information Technology, Zug-Rotkreuz, Switzerland
{m.kaufmann,andreas.waldis,
patrick.siegfried,gwendolin.wilke}@hslu.ch
[2] Institute of Information Management, University of Bern, Bern, Switzerland
Edy.portmann@unibe.ch
[3] Faculty of Mathematics and Computer Science, University of Hagen, Hagen, Germany
Matthias.Hemmje@FernUni-Hagen.de

Abstract. Individual users are overwhelmed with a flood of data. Current big-data strategies focus mainly on organizational uses of data analytics. To address this gap, we focus on personal data management (PDM) in the era of big data and cloud computing. We are developing and testing a PDM software that enables individuals to construct a cross-platform knowledge network by semi-automatically connecting new relevant data to an existing network of interlinked digital objects. Because the cloud-based services that support our knowledge work are currently fragmented, we suggest an integrated federated platform for editing and searching the personal-knowledge context as a network. This forms a directed edge-labeled property multigraph that spans over all of the cloud-based data silos. We present a design and a proof-of-concept implementation of a PDM tool that allows the creation of a personal-knowledge network that incorporates digital objects from different cloud services.

Keywords: Personal data management · Big data · Cloud computing · Knowledge network · Connectivism

1 Introduction

We live in the age of *big data* [1]. Omnipresent data increase in volume, velocity and variety, and the amount of information stored globally has doubled every three years [2]. Research in big-data management usually focuses on business optimization based on data analysis. However, big data, or the *data explosion*, also affects individuals attempting to make sense of an increasingly unmanageable digital universe. Human attention is limited. Faced with an exponentially growing amount of data, the ratio between human knowledge and available data will become smaller and smaller. Big-data overload is becoming a problem for individuals. We are motivated to provide possible solutions in the area of *personal data management* (PDM) [3]. Today, most PDM solutions such as Dropbox or Evernote are cloud-based. While these services help to give data a universal feel that is independent from time and space and enable sharing

© Springer International Publishing AG 2016
F. Lehner and N. Fteimi (Eds.): KSEM 2016, LNAI 9983, pp. 337–348, 2016.
DOI: 10.1007/978-3-319-47650-6_27

and collaboration, they also fragment personal knowledge contexts into silos. Our research question is how we can support individuals to keep track of important digital objects across multiple platforms. We hypothesize that to obtain focus in a clutter of data, it is most important to connect digital objects together to contextualize them meaningfully in a *knowledge network*. Therefore, this paper will present a design of an 'autonomic connective knowledge warehouse', a system to interconnect personal data from different cloud platforms in a knowledge network, and it will present the current state of implementation and our preliminary findings.

2   State of the Art

2.1   Conceptual and Technological Foundations

Nonaka/Takeuchi [4] explain *knowledge management* as a continuous spiral of socialization, externalization, internalization and combination. Whereas PDM [3], *personal information management* (PIM) [5] and *personal knowledge management* (PKM) [6] are closely related, PDM focuses more on the explicit aspects of externalization (storage of digital objects) and combination (creation of a knowledge network). However, the implicit aspects of socialization (sharing or collaboration) and internalization (new insights through contextualization) can also be affected by PDM. The system we propose will take the role of a *mediator*, as in Wiederhold's [7] work, because our PDM knowledge network will span many different data platforms, which makes semantic integration and interfacing necessary.

The following foundational concepts inspired our design. The term *knowledge technology* was defined by Milton et al. (1999) [8] as any method or tool focused specifically on knowledge. In an elegant way, Shadbolt (2001) [9] avoids epistemological discussion by defining the concept of knowledge simply as *useable information*. Preece et al. (2001) [10] define *knowledge engineering* as a process of automation in the formation of conceptual models and knowledge bases. They recommend the use of artificial intelligence and machine learning to automatically generate *knowledge maps*. A *knowledge warehouse* (Nemati et al. 2002) [11] is a central store of knowledge components with explicit and implicit knowledge that integrates several heterogeneous knowledge sources. Nemati et al. emphasize the use of artificial-intelligence methods for the creation of knowledge, i.e., useable information as defined by Shadbolt, to extend beyond pure data storage. *Autonomic computing* [12] aims to address the complexity of information technology by introducing self-organization, helping users to focus by autonomically handling lower-level tasks. Nilsson and Palmer (1999) [13] describe a *concept browser* that allows a user to interact directly with knowledge structures. Concepts and their connections can be graphically displayed, navigated and edited within this browser.

2.2   Connective Databases and Knowledge Networks

The project Intuitive Knowledge Connectivity (IKC) explores how individuals can generate *knowledge contexts* [14] by organizing data in a way that turns this data into the most usable information. Thus, managing the flood of data can be understood didactically as a *learning process* that can be supported with software. To achieve this, we propose applying the learning theory of *connectivism* that George Siemens postulated (2005) [15]. According to connectivism Learning is a process of constructing a knowledge network, and new knowledge is interlinked with a network of already-existing knowledge. A PDM-software that supports a connectionist learning strategy allows for interactive storage and the manipulation of concept networks. This can be called a 'connective database'. The main operation of a connective database is to connect digital objects. Data records represent any digital object that plays a role in the everyday learning process of the individual, together with its connections to other objects. For example, a network node may be a plain note referencing a concrete entity such as "my brother," or a reference to a digital object such as "my brother's email from today." Nodes in the knowledge network, the data records, are stored together with their associations to other network nodes.

Existing solutions that allow for connecting digital objects from different cloud sources are, for example, Microsoft Delve, Storify, Evernote and MediaWiki. Microsoft Delve automatically connects documents and users from different platforms such as OneDrive, SharePoint or Exchange in a graph structure. However, it is limited to Microsoft products. Storify enables linking together social content from blogs and other social media but does not incorporate personal data such as emails or contacts. Evernote is a notetaking solution that can link notes together via a hyperlink, similar to MediaWiki, but more user friendly. Both tools can store, combine, and link to anything that has a URL. Yet, both are restricted to notes/tags or wiki pages/categories, and there are no other specific PDM data types such as calendar events or places.

There exist similar database models that allow to form conceptual networks, such as the (ancient) *CODASYL network model* [16] and the (more recent) *graph-database* model [17], both of which are built on network-like structures. Additionally, the *worldwide web*, in its initial inception [18], and *project Xanadu* [19] envisioned information management by forming conceptual networks. However, our connective database model differs from existing models by emphasizing the cognitive adequacy of knowledge representation. Data is not stored in low-level data types such as String or Integer but in higher-level, personally relevant data types such as notes, emails, documents, contacts, photos, web links, or calendar events. Thus, it focuses on digital objects that an individual in the 21st century works with daily.

3   Design

3.1   The Vision of Connective User Interaction

In our vision, the implementation of a connective database for platform integration supported by machine learning can be called an 'autonomic connective knowledge warehouse'. This enables the manual construction and automatic generation of a network of digital objects from different distributed sources in the cloud, each represented by network nodes or vertices, together with associations between these nodes, represented by arrows or directed labeled edges. This knowledge network can be accessed and edited by the user through a search field or a concept browser. Being able to search on one

Fig. 1. Concept study of a user interface: digital objects such as events, emails, photos, contacts, tasks, notes, files, folders and web links from different cloud-based personal knowledge bases such as Evernote, Dropbox, and iCloud, can be interconnected to form a personal knowledge network. Data in different cloud services is accessible through one search field.

platform that integrates all cloud knowledge sources makes relevant data accessible across platform boundaries. Because knowledge connectivity should be as intuitive as possible to blend into day-to-day activities, not only should the user be able to connect new knowledge items manually but the system should also add and interconnect relevant information elements (semi-) autonomously, e.g., by learning context or user preferences, or by harvesting semantic information from structured and unstructured data in the user's information space. As illustrated in Fig. 1, a user can interact with the knowledge network by actively interconnecting photos with events, notes with tasks, contacts with files, and so on. Also, the user can search on one platform over several distributed knowledge sources to find relevant data.

3.2 Cross-Platform Knowledge Generation

As a use case, imagine a user who attends a project meeting. In the context of the meeting, several digital representations are generated:

- a calendar event,
- contacts that attended the meeting,
- photos of whiteboard notes that were created during the meeting,

- emails regarding the topic of the meeting,
- documents that were discussed in the meeting,
- tasks that were organized by the meeting manager, and
- stored notes from the discussion.

All of these knowledge elements are usually stored in different, fragmented cloud services such as iCloud, Dropbox, Google Drive and Evernote. They are logically connected, and the user would like to connect their digital representations as well, but this is currently hardly possible. The vision of project IKC is to federate the distributed digital objects of the user's information space into a single coherent view. Figure 2 shows a concept study of such a software-system architecture that allows users to interconnect everyday knowledge items from several sources. A top layer shows different cloud platforms that are integrated and federated in a coherent knowledge network (layer 2). The autonomous-reasoning component is illustrated in layer 3 as a software bot. Layer 4 represents the user interface that combines a concept browser and editor with a search engine.

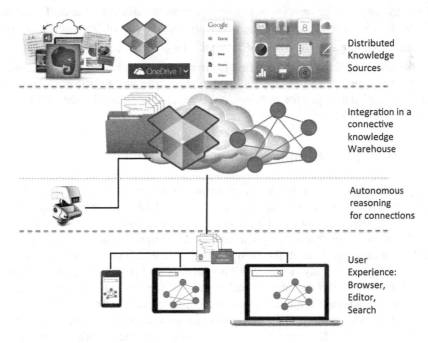

Fig. 2. Possible system architecture: distributed knowledge sources are integrated in a knowledge network. The data is interconnected and annotated either by user input or an autonomic-reasoning engine. The knowledge network is editable and accessible on mobile devices.

3.3 Mathematical Model of the Knowledge Network

We provide a mathematical model of the knowledge network that we will implement in our system. Generally, according to Knauer (2010) [20], pp. 1–2, a *general directed graph* is a triple G = (V, E, p), where V is the *vertex* set, E is the *edge* set, and p: E → V is the *incidence mapping*, which defines two more mappings o, t: E → V by (o(e), t(e)) : = p(e), where o(e) is the *origin* and t(e) is the *tail* of directed edge e. If p is not injective, G is called a *multigraph*. A mapping w: E → W or w: V → W is called a *weight function*. W is any set, called the *set of weights*, and w(x) is the *weight* of edge x or vertex x. It is important to note that this model of a multigraph using incidence and weight mapping as a blueprint for knowledge networks has the advantage that it can model directed labeled graphs with multiple distinct edges for the same vertices. Therefore, we use an instance of the general graph structure described by Knaurer as a model of the knowledge network in the IKC prototype. Also, this ensures that our structure is mathematically sound, and our terminology is clear.

Accordingly, an IKC *knowledge network* K is modeled as a directed weighted multigraph K = (N, A, δ, L, λ, P, π). The vertices in N, called *nodes*, represent digital objects. The edges in A, called *arrows*, represent relationships between them. The incidence mapping δ indicates the *direction* of arrows, their origins and tails. Indicated by the edge weight function λ, the edge weight set L consists of *relationship labels* from a domain L defined by the user. The vertex weight function π maps nodes to elements of the vertex weight set P called a node's *properties*, which are ordered sets of key-value pairs {(k$_1$, v$_1$), …, (k$_n$, v$_n$)}. For a single *property* (k$_i$, v$_i$), k$_i$ is a *key* from a set K, and v$_i$ is a *value* from a domain Δ. In K, there are keys for system properties such as id, title, description, data type, data source, creation time and modification time, and keys that are attributed to different PDM data types for websites, contacts, etc. Additionally, the user can add user-defined properties.

4   Current State of Implementation

An IKC prototype has been developed[1] to explore the proposed design synthetically. Our prototype is a technical-feasibility study that will be continuously expanded in follow-up projects. As shown in Fig. 3, our prototype allows for the formation of a knowledge network as described by our mathematical model. The nodes can be associated with digital objects from the Web and two different cloud sources. Technically, the integration of almost any cloud service is conceivable. For now, the prototype opens the possibility to connect and arrange text notes, websites and Dropbox and Evernote items together in a network, put those into meaningful context and add additional information. With the IKC prototype, the user can create, browse, edit and search a knowledge network interactively. This network consists of nodes (vertices with properties) and arrows (labeled edges), where nodes contain or represent the digital objects in the knowledge network and arrows describe the relationship between nodes. There are four types of nodes:

- *plain node:* a text node without connection to an external cloud data service,
- *link node:* a node that points to a web URL,

---

[1] Available at http://demo.ikc.today/nodeDetail.html (accessed September 1st, 2016).

- *Evernote node:* a node that points to a note on a connected Evernote account, and
- *Dropbox node:* a node that points to an item on a connected Dropbox account.

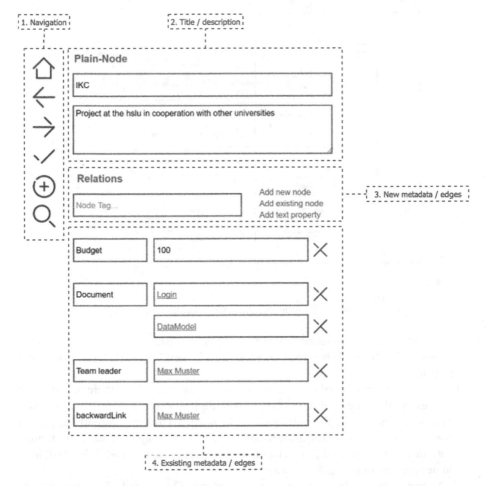

Fig. 3. User interface overview: the node is connected to three other nodes: Login, DataModel, and Max Muster (in the role of team leader). Thus, for example, there is a directed edge from IKC to DataModel labeled as "Document." Also, there is an atomic node property.

4.1 Manipulating the Knowledge Network

The database in the IKC prototype consists of a network of associated nodes. There is a special node as a point of entry for the user, the predefined Home node (the root of the network). Every node features a title and a short description Further down, atomic properties and labeled arrows that link to other nodes are shown and can be added. Each arrow is shown with a link to the associated node, its relationship label, and a delete button. The user interface (UI) uses the labels to group together similar relations.

Figure 4 shows the interface to a node, and the following points explain the numbered UI fields.

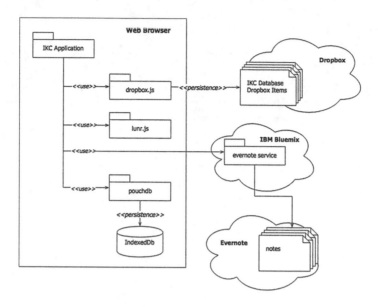

Fig. 4. Overview of the system architecture

1. For manipulation of the network, there are icons for basic functions, such as "Back to home node" and node traversal, such as "Last"/"Next", "Save", "New node" and "Search".
2. The title and description of the node are displayed, and if it exists, there is also a link to the external resource of the corresponding digital object. It is possible to add a text property, an atomic key value pair, to the node. To link the currently displayed node to other network nodes, arrows can be created, pointing from the current node as origin, to a related node as tail, together with a relationship label. It is possible to add new arrows in two different ways:
   - Adding a new node creates a new network node as the tail of the new arrow. Currently, the possibilities are plain (text only), link (URL), Evernote or Dropbox.
   - Adding an already-existing node from the knowledge network. This allows the user to choose the data object for the tail of the arrow.
3. This part of the UI shows all properties and arrows grouped by their label (the arrow label or property key), along with the possibility to delete the arrow. If there are multiple outgoing arrows with the same relationship label, the items are grouped (e.g., document).

4.2   System Architecture

A large part of the current prototype is written in JavaScript, and data is stored in Java-Script Object Notation (JSON). It runs inside the user's browser as a webpage hosted on IBM Bluemix, a commercial cloud-development platform. It holds the main part of the UI, and the application logic, which provides interfaces to the Dropbox and Evernote APIs. Furthermore, the application is responsible for the search index, metadata handling and persistence. A detailed diagram can be found in Fig. 4. The following libraries were used to implement these functionalities in the browser:

- dropbox.js: a JavaScript library that allows access to the Dropbox data of the user
- Dropbox Chooser: a Dropbox file browser,
- lunr.js: a JavaScript library for building and maintaining a full-text search index,
- Pouchdb: a JavaScript wrapper of the indexed data to save JSON document, and
- EventBus.js: a small JavaScript library for event-based programming.

Because it is not possible to connect directly from the browser to Evernote application programming interface (API) for security reasons (cross-domain scripting), a small additional interface service (a mediator) had to be developed that resides on the remote cloud server. It is written in Node.js and is also hosted on IBM Bluemix. Evernote provides a JavaScript API, which is used to communicate with the Evernote service. For data-protection reasons, the IKC server does not centrally store any user data whatsoever. All user data, such as metadata and the whole network structure, are stored in the user's Dropbox account in JSON-structured files ending with.ikc. Furthermore Dropbox, as well as Evernote, act as an information source for digital objects that can be incorporated into the knowledge network.

4.3 Data Organization

To persistently store the knowledge network, nodes and arrows are described in JSON files and saved in the user's Dropbox. Every IKC file represents a node and its arrows to other nodes. More specifically, each file contains an array of properties. Those properties store information on linked nodes, supplemented by property type and associated

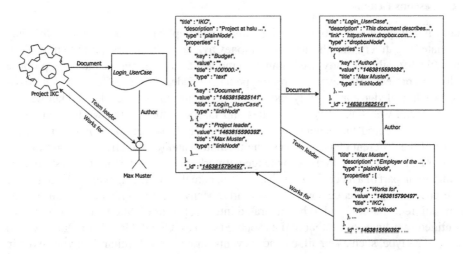

Fig. 5. Data organization in JSON (Metadata)

key. The type is necessary to distinguish between atomic aproperties and arrows that provide links to other nodes in the network. Figure 5 illustrates the described data organization, based on a simple knowledge network with three nodes and four arrows.

5 Conclusions

5.1 Discussion

IKC explores the concept of enabling individuals to master their data by building a knowledge network with integrated search, intuitive navigation, and automatically enriched knowledge connectivity. The project has developed guidelines for the design of connective knowledge technologies that put user interaction in the center of personal knowledge base design and organization and has begun implementation.

Based on design-oriented research in information systems [21], our research project has four phases: (1) analysis, (2) design, (3) evaluation and (4) diffusion. Based on an analysis of connective database technology, we have begun phase 2 of designing a software artifact. Thus far, no empirical evaluation has been performed because the development of the prototype is ongoing. Expert workshops have been planned to collect qualitative feedback on the existing implementation.

To discuss the current state, our prototype lacks many features of our conceptual design, such as an intuitive UI, automatic connections, or comprehensive interfaces to many possible cloud sources. Our current implementation shows a first step in the direction of a cross-platform personal-knowledge network. Compared to simple solutions such as mind maps with URLs or a bookmarking system, our implemented system has the advantage that it implements a knowledge network, not a hierarchy; that the network can be edited and browsed; that all network nodes can be annotated; and that it provides full-text search over all cloud sources.

5.2 Lessons Learned

There is always a gap between visions and their implementation. Usually, this gap correlates negatively with the amount of money available. In the first part of our project, a technical-feasibility study, we began with modest financing, but we intend to propose follow-up projects. However, our prototype has provided a proof-of-concept for cross-platform knowledge connectivity and search, i.e., connecting a user-defined knowledge network with web links, Evernote nodes and Dropbox files. What we learned so far concerns two central components in such a system: (1) the server that provides interfaces for all of the different cloud services (the mediator) and (2) the client that provides to the user an interaction with the knowledge network (the user application). First, there is an important part of a mediator in such a system, namely interfacing with and integrating different cloud knowledge sources coherently. Many different types of digital objects from different sources need to be integrated into a coherent PDM data model. We have identified two dimensions, that of data source (such as iCloud, Google or Outlook) and that of data type, such as email, calendar events, or contacts. Similar datatypes exist in different cloud services, and an integration layer needs to harmonize these data types over different sources to make sense to the user. We know now that the mediator will integrate the sources into a PDM data architecture. Second, there is another very important part, which is that of the client. Our browser-based client is a JavaScript AJAX application. It is clear that for truly intuitive knowledge connectivity, a significant amount of work must be invested into methods of making it easy for the user to create a knowledge network, add new nodes and add new links to existing nodes on both a large screen and a mobile device. Our simple demonstrator is far from optimal because we have concentrated on foundational technology aspects. A follow-up project with design researchers will focus on designing an intuitive interaction in the user application, and federating different cloud sources on a technical level in a mediator.

5.3 Outlook

There are three important aspects that will be addressed in the future. First, the technical mediation and interfacing will be extended. Second, the user interaction will be redesigned from the ground up with a focus on intuitiveness. Third, an empirical evaluation of the software artefact needs to be performed. The present prototype already offers a range of functions and is fully functional to create a knowledge network, including personal notes, files, and anything with a URL. The next step is to integrate other cloud services. In addition, the functionality itself can be extended. Currently, for example, Evernote nodes are only editable outside of the prototype environment and within an independent browser frame. The objective is to add, delete and edit nodes from all different cloud services directly inside of the prototype.

In a follow-up project, we plan to conduct case studies and user feedback with university students on the bachelor's and master's levels with a software development company and with a private foundation that conducts journalistic research. The use case will be for the individual knowledge workers (the students, software developers, journalists and researchers). That way, our prototype can be evaluated qualitatively through expert feedback. Based on the expert feedback, we will redesign the UI, which will be a responsive UI that works on mobile devices as well as on large screens. On our interdisciplinary research team, we have design practitioners that will explore interface concepts to design new ways of interaction with a cloud-based knowledge network.

Acknowledgements. This research is funded by the Hasler Foundation (www.haslerstifung.ch) under the project "Intuitive Knowledge Connectivity" with grant number 15026.

References

1. De Filippi, P.: Big data, big responsibilities. Internet Policy Rev. **3** (2014)
2. Hilbert, M., López, P.: The world's technological capacity to store, communicate, and compute information. Science **332**, 60–65 (2011)
3. Hildebrandt, M., O'Hara, K., Waidner, M.: Personal data management – a structured discussion. In: Digital Enlightenment Yearbook 2013: The Value of Personal Data, pp. 270–287. IOS Press (2013)

4. Lehner, F.: Wissensmanagement: Grundlagen, Methoden und technische Unterstützung. Carl Hanser Verlag GmbH & Co. KG, Munich (2012)
5. Jones, W.: Personal information management. Annu. Rev. Inf. Sci. Technol. **41**, 453–504 (2007)
6. Razmerita, L., Kirchner, K., Sudzina, F.: Personal knowledge management: the role of Web 2.0 tools for managing knowledge at individual and organisational levels. Online Inf. Rev. **33**, 1021–1039 (2009)
7. Wiederhold, G.: Mediators in the architecture of future information systems. Computer **25**, 38–49 (1992)
8. Milton, N., Shadbolt, N., Cottam, H., Hammersley, M.: Towards a knowledge technology for knowledge management. Int. J. Hum. Comput. Stud. **51**, 615–641 (1999)
9. Shadbolt, N.: Knowledge technologies. Ingenia R. Acad. Eng. **8**, 58–61 (2001)
10. Preece, A., Flett, A., Sleeman, D., Curry, D., Meany, N., Perry, P.: Better knowledge management through knowledge engineering. IEEE Intell. Syst. **16**, 36–43 (2001)
11. Nemati, H.R., Steiger, D.M., Iyer, L.S., Herschel, R.T.: Knowledge warehouse: an architectural integration of knowledge management, decision support, artificial intelligence and data warehousing. Decis. Support Syst. **33**, 143–161 (2002)
12. Kephart, J.O., Chess, D.M.: The vision of autonomic computing. Computer **36**, 41–50 (2003)
13. Nilsson, M., Palmér, M.: Conzilla - towards a concept browser. Department Computing Science, Centre for User Oriented IT Design, Royal Institute of Technology KTH, Stockholm (1999)
14. Young, R., Letch, N.: Knowledge contexts-through the theoretical lens of Niklas Luhmann. In: Proceedings of PACIS 2003 (2003)
15. Siemens, G.: Connectivism: a learning theory for the digital age. Int. J. Instr. Technol. Distance Learn. **2** (2005)
16. Codd, E.F., Date, C.J.: Interactive support for non-programmers: the relational and network approaches. In: Proceedings of the 1974 ACM SIGFIDET (Now SIGMOD) Workshop on Data Description, Access and Control: Data Models: Data-Structure-Set Versus Relational, pp. 11–41. ACM, New York (1975)
17. Angles, R., Gutierrez, C.: Survey of graph database models. ACM Comput Surv. **40**, 1–39 (2008)
18. Berners-Lee, T.: Information management: a proposal. CERN (1989)
19. Nelson, T.: Literary Machines. Mindful Press, Swarthmore (1981)
20. Knauer, U.: Algebraic Graph Theory: Morphisms, Monoids and Matrices. De Gruyter, Berlin, Boston (2011)
21. Österle, H., Becker, J., Frank, U., Hess, T., Karagiannis, D., Krcmar, H., Loos, P., Mertens, P., Oberweis, A., Sinz, E.J.: Memorandum on design-oriented information systems research. Eur. J. Inf. Syst. **20**, 7–10 (2010)

Knowledge Management

Intellectual Capital and Boundary-Crossing Management Knowledge

Shiuann-Shuoh Chen[1], Min Yu[1(✉)], and Pei-Yi Chen[2]

[1] Department of Business Administration, National Central University, No. 300, Jung-da Road, Jung-li City, Taoyuan 320, Taiwan (Republic of China)
kenchen@cc.ncu.edu.tw, m2121374@gmail.com
[2] Department of International Business, Hsin Sheng College of Medical Care and Management, No. 418, Gaoping Section, Zhongfeng Road, Longtan Township, Taoyuan 320, Taiwan (Republic of China)
peiyi01@ms47.hinet.net

Abstract. Exploring how the intellectual capital (i.e., the capability of system, coordination, and socialization, and the human capital) influences the boundary-crossing management knowledge (i.e., the syntactic transfer, semantic translation, and pragmatic transformation), this study identifies differing effects for three dimensions of boundary-crossing management knowledge. The results indicate that the coordination capability primarily enhances a team's syntactic transfer, semantic translation, and pragmatic transformation. The socialization capability primarily improves a team's semantic translation and pragmatic transformation. Our findings reveal why teams may have difficulty managing levels of syntactic transfer, semantic translation, and pragmatic transformation and vary in their ability to create value from their boundary capability.

Keywords: Intellectual capital · Boundary-crossing knowledge management · Organization learning · Product development · Innovation

1 Introduction

The fact that most innovation happens at the boundaries between domains has ensured focused attention on effective boundary-crossing management as a dominant source of competitive advantage [1, 40, 42, 43]. To survive selection pressures, the firms need to transfer knowledge, translate it, and transform it to the commercial ends. This ability, referred to as the boundary-crossing management knowledge [1], also known as the boundary capability, has emerged as an essential subject in the studies on strategy and organization.

Nowadays, an organization's value creation is largely based on the intellectual capital [44], conceptualized as the combination of all the knowledge and competences that can manifest as a company's sustained competitive advantage [45]. Prior research has shown that the intellectual capital drives firm performance, influences firm innovative capabilities, and positively affects firm capability and firm performance [46].

© Springer International Publishing AG 2016
F. Lehner and N. Fteimi (Eds.): KSEM 2016, LNAI 9983, pp. 351–363, 2016.
DOI: 10.1007/978-3-319-47650-6_28

However, the studies on the relationship between the intellectual capital and the boundary capability are scarce [47]. The link between intellectual capital and innovation capability needs more empirical research [48]. Therefore, it is necessary to identify the critical factors and their interacting relationship in order to provide a comprehensive understanding of the overall contribution to innovation. Examining the different effects of intellectual capital on the syntactic transfer, semantic translation, and pragmatic transformation would not only specify how the boundary capability can be developed, but also infer why the firms have difficulties in managing the dimensions of boundary capability successfully.

We organize this paper as follows: the next section presents a review of theory and hypotheses. The following section shows our methodology. The final section reports the implications and conclusion of our work.

2   Theory and Hypotheses

2.1   Boundary-Crossing Management Knowledge

Following Carlile [1], the boundary-crossing management knowledge denotes the efforts expended in managing knowledge across specialized domains when innovation is a desired outcome. It is distinguished into the syntactic transfer, semantic translation, and pragmatic transformation. Besides, it can be conceptualized as a combination of the capacities of common lexicons, meanings, and interests, and the abilities to use these capacities.

At a syntactic boundary [1], differences and dependencies between actors are known; the challenge is to increase capacity to process more information [49]. A common lexicon is necessary but not always sufficient. Domain-specific knowledge can be efficiently managed across the boundary if knowledge is transferred according to a common lexicon.

At a semantic boundary [1], novelty generates different translations to the differences and dependencies; the challenge is to make tacit knowledge explicit [50]. To create common meanings often requires creating new agreements. Developing the common meanings is regarded as a way to address the interpretation differences.

At a pragmatic boundary [1], novelty generates different interests between actors; the challenge is to change knowledge that is at stake [40]. Establishing the common interests affords the key to interest conflicts [1]. To create common interests requires significant practical and political effort.

2.2   Intellectual Capital

The intellectual capital has been defined as the knowledge that firms utilize for competitive advantage [41]. We draw upon and synthesize insights from prior studies to divide the intellectual capital into four dimensions: the capability of system, coordination, and socialization, and the human capital [2, 3].

The system capability formed by the routinization and formalization establish the patterns of organizational action either through the memory or the standard operational procedure respectively [5]. The formalization captures the extent to which an organization sets its rules and procedures to prevent its employees deviating from established behavior [6], while the routinization reflects how an organization establishes the grammars of action through the individuals' repeated actions to support the complex patterns of interactions between the employees [7].

The formalization enables the team members to efficiently apply a team's codified knowledge through the best practices [8]. By doing so, a team can expose its members to the jargon used in the different functional areas in turn facilitate the shared understanding about the specific terms [9]. Once the shared understanding about the specific terms is in place, there is a greater chance that the group members are able to effectively create a common lexicon, leading to the syntactic transferring knowledge across boundaries [1]. The measures of formalization are the situations where the team members had procedures to follow in dealing with any situation, or the organization kept a written record of everyone's performance.

The routinization is embodied by the repetitious behaviors guided by the experiences [11]. It reflects strict patterns of norms and rules intended for imitation, replication, and control [12]. Drawing on Cohen and colleagues' generalization [13], a critical portion of the representations of routines encompasses the memories of individuals for their respective roles, locally shared language, and general language forms such as formal oral codes and pledges. We contend that the more routines a team has, the more likely its members will increase their tendencies to use the common lexicons for effective communication [10]. Therefore, the routinization is instrumental to shape the common lexicons and thereby strengthen the syntactic transferring knowledge across boundaries [1]. The measures of routinization include the extent to which the individuals will regard their work as routine, or the volume of repeated tasks from day-to-day.

Hypothesis 1: The system capability will be positively related to the syntactic transfer of boundary-crossing management knowledge.

The high level of formalization tends to confine the individuals' thoughts into a frame of reference and in turn provide a common perspective to mitigate the interpretative discrepancies [6]. It consequently leads the members to interpret circumstances with the same path-dependent trajectory of prevailing knowledge and thereby assist the development of shared meanings [4].

A team routinizing tasks is seeking for invariably performing sequences of activities with a few exceptions [14]. Such institutionalization often steers the team members to focus on the areas closely related to the existing knowledge and on what has previously proved useful [4]. In such a case, there is a greater likelihood that the team members are prone to respond to the environmental changes in a shared perspective and interpret the circumstances according to the prevailing norms [2]. Eventually, such processes will facilitate the development of shared meanings underlying the semantic translating knowledge. Therefore, we hypothesize the following:

Hypothesis 2: The system capability will be positively related to the semantic translation of boundary-crossing management knowledge.

The coordination capability including the cross-functional interfaces, participation in decision making and job rotation captures the efforts expended in incorporating different sources of expertise and promoting the lateral interaction between the individuals [4].

The cross-functional interfaces contain the liaison personnel, task forces, teams, and so forth [15]. The liaison personnel serve as a knowledge broker to bridge the differences among functions and thus develop a consensus on the meanings of specific terms, symbols, or behaviors [4]. Consequently, such interfaces enable the common lexicons needed for the success of transferring knowledge [1]. They also foster the constructive dialogues in turn reconcile the discrepancies in interpreting the tasks [4]. They function as a knowledge translator to interpret the problems in turn enable a shared understanding about the team's goals and thus lead to a shared perspective on the specific issues [1]. In this way, they incorporate the interpretations from diverse functions into a common meaning needed for the success of translating knowledge [1]. Furthermore, they encourage the team members to reconsider the value of existing products and to review the combination of components [16]. By doing so, the team members are more willing to negotiate interests and make trade-off with one another, in turn reach a consensus to change the knowledge and interests from their own domains for the shared interests [1, 4]. Therefore, they enable the common interests that constitute the transforming knowledge across boundaries [1]. The measure of cross-functional interfaces is the extent to which the subsidiary used liaison personnel, temporary tasks forces, and permanent teams to coordinate decisions and actions with sister subsidiaries.

The participation in decision making brings the opportunities of opinion sharing [17]. The participants tend to largely use the general language instead of jargon to increase a shared understanding in turn facilitate the development of shared language [9]. A shared understanding also increases their willingness to identify the commonality of each other's notion in turn develop a shared perception [4, 17]. They consequently tend to put similar interpretations on the specific problems and in turn develop the shared meanings. Besides, a shared understanding also improves their relations in turn enables the development of shared beliefs inspiring the social climate of trust. Such climate encourages them to believe that a current trade-off will lead to the later common interests and thereby increases their tendencies to risk changing their domain-specific knowledge. The measures of participation in decision making are the situations where the team members have wide latitude in the choice of means to accomplish goals, or the managers are allowed flexibility in getting work done.

Job rotation provides the opportunities to gain the experience and learn the jargon in different fields [17]. Consequently, the team members tend to develop a shared understanding on the specific facts or terms in turn improve their efficiency to create the common lexicons [9, 10]. They also tend to develop a shared experience about the projects and thereby increase their tendencies to develop a mutual knowledge, supporting the integration of interpretations [4]. In this way, they can improve their efficiency to create the common meanings [1]. The shared experience also improves their relations in turn contributes to the shared beliefs that directly influence the trust [4]. Trustworthy social conditions increase their tendencies to believe that a current trade-off will lead to the later common interests and thereby encourage them to risk changing their domain-specific knowledge [18]. The measures of job rotation include the situations where the team members are regularly rotated between different functions or subunits. This discussion suggests the following hypotheses:

Hypothesis 3: The coordination capability will be positively related to the syntactic transfer of boundary-crossing management knowledge.

Hypothesis 4: The coordination capability will be positively related to the semantic translation of boundary-crossing management knowledge.

Hypothesis 5: The coordination capability will be positively related to the pragmatic transformation of boundary-crossing management knowledge.

The socialization capability including the connectedness and socialization tactics captures the efforts expended in developing the unspoken rules, dominant values, and common codes of communication [3].

The connectedness fosters the communication and knowledge exchange in turn contributes to a shared understanding [3], which is proposed to enable the common meanings and interests, as explained in the section of coordination capabilities above. The measures of connectedness include the situations where the relationships among team members are very close, or the team members understand the personalities of one another.

The socialization tactics has been shown to structure the shared experiences [3], which are also proposed to enable the common meanings and interests, as explained in the section of coordination capabilities above. The measures of socialization tactics include the situations where the new team members are trained by the same program to know the operation of the team, or the senior team members often provide others with many working guides. Formally, we posit that:

Hypothesis 6: The socialization capability will be positively related to the semantic translation of boundary-crossing management knowledge.

Hypothesis 7: The socialization capability will be positively related to the pragmatic transformation of boundary-crossing management knowledge.

The human capital denotes the knowledge, skills, and abilities residing with and utilized by individuals [35]. Its characteristics are creative, bright, skilled employees with expertise in their roles, who provide the main source for new ideas and knowledge in an organization [36]. Individuals and their associated human capital are crucial for exposing an organization to technology boundaries that increase its capacity to absorb and deploy knowledge domains [37].

Following Wensley [38], explicit knowledge is explicit because of the sharing of tacit knowledge about the explicit knowledge. Knowledge can be transferred because the individuals between whom it is transferred have a rich set of mutual understandings – they share a great deal of tacit knowledge that they use to interpret the explicit knowledge. In order to transfer knowledge it is necessary to ensure that an extensive foundation of shared tacit knowledge is exists. As Mclagan noted [39], the excellent senior employees can play the role of consultant, helping others to get familiar with the work environment more quickly, explaining the problems faced within a practice, sharing their experiences, and providing necessary support. Hence, we posit that:

Hypothesis 8: The human capital will be positively related to the syntactic transfer of boundary-crossing management knowledge.

Hypothesis 9: The human capital will be positively related to the semantic translation of boundary-crossing management knowledge.

Drawing from Carlile [40], knowledge is invested in practice – invested in the methods, ways, of doing things, and successes that demonstrate the value of the knowledge developed. When knowledge proves successful, individuals are inclined to use that knowledge to solve problems in the future. In this way, individuals are less able and willing to change their knowledge to accommodate the knowledge developed by another group that they are dependent on. Changing their knowledge means an individual will have to face the costs of altering what they do to develop new ways of dealing with the problems they face. The measures of human capital include the levels of employees' skill, expertise, and creativity. Under this logic, we reason the following:

Hypothesis 10: The human capital will be negatively related to the pragmatic transformation of boundary-crossing management knowledge.

3 Methodology

3.1 Measurements

The measurements used in this study were primarily derived from the previous studies and some items were modified to make them applicable to our research purposes. All of the items were measured with 7-point Likert scale (1 = strongly disagree, 7 = strongly agree).

3.2 Data Collection Procedure

The unit of analysis is the firm. To assure the validity of respond data, we call each firm to find out a representative with sufficient knowledge and ability to respond this survey. In this survey, each respondent were asked to provide two different NPD projects, one of them has to be the superior performance project and another is inferior one, and then assess their relationships respectively. Therefore, each questionnaire would consist of two samples in the survey.

Two rounds of survey were conducted by distributing the survey instrument in the form of questionnaire to the production managers of 770 electrical manufacturing firms in Taiwan from June 1 to July 31, 2013. These firms were listed in the directories of the 2012 top 2000 firms in Chinese Credit (Taiwan's leading credit company). We sent 241 questionnaires by email and fax and received questionnaires from 139 buyer firms; the response rate is about 57.67 %. Of the 149 returned questionnaires, 10 were excluded because they did not meet all sampling criteria or due to the incomplete answers, including excessive missing data and lower levels of confidence, leaving 139 usable responses.

To examine the possibility of nonresponsive bias, a Chi-square test was conducted to compare the early (79) and late respondents (60) on the research variables [21, 34]. The responses from the first mailing were 79 questionnaires. The non-response bias is not a problem in this study [19].

3.3 Analysis of Measurement Model

The partial least squares (PLS), a component-based technique for structural equation modeling, is our main data analyzing tool, because this study includes both formative (the system, coordination, and socialization capabilities) and reflective (the boundary management knowledge for transfer, translation, and transformation) constructs [22, 26, 30].

Formative Constructs: We reveal the weights of formative indicators linked with the factors for the system capability, coordination capability and socialization capability in Fig. 1. The formative items should correlate with a "global item that summarizes the essence of the construct" [28] (p. 272). PLS item weights, which indicate the impact of individual formative items [23], can be multiplied by the item values and summed, as noted by Bagozzi and Fornell [20]. In effect, this results in a modified multitrait, multimethod (MTMM) matrix of item-to-construct and item-to-item correlations similar to that analyzed by Bagozzi and Fornell as well as Loch et al. [32]. The resulting matrix, showing the item-to-construct correlations as the grayed out cells, appears as Table 1.

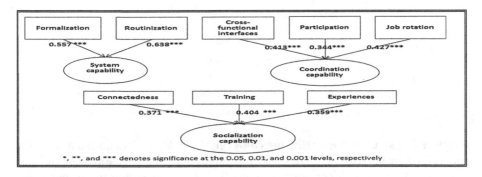

Fig. 1. Formative indicator weights

The convergent validity is demonstrated if the items of the same construct correlate significantly with their corresponding composite construct value (item-to-construct correlation) [24, 32]. This condition has been met, as all items correlated significantly (p < 0.01) with their respective construct composite value. *The discriminant validity* can be established if the item-to-construct correlations are higher with each other than with other construct measures and their composite value [32]. This condition is also met. The resulting matrix, showing the item-to-construct correlations as the grayed out cells, appears as Table 1.

Table 1. Inter-Item and item-to construct correlation matrix for formative constructs

		1	2	3	4	5	6	7	8	9	10	11
1	Formalization	1										
2	Routinization	0.330**	1									
3	System capability	0.818**	0.814**	1								
4	Cross-functional interfaces	0.752**	0.225**	0.601**	1							
5	Participation in decision making	0.284**	0.146*	0.264**	0.420**	1						
6	Job rotation	0.626**	0.164**	0.486**	0.747**	0.498**	1					
7	Coordination capability	0.676**	0.213**	0.546*	0.877**	0.729**	0.907**	1				
8	Connectedness	0.563**	0.140*	0.432**	0.638**	0.505**	0.734**	0.751**	1			
9	Training	0.715**	0.192**	0.557**	0.644**	0.369**	0.635**	0.664**	0.551**	1		
10	Experiences	0.597**	0.146*	0.457*	0.684**	0.517**	0.726**	0.771**	0.744**	0.700**	1	
11	Socialization capability	0.706**	0.180**	0.545**	0.744**	0.528**	0.793**	0.827**	0.869**	0.846**	0.929**	1

In a sense, very high reliability can be undesirable for the formative constructs because the excessive multicollinearity among the formative indicators can destabilize the model [33]. To ensure that the multicollinearity is not a significant issue, we assessed the VIF (variance inflator factor) statistic. If the VIF statistic is greater than 3.3, the conflicting item should be removed as long as the overall content validity of the construct measures is not compromised [27]. For our formative measures, we find the VIF values of both formalization and routinization in the system capability to be 2.187 and 1.145, equivalently, 2.397, 2.318 and 1.508 for the cross-functional interfaces, job rotation and participation in decision making in the coordination capability; while 2.849, 2.748, and 2.444 for the connectedness, preservice training and experiences inheritance in the socialization capability. In summary, the results suggest that all indicators have VIF statistics lower than 3.3.

Table 2. Reliability and convergent and discriminant validity for reflective constructs

		1	2	3	4
1	Syntactic transfer	0.827			
2	Semantic translation	0.723	0.858		
3	Pragmatic transformation	0.665	0.756	0.882	
4	Human capital	0.585	0.589	0.576	0.830

Square root of AVE reported along diagonal in bold type.*Significant at the .05 level; **significant at the .01 level

Reflective Constructs: Generally, *the convergent validity* is demonstrated if (1) the item loadings are in excess of 0.70 on their respective factors and (2) the average variance extracted (AVE) for each construct is above 0.50 [31]. According to our parameter estimates, these conditions have been met. Gefen and Straub also contend that the discriminant validity is demonstrated if (1) the square root of each construct's AVE is greater than the inter construct correlations (Table 2), and (2) the item loadings on their respective constructs are greater than their loadings on other constructs (Table 3) [31]. These conditions have also been met, thereby demonstrating that the independent construct indicators discriminate well.

Table 3. PLS component-based analysis: cross-loadings for reflective constructs

			1	2	3	4
1	Syntactic transfer	TR1	0.851	0.521	0.559	0.427
		TR2	0.862	0.634	0.526	0.461
		TR3	0.766	0.652	0.564	0.573
2	Semantic translation	TT1	0.714	0.875	0.651	0.562
		TT2	0.566	0.831	0.636	0.420
		TT3	0.570	0.869	0.662	0.522
3	Pragmatic transformation	TM1	0.584	0.704	0.871	0.557
		TM2	0.557	0.637	0.898	0.508
		TM3	0.616	0.658	0.878	0.457
4	Human capital	Hu1	0.531	0.525	0.471	0.848
		Hu2	0.396	0.354	0.371	0.776
		Hu3	0.510	0.552	0.562	0.857

Finally, the *reliability* for the scales was gauged via the composite reliability scores provided in the PLS output. The composite reliability scores equal to or greater than 0.70 are regarded as acceptable [29]. So the composite reliability scores of these reflective variables are acceptable. Our validation results suggest that all reflective measures demonstrated satisfactory reliability and construct validity, and all formative measures demonstrated satisfactory construct validity and no significant multicollinearity. Therefore, all of the measures were valid and reliable.

A Harman one-factor test serves to assess the potential for common method bias in the data [31]. An unrotated factor analysis using the eigenvalue-greater-than-one criterion results in a solution that accounts for 52.51 % of the total variance, and the first

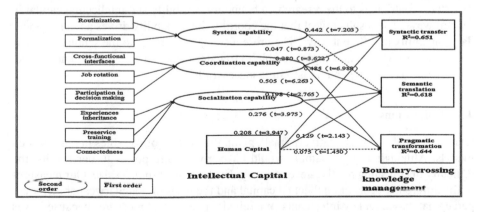

Fig. 2. Result of structural equation model (PLS) analysis

factor accounts for only 5.78 % of the variance. Therefore, common method bias is unlikely to be a serious problem.

3.4 Assessment of Structural Model

In this section, we tested the amount of variance explained and the significance of the relationships. Additionally, a bootstrap re-sampling approach is suggested in order to estimate the precision of the PLS estimates. Following this suggestion, a bootstrap analysis with 500 bootstrap samples and the original 278 cases was performed to examine the significance of the path coefficients [25]. Figure 2 shows the structural model with the coefficients for each path (hypothesized relationship).

4 Discussion

4.1 Implications

The results did not support the positive relationship between the system capability and semantic translation of Hypothesis 2. Firstly, the main reason could be that the common meanings may be formalized to such an extent that impedes the flexible incorporation of newly acquired and existing knowledge. Secondly, the routinization tends to isolate knowledge, to limit the joint dialogues, and to confine the development of new perspective by imposing the existing knowledge.

We did not find the human capital to hinder the pragmatic transformation of boundary-crossing management knowledge. It appears that the relationship between human capital and boundary capability is dependent on other elements, such as the capability of coordination, socialization, and leadership [51]. Additionally, the interaction between employees and customers also enables employees to understand customers' needs, thereby encourages employees to change their domain knowledge [51].

4.2 Contributions

Our study contributes to the research on boundary capability and intellectual capital in several ways. The results reveal that the coordination capability drives a team's boundary capability in different ways. The present study contributes to the scholars' understanding as to why certain firms are able to transfer their domain-specific knowledge, but not able to translate and transform it successfully.

4.3 Limitations

Our net sample size was relatively small, given the number of variables in our research models. Although a larger sample would have given more power to our results, the difficulty of collecting relative-level primary data imposed limits on size. Our measures of the different aspects of intellectual capital and the types of boundary capability were perceptual, based on key informants. We relied on perceptual measures because it was difficult to obtain relevant objective measures capturing the variations in intellectual capital and boundary capability with the kind of precision we required.

4.4 Future Research Directions

One area into which this study can be extended is to focus more closely on the link between social capital and boundary capability in an effort to understand why social capital influences three types of boundary capabilities. Another area is the exploration of what factors influence the link between human capital and boundary capability. These are some possible points of departure for future research.

4.5 Conclusion

To conclude, our study provides an empirically grounded framework simultaneously linking various aspects of intellectual capital and their interrelationships to different types of boundary capability. This framework shows how firms need to distinctively utilize their varied knowledge resources to achieve different types of boundary capability. It also provides a structure for future research probing of more specific questions regarding the knowledge-innovation link.

References

1. Carlile, P.R.: Transferring, translating, and transforming: an integrative framework for managing knowledge across boundaries. Organ. Sci. **15**(5), 555–568 (2004)
2. Subramaniam, M., Youndt, M.A.: The influence of intellectual capital on the types of innovative capabilities. Acad. Manag. J. **48**(3), 450–463 (2005)
3. Jansen, J.J., Van Den Bosch, F.A., Volberda, H.W.: Managing potential and realized absorptive capacity: how do organizational antecedents matter? Acad. Manag. J. **48**(6), 999–1015 (2005)
4. Gardner, H.K., Gino, F.: Dynamically integrating knowledge in teams: transforming resources into performance. Acad. Manag. J. **55**(4), 998–1022 (2012)
5. Van Den Bosch, F.A.J., Volberda, H.W., De Boer, M.: Coevolution of firm absorptive capacity and knowledge environment: organizational forms and combinative capabilities. Organ. Sci. **10**, 551–568 (1999)
6. Weick, K.E.: The Social Psychology of Organizing. Addison-Wesley, Reading (1979)
7. Grant, R.M.: prospering in dynamically-competitive environments: organizational capability as knowledge creation. Organ. Sci. **7**, 375–387 (1996)
8. Lin, X., Germain, R.: Organizational structure, context, customer orientation, and performance: lessons from Chinese state-owned enterprises. Strateg. Manag. J. **24**, 1131–1151 (2003)
9. Noe, R.A.: Employee Training and Development. Irwin/McGraw-Hill, Boston (1999)
10. Nahapiet, J., Ghoshal, S.: Social capital, intellectual capital, and the organizational advantage. Acad. Manag. Rev. **23**, 242–266 (1998)
11. Rerup, C., Feldman, M.S.: Routines as a source of change in organizational schemata: the role of trial-and-error learning. Acad. Manag. J. **54**(3), 577–610 (2011)

12. Nelson, R.R., Winter, S.J.: An Evolutionary Theory of Economic Change. Harvard University Press, Cambridge (1982)
13. Cohen, M.D., Burkhart, R., Dosi, G., Egidi, M., Marengo, L., Warglien, M., Winter, S.: Contemporary issues in research on routines and other recurring action patterns of organizations. Ind. Corp. Change **5**, 653–698 (1996)
14. Galunic, D.C., Rodan, S.: Resource recombinations in the firm: knowledge structures and the potential for schumpeterian innovation. Strategic Manag. J. **19**, 1193–1201 (1998)
15. Gupta, A.K., Govindarajan, V.: Knowledge flows within multinational corporations. Strategic Manag. J. **21**, 473–496 (2000)
16. Henderson, R., Cockburn, I.: Measuring competence? Exploring firm effects in pharmaceutical research. Strateg. Manag. J. **15**, 63–84 (1994)
17. Cohen, W., Levinthal, D.: Absorptive capacity: a new perspective on learning and innovation. Adm. Sci. Q. **35**, 128–152 (1990)
18. Collins, C.J., Smith, K.G.: Knowledge exchange and combination: the role of human resource practices in the performance of high-technology firms. Acad. Manag. J. **49**(3), 544–560 (2006)
19. Armstrong, J.S., Overton, T.S.: Estimating nonresponse bias in mail surveys. J. Mark. Res. **14**(3), 396–402 (1977)
20. Bagozzi, R.P., Fornell, C.: Theoretical concepts, measurement, and meaning. In: Fornell, C. (ed.) A Second Generation of Multivariate Analysis. Praeger, New York (1982)
21. Bailey, K.D.: Methods of Social Research. Free Press, New York (1987)
22. Barclay, D., Higgins, C., Thompson, R.: The partial least squares approach (PLS) to causal modeling, personal computer adoption and use as an illustration. Technol. Stud. **2**(2), 285–309 (1995)
23. Bollen, K., Lennox, R.: Conventional wisdom on measurement: a structural equation perspective. Psychol. Bull. **110**(2), 305–314 (1991)
24. Campbell, D.T., Fiske, D.W.: Convergent and discriminant validation by the multi-trait–multi-method matrix. Psychol. Bull. **56**(2), 81–105 (1959)
25. Chin, W.W.: The partial least squares approach for structural equation modeling. In: Marcoulides, G.A. (ed.) Methodology for Business and Management, pp. 295–336. Lawrence Erlbaum Associates, Mahwah (1998)
26. Chin, W.W.: The PLS approach to SEM. In: Marcoulides, G.A. (ed.) Modern Methods for Business Research, pp. 295–336. Erlbaum, Mahwah (1998)
27. Diamantopoulos, A., Siguaw, J.A.: Formative versus reflective indicators in organizational measure development: a comparison and empirical illustration. Br. J. Manag. **17**(4), 263–282 (2006)
28. Diamantopoulos, A., Winklhofer, H.M.: Index construction with formative indicators: an alternative to scale development. J. Mark. Res. **38**(2), 269–277 (2001)
29. Fornell, C., Larcker, D.F.: Evaluating structural equations with unobservable variables and measurement error. J. Mark. Res. **18**(1), 39–50 (1981)
30. Gefen, D., Rigdon, E.E., Straub, D.: An update and extension to SEM guidelines for administrative and social science research. MIS Q. **35**(2), iii–xiv (2011)
31. Podsakoff, P.M., Organ, D.W.: Self-reports in organizational research: problems and prospects. J. Manag. **12**(3), 531–544 (1986)
32. Loch, K.D., Straub, D.W., Kamel, S.: Diffusing the Internet in the Arab world: the role of social norms and technological culturation. IEEE Trans. Eng. Manage. **50**(1), 45–63 (2003)
33. Petter, S., Straub, D., Rai, A.: Specifying formative constructs in information systems research. MIS Q. **31**(4), 623–656 (2007)

34. Sivo, S.A., Saunders, C., Chang, Q., Jiang, J.J.: How low should you go? Low response rates and the validity of inference in is questionnaire research. J. Assoc. Inf. Syst. **7**(8), 351–414 (2006)
35. Schultz, T.W.: Investment in human capital. Am. Econ. Rev. **51**, 1–17 (1961)
36. Snell, S.A., Dean, J.W.: Integrated manufacturing and human resources management: a human capital perspective. Acad. Manag. J. **35**, 467–504 (1992)
37. Hill, C.W.L., Rothaermel, F.T.: The performance of incumbent firms in the face of radical technological innovation. Acad. Manag. Rev. **28**, 257–274 (2003)
38. Wensley, A.: Editorial: some further thoughts about knowledge transfer and understanding. Knowl. Process Manag. **8**(4), 195–196 (2001)
39. Mclagan, P.A.: Models for HRD practice. Train. Dev. **43**(9), 49–59 (1989)
40. Carlile, P.R.: A pragmatic view of knowledge and boundaries: boundary objects in new product development. Organ. Sci. **13**(4), 442–455 (2002)
41. Youndt, M.A., Subramaniam, M., Snell, S.A.: Intellectual capital profiles: an examination of investments and returns. J. Manag. Stud. **41**, 335–362 (2004)
42. Leonard-Barton, D.: Well Springs of Knowledge: Building and Sustaining the Sources of Innovation. Harvard Business School Press, Boston (1995)
43. Carlile, P.R., Rebentisch, E.S.: Into the black box: the knowledge transformation cycle. Manag. Sci. **49**(9), 1180–1195 (2003)
44. Edvinsson, L., Malone, M.: Intellectual Capital: Realising Your Company's True Value by Finding Its Hidden Brainpower. Harper Collins, New York (1997)
45. Sullivan, P.: Profiting from Intellectual Capital: Extracting Value from Innovation. Wiley, New York (1998)
46. Chen, C.J., Huang, J.W.: Strategic human resource practices and innovation performance-the mediating role of knowledge management capacity. J. Bus. Res. **62**(1), 104–114 (2009)
47. Vargas, N.M., Lloria, B.M.: Dynamizing intellectual capital through enablers and learning flows. Ind. Manag. Data Syst. **114**(1), 2–20 (2014)
48. Wu, X., Sivalogathasan, V.: Intellectual capital for innovation capability: a conceptual model for innovation. Int. J. Trade Econ. Finan. **4**(3), 139–144 (2013)
49. Galbraith, J.: Designing Complex Organizations. Addison-Wesley, Reading (1973)
50. Nonaka, I.: A dynamic theory of organizational knowledge creation. Organ. Sci. **5**, 14–37 (1994)
51. Massaro, M., Dumay, J., Bagnoli, C.: Where there is a will there is a way: IC, strategic intent, diversification and firm performance. J. Intellect. Capital **16**(3), 490–517 (2015)

KPD: An Investigation into the Usability of Knowledge Portal in DMAIC Knowledge Management

Thanh-Dat Nguyen[1(✉)], Sergiu Nicolaescu[2],
and Claudiu Vasile Kifor[2]

[1] Engineering Faculty, Lucian Blaga University of Sibiu, Sibiu City, Romania
ntdat@qnu.edu.vn
[2] Information Technology Faculty,
Quy Nhon University, Quy Nhon City, Vietnam
{sergiu.nicolaescu,claudiu.kifor}@ulbsibiu.ro

Abstract. Knowledge is considered as a resource that contributes an important role in the success of Six Sigma DMAIC methodology. However, knowledge resides brain of the individuals and exists in various forms and different places. This rise the problem of how to collect and share DMAIC knowledge everywhere all time. In this paper, we introduce a proposed Knowledge Portal named KPD that had been designed as a tool to manage DMAIC knowledge. Through the deployment of the Knowledge Portal, this paper aims at investigating its impacts on DMAIC execution based on experiments and appreciation of experts who are working in the areas of quality management and information technology. The results of the survey reveal that KPD benefits DMAIC deployment and impacts positively on the success of DMAIC through its knowledge management process.

Keywords: Knowledge portal · Ontology · Six sigma · DMAIC · Quality improvement process

1 Introduction

Six Sigma DMAIC (Define-Measure-Analyze-Improve-Control) methodology [1] with its strong points of eliminating waste, improving the satisfaction of customers, enhancing process performance [2] plays a more and more important role in manufacturing and business processes. Its ultimate objective is to optimize whole defects rates in services or transaction, manufacturing, and processes in order to improve business profits and excellence [3]. Through DMAIC execution, insights and ideas of experts, employees, customers and managers are shared and discussed in improvement or discussion sessions. They are a valuable knowledge resource that is considered as a precious resource in an organization.

In the era of information technology (IT) and knowledge economy, the precious resource of knowledge brings many benefits to an organization. It enhances competitive advantage, encourages innovation, keeps efficiencies, facilitates resource distribution, and changes the nature of investment decisions as well as work and property [4, 5].

F. Lehner and N. Fteimi (Eds.): KSEM 2016, LNAI 9983, pp. 364–375, 2016.
DOI: 10.1007/978-3-319-47650-6_29

Therefore, managing knowledge helps an organization to maximize its strategic value and competitive advantages [5].

As the importance of applying DMAIC methodology increases incessantly and knowledge management enables a knowledge resource to be used and exploited effectively, how to deliver the right knowledge to the right people at the right time is one of the important problems for an organization. In order words, knowledge management in improvement activities of DMAIC execution is one of vital tasks in Six Sigma projects.

In recent years, a knowledge portal (KP) is considered as a notable solution for knowledge management. KP is a type of portal that supports knowledge creation, knowledge storage and retrieval, knowledge transfer, knowledge integration, and distributed knowledge searching [5, 6]. It is a type of knowledge management system that struggles to provide a knowledge worker with collective knowledge at a "one-stop knowledge shop" [6] or a single point of access. Several KPs have been developed and applied in many fields. For example, a KP named "Institute for Advanced Engineering Studies" is built for managing and disseminating implicit and explicit knowledge in academic institutes [7]; KPs in [8, 9] are designed for Libraries; KPs named SkillSoft [10] and iSixSigma (http://www.isixsigma.com/) are developed for Six Sigma. The one trait that is common among the KPs is that they provide its users with functionalities to share and exchange knowledge all time and everywhere in order to enhance users' skills. However, in the KPs, knowledge is not represented structurally, therefore, the KPs do not support as well as provide its users with helpful functions of knowledge reasoning or inference that is one of important requirements to reuse and exploit available knowledge efficiently and effectively. Moreover, because they lack inference engines, the KPs do not support completely to managing DMAIC knowledge yet.

In this paper, the authors want to improve the DMAIC process through a KP approach (that is explained in Subsect. 2.2), to be deployed more successfully. The KP for DMAIC (KPD) supports organizing knowledge repositories, communication between workers and experts, collaboration, document sharing and knowledge validation. KPD is also designed for the proposed model [11], which includes a process of knowledge management, knowledge bases, techniques applied in ontology engineering, and a KP (Fig. 1).

The ultimate objective of the authors is to present a strategy in order to investigate the usability of KPD. More particularly, the authors introduce how the quality attributes for a knowledge management system, a survey questionnaire, the Likert scale-based measures, weights, weighted average scores, and a ranking and rating table are applied. Our research methodology is conducted based on the questionnaire survey and an evaluation is conducted based on quantitative analysis of the survey results. Using the strategy and methodology, the usability of KP can be ranked and rated.

In the next section, a brief description on the model of knowledge management for DMAIC, KPD, and the quality attributes are presented. Section 3 introduces the questionnaire-based survey and evaluating methodology. A discussion on weights-based measures and results is also represented in the same section. The last section is to conclude the paper.

2 A Proposed Model and Knowledge Portal

2.1 A Conceptual Model of Knowledge Management for DMAIC Processes

Knowledge management is a process of collecting, analyzing, transforming, and applying knowledge [13]. In DMAIC execution, a knowledge management process is integrated with each of DMAIC steps in a conceptual model (Fig. 1) [11] that allows to collect and analyze reports, to translate information into valuable knowledge, to represent the knowledge on the basis of Ontologies, and to apply available knowledge into innovation.

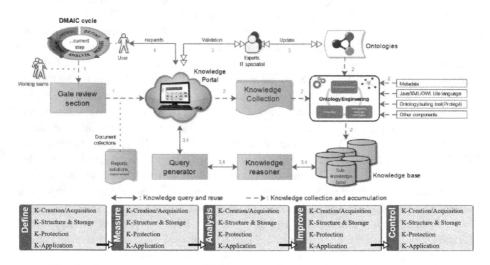

Fig. 1. A four-stage process of KM integrated with DMAIC in OKMD model.

The model illustrates a process that supports to consolidating, updating valuable knowledge, and creating new knowledge. In order to facilitate KM in Six Sigma DMAIC methodology, the process should be executed in four stages: Knowledge Creation/Acquisition, Knowledge Structure & Storage, Knowledge Protection, and Knowledge Application. More particularly, in the KM process, the discussion and reports created by members of Six Sigma project such as managers, Black Belts, and Champions in Tollgate reviews [14] or Gate Review sessions [15, 16] are collected through KPD in the first stage. They are then used to analyzed and translated into Ontologies based on Ontology engineering in the second stage. Before reusing knowledge, it is protected by every web user's an account and the security policy on the Knowledge Portal in the third stage. The knowledge in the reports would be found and inferred by KPD's users through a module of Knowledge reasoner in order to reuse in the next steps of the DMAIC process in the last stage.

2.2 Knowledge Portal for DMAIC

Knowledge Portal is an effective solution for collecting and sharing knowledge in a process of knowledge management. It allows to convert tacit knowledge to explicit knowledge and vice versa as well as to share available knowledge [7]. As we can see in Fig. 1, KPD is designed as an important component to manage effectively the knowledge of DMAIC processes. It should provide necessary functionalities to collect reports and document, share resources, exchange, search, and infer knowledge available. A successful deployment of KPD should be evaluated on the basis of the following aspects:

Aspect 1: KPD should be deployed and applied in reality DMAIC processes.

Aspect 2: DMAIC knowledge should be collected, found and reused completely through the four-stage process of knowledge management (Fig. 1) during DMAIC execution.

Aspect 3: All functionalities of KPD satisfy the requirements of a conventional knowledge portal and are evaluted by measures on the basis of the quality attributes for a knowledge management system [12] consisting of Availability, Relevance, Security, Usefulness, Procedures, Policies, Usability, Economy and Social (Fig. 2).

Fig. 2. The important quality factors for the framework of a knowledge management system. Adapted from [12]

The requirements for a conventional Knowledge portal development applied to KPD are introduced by several researchers in [5, 17–19] involving the following criteria:

– Organizing knowledge repositories
– Collaboration
– Document sharing tools
– Communication between workers and experts
– Tools to search and retrieve knowledge
– Knowledge validation
– E-learning services
– Open source web-based applications are recommended

On the basis the aforementioned requirements, the useful functionalities of KPD are proposed as the following table:

As shown in Table 1, the functionalities of KPD are divided into five groups: Content management, knowledge exchange, knowledge sharing, document, and

368   T.-D. Nguyen et al.

Table 1. The functionalities and tools for KPD

Group	Functionality	Description	Tools
Content management	News	Publish news, articles, organizational structure, events, activities, surveys, course information, search.	**NukeViet 4.0** (a free and open source CMS downloaded at http://nukeviet.vn)
	Resources	Share collected DMAIC reports, sample files of report, images, video, audio, presentation.	
	Links to	External resources, social networks, channels (Youtube), Webmail.	
Knowledge exchange	Communication	By email, chat, blog, forum, questionnaires, discussion sessions.	**Moodle 3.0** (a free and open source LMS downloaded at http://moodle.org)
	Learning	Organize courses, exercises / training, live presentation.	
Knowledge Sharing	Knowledge dissemination/ sharing	Share Ontologies, collected DMAIC reports; Query and infer knowledge using Knowledge Reasoner module.	**K-Reasoner module** (Developed using PHP, MySQL+ARC2, SPARQL, Fuseki, Protégé, and Ontology engineering), **WebVOWL**
Document	Supporting document	Provide document to web developers: PHP, SPARQL, Fuseki, Protege, Sublime, Text, ARC2, design of K-Reasoner module, Ontologies's description	A module of document management on KPD
Administration	User	Security policy, User and group definition.	Administration function from NukeViet, Moodle and K-Reasoner.
	System management	Domain, hosting, servers, files, SPARQL Endpoint, Databases.	
	Configuration Management	Interface, Modules and Plug-ins, Layout / Appearance, Users (Publishers)	

administration. They are constructed based on a free and open source CMS (content management system, i.e. Joomla, NukeViet), a LMS (learning management system, i.e. Moodle), and Ontology engineering. The most important function group used to share and infer knowledge is Knowledge Sharing. It provides virtual and visible Ontologies and collected DMAIC reports in order to evaluate and update the knowledge base. Moreover, a K-Reasoner module (Fig. 3) responsible for querying and inferring knowledge is also a core functionality of the function group. The module developed in PHP language receives user's request and creates to send SPARQL queries using Query generator to SPARQL endpoint. The server takes responsibility for analyzing the queries and extracting knowledge from a knowledge base. Finally, querying results are displayed on KPD's web-based interface by the module.

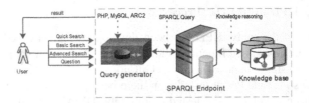

Fig. 3. The architecture of K-Reasoner module

All the rest of function groups are to collect data (i.e. reports and documents) created during DMAIC execution, to support activities of knowledge exchange, to share technical documents with programmers, and to help administrators in website management.

3 Evaluating Usability of Knowledge Portal Application

3.1 Research Methodology

The international Organizaton for Standardization (ISO) determines the usability of a product on the basis of *"the extent to which the product can be used by specified users to achieve specified goals with effectiveness, efficiency, and satisfaction in a specified context of use."* In this paper, we consider the key attributes/parameters in order to assess the usability of KPD: Relevance, Security, Usefulness, Usability, Economy, Social, User satisfaction and expert ranking.

Based on experiments and rating opinions of experts who are working in the fields of quality management and IT., data is collected for evaluation. For experiments, the authors of the paper have identified a procedure of KPD installation and deployment and its technical problems. Thereby, some recommendations and different approaches to deploy KPD have been pointed out in Subsect. 3.2. Besides, a survey questionnaire is prepared by the authors to collect the opinion of participants. The Likert-scale-based questionnaire is composed based on literature review and experience of individuals who have ever participated in Six Sigma projects.

The survey questionnaire is then sent to 19 participants who have knowledge in the fields of Six Sigma, quality improvement or engineering, and IT. In particular, the questionnaire survey is carried out in first two weeks of May 2016 in Romania and first two weeks of Jun 2016 in Vietnam with a 90 % response rate (17 persons) collected. The age of participants is between 24 years of age and 47 years of age. They are the experts (18 % of respondents), who work in 03 Romania companies, professors (18 %), Ph.Ds. (24 %), and Ph.D. candidates (40 %). 53 % of respondents belongs to the ones who are working in the fields of Six Sigma or engineering (quality improvement) while the rests work in the IT. area. In Romania, 11 respondents are collected from 03 experts (Ph.Ds.), 02 professors, 01 Ph.D., and 05 Ph.D. candidates. In Vietnam, 06 respondents come from 01 professor, 03 doctors, and 02 Ph.D. candidates. Ph.D. candidates are researching in the areas of IT. (03 respondents), Engineering (03), and Six Sigma/IT. (01).

3.2   Applying KPD in DMAIC Processes

In the first stage of the study, KPD has been developed (Fig. 4) based on aforementioned requirements, proposed functionalities and tools. A deployment procedure of six steps namely, Design, Data collection, Ontology building, K-Reasoner module development, Testing, and Portal installation have been performed. The functionalities of KPD have been developed and tested completely. DMAIC reports have been collected and used to create Ontologies. K-Reasoner module has been developed and installed on KPD and provides accurately all knowledge found in the reports. KPD is deployed on Dreamhost server, and the address of KPD at http://p.kdmaic.net is included in the questionnaire to send to the participants of the survey.

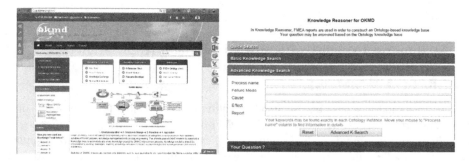

Fig. 4. Homepage and K-Reasoner module of KPD

Through the deployment, we found that a particular knowledge of IT for participants who install the applications are necessary. They should be administrators of companies and would support technically, applying KPD into their DMAIC processes. Moreover, DMAIC reports must be converted to structured text-based formats before they are used to translate into Ontologies through Parsers. K-Reasoner module should be updated immediately after updating Ontologies. Last but not least, big Ontologies should be stored in Apache Jena Fuseki servers with an advantage of fast query time while smaller Ontologies can be deployed in MySQL database.

3.3   Questionnaire Survey, Results and Discussion

In order to collect opinions of participants on the usability of KPD, a Likert scale-based questionnaire, which involves an overview of KPD's functionalities and tools, and a set of statements and questions are prepared. The questionnaire is to aim at investigating seven important quality attributes/parametters (mentioned in Subsect. 3.1) and grouped into two main issues: (1) benefits of applying KPD to DMAIC processes and (2) influences of KPD on successful aspects of Six Sigma projects. Each statement is measured by a quantitative value formatted from a scale of five levels from 1 to 5. After the questionnaire is completed, it is provided to the survey participants. The responses are then collected and analyzed quantitatively.

In the mentioned parameters, usability, usefulness and relevance are considered as the most important parameters corresponding to the highest weights, other quality parameters are identified by the lower weights. The allocated evaluation percentage of KPD quality parameters are presented in the Table 2.

Table 2. Distribution table for the quatity parameters of KPD

KPD quality parameter	Allocated evaluation percentage
Usability	60 %
Usefulness	10 %
User satisfaction and expert ranking	10 %
Relevance	5 %
Security	5 %
Economical	5 %
Social	5 %

For the first issue, a list of statements are composed in order to assess beneficial aspects of KPD deployment in an organization. The participants rated every statement/aspect by giving a quantitative value from 1 to 5, where 1-Not beneficial at all, 2-less beneficial, 3-neutral, 4-beneficial and 5-extremely beneficial. Similarly, a set of statements is formulated to measure impacts of KPD on successful aspects of a DMAIC process. The aspects are introduced by the authors in [3, 20]. Each statement is

Table 3. Weightage table for usability

KPD usability measures	Weight (60 %)	Captured rating from survey	Weighted calculation
Support employees and experts to quickly access and available knowledge	20 %	4.5	0.9
Contribute knowledge to innovation and improvement solutions	20 %	4.2	0.84
Support a quick access to organizational resources	10 %	4.4	0.44
Enhance the improving skill of new employees	10 %	4.1	0.41
Enhance customer satisfaction	10 %	4.2	0.42
Curtail the deployment time of DMAIC	10 %	3.8	0.38
Create a favourable environment to share knowledge	10 %	4.5	0.45
Enhance competitive advantage of an organization	10 %	4.1	0.41
	100 %	**Total score**	**4.25**
KPD parameter score			**2.55**

measured by marks from 1 to 5 where 1-Not influential at all, 2- Less influential, 3-Neutral, 4-Influential, 5-Extremely influential. Furthermore, the questions to measure the parameters of security, economic, social, user satisfaction, and expert ranking are also adapted in the questionnaire.

Table 3 indicates how the evaluation percentage (30 %) is distributed among several Usability measures; the captured ratings collected from the survey, and weighted calculation for the quality parameter of usability. For example, the most important measure considered is "*Support employees and experts to quickly access and available knowledge*" with the weight of 20 % and received the highest rating from the respondents (4.5). With the total score of 4.25 and its percentage (60 %), the parameter score of Usability is calculated by the highest value (2.55) compared to other parameter scores in Table 6.

Similarly, Tables 4 and 5 reveal other total scores for Relevance and Usefulness that are calculated from survey results. The total score of Usefulness is higher than that of Relevance, 4.1 and 3.87 respectively. Moreover, the scores show a significant influence of KPD on the successful deployment of a Six Sigma DMAIC process. Therefore, KPD can contribute an important role to Six Sigma success.

Table 4. Weightage table for relevance

KPD relevance measures	Weight (5 %)	Captured rating from survey	Weighted calculation
Improving quality of Six Sigma projects	20 %	4.2	0.84
Strong leadership and top management commitment	20 %	3.9	0.78
Continous education and training	10 %	3.9	0.39
Organizational infrastructure, and IT infrastructure	10 %	3.8	0.38
Encouraging and accepting cultural change	10 %	3.8	0.38
Management involvement and organizational	10 %	3.8	0.38
Stakeholders such as customers, human resources, and suppliers	10 %	3.6	0.36
Project selection, management, and control skills	10 %	3.6	0.36
	100 %	**Total score**	**3.87**
KPD parameter score			**0.1935**

In order to rate the quality parameters of KPD, a table of ranking and rating [12] is applied (Table 7). Based on the table, the higher the total score of KPD quality parameters, the better the usability of KPD. The Fig. 5 represents the overall summary weightage for the quality attributes/parameters. On the basis of the ranking and rating table (Table 7) and the parameters, it is clear that KPD is an extremely usable model

Table 5. Weightage table for usefulness

KPD usefulness measures	Weight (10 %)	Captured rating from survey	Weighted calculation
Knowledge exchange: Chat, forum, discussion, survey	30 %	4.2	1.26
Knowledge inference for reusing knowledge available	30 %	4.2	1.26
Content management (news, articles, media resources)	20 %	3.9	0.78
Course organization and management	10 %	3.9	0.39
DMAIC reports collection and sharing	10 %	4.1	0.41
	100 %	Total score	4.1
KPD parameter score			**0.41**

Table 6. The summary table of quality parameters

KPD quality parameter	Weight	Captured rating from survey	KPD parameter score
Usability	60 %	4.25	2.55
Usefulness	10 %	4.1	0.41
User satisfaction and expert ranking	10 %	4.1	0.41
Relevance	5 %	3.87	0.1935
Security	5 %	4.0	0.2
Economical	5 %	3.5	0.175
Social	5 %	4.2	0.21
Total score			**4.1485**

Table 7. Ranking and rating table. Adaped from [12]

Rank	Description	Rating
1	Outstanding	5
2	Extremely usable	4
3	Usable	3
4	Somewhat usable	2
1	Not usable	1

that can be applied effectively in DMAIC deployment, with the overall evaluation score of 4.15 (Table 6) though the aspect of economical should be improved from the lowest score (3.5).

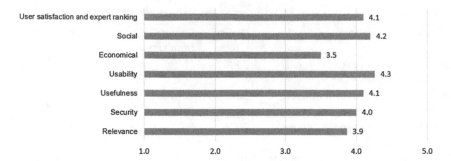

Fig. 5. The quality parameters for KPD

4   Conclusion and Future Works

Constructing a knowledge portal should be related to a particular process of knowledge management where activities and tools are identified. Thereby, the knowledge portal is developed more efficiently and effectively and its role is enhanced. In this paper, KPD has been designed for DMAIC processes in order support to knowledge management of DMAIC. Through the deployment of KPD and the questionnaire survey, positive feedbacks about its benefits and impacts on successful DMAIC execution are achieved. Based on the quality attributes/parameter calculated, KPD is considered as an important tool that brings many benefits to Six Sigma DMAIC methodology and enables the quality improvement process to be deployed more successfully. However, the proposed methodology of evaluation is easy to implement as it is based on the questionnaire survey and the weighted average scores. Evaluating the usability of KPD can be based on other measures such as Six Sigma performance, products improved by the Six Sigma projects, the amount of knowledge collected, and the number of ontologies developed. For the future work, KPD may be applied as a crucial tool in a model knowledge management for DMAIC process and should be implemented in IT projects.

Acknowledgement. This article has been done within Doctoral studies under the financial support of Erasmus Mundus Mobility with Asia (EMMA). Authors are very much thankful to the EMMA management team for their support.

References

1. Park, S.H.: Six Sigma for Quality and Productivity Promotion. Asian Productivity Organization, Tokyo (2003)
2. Chang, S.I., Chou, C.C., Wu, H.C., Lee, H.P.: Applying Six Sigma to the management and improvement of production planning procedure's performance. Total Qual. Manag. Bus. Excellence **23**(3–4), 291–308 (2012)
3. Antony, J., Banuelas, R.: Key ingredients for the effective implementation of Six Sigma program. Measuring Bus. Excellence **6**(4), 20–27 (2002)

4. Oprean, C., Kifor, C.V., Negulescu, S.C., Bǎrbat, B.E.: Innovating engineering education, to face the knowledge society. In: Proceedings of the Balkan Region Conference on Engineering and Business Education and International Conference on Engineering and Business Education, Sibiu, pp. 80–85 (2009)

5. Lee, H.J., Kim, J.W., Koh, J.: A contingent approach on knowledge portal design for R&D teams: relative importance of knowledge portal functionalities. Expert Syst. Appl. **36**(2), 3662–3670 (2009)

6. Loebbecke, C., Crowston, K.: Knowledge portals: components, functionalities, and deployment challenges. In: International Conference on Information Systems, Orlando, FL (2012)

7. Das, S., Biswas, S.: Knowledge management in academic institution through knowledge portal. In: Conference proceedings of Trends in Management of Academic Libraries in Digital Environment (TMALDEN-2014), pp. 543–551 (2014)

8. Neubauer, W., Piguet, A.: The knowledge portal, or, the vision of easy access to information. In: Library Hi Tech (27.4), pp. 594–601 (2009)

9. Jotwani, D.: Library portal: a knowledge management tool. In: INFLIBNET Centre (2005)

10. Skillsoft® Six Sigma Knowledge Center™. http://www.skillsoft.com

11. Nguyen, T.D., Kifor, C.V.: The sustainability in a quality improvement model. In: 3rd International Engineering and Technology Education Conference & 7th Balkan Region Conference on Engineering and Business Education, Sibiu, Romania (2015)

12. Subramanian, D.V., Geetha, A.: Evaluation strategy for ranking and rating of knowledge sharing portal usability. IJCSI Int. J. Comput. Sci. **9**(1), 395–400 (2012)

13. Pinto, M.: Knowledge management in higher education institutions: a framework to improve collaboration. In: 2014 9th Iberian Conference on Information Systems and Technologies (CISTI), pp. 1–4 (2014)

14. George, M.L., George, M.: Lean Six Sigma for Service. Mcgraw-Hill, New York (2003)

15. Baral, L.M., Kifor, C.V., Bondrea, I.: Assessing the impact of DMAIC-knowledge management methodology on six sigma projects: an evaluation through participant's perception. In: Knowledge Science, Engineering and Management, Sibiu, Romania, pp. 349–356 (2014)

16. Stevens, D.E.: The leveraging effects of knowledge management concepts in the deployment of Six Sigma in a health care company, Walden University (2006)

17. Zaihisma, C.C.: Nor'ashikin, A., Hidayah, S., Wan, M.I.W.M.: Islamic knowledge portal: an analysis on knowledge portal requirements. J. Eng. Appl. Sci. **10**(2), 451–456 (2015)

18. Hauke, K., Owoc, M., Pondel, M.: Knowledge portal for exclusion process services. In: Federated Conference on Computer Science and Information Systems (FedCSIS). IEEE (2014)

19. Munive-Hernandez, J.E.: Implementation of a knowledge portal as an e-learning tool to support MSc projects. In: Proceedings of the 11th International Conference on Knowledge Management and Knowledge Technologies. ACM (2011)

20. Johnson, A., Swisher, B.: How six sigma improves Research and Development. Research Technology Management, pp. 12–15 (2003)

Knowledge Management and Intellectual Capital in the Logistics Service Industry

Vincenzo Del Giudice[1], Pietro Evangelista[2],
Pierfrancesco De Paola[1(✉)], and Fabiana Forte[3]

[1] University of Naples Federico II, Piazzale Vincenzo Tecchio, 80125 Naples, Italy
{vincenzo.delgiudice,pierfrancesco.depaola}@unina.it
[2] Research Institute on Innovation and Services for Development (IRISS),
National Research Council (CNR), Via G. Sanfelice 8, 80134 Naples, Italy
p.evangelista@iriss.cnr.it
[3] Second University of Naples, Via San Lorenzo, 81031 Aversa, Italy
fabiana.forte@unina2.it

Abstract. The changing business scenario in the logistics service market is affecting the development of relationships with customers and the continuous adaptation of service offering. In this context, knowledge management and intellectual capital are potentially successful assets for developing and improving competitive capabilities of logistics service companies. In order to supply more complex and knowledge-intensive services, it is necessary to evaluate the existing IC assets to identify future needs in this area. The main aim of this paper is to investigates how to assess the intellectual capital in order to improve the management of knowledge in third-party logistics service providers. The paper reviews the main methods for assessing intellectual capital assets in the logistics service industry. It suggests the non-monetary methods as the most appropriate ones.

Keywords: Knowledge management · Intellectual capital · Third-party logistics service providers · Evaluation methods

1 Introduction

Over the last decades a number of major changes have profoundly changed the logistics service industry. Third-party logistics companies (3PLs) are under constant pressure to enhance their customer relationships and continually expand the range of services offered. In addition, the competitive scenario in the 3PL market has become much more complex as a result of the dissemination of ICT and web technologies. The net result of this changes is that many 3PLs now offer a range of value-adding services (e.g. inventory management, contract manufacturing and supply chain integration). This has fuelled the transition from the traditional "arms length" approach to the supply of integrated logistics services packages and has further facilitated the migration of 3PL companies from asset-based approach to an information and knowledge-based, value-added logistics service enterprises.

© Springer International Publishing AG 2016
F. Lehner and N. Fteimi (Eds.): KSEM 2016, LNAI 9983, pp. 376–387, 2016.
DOI: 10.1007/978-3-319-47650-6_30

This has serious implications in terms of knowledge and intellectual capital assets requirements. This situation needs to be addressed if the sector is to achieve its true competitive potential in the coming years.

In fact, in the evolving 3PL business landscape, knowledge management (KM) and intellectual capital (IC) have the potential to play an increasingly important role, providing new opportunities for logistics service providers. The more dynamic environment in which 3PLs operate should force these companies to adopt more sophisticated KM tools and practices and supply increasingly complex and knowledge-intensive services.

Recent research indicate that there is a growing interest in studying KM in logistics but the adoption of KM approaches and the related role of the IC in the logistics service industry have not been widely investigated and more research are needed in this area.

The main aim of this paper is to identify the most appropriate IC evaluation methods in the logistics service industry.

The paper is organised into seven sections. The section following this introduction provides an overview of the most relevant changes that have affected the logistics service industry over past decades. Section 3 discusses the relevance of KM and IC in the emerging business model of 3PL companies. The relationships among KM and IC is discussed in the Sect. 4, while Sect. 5 reviews the main aspects of the IC. Section 6 analyses the methods for assessing IC assets. Finally, the concluding section discusses and identifies the most appropriate IC evaluation methods for the logistics service industry. Conclusion and implication of the paper are then provided.

2 The Logistics Service Sector: An Industry in Transition

Over the last decades a number of major changes have occurred which have profoundly affected the logistics service industry. Firstly, the globalisation of production is forcing third-party logistics services providers (3PLs) to move from a regional to a global scale. Secondly, the increasing use of outsourcing by customers has progressively reduced the number of 3PLs they used. Thirdly, transportation services are requested to be more frequent, punctual, reliable, secure and cover a wider geographical area. Fourthly, greater integration with customers are demanded through the support of ICT (e.g. tracking and tracing services). Finally, customers are increasing requesting value added logistics services beyond pure transportation services.

As results of the above changes, 3PL companies are under constant pressure to enhance their customer relationships and continually expand the range of services offered. Consequently, many 3PLs have been forced to transform the scope of their business model and service offering as a result of the changing customer requirements [17]. In this process, core service offerings are being commoditized (e.g. transportation and warehousing), while value-added logistics services and technological capabilities are considered points of differentiation [14].

The changing scenario in the logistics service industry may be descried identifying tree different evolving stages over the last decades [16]. In each stage different types of companies entered the logistics service market according to the varying levels of customer logistics outsourcing.

During the first stage (1970s–1980s) commodity providers (carriers) operated in the market, and transportation was the main outsourced activity. The competitive weapons of these companies were mainly based on operational efficiency and a resulting low cost base.

In the second stage (1980s–1990s) third party logistics companies appeared on the market. The competitive abilities of these companies mainly relied on integrated transport and warehousing services provided in combination with a range of customised value-added logistics services. Such companies often outsourced basic and low margin services (such as transportation) to carriers which functioned as tier suppliers in the system.

The most recent stage (2000 and beyond) has been characterised by the entry of a new type of company: the fourth party logistics service providers (4PLs) or lead logistics providers (LLPs), offering integrated strategic supply chain orchestration. These companies typically handle all aspects of the supply chain from procurement through to inventory control, final delivery and invoicing. In this context, 4PLs are able to supply highly customised and specialised services through the control of strategic functions such as supply chain design and integration on a global scale, while value-added services and other logistics activities are outsourced to local 3PLs that act as sub-contractors.

This evolution resulted in a transition from a single-activity model toward a more complex business model based on providing a wider range of integrated services [1]. This has given 3PLs a more critical supply chain role than in the past as they assume responsibility for a growing number of activities beyond transportation and warehousing [23] and coordinating physical and information flows at multiple levels of the supply chain [6, 34]. This has fuelled the transition from the traditional "arms length" approach to the supply of integrated logistics services packages and has further facilitated the migration of 3PL companies from asset-based approach to an information and knowledge-based, value-added logistics service enterprises [35].

3   The Role of KM and IC in the New Business Model of Third-Party Logistics Service Providers

In the evolving 3PL business landscape outlined above, KM and IC have the potential to play an increasingly important role, providing new opportunities to logistics service providers [47]. In fact, in today's turbulent supply chain environment, which is characterized by time compression and the need for agility, KM capabilities and related IC assets are considered a critical variable for logistics service differentiation [20] and a tool to optimize costs and better serve clients [33]. Moreover, as supply chain processes and planning increasingly require real-time data availability and exchange, 3PLs must have the capability to manage information and knowledge flows for coordinating a wider range of supply chain activities. The more dynamic environment in which 3PLs operate should force these companies to adopt more sophisticated KM tools and practices and supply increasingly complex and knowledge-intensive services.

The management of knowledge may be considered a strategic asset for improving the performance of logistics processes and services. The logistics service industry, like other industries, has recognized the benefits of KM and started implementing KM strategies to some extent [19]. In addition the use of KM systems is considered as essential to support the specific needs of logistics service companies as well as to enhance the collaboration with other supply chain actors [38].

Given the critical role that logistics companies play in national and international markets [53] it is important to study the role of KM and the assessment of IC needed to support the ongoing change processes in logistics services companies.

Recent research has shown that there is a growing interest in the study KM in logistics [18, 20, 23, 28]. Nevertheless, despite the increasing knowledge intensity of the logistics service industry and the importance of 3PLs in contemporary supply chain configurations, the adoption of KM approaches and the related role of the IC in the logistics service industry have not been widely investigated [32] and more research are needed in this area.

In the next section, the role of intellectual capital in KM is analysed. This will allow to focus on the criteria and most used methods for evaluating IC.

4   The Importance of Intellectual Capital in Managing Knowledge

In the context of the so called "new economy", many organisations have started to pursue KM. Their objectives typically are to increase the enterprise's organizational effectiveness and to improve its short and long-term competitiveness. In modern organisation, the management of knowledge cannot leave aside from the knowledge embedded in individuals and organisations that has been termed as IC [12, 42, 46].

There is considerable overlap in the scope of IC and KM. There are, however, major differences between their perspectives. IC focuses on building and governing intellectual assets from strategic and enterprise governance perspectives with some focus on tactics. Its function is to take overall care of the enterprise's IC. KM focuses on facilitating and managing knowledge related activities such as creation, capture, transformation and use. Its function is to plan, implement, operate and monitor all the knowledge-related activities and programs required for effective IC management. Similarly to IC, the overall purpose of KM is to maximize the enterprise's knowledge-related effectiveness and returns from its knowledge assets and to renew them constantly. KM focus on and manage systematic, explicit, and deliberate knowledge building, renewal and application. From a managerial perspective systematic and explicit KM covers four areas [52]: (i) top-down monitoring and facilitation of knowledge-related activities; (ii) creation and maintenance of knowledge infrastructure; (iii) renewing, organizing and transforming knowledge assets; (iv) leveraging (using) knowledge assets to realize their value.

Ulrich identify following three main components of IC: human capital, relational capital and organisational capital [50]. Human capital represents employee knowledge, competency and brain power. Relational capital refers relations with customers, suppliers, distributors and others related parties. Organisational capital designates the organisational systems, culture, practices and processes.

In summary, IC is a key driver of innovation and competitive advantage in today's knowledge based economy [50]. Simultaneously, KM is recognized as the fundamental activity for obtaining, growing and sustaining IC in organisations [49]. This means that the successful management of IC is closely linked to the KM processes of an organization; which in turn implies that the successful implementation and usage of KM ensures the acquisition and growth of IC.

By integrating the management of intellectual assets knowledge and employees, a company will be in position to exploit effectively the internal and external resources for improving its competitive success. The opportunities to capitalize on the use of knowledge can cover a wide range of activities. In its simplest form it may mean using the best knowledge available to perform a particular task. In more complex circumstances, it may involve embedding knowledge in specific 'building blocks' such as technology platforms. In other situations, it will mean to license patents or sell technology outright. Of great importance is the need to integrate the IC management and KM objectives and perspectives. The combined system must be treated as a dynamic process.

Considering the evolving trend that are affecting logistics service providers today and the shift of their management model toward a more knowledge-based configuration, the importance of assessing and measuring intellectual capital performance is of great importance. The next sections are aimed to highlight most relevant aspects of IC and subsequently the criteria and most used methods for assessing IC are illustrated.

5   Main Aspects of Intellectual Capital

Intangible resources are oriented to knowledge-based economy and aimed to services rather than products.

Two are the ways to interpret IC. On the one hand, IC may be considered as the knowledge highway and brainpower. Here the focus is on knowledge creation with its expansion in the company. On the other, IC pursues a perspective based on resources and it is focused on creating value from the combination of existing tangible and intangible resources present in any firm.

Moreover, the IC exists in tacit form in every firm, but only with its exploitation is possible to get a reliable measure of corporate performance. According to Sullivan [46], IC consists of two main basic elements: intellectual resources and intellectual assets.

The firm's intellectual resources are difficult to exchange because have their headquarters into the minds and ability of its employees, or better, reside in the collective experiences and in the overall know-how of employees. Instead, the intellectual assets are represented by coded tangible and specific information owned by the company.

A further contribution to the identification of the main elements that constitute IC has been accomplished by Edvinsson et al. [13] with the "Value Platform Model". This model is able to indicate how the value creation is the result of positive interactions between three fundamental elements, i.e. human capital, structural or organizational capital, customer capital.

Based on this distinction, Skandia AFS (an insurance and financial services company) showed that firm market value can mainly derive from financial and intellectual capital, divided, in turn, into other subcategories (see Fig. 1).

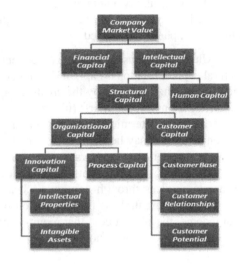

Fig. 1. Value scheme proposed by Skandia AFS.

Roos et al. [39] also classify IC by dividing it into human capital and structural capital (see Fig. 2). The first includes skills, attitudes and intellectual vitality, while the second includes relationships, organization and renovation/development policy.

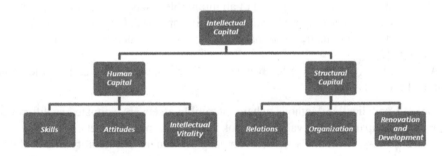

Fig. 2. Intellectual capital according to Roos et al. [39]

Three aspects differentiate IC from company's tangible assets:

– IC is re-evaluated over time and tends to not depreciate. Machinery, equipment, building structures, industrial warehouses and similar assets begin to lose their value from the time of purchase or building [7–11, 27, 29, 48]; the know-how for each individual, instead, is constantly evolving, and gives new knowledge in the future. A modern manager must develop and make fruitful the know-how of employees, transforming the intellectual capital into value for customers;

- IC is not "consumed" and decreases with its use, rather it grows with its use. It is possible to sell or exchange a product or service created from IC development, but know-how remains within the company and can be used many times;
- Unlike traditional assets, the IC is mainly owned by company workers. In the knowledge economy, the distinction between job attributable to individual workers and human capital held by the company is exceeded.

As already mentioned, Ulrich [50] considers three components for the analysis of IC: human capital, structural capital, relational capital: (i) human capital may be identified as the source of innovation characterized by the creative capacity of individuals aimed at offering innovative solutions to customers; (ii) structural capital is the company's organizational ability to meet predetermined requirements and improve them in the market; (iii) finally, relational capital, may be identified as the value of relations with external stakeholders, such as suppliers, competitors and customers.

From the intersection of these three main elements it is possible to create added value in the products and corporate services through continuous learning and knowledge circulation. Therefore, it should be highlighted that the notion of IC is an economic concept: before being an income, the IC is a cost that requires investments in human resources, technologies, software, know-how, cooperation agreements, etc.

6   Methods for Evaluating Intellectual Capital

In searching new and more appropriate methodologies to represent IC value, several approaches have been proposed that go beyond the traditional accounting approach in recent years.

Thereby it is required the budgeting of all intangible assets at both balance sheet and income level. It has been proposed a non-accounting approach for producing qualitative and quantitative systematic information (not only monetary) called "satellite accounting" [2].

Amongst the most effective IC evaluation criteria there is one that identifies four classes of methods [45]:

- Market Capitalization Methods, where intangible assets are estimated as difference between company market value and book values of company assets;
- Return on Assets Methods, where company's ordinary profitability is calculated considering the average income before taxes divided by average value of tangible assets; this ratio is compared with the average value of corresponding industry sector. The difference, multiplied by the average value of material goods, is used to calculate the return of intangible assets and their value;
- Direct Intellectual Capital Methods include the economic valuation of various company components, individually or in aggregate form;
- Scorecard Methods, where the elements of intangible assets are first identified and then measured with non-monetary indicators.

These different classes identify methods based on scientific or professional knowledge.

In general, methods that assign an economic value to intangible assets, such as IC, are generally used in case of mergers or acquisitions, where primary goal is to have a monetary quantification of value generated by these assets. These methods are adequate even in the case of comparisons between companies operating in the same industrial sector. These methods presents a weak point, as any assumptions defined in terms of discount rate or interest tend to bias the final value. Moreover, without referring to the division of IC in its main components, these methods would appear unclear and minimally useful for management. The main advantage of these methods is the possibility to discriminate IC in cardinal components and this offers a broader representation of the company structure and for this reason it is much more suitable for managerial use.

Three groups of criteria were proposed by the extant literature. The first group considers only qualitative methods, the second are based on quantitative methods, while the third includes mixed methods comprising both quantitative and qualitative approaches (see Fig. 3).

	Intellectual Capital Evaluation Methods		
Qualitative Methods	Quantitative Methods		Mixed Methods
	Monetary Methods	Non-Monetary Methods	
IC Audit Method (Brooking, 1996)	IC Management Method (Sullivan, 2000)	IC Statement Method (Mouritsen et Al., 2001)	Skandia Navigator (Edvinsson and Malone, 1997)
IC Benchmarking System (Viedma Marti, 2000)	Market-to-Book-Ratio	IC Index Method (Roos et Al., 1997)	Technology Factor (Khoury, 1994)
	Intagibles Scoreboard (Gu and Lev, 2001)	Intangible Asset Monitor (Sveiby, 1997)	Value Chain Scoreboard (Lev, 2001)
	Calculated Intangible Value (Stewart, 1997)	Balance Scorecard (Kaplan and Norton, 1992)	Konrad Group (Sveiby, 1989)
	Citation-Weighted Patents (Hall et Al., 2001)	IC Dynamic Value (Bounfour, 2002)	iValuing Factor (Standfield, 2001)
	Tobin's Q Method (Tobin, 1981)	Inclusive Value Methodology ^{TM} (M'Pherson et Al., 2002)	
	EVA ^{TM} (Stewart III, 1994)		
	HR Accounting (Bullen, 2008)		
	VAIC ^{TM} (Pulic, 2000)		

Fig. 3. Intellectual capital evaluation methods [3–5, 21, 22, 24–26, 30, 31, 36, 37, 40, 41]

Some methods are currently protected by patents or copyright, as they are the result of experience and practices of some consulting companies.

Quantitative methods can be divided into monetary and non-monetary methods with parameters characterized by economic or financial nature or not.

Qualitative methods describe intangible assets without techniques in which are inferred their monetary value.

Mixed methods use parameters of qualitative and quantitative nature.

Qualitative method may be converted into a quantitative method for particular needs (e.g. comparing companies for mergers and acquisitions). In order to achieve this goal it is recommended to use: (i) scaling techniques that help to identify opinions, attitudes or values not having a specific unit of measure (in this cases it can be used a Likert scale); (ii) techniques with percentages, which are used to determine the optimal corporate status in relative terms.

It is worth noting that all methods reviewed by the international literature, the degree of subjectivity is highly significant and it cannot be completely removed but reduced only.

7   Discussion and Concluding Remarks

As described above, the logistics service industry is undergoing a radical transition as a result of the changes that have affected the sector. One of the main effect of the this evolution consists in the increased typology of 3PL companies operating on the market (see for example the third stage in the evolving model presented in Sect. 2). A useful taxonomy to classify 3PLs may be based on the following three categories [15]:

1. Full Haulage Providers (e.g. road transport hauliers): those companies for which transport activities represent 100 % of turnover;
2. Basic Logistics Providers (e.g. 3PLs): those companies for which transport and warehousing together comprise over 50 % of turnover;
3. Advanced Logistics Providers (e.g. 4PLs): those companies for which transport and warehousing together comprise less than 50 % of turnover (e.g. where more than 50 % of the total turnover is generated by value added logistics and SCM services).

Using the above taxonomy and considering the specific characteristics of each single type of 3PL company, it is possible to identify the most suitable IC evaluation methods that can be associated to each of the above three categories:

– For Full Haulage Providers it can be suggested the use of quantitative monetary methods because these type of companies use predominantly tangible assets (such as trucks, equipments and so on);
– Basic Logistics Providers typically use both tangible and intangible assets. In fact their business is based on the provision of both physical-based services (e.g. transport and warehousing) and a number of value added services that are much more based on skill and knowledge. For this reason quantitative non monetary methods should be the most effective one as they split the IC in its macro-components: human capital, structural capital and relational.
– The business of Advanced Logistics Providers mainly consists in a consulting activity. In fact, the objective of the 4PLs is to enable customers to outsource to a single organization the entire re-engineering of their supply chain processes, beginning with the design stage through to implementation, and ending with the execution of comprehensive supply chain solutions. Considering the characteristics of the

Advanced logistics providers, the adoption of mixed methods seem to be the most effective option as these kind of methods put together both financial and non-financial aspects of the IC (e.g. customer satisfaction and consistency of corporate culture, brand awareness of sector and efficiency of processes, managerial skills and innovation degrees).

Finally, the most relevant limitation of this paper consist in the lack of a practical experimentation. For this reason, future research in this field should test the appropriateness of the evaluation methods identified in each of the three company categories identified.

References

1. Ashenbaum, B., Maltz, A.B., Rabinovich, E.: Studies of trends in third-party logistics usage: what we can conclude? Transp. J. **44**(3), 39–50 (2005)
2. Bismuth, A., Tojo, Y.: Creating value from intellectual assets. J. Intellect. Capital **9**(2), 228–245 (2008)
3. Bounfour, A.: How to measure intellectual capital's dynamic value: the IC-dVAL approach. In: Presented at the 5th World Congress on Intellectual Capital. McMaster University, Hamilton, Ontario, Canada (2002)
4. Brooking, A.: Intellectual Capital: Core Assets for the Third Millennium Enterprise. International Thomson Business Press, London (1996)
5. Bullen, M.L.: Incorporating Human Resource Accounting Value Measures in Capital Investment Decisions (2008)
6. Cooper, M.C., Lambert, D.M., Pagh, J.D.: What should be the transportation provider's role in supply chain management? In: Proceedings of the 8th World Conference on Transport Research, Antwerp (Belgium), 12–17 July 1998
7. Del Giudice, V., De Paola, P.: Undivided real estate shares: appraisal and interactions with capital markets. Appl. Mech. Mater. **584–586**, 2522–2527 (2014). Trans Tech Publications
8. Del Giudice, V., Manganelli, B., De Paola, P.: Depreciation methods for firm's assets. In: Gervasi, O., Murgante, B., Misra, S., Rocha, A.M.A.C., Torre, C., Taniar, D., Apduhan, B.O., Stankova, E., Wang, S. (eds.) ICCSA 2016. LNCS, vol. 9788, pp. 214–227. Springer, Heidelberg (2016). doi:10.1007/978-3-319-42111-7_17
9. Del Giudice, V., Manganelli, B., De Paola, P.: Spline smoothing for estimating hedonic housing price models. In: Gervasi, O., Murgante, B., Misra, S., Gavrilova, M.L., Rocha, A.M.A.C., Torre, C., Taniar, D., Apduhan, B.O. (eds.) ICCSA 2015. LNCS, vol. 9157, pp. 210–219. Springer, Heidelberg (2015)
10. Del Giudice, V., De Paola, P.: The effects of noise pollution produced by road traffic of Naples Beltway on residential real estate values. Appl. Mech. Mater. **587–589**, 2176–2182 (2014). Trans Tech Publications
11. Del Giudice, V., De Paola, P.: Geoadditive models for property market. Appl. Mech. Mater. **584–586**, 2505–2509 (2014). Trans Tech Pubblications
12. Demediuk, P.: Intellectual capital reporting: new accounting for the new economy. Asian Acad. Manage. J. **7**(1), 57–74 (2002)
13. Edvinsson, L., Malone, M.S.: IC: Realizing Your Company's True Value by Finding its Hidden Brainpower. Harper Business, New York (1997)
14. Evangelista, P., Sweeney, E.: The role of ICT in small Italian logistics enterprises. Int. J. Bus. Syst. Res. **3**(1), 1–18 (2009)

15. Evangelista, P., Sweeney, E.: Technology usage in the supply chain: the case of small 3PLs. Int. J. Logistics Manage. **17**(1), 55–74 (2006)
16. Evangelista, P.: ICT Diffusion in SMEs. An Investigation on the Italian Transport and Logistics Service Industry, Collana di Ingegneria Economico-Gestionale, n. 43, ESI, Naples (2011)
17. Evangelista, P., McKinnon, A., Sweeney, E.: Technology adoption in small and medium-sized logistics providers. Ind. Manage. Data Syst. **113**(7), 967–989 (2013)
18. Evangelista, P., Durst, S.: Knowledge management in environmental sustainability practices of third-party logistics service providers. VINE J. Inf. Knowl. Manage. Syst. **45**(4), 509–529 (2015)
19. Fugate, B.S., Stank, T.P., Mentzer, J.T.: Linking improved knowledge management to operational and organizational performance. J. Oper. Manage. **27**(3), 247–264 (2009)
20. Fugate, B.S., Autry, C.W., Davis-Sramek, B., Germain, R.N.: Does knowledge management facilitate logistics-based differentiation? The effect of global manufacturing reach. Int. J. Prod. Econ. **139**(2), 496–509 (2012)
21. Gu, F., Lev, B.: Intangible Assets: Measurement, Drivers. Usefulness. New York University, New York (2001)
22. Hall, B., Jaffe, A., Trajtenberg, M.: The NBER Patent Citation Data File: Lessons, Insights and Methodological Tools. NBER Working Paper, Cambridge, MA (2001)
23. Hertz, S., Alfredsson, M.: Strategic development of third party logistics providers. Ind. Mark. Manage. **32**(2), 139–149 (2003)
24. Kaplan, R., Norton, D.: The balanced scorecard: measures that drive performance. In: Harvard Business Review on Measuring Corporate Performance. Harvard Business School Press, Boston (1992)
25. Khoury, S.: Valuing intellectual properties. The Dow Chemical Company (1994)
26. Lev, B.: Intangibles: management, measurement and reporting. The Brookings Institution, Washington, DC (2001)
27. Manganelli, B., De Paola, P., Del Giudice, V.: Linear programming in a multi-criteria model for real estate appraisal. In: Gervasi, O., Murgante, B., Misra, S., Rocha, A.M.A.C., Torre, C., Taniar, D., Apduhan, B.O., Stankova, E., Wang, S. (eds.) ICCSA 2016. LNCS, vol. 9786, pp. 182–192. Springer, Heidelberg (2016). doi:10.1007/978-3-319-42085-1_14
28. Marr, B., Schiuma, G.: Measuring and managing intellectual capital and knowledge assets in new economy organisations. In: Bourne, M. (Ed.) Handbook of Performance measurement, Gee, London (2001)
29. Morano, P., Tajani, F., Locurcio, M.: Land use, economic welfare and property values: an analysis of the interdependencies of the real estate market with zonal and macro-economic variables in the municipalities of Apulia Region (Italy). Int. J. Agric. Environ. Inf. Syst. **6**(4), 16–39 (2015)
30. Mouritsen, J., Larsen, H.T., Bukh, P.N., Johansen, M.R.: Reading an intellectual capital statement: Describing and prescribing knowledge management strategies. J. Intellect. Capital **2**(4), 359–383 (2001)
31. M'Pherson, P.K., Pike, S.: Accounting, empirical measurement and intellectual capital. In: Presented at the 4th World Congress on the Management of Intellectual Capital. McMaster University, Hamilton, Ontario, Canada (2001)
32. Neumann, G., Tomé, E.: Knowledge management and logistics: where we are and where we might go to. LogForum **2**(3), 1–15 (2006)
33. Neumann, G., Tomé, E.: Empirical impact study on the role of knowledge management in logistics. Int. J. Electron. Customer Relat. Manage. **3**(4), 344–359 (2009)

34. Ojala, L., Andersson, D., Naula, T.: The definition and market size of third party logistics services. In: Ojala, L., Jamsa, P. (eds.) Third Party Logistics - Finnish and Swedish Experiences, Series Discussion and Working Papers n. 3. Turku School of Economics, Turku (2006)
35. Panayides, P.M.: Logistics service provider-client relationships. Transp. Res. Part E (Logistics Transp. Rev.) **41**(3), 179–200 (2005)
36. Pulic, A.: VAICTM – An accounting tool for IC management. Int. J. Technol. Manage. **20**(5–8), 702–714 (2000)
37. Tobin, J.: The monetarist counter-revolution today: an appraisal. Econ. J. **91**(361), 29–42 (1981). Wiley
38. Rajesh, R., Pugazhendhi, S., Ganesh, K.: Towards taxonomy architecture of knowledge management for third party logistics service provider. Benchmarking: Int. J. **18**(1), 42–68 (2011)
39. Roos, J., Roos, G., Dragonetti, N.C., Edvinsson, L.: Intellectual capital: navigating in the new business landscape. Macmillan, London (1997)
40. Standfield, K.: Time Capital and Intangible Accounting: The Approaches to Intellectual Capital. Idea Group Publishing, Hershey (2001)
41. Stewart, T.A.: Intellectual Capital: The New Wealth of Organizations. Doubleday/Currency, New York (1997)
42. Stewart, T.: Tom Stewart on Intellectual Capital. Retrieved (2003)
43. Stewart III, G.B.: EVA: fact and fantasy. J. Appli. Corp. Finance **7**(2), 71–84 (1994)
44. Sveiby, K.E.: The invisible balance sheet (1989)
45. Sveiby, K.E.: The New Organizational Wealth: Managing and Measuring Knowledge Based Assets. Berrett-Koehler, San Francisco (1997)
46. Sullivan, P.H.: Profitability for intellectual capital. J. Knowl. Manage. **3**(2), 132–143 (2000)
47. Sweeney, E., Evangelista, P., Passaro, R.: Putting supply-chain learning theory into practice: lesson from an Irish case. Int. J. Knowl. Learn. **1**(4), 357–372 (2005)
48. Tajani, F., Morano, P.: An evaluation model of the financial feasibility of social housing in urban redevelopment. Property Manage. **33**(2), 133–151 (2015)
49. Teece, D.J.: Managing Intellectual Capital: Organizational, Strategic, and Policy Dimensions. Oxford University Press, Oxford (2000)
50. Ulrich, W.: System thinking, system practice, and practical philosophy: a program for research. Syst. Pract. **1**, 137–163 (1998)
51. Marti, J.M.V.: ICBS intellectual capital benchmarking system. J. Technol. Manage. **20**(5–8), 799–818 (2000)
52. Wiig, K.M.: Knowledge management: Where did it come from and where will it go? Expert Syst. Appl. **13**(1), 1–14 (1997)
53. World Bank: Connecting to Compete: Trade Logistics in the Global Economy (2014)

Knowledge Retrieval

Digitalizing Seismograms Using a Neighborhood Backtracking Method

Yaxin Bi[1(✉)], Shichen Feng[1], Guoze Zhao[2], and Bing Han[2]

[1] School of Computing and Mathematics, University of Ulster,
Newtownabbey, Co., Antrim BT37 0QB, UK
`y.bi@ulster.ac.uk`, `Feng-S@email.ulster.ac.uk`
[2] Institute of Geology, China Earthquake Administration,
Hua Yan Li, Chaoyang District, Beijing, China
`zhaogz@ies.ac.cn`, `zddhb@163.com`

Abstract. In this paper, we present a new algorithm for tracing waves on seismograms and digitalising the waves into single vectors in the form of a time series data. The algorithm consists of two main components that will be used to handle the smooth and complicate cases, respectively. The underlying feature of the algorithm lies in a novel searching process based on examining the context of pixels to ascertain how the tracing moves forward. The algorithm has been evaluated on a limited number of samples and the result demonstrates its competence. The work presented can be regarded as an effort of developing a uniform earthquake archive covering both the historical and the digital device periods for future reassessment of the seismic hazard cross the world. The archive developed will serve as an effective means for discovering precursor of earthquakes by characterising the spectral seismograms and the source parameters of less active sources, whereby permitting comparative studies on earthquakes and development of prediction models.

Keywords: Analogous seismograms · Digitalization of seismic wave · Vectorization of analogous signal

1   Introduction

Many countries have a long history of using seismographs to record the ground motion with analogous measurements for monitoring earthquakes and place massive historical seismograms in the achieve storages [1]. Seismograms contain rich information including displacement, velocity, or acceleration of the ground motion, they are particularly important in zones with a low to moderate seismicity associated with seismogenic sources with medium to long recurrence intervals [2]. Although a large number of digital seismic networks have been constructed across countries in the past decades, there are no uniform seismogram archives that consist of both the historical and the digital device periods, allowing seismologists to conduct the reassessment of the historical seismic hazards. It is extremely difficult for scientists to retrieve original seismograms for a particular

© Springer International Publishing AG 2016
F. Lehner and N. Fteimi (Eds.): KSEM 2016, LNAI 9983, pp. 391–401, 2016.
DOI: 10.1007/978-3-319-47650-6_31

historic seismic event and recover seismic information contained in old records [2]. Digitalisation of analogous seismograms is therefore imperative for developing uniform seismogram achieves, permitting researchers to analyse and study earthquakes at local and global scales with different historical periods.

Several studies have recently been reported in the literature [6–9]. In [3], the authors developed an algorithm that can be used to extract waves from individual seismograms and concatenate them to 24 h single waves. A further development of this work is a waveform tracing algorithm called Waveform Mosaic Algorithm that is able to approximately vectorize analogous seismograms based on local features in tracing seismograms [4]. An interactive system designed for automatic digitization of waveform has been reported in [5], the authors convert the digitization of seismograms as an inversion problem. The system integrates automatic digitization with manual digitization and users can interactively switch between these two modality functions to complete wave tracing tasks.

In this paper, we present a new algorithm for tracing waves on seismograms and digitalising the waves into single vectors in the form of a time series data. The development of this algorithm is based on the previous work [3], but the idea of this algorithm is on the basis of backtracking within a neighborhood of any dichotomous points (white and black), and concatenating black points in fragments into single waves in the form of digital vectors. The underlying difference between the algorithm from the previous one is, when searching next black points, all points within a neighborhood will be examined instead of the points in vertical lines based on local features. In addition, we develop an effective method for counting white points along the tracked waves, which can be used to evaluate and compare the performance of algorithms. This paper presents the detailed design of the algorithm and performance comparison with the previous algorithm [3].

Seismographs record the displacement, or velocity, or acceleration of the ground motion. Only one value is recorded at each time. No matter what the recording are, the traces aligned together and give several continuous series of black pixels, rather than one series, at each time point [2].

2   Description of Seismograph Readings

Seismometers, also called seismographs, are instruments that record motions of the ground, including those of seismic waves generated by earthquakes, volcanic eruptions, and other seismic sources. They could take measurements of ground motion and draw measurements on papers, each time only one value is recorded. No matter what the recording are, the traces are aligned together to form continuous series of black pixels, Due to the size of a paper, the continuous motion will be recorded on paper with equal segment of component waves for 24 h, and then replaced with another paper. Figure 1 illustrates an example of seismogram, the numbers labelled on the paper indicate recording times. In order to vectorise seismograms, the first step process is to trace segments of waves and concatenate them into vectors in the form of a time series of format.

Definition 1: Let $S_i = \{(x_0^i, y_0^i), \cdots, (x_M^i, y_M^i)\}$ be a component wave on seismograms, where x_j^i is a time stamp and y_j^i is an amplitude value. In addition, let $S = S_0 \tau S_1 \tau \cdots S_K$ be a seismogram, where K is the number of component waves, τ is an operation that concatenates component waves to a single seismogram.

From Fig. 1, we can see that it is straightforward to concatenate the time stamps of all component waves by operator τ in a linear way, i.e. $x_M^0 + x_0^1, x_M^1 + x_0^2, \cdots, x_M^{k-1} + x_0^K$, but the alignment of amplitudes to a single seismogram is complex. This will involve converting image pixels on a seismogram to amplitude values based on pixel grey-level values across waves, placing them onto the same coordinate system along with a zero point, and then adjusting them with offsets by accounting for the zero point. Simply, suppose y_0^i be a starting reference point of wave y^i in y coordinate axis and Δy^{i-1} be difference between y^i and y^{i-1} in y coordinate axises of two traced waves, then we have

$$y_j^{1'} = y_j^1 - \Delta y_j^0,$$

$$y_j^{2'} = y_j^2 - \Delta y_j^1 - \Delta y_j^0,$$

$$\cdots$$

$$y_j^{K'} = y_j^K - \sum_{i=0}^{K-1} \Delta y_j^i$$

where $0 \leq j \leq M$. Following the above process, all component waves $S_i (0 \leq i \leq K)$ will be placed down in the same y coordinate reference and then can be concatenated in a linear way corresponding to the concatenated time stamps. After that a seismogram will be vectorised into a time series wave. However prior to vectorising seismograms above, we have to perform image processing, i.e. a binarization process of enhancing image pixels with white and black points and use black pixels to trace each component waves into vectors. We detail the design of the algorithm in following sections.

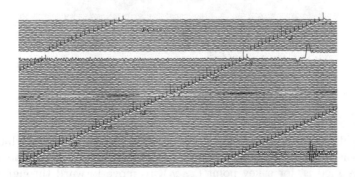

Fig. 1. A sample of original seismogram

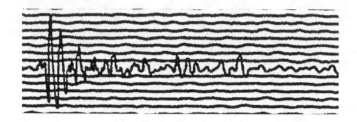

Fig. 2. A fragment of the original seismogram in Fig. 1

As a seismogram wave can be either smooth or of fluctuation. The former does not contain much changes throughout the wave, whereas the latter may include many irregular parts with large variations. Thus the tracing algorithm consists of two main functions designed on the basis of smooth and complex situations.

3   Description of Wave Tracing Algorithm – Tracing a Smooth Seismogram Wave

Zooming in a wave as shown in Fig. 2, we can take a fragment and illustrate it in Fig. 3. It can be seen that a wave is not a single pixel width line, actually it is composed of many pixels vertically. Instead tracing whole width pixel line, the process is first to search the highest and lowest point values and then move forward in the middle of the wave pixel by pixel or with a number of pixels as a step.

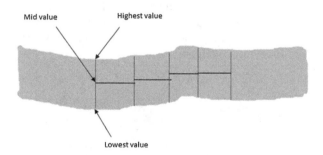

Fig. 3. A single seismograph line

Specifically assume $P_0(x_0, y_0)$ be the starting point of a wave, which is also referred to as the first key point, ε_x and ε_y be the searching steps. The searching process designed is, for a key point (x_i, y_i), to move forward through checking whether the next point $P_i(x_i + \varepsilon_x, y_i)$ is a black pixel. If yes, the process will vertically search from $P_i(x_i + \varepsilon_x, y_i)$ up and down until reaching the highest and lowest values, denoted by $\mathcal{L}(y_i)$ and $\mathcal{H}(y_i)$ as illustrated in Fig. 3, and then decide a new next key point P_{i+1} as follows:

$$x_{i+1} = x_i + \varepsilon_x = x_0 + (i+1) \times \varepsilon_x \tag{1}$$

$$y_{i+1} = \frac{\mathcal{L}(y_i) + \mathcal{H}(y_i)}{2} \tag{2}$$

The above searching process is repeated until a new key point is not a black pixel and then turn to the complicate case given in the following section.

$$\omega = \mathcal{H}(y_i) - \mathcal{L}(y_i) \tag{3}$$

where ω is the width of a wave, which will be used as a searching range for the complicate case.

4 Description of Wave Tracing Algorithm – Tracing a Seismogram Wave with Fluctuations

Following the process given in Sect. 3, if $P_i(x_i + \varepsilon_x, y_i)$ is not a black pixel, the tracing process becomes complex, and needs to account for the situations of irregular variations, wave crests and noise, etc.

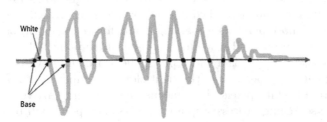

Fig. 4. Searching key points in one pass

Given $P_i(x_i + \varepsilon_x, y_i)$ being a white pixel, we assume V be a vector, holding so called base points. The first step of the searching process is to start from $P_i(x_i, y_i)$ to search all base points as shown in Fig. 4 and put them into V. This process will result in $V[0] = P_i(x_{i+0}, y_i), \cdots, V[r] = P_i(x_{i+r}, y_i)$.

For each base point, it divides a wave fragment to two parts – up and down ones, correspondingly the searching process consists of upward and downward cases. The following will detail how the searching process is conducted in tracing a wave into a vector.

4.1   Upward Case

For each base point in V, by the smoothing algorithm the searching starts from a base point up and down vertically until reaching the leftmost and rightmost points that are referred to as the left and right slope points, as illustrated in Fig. 5. For each of these slope points, it is represented by a six by six grid, denoted by $\mathcal{G}$, as shown in Fig. 6, where the central cell $\mathcal{G}_{[0][0]}$ holds either left slop or right slope point.

For the right slope point case of upward tracing, the searching process is composed of horizontal and vertical two steps. Through checking a white pixel in the upper right of the grid, the algorithm decides to turn the searching direction towards the leftmost point, denoted by $\mathcal{L_M}$, and then the searching process will turn to move up. Prior to moving up, a key point has to be generated into a vector, denoted by V^U. Specifically if $\mathcal{G}_{[1][1]}$ is a white pixel, the search turns the left direction and apply the smooth algorithm horizontally trace until reaching the leftmost point as illustrated in Fig. 5, and then a key point will be generated into $V^U_{\mathcal{R}}$ by Eq. (4),

$$V^U_{\mathcal{R}}[i] = (\frac{|\mathcal{L_M}.x - \mathcal{G}_{[0][0]}.x|}{2}, \mathcal{G}_{[0][0]}.y) \tag{4}$$

where i starts from 0 with one pixel increment. In addition, the searching from the right slope point to the leftmost point has to be stratified with an approximate certain range obtained by Eq. (4), as it is very likely for one wave line to cross another, thereby searching the leftmost point might go astray.

Once the leftmost point reached and a key point generated, the second step is to use the leftmost point as a starting point for the smooth algorithm, the searching will move up until reaching the right slope point, which will be represented by $\mathcal{G}_{[0][0]}$. If $\mathcal{G}_{[1][1]}$ is a white pixel, the searching towards the leftmost point will repeat the process as in the right slope case above until $\mathcal{G}_{[-1][1]}$, $\mathcal{G}_{[0][1]}$ and $\mathcal{G}_{[1][1]}$ are all white pixels. This means the searching arrives at a crest of the wave, the searching process stop and the last key point will be obtained by Eq. (5) below.

$$V^U_{\mathcal{R}}[i] = (\mathcal{G}_{[0][0]}.x, \mathcal{G}_{[0][0]}.y - \frac{\omega}{2}) \tag{5}$$

For the left slope point case, the searching process is almost the same as the right slope point case. Instead of checking $\mathcal{G}_{[1][1]}$, the algorithm will check if $\mathcal{G}_{[-1][1]}$ contains a white pixel. If yes, the search turns the right direction and apply the smooth algorithm horizontally trace until reaching the rightmost point as illustrated in Fig. 5. Likewise a key point will be generated into $V^U_{\mathcal{L}}$ by Eq. (6),

$$V^U_{\mathcal{L}}[i] = (\frac{|\mathcal{R_M}.x - \mathcal{G}_{[0][0]}.x|}{2}, \mathcal{G}_{[0][0]}.y) \tag{6}$$

and the rightmost point will be treated as a starting point for the smooth algorithm, the searching will move up until reaching a left slope point, which will be represented by $\mathcal{G}_{[0][0]}$ again, the subsequent searching will repeat the above process until $\mathcal{G}_{[-1][1]}$, $\mathcal{G}_{[0][1]}$ and $\mathcal{G}_{[1][1]}$ are all white pixels, that means that the searching arrives at a crest of the wave. In order to avoid generating a duplicate key point with the right slope case above, the searching process just stops and a key point will not be generated.

Once the searching process for one base point stop, $V_{\mathcal{L}}^{U}[*]$ and $V_{\mathcal{R}}^{U}[*]$ will be concatenated and then linked to $P_i(x_i, y_i)$. In the following, we describe the second part of the algorithm that the fragment wave is below the same base point.

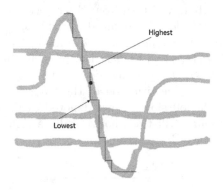

Fig. 5. Tracing a curve in the vicinity of a key point

Fig. 6. Searching black pixels in the vicinity of a key point

4.2   Downward Case

In the downward case as illustrated in Fig. 5, the searching process starts moving down vertically from the base point and move up when reaching a tough of the wave. The first point encountered is the left slope point, like the situation of the left slope point in the upward case above, this point is denoted by $\mathcal{G}_{[0][0]}$. If $\mathcal{G}_{[1][-1]}$ is a white pixel, the searching will turn the right direction and trace horizontally until reaching the rightmost point, denoted by $\mathcal{R}_{\mathcal{M}}$, and then a key point will generated by Eq. (7) below.

$$V_{\mathcal{L}}^{D}[i] = (\frac{|\mathcal{R}_{\mathcal{M}}.x - \mathcal{G}_{[0][0]}.x|}{2}, \mathcal{G}_{[0][0]}.y) \tag{7}$$

where i starts from 0 with one pixel increment.

The subsequent searching will turn from the rightmost point to move down vertically to find another the left slop point, denoted by $\mathcal{G}_{[0][0]}$ again. This searching process will repeat until $\mathcal{G}_{[-1][-1]}$, $\mathcal{G}_{[0][-1]}$ and $\mathcal{G}_{[1][-1]}$ are all white pixels, the searching processing stops and the last key point is obtained by Eq. (5).

The second part of searching in the downward case is a moving upward process, which is similar to the situation of the left slop point in the upward case. In this situation, the left slop point is denoted by $\mathcal{G}_{[0][0]}$. If $\mathcal{G}_{[0][0]}$ contains a white pixel, the searching will trace horizontally a rightmost point, denoted by $\mathcal{R}_{\mathcal{M}}$, and a key point will be generated by

$$V_{\mathcal{R}}^{D}[i] = (\frac{|\mathcal{R}_{\mathcal{M}}.x - \mathcal{G}_{[0][0]}.x|}{2}, \mathcal{G}_{[0][0]}.y) \tag{8}$$

where i starts from 0 with one pixel increment.

Noted that the downward case considers the left slop points in both moving up and moving down, which is different from those in the upward case. Here is a summary of the searching algorithm described above:

For the upward case:

– for the left slop point, turn right to reach the rightmost $\mathcal{R}_{\mathcal{M}}.x$
– for the right slop point, turn left to reach the leftmost $\mathcal{L}_{\mathcal{M}}.x$
– for the above two case, searching process is moving up

For the downward case:

– for the left slop point on the left hand side of a wave, turn to right and reach the rightmost $\mathcal{R}_{\mathcal{M}}.x$, and searching is moving down
– for the right slop point on the right hand side, turn to right and reach the rightmost $\mathcal{R}_{\mathcal{M}}.x$, searching is moving up.

5   Evaluation – Comparing Traced Waves with Original Data

Given the algorithms in Sects. 3 and 4, now we introduce how the accuracy of the algorithms for digitalising seismograms can be measured. Since there are no benchmark methods and data available for evaluation, we present the basic idea of the deigned measurement that can be used as an indicator of performance of the algorithm.

In the above, we use $P(x_i, y_i)$ to hold all the traced points, where x_i represents time stamps on x-axis and y_i are wave amplitudes. The tracing algorithm picks y value with a ten pixel interval for the smooth case, and with less than ten pixels for the complex situation that can be tuned or adjusted. The assumption here is that each of $P(x_i, y_i)$ corresponds to a black pixel on the original seismogram that was used to derive $P(x_i, y_i)$. However, if the corresponding pixel on the original seismogram is a white one, that means the tracing algorithm has not appropriately picked a black pixel. Figure 7 is a traced wave line, which has been redrawn on the original seismogram as shown in Fig. 8. It can be seen that the traced wave approximately depicts the shape of the original wave seismogram, it is extremely challenging to trace the locus of seismic waves with precision and certainty. Figure 9 presents a tracing result by the previous algorithm [3].

Additionally for each of the traced points in $P(x_i, y_i)$, the tracing process involves comparisons with all possible points within its neighbourhood. As illustrated in Fig. 6, the comparisons can involve 2^3 possible points to compare in order to ascertain tracing proceeding direction. Meanwhile from Definition 1, the number of windows is K, in which each window consists of M points. Therefore the computational complexity of the tracing algorithm can be estimated by $O(K \times M + 2^3)$.

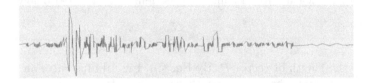

Fig. 7. Tracing on a single wave line

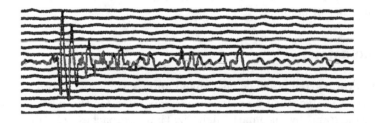

Fig. 8. Tracing accuracy on single wave line

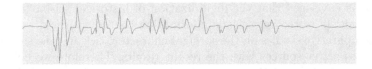

Fig. 9. Previous algorithm tracing on a single wave line

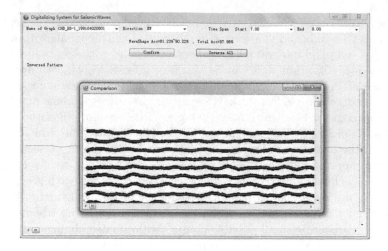

Fig. 10. An illustration of evaluation of the algorithm with 97 % tracing accuracy

In order to quantify the accuracy and make a quantitive comparison with the previous algorithm, we denote the number of white pixels held in $P(x_i, y_i)$ by N_w, then the measure for tracing error is defined as Eq. (9) below:

$$E_T = \frac{N_w}{|P|} \qquad (9)$$

where $|P|$ is the length of vector P. By Eq. (9), Fig. 10 illustrates an evaluation result, it can be seen that in s smooth case, 96 % tracing accuracy has been achieved. Note that when calculating the accuracy, we did not consider the overlapping points, thus a range is added to estimate numbers of overlapping points on total number of points.

6   Conclusions

Although several attempts have been made and considerable improvements have been achieved in digitalizing analogous seismograms for the past decades, vectorised seismograms by automatic digitalization approaches are still not perfect, which could be reviewed as approximations of the original seismograms to some extent and require human intervention to subsequently undertake corrections. Even so, it is envisaged that the digitalization of historical seismograms allow research scientists to recover seismic information contained in old records, and perform comparative studies on the spectral characteristics and the source parameters of less active sources with the recent results. Therefore we believe that the algorithm developed in this study provides a useful tool for the digitalization of analogous seismograms and contribution to improvement of seismic hazard assessment in many regions of the world.

Acknowledgements. This work is partially supported by the project funded by the National Natural Science Foundation of China (Grant No. 41374077).

References

1. Seismology Harvard Resources. http://www.seismology.harvard.edu/resources.html#HRV
2. SanDron, D., Renner, G., Rebez, A., Slejko, D.: Early instrumental seismicity recorded in the eastern Alps. Bollettino di Geofisica Teorica ed Applicata **55**(4), 755–788 (2014)
3. Wang, M., Jiang, Q., Feng, J., Feng, S.: Researches on key algorithms in analogue seismogram records vectorization. Sens. Transducers **178**(9), 209–213 (2014)
4. Wang, M., Jiang, Q., Feng, J., Yu, X., Lin, N., Feng, S., Liu, N.: A new waveform mosaic algorithm in the vectorization of paper seismograms. Sens. Transducers **182**(11), 203–206 (2014)
5. Xu, Y., Xu, T.: An interactive program on digitizing historical seismograms. Comput. Geosci. **63**, 88–95 (2014)
6. Otsu, N.: A threshold selection method from gray-level histograms. IEEE Trans. Syst. Man Cybern. **9**(1), 62–66 (1979)
7. Xu, J., Xu, K., Wei, Y., Guo, Y.: Saving and data sharing of historical seismogram in Beijing National Earth Observatory. Seismol. Geomagnetic Obs. Res. **29**(3), 100–104 (2008)
8. Pan, Z., Feng, J., Wang, M.: A base-point searching algorithm in the digitization of seismograms. Commun. Inf. Process. **289**, 699–706 (2012)
9. Chen, B.: Research and implementation of log curve vectorization methods and rebuilding technique, Harbin Institute of Technology (2007)
10. Cao, D., Tang, J., Wu, Y.: Algorithm for fast extracting human limb contours using searching NCM points. J. Huazhong Univ. of Sci. Tech. (Nature Science Edition) **35**(5), 16–18 (2007)

A Document Modeling Method Based on Deep Generative Model and Spectral Hashing

Hong Chen, Jungang Xu, Qi Wang, and Ben He[✉]

University of Chinese Academy of Sciences, Beijing, China
{chenhong113,wangqi615}@mails.ucas.ac.cn, {xujg,benhe}@ucas.ac.cn

Abstract. One of the most critical challenges in document modeling is the efficiency of the extraction of the high level representations. In this paper, a document modeling method based on deep generative model and spectral hashing is proposed. Firstly, dense and low-dimensional features are well learned from a deep generative model with word-count vectors as its input. And then, these features are used for training a spectral hashing model to compress a novel document into compact binary code, and the Hamming distances between these codewords correlate with semantic similarity. Taken together, retrieving similar neighbors is then done simply by retrieving all items with codewords within a small Hamming distance of the codewords for the query, which can be exceedingly fast and shows superior performance compared with conventional methods as well as guarantees accessibility to the large-scale dataset.

Keywords: Spectral hashing · Document modeling · Deep generative model · Hamming distance · Codeword

1 Introduction

In the domain of Information Retrieval, TF-IDF [1] is one of the most widely used weighting schemes, which measures the similarity between documents solely by comparing their respective word-count vectors. The drawbacks of this method are obvious that neither the high computational cost nor the low level of concern shown on semantic similarity can satisfy the requirements of fast inference in the large-scale dataset. To alleviate the pressures, a multitude of models for capturing low-dimensional, latent representations have been proposed, such as Latent Semantic Analysis (LSA) [2] and its probabilistic version, known as pLSA [3] as well as Latent Dirichlet Allocation (LDA) [4]. However, the exact inference in these recently introduced probabilistic models is still intractable and the time complexity of retrieving similar documents is in proportion to the dimensions of the document sets.

To remedy the foregoing drawbacks, we propose a document modeling method based on deep generative model and spectral techniques. The experimental results on 20-Newsgroups and Reuters Corpus Volume 1 demonstrate that our method can generate superior compact binary codes and is competitive in terms of retrieval efficiency.

© Springer International Publishing AG 2016
F. Lehner and N. Fteimi (Eds.): KSEM 2016, LNAI 9983, pp. 402–413, 2016.
DOI: 10.1007/978-3-319-47650-6_32

The rest part of this paper is organized as follows. The related work is described in Sect. 2. Before we describe the details of our method in Sect. 4, we briefly review Restricted Boltzmann Machines and some Spectral techniques in Sect. 3, which is a stepping stone towards the proposed Deep Spectral Hashing (DSH) method. In Sect. 5, we show the experimental results and make some empirical analysis. Finally, we conclude the paper and discuss the directions of the future work.

2   Related Works

Recently, a class of two-layer undirected graphical models, known as Restricted Boltzmann Machines (RBMs) [5], has exerted a tremendous fascination on numerous researchers. The RBMs can be viewed as a series of non-linear feature detectors, which have been manifested as good as, or in many cases better than, conventional learning methods, such as back propagation with random initialization, which is prone to convergence at local optimum. This architecture allows RBMs to model non-binary data and to use non-binary hidden variables, which can be viewed as the latent representations of the visible layer. Hinton et al. proposed a deep generative model called deep belief networks (DBNs) and a fast unsupervised learning algorithm for DBN [6]. Since then, a multitude of variants based on DBN, such as deep neural networks (DNNs), deep autoencoder (DAE), recurrent neural network (RNN), emerged and are successfully applied to a variety of machine learning problems, such as classification, regression, dimensionality reduction, collaborative filtering. More historical context and applications about deep generative models are summarized by Xu et al. [7].

Document modeling methods based on RBMs [8,9] typically perform better than directed graphical models, such as pLSA and LDA, in terms of both the log probability they assign to unseen data and their document retrieval accuracy. Although RBMs are famous for their powerful expression and tractable inference, careful hyper parameter selection is required and the retrieval time complexity is yet $O(NV)$, where N is the size of the document corpus and V is the size of vocabulary of the latent space. Semantic Hashing (SH) [10], a graphical model based on the DAE architecture, opened up the space of new deep undirected topics to explore, highlighting the values of the latent variables in the deepest layer, which is quite tractable to infer and give a much better representation of each document than topic models. Using a hamming-ball (see Fig. 1), this method can create a candidate list by varying a few bits without any search, which can be amazingly fast. However, this method assumes that the decoder network is insensitive to very small differences in the output of a code unit, and introduces Gaussian noise in the bottom-up input to make the codes binary at the expense of messing up the conjugate gradient fine-tuning process. What's more, finding the exact threshold to make the codes binary in SH is quite questionable and hence introduces uncertainty into the codewords. Weiss et al. proposed a novel hashing method [11], which is based on spectral techniques and aims at effectively seeking the best code for a given dataset and novel data point. The work demonstrated that finding the best code for a given dataset is closely related to the problem of graph partitioning, which has been proved to be NP-hard. By relaxing the binaryzation constraint, the solutions are simply a subset of threshold eigenvectors of the graph Laplacian. However, applying these spectral techniques directly to text data typically achieves extremely poor performance for the reason that spectral relaxation assumes all the input features to be obedient to multidimensional uniform distribution. Distributed representations for documents are typically high-dimensional and sparse, which violates the prerequisites for the use of spectral techniques.

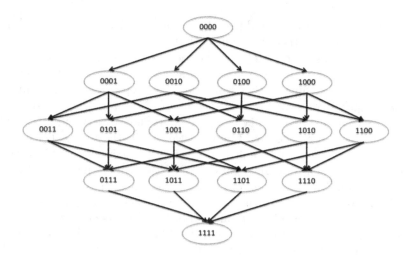

Fig. 1. The structure of a Hamming-ball with 4 bits

In order to further alleviate the problems listed above, this paper aims at seeking a better compromise between document retrieval accuracy and efficiency, guaranteeing accessibility to the large-scale dataset and improvements on the quality of the compressed codewords.

3   Restricted Boltzmann Machines and Spectral Techniques

3.1   Restricted Boltzmann Machines

A RBM is a Markov Random Field (MRF) associated with a bipartite undirected graph as shown in Fig. 2, with m visible units $v = (v_1, \cdots, v_m)$ to represent observable data and n hidden units $h = (h_1, \cdots, h_n)$ to capture dependencies between observed data.

To train a RBM, we give the probability distribution of (v, h) parameterized by θ in Eqs. 1 and 2.

$$P(v, h|\theta) = \frac{e^{-E(v,h|\theta)}}{Z(\theta)} \qquad (1)$$

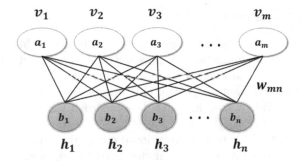

Fig. 2. A RBM with m visible and n hidden variables

$$Z(\theta) = \sum_{v,h} e^{-E(v,h|\theta)} \tag{2}$$

Where $E(v,h|\theta)$ denotes the energy function, which simplifies the inference of probability distribution and is defined in Eq. 3.

$$E(v,h|\theta) = -\sum_{i=1}^{m} a_i v_i - \sum_{j=1}^{n} b_j h_j - \sum_{i=1}^{m}\sum_{j=1}^{n} v_i W_{ij} h_j \tag{3}$$

With all visible units specified, we can infer the activation probability of the jth hidden unit in Eq. 4.

$$P(h_j = 1|v,\theta) = \sigma(b_j + \sum_i W_{ij} v_i) \tag{4}$$

Similarly, with all hidden units specified, the activation probability of the ith observed unit is quite tractable as shown in Eq. 5.

$$P(v_i = 1|h,\theta) = \sigma(a_i + \sum_j W_{ij} h_j) \tag{5}$$

Where $\sigma(x) = 1/(1 + exp(-x))$ denotes the sigmoid activation function.

With the foregoing preliminaries, the update rules of RBM parameters now can be inferred in Eqs. 6–8.

$$\Delta W_{ij} = \langle v_i h_j \rangle_{data} - \langle v_i h_j \rangle_{recon} \tag{6}$$

$$\Delta a_i = \langle v_i \rangle_{data} - \langle v_i \rangle_{recon} \tag{7}$$

$$\Delta b_j = \langle h_j \rangle_{data} - \langle h_j \rangle_{recon} \tag{8}$$

Where $\langle \cdot \rangle_{recon}$ indicates the model distribution.

3.2 Spectral Techniques

Spectral techniques are originally widely applied to clustering task and can bring several orders of magnitude improvement over conventional clustering approaches, such as K-means and Gaussian Mixture Model (GMM) [12]. Accordingly, the superiority showed in spectral clustering captures considerable attention and has been widely used in other domains of image processing, such as segmentation, retrieval [13,14].

Spectral clustering [15] treats conventional clustering problem as an issue of graph partitioning (see Fig. 3), where nodes correspond to input samples and edges are related to the similarity distance between samples.

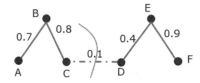

Fig. 3. Spectral graph partitioning

Given m samples $S = \{s_1, \cdots, s_m\}$ in $\mathbb{R}^n$ that we want to cluster them into k subsets, the procedure of clustering is described as follows.

- Construct the affinity matrix $A \in \mathbb{R}^{m \times m}$ defined by $A(i,j) = exp(-\|s_i - s_j\|^2/\epsilon^2)$ if $i \neq j$, and $A(i,j) = 0$ else.
- Define $D \in \mathbb{R}^{m \times m}$ to be the diagonal matrix whose (i,i) - element is defined with $D(i,i) = \sum_j A(i,j)$ and construct the normalized graph Laplacian matrix: $L = D - A$.
- Find k eigenvectors of L with minimal eigenvalues excluding the trivial eigenvector 1 with eigenvalue 0 and construct the matrix $Y = y_1 \, y_2 \, \cdots \, y_k] \in \mathbb{R}^{m \times k}$ by stacking the eigenvectors in columns.
- Renormalize each of Y's rows to have unit length, and treat each row of Y as a point in $\mathbb{R}^k$, and then cluster them into k clusters via K-means or any other algorithms.
- Finally, assign the original sample s_i to cluster j if and only if row i of matrix Y is assigned to cluster j.

4   The Deep Spectral Hashing Model

Despite the fact that applying spectral techniques directly into document set can never be a pleasurable choice, we can extract dense and low-dimensional features from top layer in the DAE architecture for training a spectral hashing model, which forms the core idea of our method called Deep Spectral Hashing (DSH) Method.

4.1   Extract Dense and Low-Dimensional Features

We extract these low-dimensional features through learning a deep generative model, where the top layer outputs these features as shown in Fig. 4 (left part).

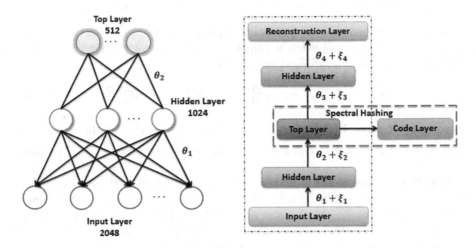

Fig. 4. The architecture of the DSH model

Our feature detector, as shown in Fig. 4 (left part) is actually a DBN, where the top two layers compose a RBM. Note that RBM shows superior ability in dealing with binary values or decimal values less than 1.0 and is weak at dealing with word count embedding, which composes the input layer in our deep architecture with decimal integers. To better model the observed "visible" word count data v, we exploit the Constraint Poisson Model, proposed in [10]. Consequently, the probability distribution of the positive phase is defined in Eq. 10 and that of the negative phase is defined in Eq. 9.

$$P(v_i = n|\mathbf{h}) = Ps(n, \frac{exp(\lambda_i + \sum_j h_j W_{ij})}{\sum_k exp(\lambda_i + \sum_j h_j W_{ij})} N) \qquad (9)$$

$$P(h_j = 1|\mathbf{v}) = \sigma(b_j + \sum_i v_i W_{ij}) \qquad (10)$$

Where $Ps(n, \tau) = e^{-\tau}/n!$, W_{ij} denotes an interaction term between word i and feature j, $N = \sum_i v_i$ is the total length of the document, and λ_i is the bias of the conditional Poisson model for word i. Equation 9 infers that the Poisson rate is normalized to make learning stable and is used to deal with documents of different lengths more appropriately. Note that we exploit the Tempered Transition method [16] instead of Gibbs sampling or Parallel Tempering to improve mixing rate and help approximate the likelihood gradient further. Taking Gibbs sampling as the base transition, we generate the states $(v_K, h_K), \cdots, (v_2, h_2), (\tilde{v}_1, \tilde{h}_1), (\tilde{v}_2, \tilde{h}_2), \cdots, (\tilde{v}_K, \tilde{h}_K)$ and the probability to accept $(\tilde{v}_K, \tilde{h}_K)$ as the next state of the Markov chain is denoted as Eq. 11.

$$\min\{1, \prod_{i=2}^{K} e^{(\beta_i - \beta_{i-1}) \times (E(v_i, h_i) - E(\tilde{v}_i, \tilde{h}_i))}\} \tag{11}$$

Where K denotes the number of Markov chains and $\beta_1 < \cdots < \beta_K = 1$ denote the inverse temperatures, extra parameters of the model. Finally, we update RBM parameters in an optimized rule, as in Eq. 12.

$$\theta_{ij}^{(t+1)} = \theta_{ij}^{(t)} + \eta \times \Delta\theta_{ij}^{(t)} - \lambda \times \theta_{ij}^{(t)} + v \times \Delta\theta_{ij}^{(t-1)} \tag{12}$$

Where η denotes the learning rate, $-\lambda \times \theta_{ij}^{(t)}$ penalizes the weights with large magnitude and $v \times \Delta\theta_{ij}^{(t-1)}$ denotes the momentum term weighted by the parameter v.

After layer-by-layer pre-training as shown in Fig. 4 (left part), we unroll each RBM as shown in Fig. 4 (right part), to generate a deep autoencoder and apply back propagation to fine-tuning the weights for optimal reconstruction of the input layer. To be specific, we maximize $p(\mathbf{v})$ while approximately marginalizing out the hidden layers during the pre-training phase, and then maximize the likelihood of correct construction of $\mathbf{v}$ given the original values of $\mathbf{v}$ during the fine-tuning phase. In addition, we assume that a vocabulary of the 2,048 most frequent words by removing common stop words and stemming in the training dataset is enough for the input layer of the DAE architecture while other layers are determined by a number of validation test (see more details in Sect. 5.2). Considering that dividing the dataset into mini-batches is more efficient during the courses of both pre-training and fine-tuning, we set the size of the mini-batches to 100 and 1000 respectively. The epoch numbers of pre-training and fine-tuning stages are 30 and 50 respectively with the fixed learning rate of 0.001. In addition, we assign the weight cost λ with 0.00002, momentum v with 0.5 and the number of Markov chains with 25.

4.2   Generate Compact Codewords with Spectral Hashing

With dense and low-dimensional features well learned from the deep generative model, training a spectral hashing model for compressing novel data is quite simple, as shown in Fig. 4 (right part).

In multivariate statistics and the clustering of data, spectral techniques make use of the spectrum of the similarity matrix of the data to perform dimensionality reduction before follow-up works in fewer dimensions. Here, we use $\{B_i\}_{i=1}^{m}$ to denote the list of binary code vectors of length k for m documents and use $A_{m \times m}$ to denote the affinity matrix, where $A(i, j) = exp(-\|s_i - s_j\|^2/\epsilon^2)$, which denotes the Gaussian similarity function indicating that Euclidean distance correlates with similarity as described in Sect. 3.2. Then the problem of seeking a good code can be described as Eq. 13.

$$minimize: \quad trace(Y^T LY)$$
$$s.t. \quad Y(i,j) \in \{-1,1\}$$
$$Y^T 1 = 0 \tag{13}$$
$$Y^T Y = I$$

This Equation is equivalent to trying to find k independent balanced partitions and which is known to be NP hard [13]. By removing the constraint that $Y(i,j) \in \{-1,1\}$ and Rayleigh-Ritz theorem, the solutions would be the k eigenvectors of L with minimal eigenvalue excluding the trivial eigenvector 1 with eigenvalue 0, and $\{B_i\}_{i=1}^m$ can be obtained just through thresholding the matrix Y by each row at zero. Notice that the solution only provides us how to calculate the codewords for the training set. For a novel sample, we adopt the Out of Sample Extension introduced by [11], and assume that the data points $x_i \in \mathbb{R}^n$ are sampled from a probability distribution $p(x)$. Then the foregoing problem can now be redeclared as Eq. 14.

$$minimize: \quad \int \|y(x_1) - y(x_2)\|^2 A(x_1, x_2) p(x_1) p(x_2) dx_1 x_2$$
$$s.t. \quad \int y(x) p(x) dx = 0 \tag{14}$$
$$\int y(x) y(x)^T p(x) dx = I$$

Where $A(x_1, x_2) = exp(-\|x_1 - x_2\|^2/\epsilon^2)$. Finally, we can obtain binary codes through thresholding the analytical eigenfunctions at zero.

5   Experiments

5.1   Experimental Design

The corpus used in our experiments include 20-Newsgroups and Reuters Corpus Volume 1, which are widely used in text applications, such as text clustering [17] and text classification [18,19]. The 20-Newsgroups is composed of 20 different topics, some of them closely related to each other while others highly unrelated. It consists of about 20,000 news documents, which are randomly split into about 12,000 documents(10 MB) for training and 8,000 documents(6.5 MB) for testing in our experiments. The Reuters Corpus Volume 1 is an archive of about 810,000 newswire stories that have been manually categorized into 103 topics, which is randomly split into about 600,000 training articles(1.8 GB) and 210,000 test articles(0.7 GB) in our experiments. Meanwhile, 20 % of the training dataset for both corpus are randomly chosen as the validation set to further determine the architecture of the feature detector, learning rate and so on.

  We design some experiments of retrieving relevant documents on the corpus to validate the performance of Deep Spectral Hashing (DSH) over other popular methods, such as LSA, TF-IDF, and Semantic Hashing (SH). Furthermore, we use the Precision-Recall curves, which are defined in Eqs. 15 and 16, to visualize experimental results of different methods.

$$Precision = \frac{n(\hat{y})}{n(y)} \tag{15}$$

$$Recall = \frac{n(\hat{y})}{n(n_t)} \tag{16}$$

Where $n(\hat{y})$ denotes the number of retrieved relevant documents, $n(y)$ denotes the total number of retrieved documents and $n(n_t)$ denotes the total number of all relevant documents. In addition, we also make comparisons among different methods in terms of time on preprocessing, single retrieval and all retrievals. We use MATLAB to do all the experiments on a server with an Intel Quad Core CPU of 2.83 GHz, 4.0 GB Memory, and 1 TB Hard Disk.

5.2 Experimental Results and Analysis

At first, we should specify the depth of deep generative model and the dimensions for each layer. Experimental results on validation set of both corpus, as shown in Fig. 5, show that the 2048-1024-512 architecture (see Fig. 4) works best (the notation "4L_256" denotes a 4-layer architecture where dimensions of top two layers are 512 and 256). Through further observation, we can conclude that neither redundant features nor inadequate features can describe a document appropriately.

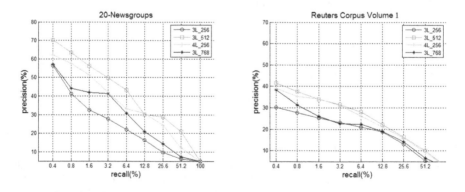

Fig. 5. Performance comparisons of different deep architectures

The performance comparisons of DSH, LSA, TF-IDF, SH are demonstrated in Figs. 6 and 7. The experimental results show that codewords generated by DSH outperform the 128 real-values produced by LSA as well as 128 codewords created by SH, and is slightly weaker than TF-IDF on 20-Newsgroups but almost as effective as TF-IDF on Reuters Corpus Volume 1. Futhermore, using 20-bit codewords to do pre-filtering (return codewords within a small Hamming distance) on TF-IDF achieves better accuracy rate. Note that unlike the method of adding deterministic Gaussian noise to the bottom-up input by each code unit to make codes binary [10], we use the real-valued activation probabilities produced by top layer as the input data for training our spectral model, which reduces the risk of messing up the conjugate gradient fine-tuning and promises to generate better codewords.

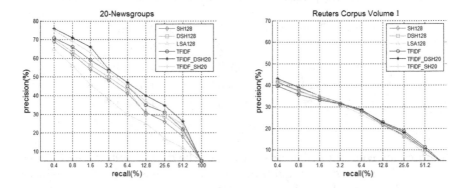

Fig. 6. Precision-Recall curves for 20-Newsgroups and Reuters Corpus Volume 1

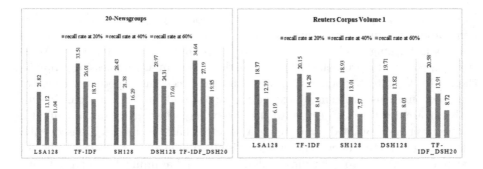

Fig. 7. Precision rates with recall rates of 20 %, 40 %, 60 % respectively

Tables 1 and 2 show the time comparison on 20-Newsgroups and Reuters Corpus Volume 1 among LSA, TF-IDF, SH and DSH. It is obvious that DSH is superior to TF-IDF, LSA and SH. In addition, some conclusions about efficiency can be obtained by comparing either the time of retrieving one document (single retrieval) or that of retrieving all the documents (all retrievals) in the dataset. Also, note that both SH and DSH are based on hashing techniques, which means that there is little difference on the retrieval stage excluding different strategies

Table 1. Time comparison on 20-Newsgroups (in second)

Algorithm	Preprocessing	Single retrieval	All retrievals
LSA	2.53	0.0914	6.58×10^2
TF-IDF	56.18	0.6351	4.77×10^3
SH	5.41	0.0065	54.6
DSH	5.14	0.0058	51.4

Table 2. Time comparison on Reuters Corpus Volume 1 (in second)

Algorithm	Preprocessing	Single retrieval	All retrievals
LSA	2.84	1.8971	5.9×10^5
TF-IDF	55.43	9.3138	2.1×10^6
SH	5.34	0.0063	1.6×10^3
DSH	5.25	0.0055	1.5×10^3

for compressing compact codewords. However, the deep architecture of our generative model is much simpler and training a spectral hashing model (mainly for generating the range of uniform distribution for each dimension) proves to be much faster than training a RBM.

6   Conclusions

In this paper, we proposed a novel document modeling approach which extracts dense and low-dimensional features first and then exploits these features to train a spectral hashing model for compressing novel data into binary codes. While spectral techniques are well studied in the domain of image processing, applying them directly into text data typically achieves exceedingly poor performance. We overcome this problem by extracting dense and low-dimensional features with an architecture of deep autoencoder, where the activation probabilities in top layer are obtained for training a spectral hashing model. Consequently, retrieving correlative documents is exceedingly simple and particularly efficient, which is equivalent to retrieving codewords in a hamming ball of predefined radius without computing any similarity measures, such as Euclidean distance. Experimental results on two kinds of corpus show that our method generates compressed codewords of high quality and outperforms the conventional retrieval methods including LSA, TF-IDF and SH as well as guarantees accessibility to the large-scale dataset. In addition, we also admit that shorter codewords may be faced with hash collisions, which is more obvious in large-scale dataset.

Acknowledgments. This work is supported in part by the Beijing Natural Science Foundation under Grant No. 4162067/4142050 and the National Science Foundation of China under Grant No. 61472391/61372171.

References

1. Salton, G., Buckley, C.: Term-weighting approaches in automatic text retrieval. Inf. Process. Manag. **24**(5), 513–523 (1988)
2. Deerwester, S.C., Dumais, S.T., Landauer, T.K., Furnas, G.W., Harshman, R.A.: Indexing by latent semantic analysis. J. Am. Soc. Inf. Sci. **41**(6), 391–407 (1990)
3. Hofmann, T.: Probabilistic latent semantic indexing. In: Proceedings of the 22nd Annual International ACM SIGIR Conference on Research and Development in Information Retrieval, pp. 50–57. ACM, New York (1999)
4. David, M.B., Andrew, Y.N., Michael, I.J.: Latent Dirichlet allocation. J. Mach. Learn. Res. **3**, 993–1022 (2003)
5. Hinton, G.E.: Training products of experts by minimizing contrastive divergence. Neural Comput. **14**(8), 1711–1800 (2002)
6. Hinton, G.E., Osindero, S.: A fast learning algorithm for deep belief nets. Neural Comput. **18**(7), 1527–1554 (2006)
7. Xu, J., Li, H., Zhou, S.: An overview of deep generative models. IETE Techn. Rev. **32**(2), 131–139 (2015)
8. Li, J., Luong, M.T., Dan, J.: A hierarchical neural autoencoder for paragraphs and documents. In: Proceedings of the 53rd Annual Meeting of the Association for Computational Linguistics and the 7th International Joint Conference on Natural Language Processing, pp. 1106–1115. Association for Computational Linguistics, Stroudsburg (2015)
9. Le, Q.V., Tomas, M.: Distributed representations of sentences and documents. In: Proceedings of the 31st International Conference on Machine Learning, pp. 1188–1196 (2014)
10. Salakhutdinov, R.R., Hinton, G.E.: Semantic hashing. Int. J. Approximate Reasoning **50**(7), 969–978 (2009)
11. Weiss, Y., Torralba, A., Fergus, R.: Spectral hashing. In: Advances in Neural Information Processing Systems, vol. 21, pp. 1753–1760 (2009)
12. Yu, G., Sapiro, G., Mallat, S.: Solving inverse problems with piecewise linear estimators: from Gaussian mixture models to structured sparsity. IEEE Trans. Image Process. **21**(5), 2481–2499 (2012)
13. Shi, J., Malik, J.: Normalized cuts and image segmentation. IEEE Trans. Pattern Anal. Mach. Intell. **22**, 888–905 (1997)
14. Kannan, R., Vempala, S., Vetta, A.: On clusterings-good, bad and spectral. J. ACM **51**(3), 497–515 (2004)
15. Andrew, Y.N., Michael, I.J., Yair, W.: On spectral clustering: analysis and an algorithm. In: Advances in Neural Information Processing Systems, vol. 14, pp. 849–856 (2002)
16. Xu, J., Li, H., Zhou, S.: Improving mixing rate with tempered transition for learning restricted Boltzmann machines. Neurocomputing **139**, 328–335 (2014)
17. Bekkerman, R., Yaniv, R.E., Tishby, N., Winter, Y.: On feature distributional clustering for text categorization. In: Proceedings of the 24th Annual International ACM SIGIR Conference on Research and Development in Information Retrieval, pp. 146–153. ACM, New York (2001)
18. Li, B., Vogel, C.: Improving multiclass text classification with error-correcting output coding and sub-class partitions. Adv. Artif. Intell. **6085**, 4–15 (2010)
19. Nigam, K., McCallum, A.K., Thrun, S., Mitchell, T.: Text classification from labeled and unlabeled documents using EM. Mach. Learn. **39**(2–3), 103–134 (2000)

Best Guided Backtracking Search Algorithm for Numerical Optimization Problems

Wenting Zhao[1], Lijin Wang[1,2], Bingqing Wang[1], and Yilong Yin[1,3(✉)]

[1] School of Computer Science and Technology,
Shandong University, Jinan 250101, China
ylyin@sdu.edu.cn
[2] College of Computer and Information Science,
Fujian Agriculture and Forestry University, Fuzhou 350002, China
[3] School of Computer Science and Technology,
Shandong University of Finance and Economics, Jinan 250014, China

Abstract. Backtracking search algorithm is a promising stochastic search technique by using its historical information to guide the population evolution. Using historical population information improves the exploration capability, but slows the convergence, especially on the later stage of iteration. In this paper, a best guided backtracking search algorithm, termed as BGBSA, is proposed to enhance the convergence performance. BGBSA employs the historical information on the beginning stage of iteration, while using the best individual obtained so far on the later stage of iteration. Experiments are carried on the 28 benchmark functions to test BGBSA, and the results show the improvement in efficiency and effectiveness of BGBSA.

Keywords: Backtracking search algorithm · Best guided · Historical information · Numerical optimization problems

1 Introduction

Optimization plays a vital role in various fields, including decision science and physical system. Generally speaking, the first step of solving the optimization problem is to specify the objective function which describes the relationship between the variables and constraints. Then, we select the appropriate optimization method to reach the global optimum according to the characteristics of the objective function. When the objective function turns out to be complex, non-linear or non-differential, evolutionary algorithm (EA) is chosen to find the global optimum. EA is expected to reach a problems global minimum value quickly with a small number of control parameters and low computational cost. Experts have come up with numerous EAs, for example, genetic algorithm(GA) [1], differential evolutionary algorithm (DE) [2], ant colony optimization algorithm (ACO) [3], particle swarm optimization algorithm (PSO) [4], artificial bee colony algorithm (ABC) [5], biogeography-based optimization algorithm (BBO) [6] and backtracking search optimization algorithm (BSA) [7].

© Springer International Publishing AG 2016
F. Lehner and N. Fteimi (Eds.): KSEM 2016, LNAI 9983, pp. 414–425, 2016.
DOI: 10.1007/978-3-319-47650-6_33

BSA is a novel population-based nature inspired optimization algorithm with a simple structure which only needs one input parameter and easy to be implemented. The experiments has been carried out on benchmark CEC-2005 and CEC-2011 and BSA shows the promising performance [7]. In addition, BSA has been applied to different kinds of engineering optimization problems recently due to its strong solving ability [8–10].

Similar to common evolutionary computation method, BSA employs mutation, crossover and selection three basic genetic operators. BSA is a double-population algorithm that employs both the historical and current populations. In the process of trial vectors generation, BSA maintains a block of memory to store the information derived from history population. During selection stage, BSA selects better individuals based on a greedy strategy guiding population towards the global optimum. BSA has a powerful exploration capability and achieves good results in solving multi-model problems. However, influenced by historical experience, the convergence speed of BSA slows down and prejudice exploitation on later evolution period.

Therefore, researches have paid more and more attention to improve the performance of BSA. Opposition-based BSA was presented by Xu et al. in [11] which incorporates the opposition-based learning method to accelerate the convergence speed. In ABSA [12] proposed by Duan et al., the parameters of mutation and crossover are varied depending on the fitness values of the solutions to refine the convergence performance. Askarzadeh et al. introduced Burger's chaotic map into BSA and proposed BSABCM [13] to adjust the scaling factor during mutation operation. Thus, the convergence rate and population diversity achieve a balance. Wang et al. proposed a hybrid algorithm HBD [14] which employs DE mutation strategy to enhance exploitation ability of BSA by optimizing one worse individual according to its probability at each iteration process. IBSA was proposed by Zhao et al. in [15] by incorporating differential evolution mutation into BSA to increase the solving speed for constrained optimization problems.

In the lights of the slow convergence on the stage of iteration, we propose the best guided backtracking search algorithm (BGBSA) which utilizes historical experience or the information derived from the best individual at different iteration process to balance the exploration and exploitation capabilities. The iteration period is divided into early stage and later stage in BGBSA. At early stage, BGBSA takes full advantage of historical experience to maintain a powerful exploration capability. During later evolution process, the efficient information provided by the best individual is utilized to speed up the convergence. The major advantage of our approach are as follows: (i) The information derived from the best individual helps BGBSA converge fast, and keep the balance between exploration and exploitation. (ii) The best guided operator is similar to the historical guided operator in BSA, therefore, BGBSA is still very simple.

The remainder of this paper is organized as follows. Section 2 introduces backtracking search algorithm. Section 3 gives the description of the proposed method. Results are presented in Sect. 4. Section 5 concludes this paper.

2   Backtracking Search Optimization Algorithm

The backtracking search optimization algorithm was first proposed by Civicioglu [7]. BSA is a novel stochastic population-based algorithm for real-valued numerical optimization problems. The structure of BSA is described as follows.

Initialization. At the first step, BSA initializes the current population P and the historical population $OldP$ containing N individuals according to the following form:

$$p_{ij} = p_j^{\min} + (p_j^{\max} - p_j^{\min}) \times rand(0,1) \tag{1}$$

$i = 1, 2, ..., N, j = 1, 2, ..., D$ where N is the population size and D is the problem dimension; p_i is the i-th individual in the current population P, $p_j^{\max}$ and $p_j^{\min}$ are the upper bound and lower bound of dimension j, and $rand(0,1)$ generates a random number distributed uniformly from 0 to 1.

Selection-I. At this step, the selection strategy is used to select the historical population which will guide the population evolution in mutation step. Before the historical population is selected, BSA has a option of updating the historical population according to Eq. (2) where a and b are randomly generated numbers distributed uniformly over the range $(0,1)$. Then, the order of the individuals in $OldP$ is changed randomly according to Eq. (3). The mechanism ensures that BSA is able to utilize randomly selected previous generation as the historical population.

$$OldP := \begin{cases} P, & a < b \\ OldP, & \text{otherwise} \end{cases} \tag{2}$$

$$OldP := permuting(OldP) \tag{3}$$

Mutation. BSA's mutation process generates the initial form of trial vectors by Eq. (4). The search-direction matrix $(OldP - P)$ is calculated and the amplitude is controlled by F which is generated randomly over the standard normal distribution range $(0,3)$. Due to the utilization of $OldP$, BSA takes partial advantage of experiences from previous generations.

$$Mutant = P + F \times (OldP - P) \tag{4}$$

Crossover. A nonuniform and more complex crossover strategy is designed in BSA. The crossover process generates the final form of the trial population V. Firstly, a binary integer-valued matrix map of size $N * D$ is obtained. Secondly, it depends on the value of map for BSA to update the relevant dimensions of mutant individual by using the relevant individual in P. This crossover process can be presented in Eq. (5).

$$v_{ij} = \begin{cases} p_{ij}, & map_{ij} = 1 \\ mutant_{ij}, & \text{otherwise} \end{cases} \tag{5}$$

Algorithm 1. Best Individual Guided Backtracking Search Algorithm

1: **Input** N- population size, $dimRate$- crossover parameter,
2:        $MaxFes$- total number of function evaluation, α- stage control parameter
3: **Output** X_{best}- the best individual
4: **Step 1: Initialization**
5:    initialize $P = \{p_1, ...p_N\}$, $OldP = \{oldp_1, ...oldp_N\}$,
6:    and current number of function evaluation $FES = 0$
7: **Step 2: Selection-I**
8:    redefine $OldP$ using Eqs. (2) and (3)
9: **Step 3: Mutation**
10:    **Step 3.1** generate $mutant$ according to Eq. (4)
11:            if $FES < \alpha * MaxFes$
12:    **Step 3.2** generate $mutant$ according to Eq. (7)
13:            if $FES \geq \alpha * MaxFes$
14: **Step 4: Crossover**
15:    generate trial vector V using Eq. (5)
16: **Step 5: Selection-II**
17:    update current population P according to Eq. (6)
18: **Step 6: Variable Update**
19:    select the best individual X_{best} and update FES
20: **Step 7: Stopping Criteria**
21:    If stopping criteria is fulfilled, then output X_{best}; otherwise go to **Step 2**

Selection-II. At this step, the individuals with better fitness value are selected and evolved in the next iteration until the stop condition is satisfied. The greedy selection mechanism is shown in Eq. (7) where $f(v_i)$ and $f(p_i)$ represents the fitness value of v_i and p_i.

$$p_i = \begin{cases} v_i, & if \ f(v_i) \leq f(p_i) \\ p_i, & \text{otherwise} \end{cases} \tag{6}$$

3   Best Guided Backtracking Search Algorithm

BSA has a powerful exploration capability but a relatively slow convergence speed, since the algorithm makes full use of historical experiences to guide the evolution. The enhanced backtracking search algorithm, combining the historical experience and the experience from best individual, is proposed. The character of the best individual obtained at later stage of iteration is able to provide efficient information contributing to convergence acceleration. The pseudo-code of BGBSA is summarized in Algorithm 1.

The whole iteration period is divided into two stages: early stage and later stage. To do this, a control parameter α over the range $(0, 1)$ is used. If the sum of function evaluations till the current iteration are less than the α scale of the maximum function evaluations, then BGBSA is regarded on the early stage of iteration, else, on the later stage of iteration. On the early stage, the population is still guided by the historical information. In this case, the exploration capacity is kept. On the later stage, to improve the convergence, the population is guided by the information derived from the best individual obtained so far, resulting in the improvement on the exploitation capacity. As a result, the balance of exploration and exploitation is kept. The best guided operator is presented in Eq. (7), where P_{best} is the best individual obtained so far, and F is the same as Eq. (4).

$$BGMutant = P + F \times (P_{best} - P) \tag{7}$$

4   Experiments

4.1   Experimental Settings

In this section, BGBSA is verified on CEC-2013 benchmark test suite including unimodal functions $F_1 - F_5$, basic multimodal functions $F_6 - F_{20}$ and composition functions $F_{21} - F_{28}$ [16]. The parameter of crossover rate $dimRate$ is set to 1.0, stage control parameter α is equal to 0.75, the population size N is 30 and the dimension is set to 30. For each problem, 25 independent runs are performed. The algorithm is stopped if a maximum of 300000 iterations for each run is met. In this paper, the amplitude control parameter F mentioned in Eq. (4) and in Eq. (7) which is subject to lévy distribution is implemented using matlab statistics and machine learning toolbox. To evaluate the performance of BGBSA, the average and standard deviation of the best error values, presented as "$AVG_{ER} \pm STD_{ER}$", are used in the following result tables. The results with high quality are marked in bold. In addition, we examine the significant differences between two algorithms utilizing Wilcoxon signed-rank test at the 5 % significance level. The statistical result is listed at the bottom of tables.

4.2   The Effect of BGBSA

Table 1 summarizes the results obtained by BSA and BGBSA. For uni-modal functions $F_1 - F_5$, BGBSA gains global optimum on F_1, F_5 and brings superior solutions on F_2 and F_4. For F_3, BGBSA exhibits a little inferior to BSA. However, it is not significant according to the Wilcoxon test results. For basic multimodal functions $F_6 - F_{20}$, BGBSA overall outperforms BSA. BGBSA obtains solutions with high quality on F_7, $F_9 - F_{18}$ and F_{20} with the help of average error values. BGBSA behaves poor solving ability on F_{19}, but these two methods are not significant. For composition functions $F_{21} - F_{28}$, BGBSA gains superior solutions on F_{22}, F_{23}, F_{25}, F_{27}, equal ones on F_{26}, F_{28}, and inferior ones on F_{21}, F_{24}. However, based on the Wilcoxon test results, BGBSA shows similar performance on F_{21}, F_{24} compared with BSA. Summarily, BGBSA wins and ties BSA on 12 and 16 out of 28 benchmark functions according to "+/ = /−", respectively. Particularly, the proposed method shows promising performance on basic multimodal functions.

Table 1. Error values obtained by BSA and BGBSA for 30-dimensional CEC-2013 benchmark functions

	BSA	BGBSA		P value
	$AVG_{Er} \pm STD_{Er}$	$AVG_{Er} \pm STD_{Er}$		
F_1	1.01e$-$30 $\pm$ 3.49e$-$30	1.01e$-$30 $\pm$ 3.49e$-$30	=	1.000000
F_2	1.37e+06 $\pm$ 5.35e+05	**4.26e+05 $\pm$ 2.13e+05**	+	0.000020
F_3	**4.54e+06 $\pm$ 4.60e+06**	6.44e+06 $\pm$ 9.98e+06	=	0.861162
F_4	1.27e+04 $\pm$ 3.58e+03	**4.95e+03 $\pm$ 1.93e+03**	+	0.000018
F_5	0.00e+00 $\pm$ 0.00e+00	5.05e$-$31 $\pm$ 2.52e$-$30	=	1.000000
F_6	2.74e+01 $\pm$ 2.47e+01	2.74e+01 $\pm$ 2.64e+01	=	0.492633
F_7	6.82e+01 $\pm$ 1.35e+01	**5.94e+01 $\pm$ 1.36e+01**	=	0.082653
F_8	2.09e+01 $\pm$ 6.72e$-$02	2.09e+01 $\pm$ 3.98e$-$02	=	0.618641
F_9	2.73e+01 $\pm$ 2.75e+00	**2.58e+01 $\pm$ 2.86e+00**	=	0.078001
F_{10}	1.90e$-$01 $\pm$ 1.42e$-$01	**1.49e$-$01 $\pm$ 1.59e$-$01**	=	0.287862
F_{11}	7.96e$-$02 $\pm$ 2.75e$-$01	**3.98e$-$02 $\pm$ 1.99e$-$01**	=	1.000000
F_{12}	8.71e+01 $\pm$ 2.14e+01	**8.36e+01 $\pm$ 1.74e+01**	=	0.492633
F_{13}	1.49e+02 $\pm$ 2.53e+01	**1.42e+02 $\pm$ 2.19e+01**	=	0.287862
F_{14}	3.56e+00 $\pm$ 1.73e+00	**1.52e+00 $\pm$ 1.28e+00**	+	0.000157
F_{15}	3.81e+03 $\pm$ 4.16e+02	**3.48e+03 $\pm$ 4.60e+02**	+	0.002259
F_{16}	1.26e+00 $\pm$ 1.66e$-$01	**1.10e+00 $\pm$ 3.01e$-$01**	+	0.021418
F_{17}	3.09e+01 $\pm$ 1.75e$-$01	**3.06e+01 $\pm$ 1.06e$-$01**	+	0.000029
F_{18}	1.16e+02 $\pm$ 1.99e+01	**9.78e+01 $\pm$ 1.90e+01**	+	0.002947
F_{19}	**1.07e+00 $\pm$ 2.11e$-$01**	1.13e+00 $\pm$ 2.38e$-$01	=	0.312970
F_{20}	1.14e+01 $\pm$ 4.91e$-$01	**1.10e+01 $\pm$ 6.39e$-$01**	+	0.017253
F_{21}	**2.67e+02 $\pm$ 8.00e+01**	2.90e+02 $\pm$ 4.91e+01	=	0.142970
F_{22}	4.33e+01 $\pm$ 1.72e+01	**2.59e+01 $\pm$ 1.12e+01**	+	0.000602
F_{23}	4.36e+03 $\pm$ 5.00e+02	**4.09e+03 $\pm$ 3.81e+02**	+	0.042207
F_{24}	**2.33e+02 $\pm$ 1.03e+01**	2.35e+02 $\pm$ 1.16e+01	=	0.396679
F_{25}	2.89e+02 $\pm$ 8.80e+00	**2.81e+02 $\pm$ 1.44e+01**	+	0.028314
F_{26}	2.00e+02 $\pm$ 1.32e$-$02	2.00e+02 $\pm$ 7.07e$-$03	+	0.000029
F_{27}	8.89e+02 $\pm$ 1.45e+02	**8.85e+02 $\pm$ 1.10e+02**	=	0.798248
F_{28}	3.00e+02 $\pm$ 1.95e$-$13	3.00e+02 $\pm$ 1.62e$-$13	=	0.637352
+/=/$-$				12/16/0

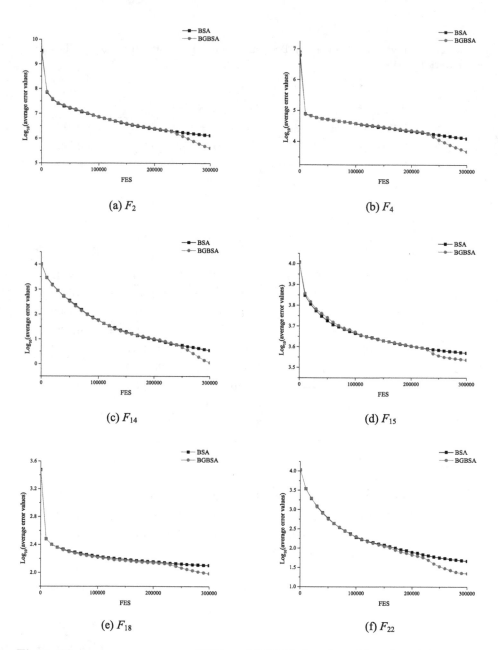

(a) F_2

(b) F_4

(c) F_{14}

(d) F_{15}

(e) F_{18}

(f) F_{22}

Fig. 1. The convergence curves of BSA and BGBSA for selected benchmark functions.

Table 2. Average ranking obtained by BGBSA and three variants of BSA for 10, 30 and 50-dimensional CEC-2013 benchmark functions

D	BGBSA	HBD	IBSA	COOBSA
10	1.93	1.93	2.14	4
30	1.84	1.89	2.27	4
50	1.75	1.93	2.32	4

Furthermore, convergence curves for six selected functions F_2, F_4, F_{14}, F_{15}, F_{18}, and F_{22} are plotted in Fig. 1 to investigate the convergence speed. It can be observed from Fig. 1 that BGBSA has a better convergence performance than BSA on the later stage of iteration, and shows the exploitation capacity.

In terms of the accuracy of solutions and convergence, BGBSA is overall superior to BSA. This is because that the information dervied from the best solution improves the exploitation capacity of BGBSA, and keeps the balance of exploration and exploitation.

4.3   Compared with Other Variants of BSA

BGBSA is compared with three BSA approaches, called COOBSA [11], HBD [14] and IBSA [15]. For each algorithm, parameter settings are kept the same. The population size N is equal to the dimension D at 30 or 50, while it is 30 in the case of $D = 10$. The stopping criterion is that a maximum of function evaluation times equal to $10000 * D$ is reached. The average rankings of the four algorithms by the Friedman test for CEC-2013 test suite at $D = 10, 30, 50$ are presented in Table 2.

In Table 2, each row shows the results obtained by the Friedman test at different dimension $D = 10, 30, 50$ respectively. In the top row of the table, it can be seen that BGBSA and HBD offer the best performance, followed by IBSA and COOBSA at $D = 10$. At $D = 30$, BGBSA is slightly better than the second best performance algorithm HBD, followed by IBSA and COOBSA. When the dimension is increased to 50, BGBSA is the best and superior to the other three algorithms. It can be concluded that the best guided BSA shows better performance than HBD, IBSA and COOBSA, and exhibits higher stability.

4.4   Compared with Other Algorithms

BGBSA is compared with five state-of-art algorithms which do not combine with BSA named NBIPOP-aCMA [17], fk-PSO [18], SPSO2011 [19], SPSOABC [20], and PVADE [21] which were proposed during CEC-2013 Special Session &Competition on Real-Parameter Single Objective Optimization. To compare fair and conveniently, the maximum of fitness evaluations is set to 300000 for each algorithm. Comparison results are listed in Table 3. In addition, the results of Friedman test similarly done in [22] for six problems are presented in Table 4.

Table 3. Error values obtained by BGBSA and 5 compared algorithms for CEC-2013 benchmark functions at $D = 30$

	NBIPOP-aCMA	fk-PSO	SPSO2011	SPSOABC	PVADE	BGBSA
	$AVG_{Er} \pm STD_{Er}$	$AVG_{Er} \pm STD_{Er}$	$AVG_{Er} \pm STD_{Er}$	$AVG_{Er} \pm STD_{Er}$	$AVG_{Er} \pm STD_{Er}$	$AVG_{Er} \pm STD_{Er}$
F_1	0.00E+00 ± 0.00E+00	0.00E+00 ± 0.00E+00	0.00E+00 ± 0.00E+00	0.00E+00 ± 0.00E+00	0.00E+00 ± 0.00E+00	**1.01e-30 ± 3.49e-30**
F_2	0.00E+00 ± 0.00E+00	1.59E+06 ± 8.03E+05	3.38E+05 ± 1.67E+05	8.78E+05 ± 1.69E+06	2.12E+06 ± 1.56E+06	4.26e+05 ± 2.13e+05
F_3	0.00E+00 ± 0.00E+00	2.40E+08 ± 3.71E+08	2.88E+08 ± 5.24E+08	5.16E+07 ± 8.00E+07	1.65E+03 ± 2.83E+03	6.44e+06 ± 9.98e+06
F_4	0.00E+00 ± 0.00E+00	4.78E+02 ± 1.96E+02	3.86E+04 ± 6.70E+03	6.02E+03 ± 2.30E+03	1.70E+04 ± 2.85E+03	4.95e+03 ± 1.93e+03
F_5	0.00E+00 ± 0.00E+00	0.00E+00 ± 0.00E+00	5.42E-04 ± 4.91E-04	0.00E+00 ± 0.00E+00	1.40E-07 ± 1.86E-07	**5.05e-31 ± 2.52e-30**
F_6	0.00E+00 ± 0.00E+00	2.99E+01 ± 1.76E+01	3.79E+01 ± 2.83E+01	1.09E+01 ± 1.09E+01	8.29E+00 ± 5.82E+00	2.74e+01 ± 2.64e+01
F_7	2.31E+00 ± 6.05E+00	6.39E+01 ± 3.09E+01	8.79E+01 ± 2.11E+01	5.12E+01 ± 2.04E+01	1.29E+00 ± 1.22E+00	5.94e+01 ± 1.36e+01
F_8	2.09E+01 ± 4.80E-02	2.09E+01 ± 6.28E-02	2.09E+01 ± 5.89E-02	2.09E+01 ± 4.92E-02	2.09E+01 ± 4.82E-02	2.09e+01 ± 3.98e-02
F_9	3.30E+00 ± 1.38E+00	1.85E+01 ± 2.69E+00	2.88E+01 ± 4.43E+00	2.95E+01 ± 2.62E+00	6.30E+00 ± 3.27E+00	2.58e+01 ± 2.86e+00
F_{10}	0.00E+00 ± 0.00E+00	2.29E-01 ± 1.32E-01	3.40E+01 ± 1.48E-01	1.32E-01 ± 6.23E-02	2.16E+00 ± 1.36E-02	1.49e-01 ± 1.55e-01
F_{11}	3.04E+00 ± 1.41E+00	2.36E+01 ± 8.76E+00	1.05E+02 ± 2.74E+01	0.00E+00 ± 0.00E+00	5.84E+01 ± 1.11E+01	8.36e-01 ± 1.74e+01
F_{12}	2.91E+00 ± 1.38E+00	5.64E+01 ± 1.51E+01	1.04E+02 ± 3.54E+01	6.44E+01 ± 1.48E+01	1.15E+02 ± 1.24E+01	1.42e+02 ± 2.16e+01
F_{13}	2.78E+00 ± 1.45E+00	1.23E+02 ± 2.19E+01	1.94E+02 ± 3.86E+01	1.15E+02 ± 2.24E+01	1.31E+02 ± 4.38E+01	1.52e+02 ± 1.28e+00
F_{14}	8.10E+02 ± 3.60E+02	7.04E+02 ± 2.38E+02	3.99E+03 ± 6.19E+02	1.55E+01 ± 6.13E+01	3.20E+03 ± 3.19E+02	3.48e+03 ± 4.60e+02
F_{15}	7.65E+02 ± 2.95E+02	3.42E+03 ± 5.16E+02	3.81E+03 ± 6.94E+02	3.55E+03 ± 3.04E+02	5.16E+03 ± 3.19E-02	1.52e+00 ± 1.28e+00
F_{16}	4.40E+01 ± 9.26E-01	8.48E-01 ± 2.20E-01	1.31E+00 ± 3.59E-01	1.03E+00 ± 2.01E-01	2.39E+00 ± 2.66E+00	3.48e+03 ± 3.01e-01
F_{17}	3.44E+01 ± 1.87E+00	5.26E+01 ± 7.11E+00	1.16E+02 ± 2.02E+01	3.09E+01 ± 1.23E-01	1.02E+02 ± 1.17E+01	3.06e+01 ± 1.06e-01
F_{18}	6.23E+01 ± 4.56E+01	6.81E+01 ± 9.68E+00	1.21E+02 ± 2.46E+01	9.01E+01 ± 8.95E+00	1.82E+02 ± 1.20E+01	9.78e+01 ± 1.90e+01
F_{19}	2.23E+00 ± 3.41E-01	3.12E+00 ± 9.83E-01	9.51E+00 ± 4.42E+00	1.71E+00 ± 4.68E-01	5.40E+00 ± 8.10E-01	**1.13e+00 ± 2.38e-01**
F_{20}	1.29E+01 ± 5.98E-01	3.11E+01 ± 9.26E-01	1.35E+01 ± 1.11E+00	1.11E+01 ± 7.60E-01	1.13E+00 ± 3.28E-01	1.10e+00 ± 6.39e-01
F_{21}	1.92E+02 ± 2.72E+01	3.11E+02 ± 7.92E-01	3.09E+02 ± 6.80E+01	3.18E+02 ± 7.53E+01	3.19E+02 ± 6.26E+01	**2.59e+01 ± 1.12e+01**
F_{22}	8.38E+02 ± 4.60E+02	8.59E+02 ± 3.10E+02	4.30E+03 ± 7.67E+02	8.41E+01 ± 3.90E+01	2.50E+03 ± 3.86E+02	4.09e+03 ± 3.81e+02
F_{23}	6.67E+02 ± 2.90E+02	3.57E+03 ± 5.90E+02	4.83E+03 ± 8.23E+02	4.18E+03 ± 5.62E+02	5.81E+03 ± 5.04E+02	2.35e+02 ± 1.1e+01
F_{24}	1.62E+02 ± 3.00E+01	2.48E+02 ± 8.11E+00	2.67E+02 ± 1.25E+01	2.51E+02 ± 1.43E+00	2.02E+02 ± 1.40E+00	2.81e+02 ± 1.44e+01
F_{25}	2.20E+02 ± 1.11E+01	2.49E+02 ± 7.82E+00	2.99E+02 ± 1.05E+01	2.75E+02 ± 9.76E+00	2.30E+02 ± 2.08E+01	2.00e+02 ± 7.07e-03
F_{26}	1.58E+02 ± 3.00E+01	2.95E+02 ± 7.06E+01	2.86E+02 ± 8.24E+01	2.60E+02 ± 7.62E+01	2.18E+02 ± 4.01E+01	8.85e+02 ± 1.1e+02
F_{27}	4.69E+02 ± 7.38E+02	7.76E+02 ± 7.11E+01	1.00E+03 ± 1.12E+02	9.10E+02 ± 1.62E+02	3.26E+02 ± 1.14E+01	3.00e+02 ± 1.62e-13
F_{28}	2.69E+02 ± 7.35E+01	4.01E+02 ± 3.48E+02	4.01E+02 ± 4.76E+02	3.33E+02 ± 2.32E+02	3.00E+02 ± 2.24E+02	

Table 4. Average ranking of six algorithms by the Friedman test for CEC-2013 functions at $D = 30$

Methods	NBIPOP-aCMA	BGBSA	SPSOABC	fk-PSO	PVADE	SPSO2011
Ranking	1.8	3.11	3.3	3.57	3.93	5.29

Table 5. Average ranking on 28 benchmark functions with varying stage control parameter

α	0.55	0.65	0.75	0.85	0.95
Ranking	3.09	2.71	2.61	3.05	3.54

As depicted in Table 3, NBIPOP-aCMA achieves the best performance since it is one of top three algorithm during CEC-2013. NBIPOP-aCMA, fk-PSO, SPSO2011, SPSOABC, PVADE and BGBSA perform better in 21, 3, 2, 4, 3, and 8 out of 28 functions respectively. According to the average ranking value of different algorithms by Friedman test, it can be concluded that NBIPOP-aCMA shows the best performance, and BGBSA offers the second overall performance, followed by SPSOABC, fk-PSO, PVADE and SPSO2011.

4.5 The Effect of the Parameter α

In BGBSA, the stage control parameter α is set to 0.75 which means the function evaluation times in early stage accounts for 75 percent of total evolutionary iteration times. It's necessary to choose a suitable value for α which determines the longitude of early evolutionary stage. The convergence speed of the proposed algorithm declines with bigger value of α. And BGBSA suffers from premature convergence if α is set to a smaller value.

To verify the effectiveness of α, five different values are tested and the Friedman test results are shown in Table 5. It can be seen clearly from Table 5, the algorithm with α equal to 0.75 offers the best performance, followed by 0.65, 0.55, 0.85 and 0.95. We infer that global search ability is gradually enhanced and local search ability can be ensured to bring high quality solutions as α is increased from 0.55 to 0.75. However, the performance of BGBSA suffers rapid decline when α is increased from 0.75 to 0.95. It can be analyzed that there is not enough time for best guided operator to exhibit exploitation ability. Therefore, $\alpha = 0.75$ is a reasonable choice for BGBSA.

5 Conclusions

In this work, we suggested a revised version of BSA for numerical optimization problems. The proposed algorithm combined the historical experience and the experience from the best individual obtained so far to enhance the convergence speed on the later stage of iteration. Our experimental results indicate that

BGBSA is effective and shows promising performance, especially when solving basic multimodal functions. For composition problems, BGBSA still leaves a large promotion space. In the future work, we plan to design BSA with adaptive experience guided operator.

Acknowledgments. This work was supported by the NSFC Joint Fund with Guangdong of China under Key Project U1201258, the National Natural Science Foundation of China under Grant No. 61573219, the Shandong Natural Science Funds for Distinguished Young Scholar under Grant No. JQ201316, the Natural Science Foundation of Fujian Province of China under Grant No. 2016J01280 and the Fostering Project of Dominant Discipline and Talent Team of Shandong Province Higher Education Institutions.

References

1. Holland, J.H.: Adaptation in Natural and Artificial Systems. MIT press, Cambridge (1992)
2. Storn, R., Price, K.: Differential evolution-a simple and efficient adaptive scheme for global optimization over continuous spaces. ICSI, Berkeley (1995)
3. Dorigo, M., Maniezzo, V., Colorni, A.: Ant system: optimization by a colony of cooperating agents. IEEE Trans. Syst. Man Cybern. Part B Cybern. **26**, 29–41 (1996)
4. Eberhart, R., Kennedy, J.: A new optimizer using particle swarm theory. In: 6th International Symposium on Micro Machine and Human Science, pp. 39–43. IEEE Press, New York (1995)
5. Karaboga, D.: An idea based on honey bee swarm for numerical optimization. Technical report, Erciyes University (2005)
6. Simon, D.: Biogeography-based optimization. IEEE Trans. Evol. Comput. **12**, 702–713 (2008)
7. Civicioglu, P.: Backtracking search optimization algorithm for numerical optimization problems. Appl. Math. Comput. **219**, 8121–8144 (2013)
8. Agarwal, S.K., Shah, S., Kumar, R.: Classification of mental tasks from eeg data using backtracking search optimization based neural classifier. Neurocomputing **166**, 397–403 (2015)
9. Yang, D.D., Ma, H.G., Xu, D.H., Zhang, B.H.: Fault measurement for siso system using the chaotic excitation. J. Franklin Inst. **352**, 3267–3284 (2015)
10. Mallick, S., Kar, R., Mandal, D., Ghoshal, S.: CMOS analogue amplifier circuits optimisation using hybrid backtracking search algorithm with differential evolution. J. Exp. Theor. Artif. Intell. **28**(4), 719–749 (2016)
11. Xu, Q., Guo, L., Wang, N., Li, X.: Opposition-based backtracking search algorithm for numerical optimization problems. In: He, X., Gao, X., Zhang, Y., Zhou, Z.-H., Liu, Z.-Y., Fu, B., Hu, F., Zhang, Z. (eds.) 5th International Conference on Intelligence Science and Big Data Engineering. LNCS, vol. 9243, pp. 223–234. Springer, Switzerland (2015)
12. Duan, H., Luo, Q.: Adaptive backtracking search algorithm for induction magnetometer optimization. IEEE Trans. Mag. **50**(12), 1–6 (2014)
13. Askarzadeh, A., Coelho, L.D.S.: A backtracking search algorithm combined with Burger's chaotic map for parameter estimation of PEMFC electrochemical model. Int. J. Hydrogen Energy **39**, 11165–11174 (2014)

14. Wang, L., Zhong, Y., Yin, Y., et al.: A hybrid backtracking search optimization algorithm with differential evolution. Math. Probl. Eng. **2015**, 1 (2015)
15. Zhao, W., Wang, L., Yin, Y., Wang, B., Wei, Y., Yin, Y.: An improved backtracking search algorithm for constrained optimization problems. In: Buchmann, R., Kifor, C.V., Yu, J. (eds.) KSEM 2014. LNCS (LNAI), vol. 8793, pp. 222–233. Springer, Heidelberg (2014). doi:10.1007/978-3-319-12096-6_20
16. Liang, J., Qu, B., Suganthan, P., Hernndez-Daz, A.G.: Problem definitions and evaluation criteria for the CEC 2013 special session on real-parameter optimization. Technical report, Singapore (2013)
17. Loshchilov, I.: CMA-ES with restarts for solving CEC 2013 benchmark problems. In: 2013 IEEE Congress on Evolutionary Computation, pp. 369–376. IEEE Press (2013)
18. Nepomuceno, F.V., Engelbrecht, A.P.: A self-adaptive heterogeneous PSO for real-parameter optimization. In: 2013 IEEE Congress on Evolutionary Computation, pp. 361–368. IEEE Press (2013)
19. Zambrano-Bigiarini, M., Clerc, M., Rojas, R.: Standard particle swarm optimisation 2011 at CEC-2013: a baseline for future PSO improvements. In: 2013 IEEE Congress on Evolutionary Computation, pp. 2337–2344. IEEE Press (2013)
20. El-Abd, M.: Testing a particle swarm optimization and artificial bee colony hybrid algorithm on the CEC13 benchmarks. In: 2013 IEEE Congress on Evolutionary Computation, pp. 2215–2220. IEEE Press (2013)
21. Coelho, L.D.S., Ayala, V.H., Freire, R.Z.: Population's variance-based adaptive differential evolution for real parameter optimization. In: 2013 IEEE Congress on Evolutionary Computation, pp. 1672–1677. IEEE Press (2013)
22. Gong, W.Y., Cai, Z.H.: Differential evolution with rankingbased mutation operators. IEEE Trans. Cybern. **43**, 2066–2081 (2013)

Domain Specific Cross-Lingual Knowledge Linking Based on Similarity Flooding

Liangming Pan[✉], Zhigang Wang, Juanzi Li, and Jie Tang

Tsinghua National Laboratory for Information Science and Technology,
Department of Electronic Engineering, Tsinghua University, Beijing 100084, China
{plm,wzhigang,ljz,tangjie}@keg.cs.tsinghua.edu.cn

Abstract. The global knowledge sharing makes large-scale multi-lingual knowledge bases an extremely valuable resource in the Big Data era. However, current mainstream multi-lingual ontologies based on online wikis still face the limited coverage of cross-lingual knowledge links. Linking the knowledge entries distributed in different online wikis will immensely enrich the information in the online knowledge bases and benefit many applications. In this paper, we propose an unsupervised framework for cross-lingual knowledge linking. Different from traditional methods, we target the cross-lingual knowledge linking task on specific domains. We evaluate the proposed method on two knowledge linking tasks to find English-Chinese knowledge links. Experiments on English Wikipedia and Baidu Baike show that the precision improvement of cross-lingual link prediction achieve the highest 6.12 % compared with the state-of-art methods.

Keywords: Knowledge linking · Cross-lingual · Similarity flooding

1 Introduction

In the era of information globalization, sharing knowledge across different languages becomes an important and challenging task. Online encyclopedias, which have already become an indispensable part in people's life for knowledge acquisition, are the primary focus of globalized knowledge sharing. One fundamental research, namely *cross-lingual knowledge linking*, aims at automatically discovering *cross-lingual links* (CLs), i.e., links between articles describing the same subjects in different languages. Inter-wiki cross-lingual links can largely enrich the cross-lingual knowledge and facilitate knowledge sharing across different languages. These CLs also serve as a valuable resource for many applications, including machine translation [16], cross-lingual information retrieval [10,15], and multilingual semantic data extraction [1,3], etc.

However, the problem of cross-lingual knowledge linking is non-trivial and poses a set of challenges. One of the most serious challenges is that lexical similarities, such as the edit distance between articles, are impracticable to be utilized because of the language gap. Recognizing this problem, designing language-independent features (e.g. the link structure of articles) forms the basis of recent

© Springer International Publishing AG 2016
F. Lehner and N. Fteimi (Eds.): KSEM 2016, LNAI 9983, pp. 426–438, 2016.
DOI: 10.1007/978-3-319-47650-6_34

related works for cross-lingual knowledge linking [9,11,13,14]. However, several problems for cross-lingual knowledge linking are still in need of further investigation. First, large number of known CLs are required in all the aforementioned approaches for serving as either training data or seed set. However, in most cases, there are less existing CLs or none at all between different wikis. Clearly, it is time consuming and tedious to annotate all these training CLs. Thus, unsupervised methods need to be developed for cross-lingual knowledge linking. Second, the methods to date have been entirely focus on the general framework for knowledge linking, with less focus on finding CLs in specific domains. In fact, if we focus the knowledge linking task on one specific domain, many domain-specific features (e.g. the properties in the infoboxes of wiki) can be utilized to discover new CLs more accurately.

Based on these considerations, we propose an unsupervised domain-specific framework for cross-lingual knowledge linking. Our model takes the articles of two cross-lingual wikis K and K' as input, and the articles are all from one specific domain D. Each wiki article is a single web page and describes a realworld entity in domain D. The semantic relations between these entities are mostly contained in the wiki infoboxes and hyperlinks between articles. After extracting these semantic relations, we first build up a domain-specific knowledge graph (KG) for each wiki, and then we match the entities between the two KGs via an adaptive variation of the similarity flooding algorithm (SF) [8]. Specifically, we make the following contributions:

1. We propose a novel unsupervised knowledge linking framework for cross-lingual wikis in specific domains. The method does not dependent on any pre-given CLs, which is more practically applicable than previous methods.
2. We propose a variation of the similarity flooding algorithm for entity matching between cross-lingual wikis. The SF is commonly used for ontology alignment and can hardly be applied to entity matching because of the computational challenge. We tackle with this problem via reducing the number of nodes in the pairwise connectivity graph (PCG).
3. We conduct experiments on different wiki data sets. Experimental results show that our method outperforms the state-of-the-art framework for cross-lingual knowledge linking.

The rest of this paper is organized as follows. Section 2 presents some related works. Section 3 presents some basic concepts and the problem formulation. In Sect. 4 we present our detailed approaches. The experimental results are reported in Sect. 5. Finally we conclude our work in Sect. 6.

2 Related Work

Our work is relevant to cross-lingual knowledge linking, which concerns the discover of missing cross-lingual links across online wikis. There exist several related works. Sorg and Cimiano [11] proposed a classification-based approach to infer new CLs between German Wikipedia and English Wikipedia.

Erdmann et al. [4] extracted a dictionary from Wikipedia by analyzing the link structure of Wikipedia. Hassan et al. [6] address the task of cross-lingual semantic relatedness by exploiting the cross-lingual links available between Wikipedia versions in multiple languages. Wang et al. [14] employed a factor graph model which leverages link-based features to find CLs between English Wikipedia and Chinese Wikipedia. All the aforementioned works intend to propose a general framework for cross-lingual knowledge linking. To our best knowledge, our work is the first to focus on discovering CLs in specific domains.

Ontology and instance matching is another related problem. The goal of ontology and instance matching is to find equivalent elements between two heterogeneous semantic data sources. Currently, there exist several systems for ontology matching, such as Silk [12], idMesh [2], SOCOM [5] and RiMOM [7]. The Silk and idMesh focus on monolingual matching tasks, while SOCOM and RiMOM can deal with ontology matching across languages.

3   Preliminaries

In this section, we introduce some basic concepts, and formally define the key problem of domain-specific cross-lingual knowledge linking.

Definition 1. **Knowledge Graph.** Let E be a set of entities and R be a set of binary relations. A knowledge graph G is defined as a directed graph whose nodes correspond to entities in E and edges of the form (s, r, t), where $s, t \in E$ and $r \in R$. Each edge (s, r, t) indicates that there exists a relationship r from the entity s to entity t.

Definition 2. An **Online Wiki** can be represented as $K = \{a_i\}_{i=1}^{p}$, where a_i is a disambiguated article in K and p is the size of K. A wiki article $a \in K$ is formally defined as a 4-tuple $a = (title, text, info, link)$, where $title$ denotes the title of the article a, $text$ denotes the unstructured text description of a, $info$ is the infobox associated with a and $link$ is the set of hyperlinks in article a (hyperlinks in infoboxes are not count). Specifically, $info = \{(attr_i, value_i)\}_{i=1}^{q}$ represents the list of attribute-value pairs for the article a.

Figure 1 gives an example of these four important elements concerning the article named "Steve Jobs". Given two online wikis, K and K', a *correspondence* between entities $e \in K$ and $e' \in K'$, denoted as $\langle e, e' \rangle$, signifies that e and e' are equivalent. Cross-lingual knowledge linking is the task of finding correspondences between multi-language online wikis, which is formally defined as follows.

Definition 3. **Cross-lingual knowledge linking.** Given two online wikis, K and K', *knowledge linking* is the process of finding correspondences between K and K'. If K and K' are in different languages, we call it the problem of *cross-lingual knowledge linking.*

In our problem, we further choose a domain D and extract all articles of domain D from K and K' to form two new online wikis, denoted as K_D and K'_D. We then define the problem of cross-lingual knowledge linking between K_D and K'_D as *domain-specific cross-lingual knowledge linking.*

title →
link →
text →
←info
attr
value

Fig. 1. An example of online wiki articles

4 The Proposed Approach

Figure 2 shows the framework of our proposed approach. There are two major components: *Knowledge Graph Construction* and *Graph-based Knowledge Linking*. *KG Construction* aims to build two knowledge graphs, denoted as G_D and G'_D, for K_D and K'_D. Specifically, based on a common relation set, we extract semantic relations in a structured form of subject-predicate-object triples from infoboxes and hyperlinks of K_D and K'_D. Then, G_D and G'_D are constructed by these triples. The goal of *Graph-based Knowledge Linking* is to discover CLs between G_D and G'_D based on the variation of SF algorithm. Two main processes in this algorithm are PCG construction and similarity propagation. In the following subsections, the *KG Construction* and *Graph-based Knowledge Linking* are described in detail.

4.1 Knowledge Graph Construction

Online wiki's infoboxes contain rich structured information of various entities. Among all the infobox attributes, those attributes having hyperlinks in its values identify semantic relations between entities, which are important for creating domain knowledge graph. Because attribute names are usually annotated by human editors, an attribute often have many surface names in infoboxes of online wiki. For example, in the movie domain, the attributes "Starring" and "Actor List" both refer to the actors of a movie. Furthermore, attributes across wikis may also have same semantic meanings (e.g. "Starring" and "演员表"). Therefore, we need to unify all synonymous attribute names of K_D and K'_D to a

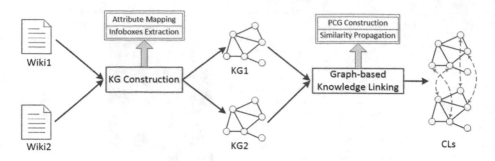

Fig. 2. The framework of the proposed method

disambiguated attribute. We call it the construction of *Attribute Mapping*, which is formally defined as follows:

Definition 4. **Attribute Mapping.** Given two online wikis K_D and K'_D in domain D, let us denote A_D and A'_D as the set of infobox attribute names of K_D and K'_D. An *attribute mapping* is a set of disambiguated attributes, denoted as $AM = \{r_i\}_{i=1}^q$. Each $r_i \in AM$ can be represented as a set $r_i = s_i \cup s'_i$, where $s_i \subset A_D$ and $s'_i \subset A'_D$. Attribute names in r_i have identical semantic meanings.

Figure 3 shows an example of attribute mapping of movie domain between English Wikipedia and Baidu Baike[1]. Based on the attribute mapping AM, we can define a function Map to map an attribute name to its corresponding disambiguated attribute in AM.

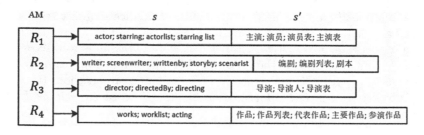

Fig. 3. An example of attribute mapping in movie domain

Given the Map function, we can build a domain knowledge graph for an online wiki using semantic relations contained in infoboxes. The knowledge graph construction algorithm is presented in Algorithm 1, where K is the input online wiki and G is the output knowledge graph. V and E are the vertex set and edge set of G, respectively. Notice that hyperlinks also represent relations between

---

[1] http://baike.baidu.com/.

entities. If article a has a hyperlink to article b, there exists some kind of relationship between a and b, thus we also add an edge $(a, <relatedTo>, b)$ to the knowledge graph.

Algorithm 1. Knowledge Graph Construction

Input: K, AM
Output: $G = (V, E)$
$V = \varnothing, E = \varnothing$;
foreach *article a* **in** K **do**
   V.add(a);
   foreach $(attr, value)$ **in** $a.info$ **do**
      $r = Map(attr)$;
      if $r \neq None$ **and** $value \in K$ **then**
         $E.add(< a, r, value >)$;
      end
   end
end
foreach *hyperlink h* **in** $a.link$ **do**
   $E.add(< a, relatedTo, h >)$;
end

4.2 Graph-Based Knowledge Linking

After the construction of knowledge graphs for two input wikis, we match equivalent entities between the two graphs. Because semantic relations between entities are language-independent features [14], we assume that the two knowledge graphs share similar structures. Similarity flooding is an efficient algorithm for alignment between two similar ontologies. Two main processes in SF are pairwise connectivity graph (PCG) construction and similarity propagation. If the number of nodes for two input graphs are m and n, the scale of nodes number for PCG will be $O(m \times n)$. Thus, the SF algorithm is possible for relatively small m and n (e.g. in schema matching tasks), but clearly infeasible when m and n are too large. The knowledge graph of one input wiki may contain thousands of nodes, which makes the nodes number of the PCG become billions. Thus, SF can not be directly applied to the knowledge linking because of the computational challenge. To tackle with this problem, we propose a variation of SF algorithm which reduce the scale of the propagation graph. The algorithm are described in detail in the following subsections.

Initial Similarity Computation. The two KGs are construed from link-based structural information such as Infoboxes and innerlinks of wikis. The text descriptions of entities in KG are not considered. In our method, we calculate the initial similarities of entities based on their text description. In order to calculate texture similarities across languages, some priori knowledge about the two languages should be given. The priori knowledge is defined as the *Domain Dictionary* in our model. Specifically, the domain dictionary is a set of word pairs between two languages. More specifically, given two different languages A and B and the domain D, the domain dictionary Dic_D is defined as:

$$Dic_D = \{<w_i^A, w_i^B>\}_{i=1}^L \qquad (1)$$

where $<w_i^A, w_i^B>$ is a translation equivalent pair between language A and B (e.g. $<$"China", "中国"$>$). To provide sufficient priori knowledge for initial similarity computation, word pairs in Dic_D should be relevant to the domain D. Given the domain dictionary Dic_D, we represent each entity in KGs as a L dimensional vector. Without loss of generality, for an entity e from a wiki written in language A, the *i-th* dimension of e's vector is the frequency of w_i^A appeared in its corresponding wiki article. Finally, the initial similarity between entity e and e' is the cosine similarity of their corresponding vectors.

Propagation Graph Construction. Based on the assumption that a part of the similarity of two elements should propagate to their respective neighbors [8], we convert the two knowledge graphs to a *Similarity Propagation Graph* (SPG) as follows:

Definition 5. **Similarity Propagation Graph.** Given two knowledge graphs G and G', $((e, e'), r, (o, o')) \in SPG(G, G')$ if and only if: (1) $(e, r, o) \in G$, (2) $(e', r, o') \in G'$, (3) $Sim(e, e') > \theta$, and (4) $Sim(o, o') > \theta$. The $Sim(e, e')$ indicates the initial similarity between entity e and e'. θ is a pre-given threshold.

Each node in the SPG represents a candidate alignment pair between the two KGs. To reduce the scale of SPG, the conditions (3) and (4) in Definition 5 remove the entities pairs which have low initial similarities.

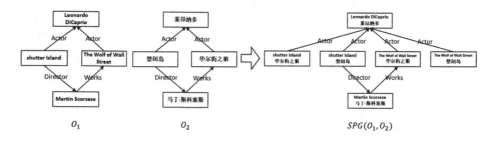

Fig. 4. An example of the construction of SPG

Figure 4 shows an example of SPG construction. The left side of the figure are two small cross-lingual KGs of movie domain, denoted as O_1 and O_2. The right side shows the SPG of O_1 and O_2. In $SPG(O_1, O_2)$, nodes are entity pairs from two KGs that have some structural relationship in common. For example, "shutter Island" and "禁闭岛" are two entities in O_1 and O_2. They are constructed into a node in $SPG(O_1, O_2)$ because they share the same relationship "Director".

Similarity Propagation. The similarity propagation starts from initial similarities between nodes of two KGs and runs an iterative propagation in the SPG. In each iteration, the similarity of a given matching pair would be propagated to the neighborhood matching pairs. The iteration stops when no similarity changes or after a predefined number of steps. Formally, let us denote $\sigma^i(e, e')$ as the similarity between e and e' after the i-th iteration. The iteration equation to perform similarity propagation is defined as follows:

$$\sigma^{i+1}(e, e') = \frac{1}{Z} \left(\sigma^0(e, e') + \sigma^i(e, e') + \varphi^i(e, e') \right) \tag{2}$$

$$\varphi^i(e, e') = \sum_{(o, o') \in IN(e, e')} \omega(o, o') \cdot \sigma^i(e, e') \tag{3}$$

$$Z = \max_{(e, e') \in SPG(G, G')} (\sigma^{i+1}(e, e')) \tag{4}$$

As defined in Eq. 2, for any entity pair (e, e') in SPG, the similarity of (e, e') in the $(i + 1)$-th iteration is dependent on its similarity of the i-th iteration $(\sigma^i(e, e'))$, its initial similarity $(\sigma^0(e, e'))$ and the similarity gain from its neighbors $(\varphi^i(e, e'))$. In Eq. 3, $\varphi^i(e, e')$ is defined as the weighted sum of similarities of (e, e')'s adjacent nodes. $IN(e, e')$ is the set of incoming neighbors of the node (e, e') in SPG. $\omega(o, o')$ is the propagation weight which is simply defined as the inverse of the number of out-linking relationships for the node (o, o'). Z is a normalization factor defined in Eq. 4.

5   Experiments

5.1   Datasets

The proposed method can be used to find cross-lingual links between any online wikis in different languages. To evaluate the performance of our method, we extract the articles of the movie domain from three different online wikis. Specifically, one English language online wiki—English Wikipedia—and two Chinese language online wikis—Baidu Baike and Chinese Wikipedia—are chosen for our experiments. For English Wikipedia and Chinese Wikipedia, we employ the latest publicly available Wikipedia dump[2], which includes **9,834,664** articles for English Wikipedia and **886,437** articles for Chinese Wikipedia. As for the Baidu Baike, which is the largest web-based encyclopedia in China, we crawled **6,223,649** web pages from the latest Baidu encyclopedia.

---

[2] https://dumps.wikimedia.org/enwiki/20160113/.

To further create datasets of movie domain, we utilize the category system of online wikis. Specifically, we select several typical categories for the movie domain such as "film", "actor" and "director", and we extract articles with these categories from the aforementioned three online wikis. As a result, a total of **222,022** movie domain articles are extracted from English Wikipedia. As for the Baidu Baike and Chinese Wikipedia, we extract **112,164** and **58,638** articles for the movie domain, respectively. We refer to these three datasets as **EWM** (English Wikipedia of movie domain), **ZWM** (Chinese Wikipedia of movie domain) and **BBM** (Baidu Baike of movie domain).

To create gold standard for evaluation, we construct a evaluation dataset that contains equivalent article pairs between EWM and ZWM. These article pairs are acquired from the existing Chinese-English cross-lingual links within Wikipedia. As a result, we obtain **2,678** CLs between EWM and ZWM. We also create a dataset of CLs between EWM and BBM. Firstly, a total of 10,000 articles are randomly selected from EWM. Then, each selected article is sent to a human annotator to find its corresponding article in BBM. However, not all articles in English Wikipedia have CLs to Baidu Baike. Finally, we obtain a total of **4,022** CLs from the 10,000 sampled articles as our evaluation dataset.

Knowledge Graph Construction. As for the knowledge graph construction, we use the attribute mapping shown in Fig. 3 for all the three datasets. In other words, we consider four kinds of relations—<Actor>,<Director>,<Writer> and <Works>—, which are frequently occurred among entities in the movie domain. Then, we apply the Algorithm 1 defined in Sect. 4.1 to construct a KG for each wiki dataset. In addition with the <relatedTo> relation defined in Algorithm 1, there exist a total of five kind of edges in the constructed KGs. Detailed statistics of the three constructed KGs are presented in Table 1. For example, in the KG for the EWM dataset, there exist 185,453 edges with the relation <Actor>.

Table 1. Statistics of knowledge graphs for three datasets

Dataset	#Nodes	#Edges				
		<Actor>	<Director>	<Writer>	<Works>	<relatedTo>
EWM	220,989	185,453	73,705	48,500	373,550	681,208
ZWM	57,842	81,717	23,544	11,730	93,257	151,299
BBM	111,768	154,112	18,921	9,603	180,370	363,006

5.2   Methods for Comparison

We define four cross-lingual linking methods as the comparison methods.

- **Title Edit Distance (TED).** This method first translates the titles of Chinese articles into English by Google Translation API[3], then we calculate the similarity between all article pairs using the edit distance of their titles.
- **Initial Similarity (IS).** This method directly regards the initial similarities defined in Sect. 4.2 as the final result. Due to the removal of the similarity propagation step, the essence of this method is a translation-based method which calculates the similarities between article texts.
- **Simple Similarity Propagation (SSP).** This method conducts the similarity propagation process without the influence of initial similarity. We initialize all nodes in the SPG with a unified initial similarity (set to 0.5 in our experiments). Accordingly, the Eq. 2 is rewritten as:

$$\sigma^{i+1}(e, e') = \frac{1}{Z} \left(\sigma^i(e, e') + \varphi^i(e, e') \right) \tag{5}$$

- **Linkage Factor Graph (LFG).** The LFG model [14] is a state-of-art method for cross-lingual knowledge linking. The method first calculates several language-independent features from input wikis and then proposes a factor graph model to discover cross-lingual links.

Implementation Details. For the construction of domain dictionary for the IS method and the proposed method, we first rank all the words in the EWM dataset by their TF-IDF, then we select the top 1000 ranked words to form a set of keywords. Finally, we translate each of the keyword into Chinese by Google Translation API to get 1000 English-Chinese word pairs. These word pairs are employed as our domain dictionary. For the LFG, we use 0.001 learning rate and run 2500 iterations in all the experiments to get its best performance.

Evaluation Metrics. We evaluate our approach on the Chinese-English cross-lingual links constructed in Sect. 5.1, i.e. the 2,678 EWM-ZWM cross-lingual links and the 4,022 CLs between the EWM dataset and the BBM dataset. For an arbitrary candidate matching pair, all comparison methods are able to predict its similarity, indicating its confidence level of being equivalent. Intuitively, for an article e in the source wiki, the article in the target wiki which has the highest similarity with e is regarded as its predicted CL. Thus, we use the prediction accuracy on the evaluation CLs to evaluate different knowledge linking methods, denoted as $P@1$. In addition, we also evaluate the methods by $P@5$, which is defined as the percentage of articles that have correct equivalent articles in its Top-5 candidates.

5.3 Influences of Parameters

There are two parameters in our method that may influence the performance including: (1) pruning threshold θ, (2) iteration time T. In this section, we look into the influences of these parameters.

---

[3] http://code.google.com/intl/zhcn/apis/language/translate/overview.html.

The Parameter θ. The parameter θ defined in Definition 5 is used for pruning the SPG. The entity pair having a lower initial similarity than θ is not considered as a candidate CL, and will be removed from the SPG. Figure 5(a) demonstrates the influence of θ on the knowledge linking between EWM and BBM. In the figure, the *Rec.* denotes the coverage rate of the PCG for the entity pairs in the evaluation set, indicating the recall of our method. *NodeP* is defined as the number of nodes in PCG divided by the nodes number of PCG when $\theta = 0$. From the figure, we observe that the *Rec.* declines rapidly with the increasing of θ. The reason is: if θ is set too high, we may prune the SPG too much. As a result, many correct entity pairs may be excluded from the SPG, which leads to a poor recall. However, if θ is too low, the number of nodes in PCG will increase rapidly, which makes the algorithm computationally challenging. We observe that the method reaches its best F1 score when $\theta = 0.3$. The similar observations are also got on the experiment between EWM and ZWM.

The Iteration Time T. Figure 5(b) shows the $P@1$ of the proposed method with different iteration time T on the two knowledge linking tasks. The parameter θ is set to 0.3 for both tasks. From the figure, we observe that the proposed method reaches its best performance through 4 to 6 iterations. In the knowledge linking task between EWM and BBM, our method converges to the best performance (89.89 % in terms of $P@1$) after the 5-th iteration. Similarly, for the datasets of EWM and ZWM, the method converges to the $P@1$ of around 83 % after 6 iterations.

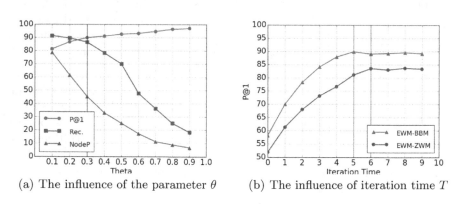

(a) The influence of the parameter θ (b) The influence of iteration time T

Fig. 5. The study of parameter influence on the proposed method (%).

5.4   Results Analysis

After we explore the influences of parameters, we further employ baseline methods to compare with the proposed method. Table 2 summarizes the performance of 5 different methods on different knowledge linking tasks. From the table, we find that the proposed method outperforms all baselines on both two tasks.

According to the result, the TED method gets the lowest $P@1$ of 55.32 % and 54.79 %, because only the entity titles are used in this method. The IS method utilizes the article texts of entities, while the SSP method take advantage of the semantic information contained in the infoboxes. The two methods have better performances than TED because of utilizing more information of the wiki article. The proposed method can be regraded as the combination of the IS and the SSP. By combining the texture information and the structural information in a synergistic way, our method outperforms the IS and the SSP by 21.75 % and 12.28 % with regards to $P@1$, respectively. Compared with the LFG method, our method focuses on the knowledge linking task in specific domains. The experimental results show that our method outperforms the LFG (+4.68 % in terms of $P@1$ in average) with regards to both $P@1$ and $P@5$.

Table 2. Performance of knowledge linking with different methods (%).

Tasks	Metrics	Methods				
		TED	IS	SSP	LFG	Proposed
EWM-BBM	$P@1$	55.32	68.14	77.61	83.73	**89.89**
	$P@5$	62.91	75.53	86.56	88.21	**93.28**
EWM-ZWM	$P@1$	54.79	61.88	70.11	80.26	**83.51**
	$P@5$	61.53	67.03	80.35	82.29	**87.33**

6   Conclusion and Future Work

In this paper, we propose a cross-lingual knowledge linking approach for discovering domain-specific cross-lingual links across online wikis. Our approach combines both domain semantic relations and texture features of the wiki article, and employs a variation of similarity flooding to predict new cross-lingual links. Evaluations on two cross-lingual linking tasks show that our approach can outperform the state-of-the-art method by an average precision of 4.68 %. Our future work is to studying the automatical construction method for the attribute mapping, to extend our approach to a more general one.

References

1. Bizer, C., Lehmann, J., Kobilarov, G., Auer, S., Becker, C., Cyganiak, R., Hellmann, S.: Dbpedia - a crystallization point for the web of data. Web Semant. Sci. Serv. Agents World Wide Web **7**(3), 154–165 (2009)
2. Cudré-Mauroux, P., Haghani, P., Jost, M., Aberer, K., De Meer, H.: idMesh: graph-based disambiguation of linked data. In: Proceedings of WWW, pp. 591–600 (2009)
3. De Melo, G., Weikum, G.: MENTA: inducing multilingual taxonomies from wikipedia. In: Proceedings of CIKM, pp. 1099–1108 (2010)

4. Erdmann, M., Nakayama, K., Hara, T., Nishio, S.: Improving the extraction of bilingual terminology from wikipedia. Int. J. TOMM **5**(4), 31 (2009)
5. Fu, B., Brennan, R., O'Sullivan, D.: Cross-lingual ontology mapping – an investigation of the impact of machine translation. In: Gómez-Pérez, A., Yu, Y., Ding, Y. (eds.) ASWC 2009. LNCS, vol. 5926, pp. 1–15. Springer, Heidelberg (2009). doi:10.1007/978-3-642-10871-6_1
6. Hassan, S., Mihalcea, R.: Cross-lingual semantic relatedness using encyclopedic knowledge. In: Proceedings of EMNLP, pp. 1192–1201 (2009)
7. Li, J., Tang, J., Li, Y., Luo, Q.: RiMOM: a dynamic multistrategy ontology alignment framework. Int. J. of TKDE **21**(8), 1218–1232 (2009)
8. Melnik, S., Garcia-Molina, H., Rahm, E.: Similarity flooding: a versatile graph matching algorithm and its application to schema matching. In: Proceedings of ICDE, pp. 117–128 (2015)
9. Oh, J.H., Kawahara, D., Uchimoto, K., Kazama, J., Torisawa, K.: Enriching multilingual language resources by discovering missing cross-language links in wikipedia. In: Proceedings of WI-IAT, pp. 322–328 (2008)
10. Potthast, M., Stein, B., Anderka, M.: A wikipedia-based multilingual retrieval model. In: Macdonald, C., Ounis, I., Plachouras, V., Ruthven, I., White, R.W. (eds.) ECIR 2008. LNCS, vol. 4956, pp. 522–530. Springer, Heidelberg (2008). doi:10.1007/978-3-540-78646-7_51
11. Sorg, P., Cimiano, P.: Enriching the crosslingual link structure of wikipedia - a classification-based approach. In: Proceedings of the AAAI Workshop on Wikipedia and Artificial Intelligence (2008)
12. Volz, J., Bizer, C., Gaedke, M., Kobilarov, G.: Discovering and maintaining links on the web of data. In: Bernstein, A., Karger, D.R., Heath, T., Feigenbaum, L., Maynard, D., Motta, E., Thirunarayan, K. (eds.) ISWC 2009. LNCS, vol. 5823, pp. 650–665. Springer, Heidelberg (2009). doi:10.1007/978-3-642-04930-9_41
13. Wang, Z., Li, J., Tang, J.: Boosting cross-lingual knowledge linking via concept annotation. In: Proceedings of IJCAI, pp. 2733–2739 (2013)
14. Wang, Z., Li, J., Wang, Z., Tang, J.: Cross-lingual knowledge linking across wiki knowledge bases. In: Proceedings of WWW, pp. 459–468 (2012)
15. Wang, Z., Li, Z., Li, J., Tang, J., Pan, J.Z.: Transfer learning based cross-lingual knowledge extraction for wikipedia. In: Proceedings of ACL, pp. 641–650 (2013)
16. Wentland, W., Knopp, J., Silberer, C., Hartung, M.: Building a multilingual lexical resource for named entity disambiguation, translation and transliteration. In: Proceedings of ICLRE, pp. 3230–3237 (2008)

LSSL-SSD: Social Spammer Detection with Laplacian Score and Semi-supervised Learning

Wentao Li[1], Min Gao[2](✉), Wenge Rong[3], Junhao Wen[2],
Qingyu Xiong[2], and Bin Ling[4]

[1] Center for Quantum Computation and Intelligent Systems, Faculty of Engineering
and Information Technology, University of Technology Sydney, Ultimo, Australia
[2] School of Software Engineering, Chongqing University, Chongqing, China
gaomin@cqu.edu.cn
[3] School of Computer Science and Engineering, Beihang University, Beijing, China
[4] School of Engineering, University of Portsmouth, Portsmouth, UK

Abstract. The rapid development of social networks makes it easy for people to communicate online. However, social networks usually suffer from social spammers due to their openness. Spammers deliver information for economic purposes, and they pose threats to the security of social networks. To maintain the long-term running of online social networks, many detection methods are proposed. But current methods normally use high dimension features with supervised learning algorithms to find spammers, resulting in low detection performance. To solve this problem, in this paper, we first apply the Laplacian score method, which is an unsupervised feature selection method, to obtain useful features. Based on the selected features, the semi-supervised ensemble learning is then used to train the detection model. Experimental results on the Twitter dataset show the efficiency of our approach after feature selection. Moreover, the proposed method remains high detection performance in the face of limited labeled data.

Keywords: Social networks · Spammer detection · Laplacian score · Feature selection · Semi-supervised learning

1 Introduction

With the development of Web 2.0, online social networks have gained increasing attention [1]. As an open platform, social networks enable people to maintain social relationships and find common interests with each other online [2]. Though social networks bring great convenience to people, their open characteristics make them vulnerable to attacks issued by social spammers [3,4].

Social spammers refer to those people who inject false information (1.e. advertisements, pornography) into social networks for economic purposes [5]. Because social relationships normally represent certain kind of trust, social spammers

© Springer International Publishing AG 2016
F. Lehner and N. Fteimi (Eds.): KSEM 2016, LNAI 9983, pp. 439–450, 2016.
DOI: 10.1007/978-3-319-47650-6_35

pose more threats to social networks than other types of spammers [6]. For example, it indicates that advertising links in Twitter clicked by more than twice people than those in e-mails [7].

Social spammers bring economic losses to normal people and hinder the long-term development of social networks [8]. To alleviate the effect of social spammers, many notable works have been done. The purpose of these detection methods is to distinguish spammers from normal users [9]. According to the amount of needed labeled data, these methods can be classified into three categories: supervised methods that train a classifier based on features derived from relationship or content information [10], unsupervised methods that cluster users into different groups [11], and semi-supervised learning methods based on the label propagation process [12].

Among these methods, supervised methods need a large number of labeled data, which become impractical in the real-world situation because of the high cost of labeling. Unsupervised methods have low detection accuracy due to the lack of labels. In addition, they are susceptible to the interference of noise data. Existing semi-supervised methods make use of the random walk process to obtain users' credibility, but this process brings high time cost. Moreover, all these methods train models based on high dimension features because of the large scale of social networks, which reduces the detection performances.

To solve these problems, in this paper, we combine the unsupervised feature selection method and the semi-supervised ensemble learning method to get our detection method, which is called LSSL-SSD. More specifically, we first select features through their ability to maintain the local geometrical information in the original data space, or Laplacian score, without the use of label information. After selecting useful features, a semi-supervised random forest approach is used to train the detection model to make use of both labeled and unlabeled data.

Note that the feature selection process in LSSL-SSD is an unsupervised one, so it can be combined with the process of semi-supervised classification. Experiments on the Twitter data set show that the proposed method outperforms state-of-the-art methods in term of detection rates when the amount labeled data is limited. Moreover, the operation of feature extraction reduces the dimension of features, resulting in better generalization ability for spammer detection.

The next of this paper is organized as follows. In Sect. 2, we introduce some related work about social spammer detection and feature extraction. The description of our proposed method is shown in Sect. 3. Section 4 introduces experimental results and discussion. Finally is the conclusion and future work.

2   Related Work

In this section, we first introduce current research about social spammer detection. Then, background knowledge about feature selection is described.

2.1   Social Spammer Detection Methods

Due to the open nature of social networks, social spammers are able to inject false information and spread them through social networks [4]. To alleviative the harness brought by social spammers, increasing attentions have been paid to detect them [6]. According to the way of training detection models, three kinds of methods can be summarized.

Supervised Detection Methods. These methods mainly find features to distinguish spammers from normal users, then train classifiers based on these features. For example, Aggarwal etc. [10] detected spammers by using features from a user's registered information or content information. Lee etc. [13] did that by extracting features from behavior information crawled by honeypots. In [14], the authors proposed detection method based on social network structures. In [2], content information and structure information are used together to train models. Supervised methods performance well in detecting social spammers, but the need of labeled data makes them hard to work well in the real-world situation due to the high cost of labeling data [15].

Unsupervised Detection Methods. Supervised methods need a large number of labeled samples, so some researchers put forward unsupervised detection methods. These methods mainly find spammers by using social network topology. For example, in [11], similarities of text content and URLs are used to cluster users into different groups. The intuition behind this method is that spammers have fewer similarities with normal users in terms of content information. By contrast, Tan [16] first located normal users by social relationship graph, then detect spammers through relationships between different users.

Semi-supervised Detection Methods. Compared with supervised detection methods, unsupervised detection methods do not require manually labeled data. But the false positive rates of them are high due to the lack of labels, and their robustness is low when noisy data exist. In order to solve these problems, Li [12] proposed a semi-supervised detection framework based on trust propagation, which uses PageRank to propagate labels to find spammers. This method works well in practice, but the process of trust propagation needs high time cost.

   To summarize, all these methods rely on features extracted from user behavior or relationship information, resulting in high-dimension of features. Among them, semi-supervised methods are suitable for the real-world application while the time cost is high. To improve the detection effectiveness and accuracy, unsupervised feature selection method is used together with a novel semi-supervised learning method to form the proposed detection model.

2.2   Feature Selection Methods

The size of social networks is so large, which makes the dimension of features derived from them so high [4]. These high dimensional features reduce the detection performances of current methods. Feature selection methods are often used to remove useless features, thus improving the detection performance [17].

There are two types of feature selection methods, i.e. wrapper and filter methods. Wrapper methods are used with a particular learning algorithm, so these methods are limited to some specific learning tasks. Filtering methods make use of the intrinsic characteristics of data to evaluate features. Filtering methods generally require the relationship between features and labels, such as the use of Pearson correlation similarity or fisher score for feature selecting [18].

Here, we use feature selection rather than feature transformation because the latter will change the original feature space, thus reduce the diversity of features. Moreover, while little labeled data can be obtained, unsupervised feature selection method can provide suitable inputs for the next classification process.

3   Proposed Method

In this section, we introduce the proposed detection method (LSSL-SSD). We treat the problem of social spammer detection as a classification, that is to make a difference between normal users and spammers. Before the specific algorithm is given, the overall process of LSSL-SSD is explained in Fig. 1.

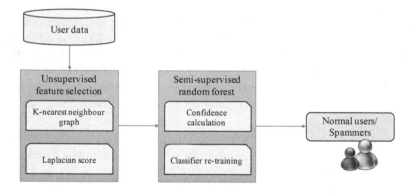

Fig. 1. The framework of LSSL-SSD

The whole process includes two main modules. The first one is to select features. Here we use unsupervised feature selection method, the reason is that we want to pick up useful features when the labels are lacking. The second module is to use the semi-supervised random forest learning based on the selected features to train a detection model. We will introduce the two modules in detail below.

3.1   Unsupervised Feature Selection Based on Laplacian Score

In social networks, each user has relationship information and content information. The large scale of social networks leads to a high dimension representation of users. Training a model on these high dimension data will result in low detection results [2]. Therefore, a feature selection method is useful for effectively spammer detection.

At the same time, getting enough labeled data needs high cost while traditional unsupervised and semi-supervised feature selection methods fail to remove redundant features for general tasks [4]. In order to get a better feature selection effect in the absence of labeled data, we apply the Laplacian score method that was proposed by Hu [17] for feature selection. This method is an unsupervised one, but it can achieve a fantastic effect as supervised ones.

The basic idea of the Laplacian score method is to use the feature's ability to maintain neighbor information as a selection standard. The intuition behind is that the discriminant effect of features represents in their local geometric relationships in the original data space. The key steps of this method include two steps, i.e. the construction of k-nearest neighbor graph and the calculation of the Laplacian score.

Construction of the K Nearest Neighbor Graph. To construct a neighborhood graph G, we need to calculate the similarity between user data in the original data space. Here we use Euclidean similarity as the basic measure. Assuming there are M users in social networks, denoted by $(x_1, x_2, \cdots, x_i, \cdots, x_M)$, where x_i is user i's feature vector. For user x_i, we find its k nearest neighbor set denoted by $n(x_i)$. Then, we add edge between x_i and users in $n(x_i)$. Note that the edge is directed, that is to say, the neighbor relationship is not symmetrical.

Calculation of the Laplacian Score. When the neighbor graph G is got, the Laplacian score of each feature can be calculated. Firstly, a weighted matrix S is constructed based on the graph G. S quantifies the local geometric relationships in the original data space and it is calculated according to Formula 1.

$$S_{ij} = \begin{cases} e^{-\frac{\|x_i - x_j\|^2}{t}}, & if\ e(x_i, x_j) \in G \\ 0, & otherwise \end{cases} \tag{1}$$

Here, t is the adaptive constant. When x_i and x_j exist an edge, then S_{ij} is obtained by the similarity between x_i and x_j, otherwise the value is 0. S is a weighted graph, the Laplacian matrix of S is $L = D - S$, where $D = diag(S)$ is the main diagonal matrix of S.

Then, assuming there are R features in total, for the r-th feature, values of M users on this features form a vector $f_r = [x_{1r}, x_{2r}, \cdots, x_{Mr}]$, and the Laplacian score of feature r can be calculated by Formula 2.

$$L_r = \frac{\widetilde{f_r^T} L \widetilde{f_r}}{\widetilde{f_r^T} D \widetilde{f_r}} \tag{2}$$

where L_r is the Laplacian score of feature r, L is the Laplacian matrix. The calculation of F_r^T is shown in Formula 3.

$$\widetilde{f_r^T} = f_r - \frac{f_r^T D1}{1^T D1} 1 \tag{3}$$

Finally, when the Laplacian score of each feature is calculated, these features can be sorted according to their scores. The features with high scores are selected. L_r does not use the label information but it has a good effect on the selection of useful features. Details of this method can be found in [17].

3.2 Semi-supervised Random Forest

After the feature selection process, the next step is to train a detector based on the selected features. To solve the problem of insufficient labeled data, an intuitive way is to make use of both labeled and unlabeled data. In this paper, we apply the semi-supervised random forest method. This method integrates ensemble learning and co-training to get the final detection model.

The semi-supervised random forest [19] first learns multiple basic classifiers on labeled data, and then unlabeled data are used to improve the performance of classifiers at each iteration. The whole process consists of two parts, namely, the confidence calculation process and the classifier re-training process.

Confidence Calculation. The whole training set in the system can be divided into labeled set L and unlabeled set U. L includes $|L|$ labeled data, denoted by $(x_1, y_1), (x_2, y_2), \cdots, (x_{|L|}, y_{|L|})$, where x_i represents user i's feature vector, y_i is user i' label. U includes $|U|$ unlabeled data, denoted by $x_1, x_2, \cdots, x_{|U|}$, the datum in this set is unlabeled.

The semi-supervised random forest method first applies resampling technology to get N data subsets from labeled set, which can be denoted by L_i, $(i = 1, 2, \cdots, N)$. And then N decision tree classifiers f_i can be trained on each subset. In order to make use of unlabeled data, the confidence of each unlabeled datum in U_i is calculated. For each base classifier, the unlabeled data with high confidence will then be moved into the corresponding labeled data subset L_i.

For a base classifier f_i, to get the confidence of unlabeled data, we use the prediction results of $N - 1$ classifiers except f_i. If there are two groups of users, i.e. normal users whose labels are -1 and spammers whose labels are 1, then the confidence of each datum x_i can be calculated by Formula 4.

$$con(x_i) = \max(\sum\nolimits_{f(x_i)=1} 1, \sum\nolimits_{f(x_i)=-1} 1) \qquad (4)$$

The first term $(f_i(x) = 1)$ means how many base classifiers predict x_i as spammers, the second term means how many base classifiers predict x_i as normal users. $con(x_i)$ reflects the consistency of classifiers to predict x_i. The data with top confidence values will be selected to moved from U_i into L_i, $(i = 1, 2, \cdots, N)$.

Classifier Re-training. When new data are added into labeled data set, N classifiers will be re-trained on the augmented labeled data. The process will continue until the output of N classifiers remains the same. When the process is over, N classifiers are obtained, and the label of new user data x is determined by voting, as shown in Formula 5.

$$f(x) = \begin{cases} 1 \ if \ \sum_{f_i(x)=1} 1 > \sum_{f_i(x)=-1} 1 \\ -1 \ if \ \sum_{f_i(x)=1} 1 < \sum_{f_i(x)=-1} 1 \end{cases} \qquad (5)$$

The label with most votes will be assigned to x, $i \in (1, 2, \cdots, N)$. If the votes are equal, the label can be assigned randomly. After the two stages, LSSL-SSD can be obtained. The process of the LSSL-SSD algorithm is shown in Table 1.

Table 1. The process of LSSL-SSD algorithm

Input:
Labeled data set L
Unlabeled data set U
M users, each represented by a R-dimension vector, $x_i = (x_{i1}, x_{i2}, \cdots, x_{iR})$
the number of nearest neighbor k
the number of selected features t
the number of base classifiers N
Output:
N classifiers $F = [f_1, f_2, \cdots, f_N]$

Steps:
1. Get the k nearest neighbor graph G of M users according to Formula 1.
2. Get the Laplacian scores of R features according to Formula 2.
3. Choose features with the top-t largest Laplacian scores.
4. Re-sample the original labeled data to get N subsets.
5. Train N initial base classifiers f_i based on each subset L_i, $i = 1, 2, \cdots, N$.
6. iterate until the output of N base classifiers remain the same.
    6.1 For each base classifier f_i.
    6.2 Calculate the confidence of each unlabeled datum according to Formula 4.
    6.3 Choose data with top confidence values and add them to L_i.
    6.4 Re-train f_i using updated L_i.
6. Output N classifiers, and predict each new-coming datum according to Formula 5.

---

4 Experiment Results

To analyze the performance of LSSL-SSD, in this section we conduct three groups of experiments on a real-world data set. The first one is to compare LSSL-SSD with related methods. The second one is to check the function of the feature selection process. The third one is about parametric sensitivity analysis.

4.1 Experiment Setup

Data Set. In this paper, we use the Twitter data set provided by Benevenuto [20]. This data set is collected since August 2009, which includes eight million users. 1065 users in this data set have been labeled, including 710 normal users and 355 spammers. Each user has 62 features, which are derived from behavior and content information. More details about this data set can be found in [20].

Evaluation Metrics. To evaluate the accuracy of the proposed method, Precision, Recall and F1-measure are used as evaluation metrics. We denote N_a as the number of spammers who are correctly detected, N as the number of spammers predicted by the algorithm, and N_t as the number of spammers in the systems. Precision, Recall and F1-measure are calculated by Formulas 6–8.

$$Precision = \frac{N_a}{N} \qquad (6)$$

$$Recall = \frac{N_a}{N_t} \qquad (7)$$

$$F1 - measure = \frac{2 \times Precision \times Recall}{Precision + Recall} \qquad (8)$$

Precision is the ratio between the number of correctly predicted spammers and those who are predicted as spammers. Recall is the ratio between the number of correctly predicted spammers and the total number of spammers. F1-measure is the weighted average of precision and recall. The range of both three metrics is 0 to 1 and the best value is 1 while the worst one is 0.

Experimental Settings. LSSL-SSD is a semi-supervised algorithm. We hope our method outperforms supervised methods in terms of detection accuracy. For fair comparison, we use a common subset for testing. The size of this public testing set is 20 % of all the data. For supervised algorithms, the remaining 80 % data is used for training. For the semi-supervised algorithm, we divided the training set into the labeled set L and unlabeled set U. The experiment was conducted 100 times, and the average results are used to report results.

4.2 Experimental Results and Discussion

Comparison of Detection Performances Between LSSL-SSD and Supervised Methods. To show that the proposed method has better detection performance, we compare LSSL-SSD with traditional methods. Naive Bayes, decision tree, logistic regression, support vector machine (SVM) and random forest are used to compare. We change the size of labeled training data from 10 %, 20 % to 40 %. Here 10 % means 10 % of the original data is labeled training data. Comparison results are shown in Fig. 2.

From Fig. 2, it can be found that the F1-measure values of LSSL-SSD are higher than those of all supervised algorithms. In addition, with the increase in the labeled training data size, F1-measure of LSSL-SSD becomes better. Even when only 10 % labeled data is obtained, LSSL-SSD outperforms other methods.

In terms of precision, the supervised random forest has a good performance. When the amount of labeled data is small, LSSL-SSD does not perform well, but with the increase in the labeled data size, prediction becomes good. In terms of recall, the decision tree has a good performance. But recall rates of LSSL-SSD are better than those of other algorithms.

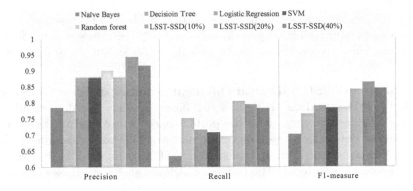

Fig. 2. Comparison of detection performance between LSSL-SSD and others

From the above results, it can be concluded that LSSL-SSD has better detection performances than traditional methods. Moreover, only a small fraction of labeled data is used to get our model, which means little labeling cost. This justifies the real-world value of our proposed method.

Performance of LSSL-SSD with Different Number of Features. The first step of LSSL-SSD is to use the Laplacian score method to select features. To verify the effectiveness of this process, we vary the number of selected features from 10 to all (62) for analysis. Supervised random forest is used as a comparison. Here 10 % labeled training data is used and the number of base classifiers changes from 3 to 100 to report results. The results are shown in Fig. 3.

As seen from Fig. 3, LSSL-SSD achieves better detection performances than random forest in all settings. Also, we can observe that with the increase in the number of selected features, F1-measure of LSSL-SSD first increases then

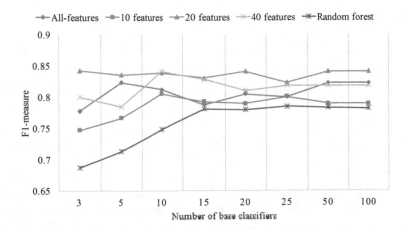

Fig. 3. Comparison of detection performance between different number of features

decreases. When the number is 20, F1-measure is the best. This result shows the importance of feature selection process because many features are redundant. In practice, we can use cross-validation to find the suitable number of features.

Performance of LSSL-SSD with Different Number of Base Classifiers. Since LSSL-SSD uses decision tree as the base classifier, the number of base classifiers may have an impact on the performance. So we discuss the effect of the number of base classifiers on detection accuracy. We vary this number from 3 to 100. We change the labeled data size of LSSL-SSD from 10%, 20% to 40% of the original data to report the results, which are shown in Tables 2, 3 and 4 respectively.

Table 2 shows the case of 10% labeled training data. With the increase in the number of base classifiers, precision increases gradually, recall remains the same and F1-measure shows some fluctuates.

Results in Tables 3 and 4 show some similarities. The results mean that the number of base classifiers has little impact on the detection performance. The possible reason is that LSSL-SSD needs the assumption of Classifier diversity. Therefore, when the number of base classifiers is in a certain range, this condition is satisfied and the detection effect will have no big change.

To summarize, LSSL-SSD has a perfect detection performance in detecting social spammers compared with traditional methods. Even when limited labeled

Table 2. Performance of LSSL-SSD with different number of base classifiers (10%)

	3	5	10	15	20	25	50	100
Precision	**0.9142**	0.86884	0.8684	0.8571	0.88	0.8552	0.88	0.88
Recall	0.7804	**0.8048**	**0.8048**	**0.8048**	**0.8048**	0.7926	**0.8048**	**0.8048**
F1-measure	**0.8421**	0.8354	0.8385	0.8301	0.8407	0.8227	0.8407	0.8407

Table 3. Performance of LSSL-SSD with different number of base classifiers (20%)

	3	5	10	15	20	25	50	100
Precision	0.8533	0.8684	0.8783	**0.9027**	0.8904	0.8767	0.8918	0.88
Recall	0.771	**0.7951**	0.7831	0.7831	0.7831	0.771	**0.7951**	**0.7951**
F1-measure	0.8101	0.8301	0.828	0.8387	0.8333	0.8205	**0.8407**	0.8354

Table 4. Performance of LSSL-SSD with different number of base classifiers (40%)

	3	5	10	15	20	25	50	100
Precision	0.8552	0.8918	0.9041	0.8918	**0.9166**	0.9041	**0.9166**	0.9041
Recall	0.7831	**0.7951**	**0.7951**	**0.7951**	**0.7951**	**0.7951**	**0.7951**	**0.7951**
F1-measure	0.8176	0.8407	0.8461	0.8407	**0.8516**	0.8461	**0.8516**	0.8461

data is available, LSSL-SSD performs well. Moreover, as a common method, LSSL-SSD can be applied for other social networks such as Facebook.

5   Conclusion and Future Work

The open nature of social networks makes them vulnerable to social spammers. To fight against social spammers, in this paper, we proposed a novel method, LSSL-SSD, for social spammer detection. It first calculates the Laplacian score of each feature for feature selection. Then, based on these selected features, the semi-supervised random forest method is used to get the final detection model. Experimental results show that the proposed method not only has a strong generalization ability due to the process of feature selection, but also has a good detection accuracy in the face of limited labeled data.

As further work, we will incorporate other information, such as the labels of data, to select features. This may improve the detection accuracy because we only make use of local information in this paper. Moreover, for the problem of limited labeled data, active learning can be used together with semi-supervised learning to improve the detection performance.

Acknowledgments. This work is supported by the Basic and Advanced Research Projects in Chongqing under Grant No. cstc2015jcyjA40049, the National Key Basic Research Program of China (973) under Grant No. 2013CB328903, the National Natural Science Foundation of China under Grant Nos. 61472021 and 61602070, the Fundamental Research Fund for the Central Universities under Grant No. 106112014CDJZR095502, and the China Scholarship Council.

References

1. Borge-Holthoefer, J., Rivero, A., Moreno, Y.: Locating privileged spreaders on an online social network. Phys. Rev. E **85**(6), 066123 (2012)
2. Hu, X., Tang, X., Liu, H.: Online social spammer detection. In: AAAI, pp. 59–65 (2014)
3. Guille, A., Hacid, H., Favre, C., Zighed, D.A.: Information diffusion in online social networks: a survey. ACM SIGMOD Rec. **42**(2), 17–28 (2013)
4. Wu, F., Shu, J., Huang, Y., Yuan, Z.: Social spammer and spam message co-detection in microblogging with social context regularization. In: Proceedings of the 24th ACM International on Conference on Information and Knowledge Management, pp. 1601–1610. ACM (2015)
5. Hu, X., Tang, J., Gao, H., Liu, H.: Social spammer detection with sentiment information. In: 2014 IEEE International Conference on Data Mining, pp. 180–189. IEEE (2014)
6. Zhu, X., Nie, Y., Jin, S., Li, A., Jia, Y.: Spammer detection on online social networks based on logistic regression. In: Xiao, X., Zhang, Z. (eds.) WAIM 2015. LNCS, vol. 9391, pp. 29–40. Springer, Heidelberg (2015). doi:10.1007/978-3-319-23531-8_3
7. Heymann, P., Koutrika, G., García-Molina, H.: Fighting spam on social web sites: a survey of approaches and future challenges. IEEE Internet Comput. **11**(6), 36–45 (2007)

8. Stringhini, G., Kruegel, C., Vigna, G.: Detecting spammers on social networks. In: Proceedings of the 26th Annual Computer Security Applications Conference, pp. 1–9. ACM (2010)
9. Ren, Y., Ji, D., Yin, L., Zhang, H.: Finding deceptive opinion spam by correcting the mislabeled instances. Chin. J. Electron. **24**(1), 52–57 (2015)
10. Aggarwal, A., Almeida, J., Kumaraguru, P.: Detection of spam tipping behaviour on foursquare. In: Proceedings of the 22nd International Conference on World Wide Web, pp. 641–648. ACM (2013)
11. Gao, H., Hu, J., Wilson, C., Li, Z., Chen, Y., Zhao, B.Y.: Detecting and characterizing social spam campaigns. In: Proceedings of the 10th ACM SIGCOMM Conference on Internet Measurement, pp. 35–47. ACM (2010)
12. Li, Z., Zhang, X., Shen, H., Liang, W., He, Z.: A semi-supervised framework for social spammer detection. In: Cao, T., Lim, E.-P., Zhou, Z.-H., Ho, T.-B., Cheung, D., Motoda, H. (eds.) PAKDD 2015. LNCS (LNAI), vol. 9078, pp. 177–188. Springer, Heidelberg (2015). doi:10.1007/978-3-319-18032-8_14
13. Lee, K., Caverlee, J., Webb, S.: Uncovering social spammers: social honeypots + machine learning. In: Proceedings of the 33rd International ACM SIGIR Conference on Research and Development in Information Retrieval, pp. 435–442. ACM (2010)
14. Song, J., Lee, S., Kim, J.: Spam filtering in twitter using sender-receiver relationship. In: Sommer, R., Balzarotti, D., Maier, G. (eds.) RAID 2011. LNCS, vol. 6961, pp. 301–317. Springer, Heidelberg (2011). doi:10.1007/978-3-642-23644-0_16
15. Zhang, Y., Jianguo, L.: Discover millions of fake followers in Weibo. Soc. Netw. Anal. Min. **6**(1), 1–15 (2016)
16. Tan, E., Guo, L., Chen, S., Zhang, X., Zhao, Y.: UNIK: unsupervised social network spam detection. In: Proceedings of the 22nd ACM International Conference on Information and Knowledge Management, pp. 479–488. ACM (2013)
17. He, X., Cai, D., Niyogi, P.: Laplacian score for feature selection. In: Advances in Neural Information Processing Systems, pp. 507–514 (2005)
18. Guyon, I., Elisseeff, A.: An introduction to variable, feature selection. J. Mach. Learn. Res. **3**, 1157–1182 (2003)
19. Li, M., Zhou, Z.-H.: Improve computer-aided diagnosis with machine learning techniques using undiagnosed samples. IEEE Trans. Syst. Man Cybern. Part A Syst. Hum. **37**(6), 1088–1098 (2007)
20. Benevenuto, F., Magno, G., Rodrigues, T., Almeida, V.: Detecting spammers on Twitter. In: Collaboration, Electronic Messaging, Anti-abuse and Spam Conference (CEAS), vol. 6, p. 12 (2010)

Knowledge Systems and Security

i-Shield: A System to Protect the Security of Your Smartphone

Zhuolong Yu$^{(\boxtimes)}$, Liusheng Huang, Hansong Guo, and Hongli Xu

University of Science and Technology of China, Hefei, China
{yzl123,guohanso}@mail.ustc.edu.cn, {lshuang,xuhongli}@ustc.edu.cn

Abstract. Losing smartphones is a troublesome thing as smartphones
are playing an important role in our daily lives. As smartwatches become
popular, we argue that smartwatches can play a role in smartphone
antitheft design. In this paper, we propose i-Shield, a real-time antitheft
system that leverages accelerometers and gyroscopes of smartphones and
smartwatches to prevent smartphone being stolen. As opposed to existing
solutions which are based on Bluetooth, NFC, or GPS tracking, i-Shield
follows a practical manner to achieve the goal of real-time antitheft for
smartphones. i-Shield recognizes taken-out events of smartphones using a
supervised classifier, and applies a dynamic time warping (DTW) scheme
to recognize whether the events are caused by users themselves. We conduct a series of experiments on iPhone6 and iPhone4s, and the evaluation results show that our system can achieve 97.4 % true positive rate of
recognizing taken-out actions, and classify taken-out actions with misclassification rate of 1.12 %.

Keywords: Antitheft · Smartphone · Smart wearable device · Supervised classification · Dynamic time warping

1 Introduction

Smartphones are now becoming very ubiquitous and extremely important in our
daily lives. As smartphones always store lots of important personal information
(even including credit card information), the security of smartphones is receiving more and more attention. A survey of smartphone theft victims conducted
by IDG Research [3] shows that 1 in 10 U.S. smartphone owners are victims of
phone theft, whose amount is 2.1 million in 2014. Smartphone producers have
also brought out some solutions to smartphone theft [1,2,14], such as "Find
My iPhone" of Apple and "Android Device Manager" of Android. Nonethe-
less, these approaches including password mechanism and fingerprint identification are effective only after the crime. Besides, Bluetooth technology is also
employed for antitheft design, in order to detect the crimes when they are happening. But since Bluetooth can detect crimes only after criminals get about 6 m
away (can be too late), it is not very effective in this scenario. Meanwhile, the
majority of smartphones and smartwatches contain abundant advanced built-
in sensors to sense users' motions. Sensor-based activity recognition has been

© Springer International Publishing AG 2016
F. Lehner and N. Fteimi (Eds.): KSEM 2016, LNAI 9983, pp. 453–464, 2016.
DOI: 10.1007/978-3-319-47650-6_36

studied extensively, especially for smartphones [12,13]. These researches focus on multiple fields, including health care [4,8,12], localization [11] and human computer interaction [6], but the field of smartphone antitheft has been paid little attention.

In this paper, we focus on the theft problem of smartphones. We propose i-Shield, a system employing sensors on smartphones and smartwatches to guard smartphones from being stolen. i-Shield can recognize taken-out actions of smartphones and check the sensor data of both smartphones and smartwatches to judge if the actions are caused by user themselves. The smartwatches will alarm users when i-Shield finds that the taken-out actions with high probability are not caused by users themselves.

The rest of this paper is organized as follows. In Sect. 2, we provide a system overview of our i-Shield system. Section 3 introduces the taken-out actions recognition part of i-Shield system, and Sect. 4 introduces how we judge whether the actions are secure. Section 5 reports the evaluation of our i-Shield system. We present the related works in Sect. 6, and give a discussion in Sect. 7.

2   System Overview

In this section, we give a brief introduction of i-Shield, our proposed smartphone antitheft system. Our system includes two parts which are based on smartphone and smartwatch respectively.

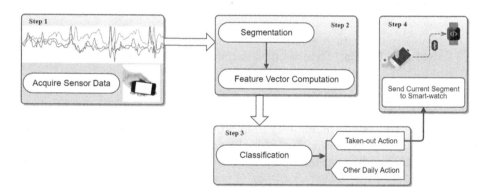

Fig. 1. System overview of i-Shield on smartphone side.

At the smartphone side, i-Shield firstly acquires the acceleration data and rotation-rate data from accelerometers and gyroscopes of smartphones, which are organized in the form of triples (x, y, z) with corresponding timestamps respectively. In the second step, we extract segments from the sensor data time series we obtain in the former step. Then, we compute a feature vector for each segment, which consists discriminative cues both in time-domain and frequency-domain. In the third step, i-Shield constructs a classifier on smartphone, it tells if there

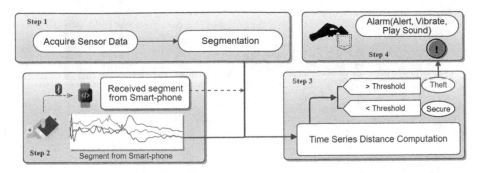

Fig. 2. System overview of i-Shield on smartwatch side.

is a Taken-out Action. When a Taken-out Action is recognized, smartphone side i-Shield will send current data segment to smartwatch side (Step 4) since there is a potential risk that smartphone is being stolen (Fig. 1).

Meanwhile, i-Shield does the same acquisition job and segmentation job at smartwatch side. Right after receiving data segment from smartphone side, i-Shield computes distance between two series of sensor data sequences from smartphone and smartwatch. Based on the distance measured, i-Shield judges if the taken-out actions are with high probability caused by users themselves. At last, i-Shield will arouse an alarm event (alert, vibrate, or play sound) when a risky action is detected. The user can choose to turn the alarm off manually (Fig. 2).

3   Recognize Taken-Out Actions

In this section, we specifically describe the recognition process i-Shield implement on smartphone side. The recognition process mainly needs three steps to recognize an action, which are Segmentation, Feature Vector Computation and Classification respectively. Firstly, i-Shield divides sensor data time series into segments, then feature vectors are computed based on segments. After feature vectors computed, classification algorithms are applied, segments can be classified into specific actions.

3.1   Segmentation

To detect taken-out actions precisely and efficiently, sensor data time series are divided into segments of appropriate length.

Figure 3 plots the waveform of accelerometer data generated by 6 min of daily activities including walking, going upstairs and downstairs, and having short rest. As shown in Fig. 3, we extract two segments from the original data. Segment A shows a intense wave (actually, taking out phone while walking), while Segment B shows a flat wave (having a short rest).

In this paper, we focus on potentially dangerous actions, namely Taken-out Actions, which will cause waveforms of a certain shape, e.g. waveform of

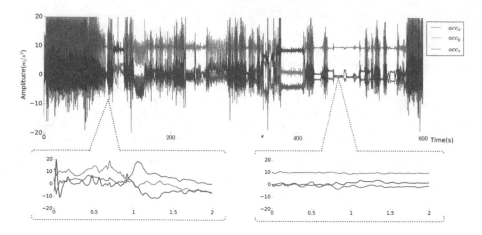

Fig. 3. Waveform of accelerometer data.

Segment A. In the meantime, Segment B indicates an almost dormant state that we regard as a safe action. Accordingly, we do not need to purchase all available segments, but only the segments containing waveforms at least of a certain intensity.

Sliding Windows [15] and End-Points Detection [7] are two of the most popular segmentation methods. As Sliding Window-based segmentation algorithm will collect massive segments including ones we will not need, and cause a enormous consumption of computing resources, memories and energy. We apply End-Point Detection-based algorithm in our system, since it can extract specific segments we are interested in.

3.2 Feature Vector Computation

For later use of classification, we compute a set of features for each segment, which are organized as a feature vector. In an attentive manner, we choose 24 features for each segment, which can be divided into three categories: Time Domain Features, Frequency Domain Features, and Statistics Features respectively. These features are listed below in Table 1 (For simplicity, we use Acc. to denote acceleration value of 3 axes, and Rtr. to denote rotation-rate value).

- **Time Domain Features.** Time Domain Features are intuitional and they can be obtained with low computational complexity.

    We extract minimum, maximum and mean value of acceleration data and rotation-rate data on 3 axes. These values can describe approximate shape of sampled data, and have been exploited generally.
- **Frequency Domain Features.** In order to describe periodic characteristics of sampled data, we leverage Fast Fourier Transform (FFT) to transform our time series segment into frequency domain.

    We extract **Peak Amplitude**, **Peak Frequency** and **Spectral Slope** of Acc.Z.

Table 1. Feature set for each sensor data segment

Time Domain Features	
Accelerometer:	$min(Acc.X)$, $max(Acc.X)$, $mean(Abs(Acc.X))$, $min(Acc.Y)$, $max(Acc.Y)$, $mean(Abs(Acc.Y))$, $min(Acc.Z)$, $max(Acc.Z)$, $mean(Abs(Acc.Z))$, $mean(\sqrt{Acc.X^2 + Acc.Y^2 + Acc.Z^2})$
Gyroscope:	$min(Rtr.X)$, $max(Rtr.X)$, $mean(Abs(Rtr.X))$, $min(Rtr.Y)$, $max(Rtr.Y)$, $mean(Abs(Rtr.Y))$, $min(Rtr.Z)$, $max(Rtr.Z)$, $mean(Abs(Rtr.Z))$, $mean(\sqrt{Rtr.X^2 + Rtr.Y^2 + Rtr.Y^2})$
Frequency Domain Features	
Accelerometer:	$PeakAmplitude(Acc.Z)$, $PeakFrequency(Acc.Z)$, $SpectralSlope(Acc.Z)$
Statistics Features	
Accelerometer:	$Kurtosis(Acc.Z)$

- **Statistics Features.** We calculate **Kurtosis** of acceleration value on Z-axis. This feature weighs how the amplitude decays near the extreme points, namely the peakedness and flatness. The Kurtosis of accelerometer value on Z-axis is calculated as:

$$Kurtosis_i = \frac{n \sum\limits_{j=1}^{n} (z_j - \bar{z})^4}{(\sum\limits_{j=1}^{n} (z_j - \bar{z})^2)^2}$$

where z_j indicates the value of the j-th sampling point in segment $Acc.Z$ and $\bar{z}$ indicates the mean value of all sampling points in segment $Acc.Z$. n is the length of segment $Acc.Z$.

3.3 Classification

We divide all actions into three categories: taken-out when user is still, taken-out when user is walking, other daily activities. We mark the three categories as TOS, TOW, and OTHER respectively. Note that OTHER includes all actions of daily activities except for taking smartphone out of pocket, such as walking, jogging, riding and so on (Table 2).

We construct four classifiers based on Hoeffding Tree, Logistic, Naive Bayes, and Random Forest respectively. The result of classification is shown in Sect. 5.1. When an action is recognized as a Taken-out Action (TOS or TOW), i-Shield will send the current segment to smartwatch right away through Bluetooth.

4   Check If Actions Are Secure

In this section, we describe how i-Shield works at smartwatch side. i-Shield will continuously acquire acceleration and rotation-rate data of smartwatch, and

Table 2. Three categories of actions.

Catagory	Actions
TOS	taken-out when user is still
TOW	taken-out when user is walking
OTHER	still, walking, jogging, riding, sitting down, standing up, going upstairs and downstairs, working out at the gym, etc. (phone in hand and phone in pocket respectively)

keep a constant length log using a circular queue. When a data segment comes from smartphone through Bluetooth, the time-series distance computation procedure will be activated. Depending on the distance of time-series, i-Shield judges whether the action is caused by user, if yes, the action is secure. At last, i-Shield will alarm if the action is insecure. We apply dynamic time warping (DTW) algorithm to accomplish the task of time-series distance computation.

Dynamic time warping (DTW) has been wildly used in the field of speech recognition, signature recognition, shape matching and etc. DTW measures similarity between two temporal sequences that may vary in time or speed. By applying DTW, we don't need to worry difference in both absolute-time (always different between devices) and sample-rate between smartwatches and smartphones.

Given two time series $S = [s_1, s_2, ..., s_n]$ and $T = [t_1, t_2, ..., t_m]$, let $Dist[i, j]$ denotes the distance between symbol s_i and t_j, that is

$$Dist[i, j] = (s_1.X - t_1.X)^2 + (s_2.Y - t_2.Y)^2 + (s_3.Z - t_3.Z)^2 \qquad (1)$$

Following DTW, we define $F[i, j]$ which satisfies:

$$F[i, j] = Dist[i, j] + minimum(F[i - 1, j - 1], F[i, j - 1], F[i - 1, j]) \qquad (2)$$

Finally, we judge if the actions are caused by users themselves in terms of $F[n, m]$ of acceleration and rotation-rate. When $F[n, m]$ is below the threshold line, we judge that action is caused by user safely. The detail figure is shown in Figs. 7 and 8 in Sect. 5.2.

Fig. 4. Coordinate system of smartwatch and smartphone are opposite.

Note that in most cases, coordinate system of smartwatch and smartphone will not be the same. We consider the most natural way shows in Fig. 4, that the screens of watches and phones should be facing users (not hands or wrists), top and bottom should not be upside down. In i-Shield, we conform smartwatch's X, Y, Z-axis value (acceleration and rotation-rate) to smartphone's coordinate system. We denote before-conform X, Y, Z-axis value of smartwatch as (x, y, z), and after-conform value as $(\tilde{x}, \tilde{y}, \tilde{z})$, we have:

$$(\tilde{x}, \tilde{y}, \tilde{z}) = (x, y, z) \begin{pmatrix} 0 & -1 & 0 \\ -1 & 0 & 0 \\ 0 & 0 & -1 \end{pmatrix} \tag{3}$$

5   Evaluation

In order to ease the development process, we make use of a smartphone tied to a user's wrist as a substitute for a smartwatch. We packed i-Shield's functionality of smartphone side and smartwatch side together as an application, and implemented our system on iOS9.

We evaluate our proposed system using an iPhone6 Plus as our to-be-protected smartphone, an iPhone4s (small enough to be tied on a wrist) as our smartwatch simulator. Our evaluation has three parts. We start with evaluating how well we recognize taken-out actions on smartphones. Then, we report the result of smartwatch side actions checking. Finally, a real world evaluation is presented.

5.1   Recognizing Taken-out Actions

Our dataset consists of 511 TOS actions, 391 TOW actions and 35756 OTHER segments, collected by 6 volunteers for over 2 days. We construct six different classifiers, respectively based on Hoeffding Tree, Logistic, Naive Bayes, Random Forest, k-Nearest Neighbors, and Multilayer Perceptron. We conduct a series of 10-fold cross-validation experiments, and the results are shown in Table. 3, Figs. 5 and 6. Table. 3 presents the confusion matrixes of the six classifiers. Figure 5 illustrates two histograms of true positive rate (TPR) and false positive rate (FPR). Figure 6 reports time consumption of the six classifiers on our dataset. Table. 3 and Fig. 5 indicates that, k-Nearest Neighbors achieves the largest true positive rate of taken-out actions (TOS and TOW) which guarantees reliable safety for smartphones. While Random Forest will barely cause a false alarm on smartwatch as it outputs only 3 times of false positive recognition of taken-out actions which is the least among the six classifiers. In the meanwhile, the true positive rate of Random Forest to recognize taken-out actions is as high as 97.4 %. As presented in Fig. 6, Multilayer Perceptron and k-Nearest Neighbors both consume much more time than other four classifiers. Even though k-Nearest Neighbors obtains the best accuracy on recognizing taken-out actions, its total time consumption is 104.95 s that is 12 times as large as Random Forest consumes (8.78 s). Note that Random Forest consumes most of its total time on training part (7.97 s), while test time is only 0.81 s on our dataset (36658 items in total).

Table 3. Time consumption and Confusion matrixes of six classifiers. Note that a, b, c stands for TOS, TOW and OTHER respectively.

Classified as:	Hoeffding Tree			Logistic			Naive Bayes		
	a	b	c	a	b	c	a	b	c
a	478	0	33	494	2	15	478	0	33
b	6	380	5	23	361	7	7	382	2
c	32	1	35723	26	23	35707	32	1	35723

Classified as:	Random Forest			KN-Neighbors			ML-Perceptron		
	a	b	c	a	b	c	a	b	c
a	490	0	21	509	0	2	480	0	31
b	1	388	2	3	388	0	0	391	0
c	0	3	35753	6	1	35749	4	1	35751

Since our system runs training process only once during system initialization, Random Forest is high-efficient in our system. So we implement Random Forest in our proposed system, it achieves approximately 0.0 % of false positive rate of recognizing taken-out actions, and the true positive rate is 97.4 %.

5.2 Check If Actions Are Secure

In this section, we present the evaluation of i-Shield on smartwatch side. We collected 100 time-series pairs of phones and watches which generated when phones were taken out by users themselves, and 200 time-series pairs generated when phones were taken out by others. Here we define the former 100 time-series pairs as Safe Actions, the latter 200 time-series pairs as Unsafe Actions. We obtained distances of acceleration & rotation-rate for both Safe Actions and Unsafe Actions using DTW mentioned in Sect. 4. Figure 7 shows the distribution of Safe Actions and Unsafe Actions in terms of the distances we got.

As shown in Fig. 7, Safe Actions aggregate at bottom-left corner, which means Safe Actions cause very small-scale distances compared to Unsafe Actions. We

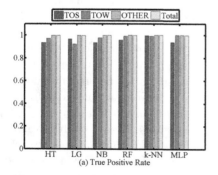
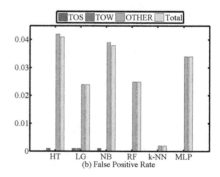

Fig. 5. TP rate (left) and FP rate (right) of six classifiers.

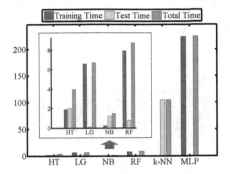

Fig. 6. Time consumption of six classifiers.

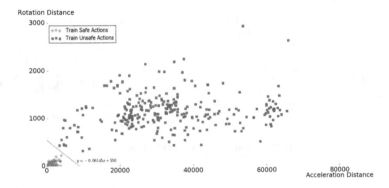

Fig. 7. Distribution of Safe Actions and Unsafe Actions of training set.

graph line $y = -0.06145x + 550$ using a linear classifier. Time-series pairs whose distances appear below $y = -0.06145x + 550$ are considered as Safe Actions, others are considered as Unsafe Actions.

We then calculated distances of other 178 Safe Actions and 424 Unsafe Actions using DTW as a test set and plot the result in Fig. 8. It's shown that only two of the Safe Actions are recognized as Unsafe Actions with misclassification rate of 1.12 %, while no Unsafe Actions are misclassified. Commonly speaking, there may be false alarms with very small probability, but will seldom be any theft situations regarded as safe situations.

5.3   Real World Evaluation

To study how well would i-Shield work in real world situation, we conducted a real world evaluation with eight volunteers. They are all smartphone users, and five of them are smartwatch users.

In the study, we let four of them act as users, carrying smartphone in the pocket and "smartwatch" on the wrist, while other four act as thieves trying to steal users' smartphones. We conducted 120 theft actions, 80 safe taken-out

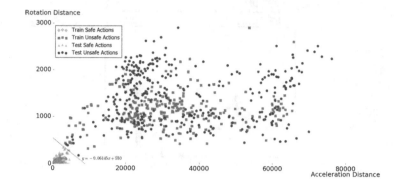

Fig. 8. Distribution of Safe Actions and Unsafe Actions.

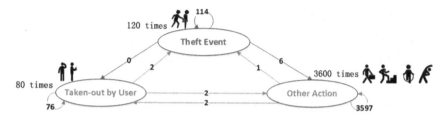

Fig. 9. Result of real world evaluation.

actions and 3600 other daily action segments among these four user-thief pairs. As shown in Fig. 9, 114 of theft actions were recognized correctly with other 6 are missed by i-Shield. False alarm appeared 3 times as two safe actions and one other daily action were recognized as theft action.

After the study, a volunteer who had once been a victim of smartphone theft said, "This system works so sensitively and naturally, it's just like an invisible shield of my cell phone!"

6   Related Work

Smartphone related works have been studied extensively in recent years. Great progress has been made in various fields about smartphone including but not limited to motion detection [4,12,13], location based service [11], security [10], energy consumption saving [5]. In this section, we describe two areas of related work.

Motion sensor related works. Motion sensor has been studied a lot, especially after smartphones appear. As for smartphones, Thompson et al. [12] proposed a system detecting car accidents using sensors of smartphones. Abbate et al. [4] proposed a smartphone-based fall detection system with concern for health of elderly people. As for smartwatches, Gouthaman et al. [6] designed a system to control computers using accelerometer, gyroscope and gravitational sensors of smartwatches. Parate et al. [8] proposed a trajectory-based method that extracts hand-to-mouth gestures to recognize smoking gestures. In addition, various commercial applications based on motion detection are available on both smartphones and smartwatches, and provide helpful and delightful functionalities, such as step counting and sleep tracking. However, motion detection hasn't been applied to the field of smartphone antitheft.

Mobile phone antitheft related works. Mobile phone antitheft design has been studied for many years, since mobile phone theft becomes a serious problem. Whitehead et al. [14] gave a review of mobile phone antitheft designs. They reviewed a great many antitheft designs, but most of them are traditional ways, which are not very practical nowadays. Ren et al. [9] proposed a model that PC and smartphone would form a loop to track each other. Yu et al. [16] leveraged emergency call mechanism to achieve remote deletion on stolen phones. These models works only after our devices are missing, in which condition the devices are probably turned off by criminals and not trackable.

7   Discussion

We plan to extend our current system to support other situations with more comprehensive functions.

Location based. We are planning to take location information into consideration to make our i-Shield system more intelligent. For example, crowded places like subway stations can be considered as high-risky places, i-Shield needs to lower the threshold line (be more sensitive), to guarantee invulnerably safety. Likewise, i-Shield needs to lower its sampling rate to reduce energy consumption when at home.

Smart wallets. Our work can be extended to the field of smart wearable devices. Wallets which are as popular as smartphones in crime of theft, can be protected like the way smartphones are, after embedded with sensor-based computation chips.

Acknowledgements. This paper is supported by the National Science Foundation of China under No. U1301256 and 51274202, Special Project on IoT of China NDRC (2012-2766).

References

1. Apple. http://www.apple.com/icloud/find-my-iphone.html
2. Google. https://www.google.com/android/devicemanager
3. Lookout. https://www.lookout.com/resources/reports
4. Abbate, S., Avvenuti, M., Bonatesta, F., Cola, G., Corsini, P., Vecchio, A.: A smartphone-based fall detection system. Pervasive Mob. Comput. 8(6), 883–899 (2012)

5. Chen, X., Jindal, A., Ding, N., Hu, Y.C., Gupta, M., Vannithamby, R.: Smartphone background activities in the wild: origin, energy drain, and optimization. In: Proceedings of the 21st Annual International Conference on Mobile Computing and Networking, pp. 40–52. ACM (2015)

6. Gouthaman, S., Pandya, A., Karande, O., Kalbande, D.: Gesture detection system using smart watch based motion sensors. In: 2014 International Conference on Circuits, Systems, Communication and Information Technology Applications (CSCITA), pp. 311–316. IEEE (2014)

7. Gu, T., Chen, S., Tao, X., Lu, J.: An unsupervised approach to activity recognition and segmentation based on object-use fingerprints. Data Knowl. Eng. **69**(6), 533–544 (2010)

8. Parate, A., Chiu, M.C., Chadowitz, C., Ganesan, D., Kalogerakis, E.: RisQ: recognizing smoking gestures with inertial sensors on a wristband. In: Proceedings of the 12th annual international conference on Mobile systems, applications, and services, pp. 149–161. ACM (2014)

9. Ren, B., Sun, Y., Lin, Y.: Anti-theft and tracking loop model based on PC and smart phone. In: 2013 Fifth International Conference on Computational and Information Sciences (ICCIS), pp. 1943–1946. IEEE (2013)

10. Shao, J., Lu, R., Lin, X.: Fine: a fine-grained privacy-preserving location-based service framework for mobile devices. In: IEEE Conference on Computer Communications, IEEE INFOCOM 2014, pp. 244–252. IEEE (2014)

11. Shu, Y., Shin, K.G., He, T., Chen, J.: Last-mile navigation using smartphones. In: Proceedings of the 21st Annual International Conference on Mobile Computing and Networking. pp. 512–524. ACM (2015)

12. Thompson, C., White, J., Dougherty, B., Albright, A., Schmidt, D.C.: Using smartphones to detect car accidents and provide situational awareness to emergency responders. In: Cai, Y., Magedanz, T., Li, M., Xia, J., Giannelli, C. (eds.) MOBILWARE 2010. LNICST, vol. 48, pp. 29–42. Springer, Heidelberg (2010). doi:10.1007/978-3-642-17758-3_3

13. Weiss, G.M., Timko, J.L., Gallagher, C.M., Yoneda, K., Schreiber, A.J.: Smartwatch-based activity recognition: a machine learning approach. In: 2016 IEEE-EMBS International Conference on Biomedical and Health Informatics (BHI), pp. 426–429. IEEE (2016)

14. Whitehead, S., Mailley, J., Storer, I., McCardle, J., Torrens, G., Farrell, G.: In safe hands: a review of mobile phone anti-theft designs. Eur. J. Crim. Policy Res. **14**(1), 39–60 (2008)

15. Wu, W., Dasgupta, S., Ramirez, E.E., Peterson, C., Norman, G.J.: Classification accuracies of physical activities using smartphone motion sensors. J. Med. Internet Res. **14**(5), e130 (2012)

16. Yu, X., Wang, Z., Sun, K., Zhu, W.T., Gao, N., Jing, J.: Remotely wiping sensitive data on stolen smartphones. In: Proceedings of the 9th ACM Symposium on Information, Computer and Communications Security, pp. 537–542. ACM (2014)

Rule Management in Clinical Decision Support Expert System for the Alzheimer Disease

Firas Zekri[✉] and Rafik Bouaziz

MIR@CL Laboratory, Computer Science Department, FSEGS,
Airport Road, km 4 B.P. 1088, 3018 Sfax, Tunisia
firas.zekri@fsegs.rnu.tn, rafik.bouaziz@usf.tn

Abstract. The explosion of medical knowledge and the uncertainty of some patient information in several diseases cause many clinical errors in medical decision support systems. Therefore, we aim to reduce these deficiencies using new decision support approach and new improvements. In this paper, we present a new way to manage rules in expert systems. First, we propose a knowledge specialization process that uses three types of rule bases. Then we explain how we can manage rules across these three types of rule bases. Finally, we present an implementation of a rule management system for the Alzheimer disease. We think that this improvement can enhance clinical decision support expert system performances to better support medical decision support.

Keywords: Clinical decision support system · Expert system · Rule management · Fuzzy logic · Alzheimer disease

1 Introduction

Alzheimer's disease (AD) is a degenerative brain disease and the most common cause of dementia [1]. Because of its socio-economic impact, various works that deal with the implementation of a medical decision support system for the diagnosis and detection of AD, have been the subject of many accomplishments. Sanchez *et al.* proposed a decision-making system for an early detection of AD [2]. This system is designed to assist physicians in the early stages of the AD and a possible early detection of this disease. Toro *et al.* proposed a system which consists in the evolution of the previous work presented above by Sanchez *et al.* [3]. They developed an experience-based CDSS for the diagnosis of AD. Obi *et al.* developed a neuron fuzzy system for the diagnosis of the AD [4]. In their work, the traditional procedure for the medical diagnosis of Alzheimer employed by physicians is analyzed using a neuro-fuzzy inference procedure. Finally, NES (Neurologist Expert System) is a rule-based expert system that uses an interactive sequence of questions and answers for the diagnosis of neurological diseases [5]. All related studies were oriented to the early detection of the AD. Some of them are designed only to help a physician to predict if Alzheimer is probable or not [4, 5]. Some others are designed to assist physicians only in the early stages of the AD [2, 3]. Therefore, there is no work which supports the AD throughout all its various phases, although relatives and family members of

© Springer International Publishing AG 2016
F. Lehner and N. Fteimi (Eds.): KSEM 2016, LNAI 9983, pp. 465–477, 2016.
DOI: 10.1007/978-3-319-47650-6_37

the patient are very important actors that must be taken into consideration. Their information and observations about the patient throughout all the phases of the disease are important for a good diagnosis and support, even in the final stage of the disease. Moreover, the patient, who is still in the mild stage of the AD, can be regarded as an important information source since he is aware of the difficulties in this stage and can express and describe his state. As a consequence, aid decision must be addressed to physicians, to the Alzheimer first stage patient and to persons regarded as caregivers (nurses, housekeepers, spouses, family members). To support all these kinds of users, we have proposed a new medical decision support approach for the AD. We have called the concretization of this approach Clinical Decision Support Expert System (CDSES) [6]. Then we made improvements to this system to handle the problem of knowledge acquisition uncertainty. In fact, AD symptoms vary among individuals. Hence, a decision support system could not be very cooperative and effective in front of this variety in AD. Therefore, we have proposed a knowledge specialization process which consists on the generation of a specific knowledge base to each enrolled patient in the system. We specialize knowledge (rules) using three different rules bases: general rules base (GRB), phase rule base (PhRB) and patient rules base (PaRB). In this paper we present how we can manage rules in the three different bases.

This paper is structured as follows: in the next section, we discuss the knowledge specialization process. Section 3 explains our new way to manage rules in expert systems. Section 4 presents an implementation of the rule management system for the AD. Finally, Sect. 5 concludes the paper and gives some perspectives for our future work.

2   Knowledge Specialization Process

Many studies on AD have shown that symptoms can be expressed differently depending on the patient [1, 7]. It is possible that a patient presents symptoms that are specific to him: some signs may appear strongly at the beginning of the disease or, conversely, stay very discrete even after several years. Moreover, the choice of a treatment is determined by the severity of the illness. Hence, a system could not be effective in front of this heterogeneity, especially in chronic progressive diseases like AD. Furthermore, the fact that new medical advances on AD are frequent involve an exponential grow of new knowledge [8]. Therefore, over time, the amount of rules will definitely increase too much since they are provided in a continuous manner by domain experts. Consequently, inference time will increase proportionally with the increase of the number of rules. Thus, the system cannot respond quickly to calls for decision aid and then we run the risk that the results come too late. We should then reduce the rule number, so that the inference becomes more efficient and does not take too much time. Tools easing the handling of such information and knowledge are needed [8]. We then propose the solution "knowledge specialization" which means that the system must accord to each patient a knowledge base containing only his pathological specific knowledge. Indeed, if knowledge is more specific, then inference will be more precise, the execution time will be shorter and therefore the results will be better. To specialize an expert system (ES), we must specialize its knowledge base and therefore specialize its rule base and its fact base. For this purpose, we propose a knowledge base specialization process (cf. Fig. 1) which is composed of three steps. In the first step, we gather all the rules gained from experiences lived by domain experts in a general rule base (GRB). In the second step, these rules must be classified and distributed on x bases, each base correspond to one disease phase (PhRB). In the third step, a rule base must be generated to each patient enrolled in the system which will be specific to him and contains rules suitable to his pathological case (PaRB). Experts responsible for the pathological case, who are the only actors able to assess the situation of the considered patient, must validate the rule base of this patient. This will more ensure the use of the appropriate rules. Then, a fact base that contains variables describing the state of the considered patient is selected for each rule base. This fact base is fueled by a data collector module. Finally, as a result we get a specific knowledge base for each enrolled patient. Once knowledge base is specialized, an inference engine is paired with this last to reason about the knowledge and gives the results. The inference is now faster and more efficient since it will run only on the knowledge of the considered patient.

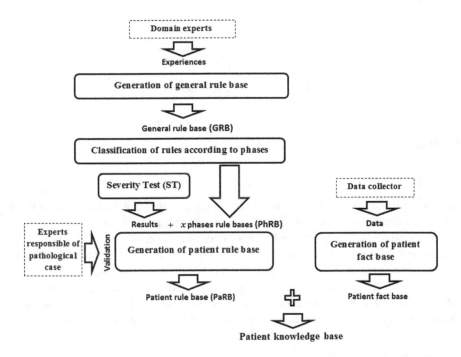

Fig. 1. Knowledge base specialization process

Based on knowledge base specialization process, we propose an ES architecture, which is presented in Fig. 2. The rule base in the knowledge base is composed of three base types (GRB, PhRB and PaRB). The fact base contain patients data as many fact bases where each base matches one enrolled patient. The fuzzy inference engine reasons only on the rule and fact base of the selected patient.

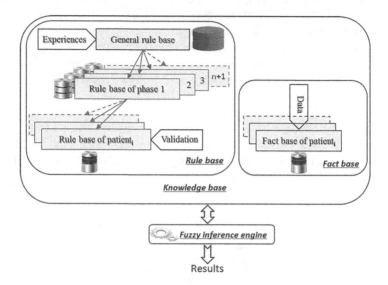

Fig. 2. The proposed expert system architecture

3   Rule Management in CDSES

3.1   Rule Addition

As previously announced, Many studies on AD have shown that the onset of the symptoms and treatments varies from one patient to another [1, 7]. A symptom may appear in the first phase as it may appear in the second phase even with different occurrence frequencies. With respect to treatments, for example in case of intolerance or failure to achieve the maximum recommended doses, it is possible to substitute AChEI[1] drug by another. For moderate and moderately severe stages, Memantine[2] drug may be an alternative to AChEI with some patients. However, in moderate forms of the disease, its efficacy is less well established than AChEI and its place must be discussed by the experts

---

[1] Acetylcholinesterase Inhibitors: restore normal levels of acetylcholine (a chemical messenger in the brain) in areas affected by the disease.

[2] NMDA receptor antagonist: protect neurons from glutamate excess, a chemical messenger in the brain that can damage nerve cells.

[9]. Therefore, the rules based on these symptoms and treatments cannot be classified in x bases with a crisp way. We cannot require that a rule belongs to one single base. This means that the system must be adaptive to different cases of the disease. Hence, we can ask how these rules will be classified with the presence of these uncertainties.

Rule addition must be done in an uncertain way. There are several logics to handle uncertainties. Nevertheless, we choose to use two methods to add a rule. (i) The first method is to add a rule in a fuzzy way using the fuzzy logic. We choose this logic because it allows taking into account the linguistic typologies, and therefore all the nuances that were created to best explain the reality. Besides, in the fuzzy logic, it is often possible to sufficiently reduce the number of rules to allow a higher speed inference [10]. One fuzzy rule can replace many (often very many) conventional rules [11]. This allows its use in most developed control systems. We denote all rule added in fuzzy way by R. (ii) If it is difficult to approach the efficiency of the rule in each phases of the disease using membership functions, we add a rule with a weight which indicate its occurrence in a particular base. We denote all rules added with weight by R'.

Rule Fuzzy Addition. In the first step "Generation of general rule base" of the knowledge base specialization process (cf. Fig. 1), the rules acquired from the domain experts must be accompanied by their membership functions. Each rule must have one of the four fuzzy membership functions (Trapezoidal Function, Triangle Function, R-function and L-Function) [12]. We can add other functions if needed. Each membership function represents degrees of the rule efficiency in each phase of the disease. The domain experts, with the aid of the knowledge engineers, must try to approach the efficiency of the rule in each phases of the disease using membership functions, which must be based on Severity Test (ST) values. These values are divided into intervals. Each interval represents a phase of the disease. To separate between disease phases we used Phase Discriminating Values (PDV). We call this step "Classification Fuzzification". To better explain our idea, let's consider the example presented in Fig. 3. In this example, and according to the membership function, rule R_i is efficient in bases $PhRB_1$, $PhRB_2$ and $PhRB_3$, but with different degrees of membership based on the ST values.

In the second step "Classification of rules according to phases", the rules must be classified according to the parameters of each membership function type and according to different PDVs. We call this step "Fuzzy Classification". We developed a complex algorithm to deal with four types of membership functions: trapezoidal function, triangle function, R-function and L-function. For example, rule R_i, presented in Fig. 4, has L-function membership function. Then, it must be classified as follow:

$$a = 5,\ b = 25,\ PDV_1 = 20,\ PDV_2 = 10$$

$$\exists R_i \in PhRB_1 / 0 < \mu_{(R_i \backslash PhRB_1)} < \frac{-PDV_1 + b}{b - a}$$

$$\exists R_i \in PhRB_2 \left/ \frac{-PDV_1 + b}{b - a} < \mu_{(R_i \backslash PhRB_2)} < \frac{-PDV_2 + b}{b - a} \right.$$

$$\exists R_i \in PhRB_3 \Big/ \frac{-PDV_2 + b}{b - a} < \mu_{(R_i \setminus PhRB_3)} < 1$$

Now, in the third step "Generation of patient rule base", the system must first select a PhRB for each patient among those previously defined (x bases). The system must make the selection based on the Severity Test (ST) and PDVs. We leave the choice of these values to the experts, because it varies from one expert to another. Assuming that the number of PDV is equal to n, then we must have "n + 1" PhRB. The formula is then:

$$\forall n \in N; IF \ number \ of \ PDV = n \ THEN \ Number \ of \ PhRB = n + 1$$

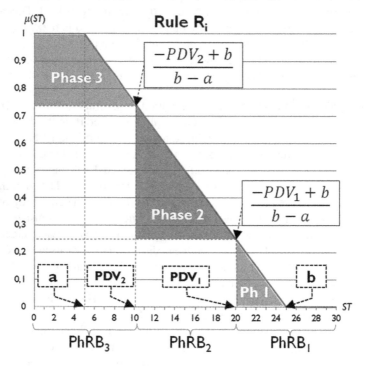

Fig. 3. Rule R_i membership function

Therefore, when selecting the appropriate PhRB, we must apply the following algorithm:

```
If (n=0) then
  select GRB;
Else If (the severity test
is descending) then
While (n>0 and PhRB is not
selected)
{
  If (ST< PDVn) then
    {
      select (PhRBn+1);
      PhRB is selected;
    }
    Else
    {
      n=n-1;
      If (n=0) then select
      (PhRB1);

    }

}
```

```
Else If (the severity test
is ascending) then
While (n>0 and PhRB is not
selected)
{
  If (ST> PDVn) then
    {
      select (PhRBn+1);
      PhRB is selected;
    }
    Else
    {
      n=n-1;
      If (n=0) then select
      (PhRB1);
    }
}
```

Then, the system must affect a degree of membership to PaRB$_n$ for each rule in the selected PhRB, based on the following formula:

$$\forall R_i \in PhRB_{selected}; \ \mu_{(R_i \setminus PaRB_n)} = F_i^j(ST)$$

Where i is the index of the rule and j is the index of the membership sub-function. We call this step "Classification Defuzzification". In the example presented in Fig. 3, if the ST value of the patient is equal to 19, the system must choose PhRB$_2$ because, in this case, $PDV_2 < ST < PDV_1$. Therefore rule R$_i$, belonging to PhRB$_2$, must have a degree of membership to the PaRB$_n$ equal to:

$$\mu_{(R_i \setminus PaRB_n)} = \frac{-ST + b}{b - a} = \frac{-19 + 25}{25 - 5} = 0,3.$$

Weighted Rule Addition We can manually use weight to specify the rule membership degree (weight). In this case weight is independent from the severity test ST, and adding rule vary depending on which type of rule base used among the three types defined in the knowledge specialization process. We present three case of weighted rule addition:

- Adding manually a rule to the general rules base: If a rule R$'$ is added to the general rule base GRB with a weight $W_{R'}$, then R$'$ must be added to all PhRB and all PaRB with a weight $W_{R'}$. So, to add a rule to all rules bases regardless of their type, the addition must be done at the general rules base.

$$\forall R' \in GRB/W_{R'} = w; \ R' \in all \ PhRB \ and \ R' \in all \ PaRB/W_{R'} = w$$

- Adding manually a rule to the phase rules base: If a rule R' is added to a particular phase rule base $PhRB_i$ with a weight $W_{R'}$, then R' must be added to all patient rules bases $PaRB_{ij}$ which inherited their rules from $PhRB_i$. In this case, R' is specific only for patients who are in phase i of the disease.

$$\forall R' \in PhRB_i/W_{R'} = w; \ \forall PaRB_{ij}/PhRB_i = \int PaRB_{ij}; \ R' \in PaRB_{ij}/W_{R'} = w$$

- Adding manually rule to patient rules base: If a rule R' is added to a particular patient rule base $PaRB_j$ with a weight $W_{R'}$, then R' is specific only for the patient j.

3.2 Rule Removal

Removing a rule is not simple since we have used three types of rule bases. Like addition, removal varies depending on the type of the used rule base. We present three cases of rule removal:

- Removing a rule from the general rules base: removing a rule R or R' from GRB require removing it from all rules bases (PhRB and PaRB). So if we want to remove a rule completely from the inference system we must do it from the GRB.
- Removing a rule from the phase rules base: removing a rule R or R' from a particular $PhRB_i$ require removing it from all $PaRB_{ij}$ which inherited their rules from $PhRB_i$. This removal cannot be spread to the GRB if it contains a copy of the removed rule.
- Removing a rule from patient rules base: removing a rule R or R' from a particular $PaRB_i$ is restricted to this rules base. This removal cannot be spread to the GRB or PhRB if it contains a copy of the removed rule.

3.3 Rule Modification

The rule modification is not as simple as the removal. The rule can be modified from any rule base, but it differs from one type to another. Modifying a rule from GRB is not like modifying it from PhRB or PaRB and it depend in how this rule is added to the rule base in fuzzy way or added with weight. We present three cases of rule modification:

- Modifying a rule in the general rules base: to modify a rule in GRB we must know if it is added in a fuzzy way or added with a weight. If it is a fuzzy addition, two cases are present. (i) The first case is to modify the fuzzy parameters of the rule. This kind of modification requires firstly a modification in all $PhRB_i$ that contains a copy of the appropriate rule, and secondly a modification in all $PaRB_{ij}$ that inherited their rules from $PhRB_i$ concerned by the modification. With respect to the modification in the PhRB, fuzzy parameters and degrees of membership intervals of the appropriate rule must be modified. For the PaRB, the degree of membership or the weight of the appropriate rule must be modified according to the new fuzzy parameters in the PhRB. (ii) The second case is to modify the rule text. This kind of modification requires that the modification must be applied in all rules bases that contain the concerned rule. Now, if the rule is added to GRB with a weight, rule can be modified in text and\or weight. This modification must be spread to all rule bases that contained the concerned rule.

- Modifying a rule in the phase rules base: Like in GRB, to modify a rule in PhRB it is necessary to know if it is added in a fuzzy way or added with a weight. If it is a fuzzy addition, then only modifying the rule text is allowed. If the rule is added originally in the PhRB with weight then the rule text and/or the weight can be modified. In two cases, modifications must be spread to all PaRB that inherited theirs rules from the concerned PhRB.

- Modifying a rule in the patient rule base: In this case there is no need to know if the rule is added in fuzzy way or added with weight because weight in this final step is equivalent to degree of membership in fuzzy addition. Here, rule text and/or the weight can be modified.

3.4 Rule Movement

Only weighted rule can be moved from one base to any other base, because rules, added in fuzzy way, are classified and fixed with parameters and functions. Moving a weighted rule require many modifications in related bases. We identify five cases of rule movement:

- Moving a weighted rule from GRB to PhRB: three steps must be flowed to move a weighted rule from GRB to a particular $PhRB_i$ (cf. Fig. 4). (i) The first step is to make a copy of the rule we want to move. (ii) Then, we need to delete the concerned rule from GRB, all PhRB and all PaRB. (iii) Finally, rule copy must be added to the selected PhRB in the some way we add manually a rule to PhRB (cf. Section weighted rule addition).

- Moving a weighted rule from GRB to PaRB: like the previous case, three steps must be flowed to move a weighted rule from GRB to a particular $PaRB_j$. (i) The first step is to make a copy of the rule we want to move. (ii) Then, we need to delete the concerned rule from GRB, all PhRB and all PaRB. (iii) Finally, rule copy must be added to the selected $PaRB_j$.

- Moving a weighted rule from PhRB or PaRB to GRB: In this case rule must be copied in all rule bases (GRB, all PhRB and all PaRB).

- Moving a weighted rule from PhRB to PaRB: In this case rule must be deleted from $PhRB_i$ and from all $PaRB_{ij}$ that inherited their rule from $PhRB_i$ except the selected PaRB that we want to move in the rule.

- Moving a weighted rule from PaRB to PhRB: In this case rule must be copied in the selected $PhRB_i$ and in all $PaRB_{ij}$ that inherited their rule from $PhRB_i$.

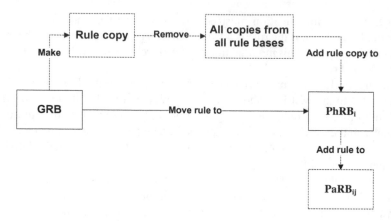

Fig. 4. Moving a weighted rule from GRB to PhRB

3.5 Inference Limiter

Rules are now classified in PaRB with different weights and the system must manipulate them in a flexible manner. The system must be able to use rules with the highest weight in a given situation and use lower weight in another situation if needed. For this purpose, we propose to use a limiter, which allows applying inference only on the rules that have a weight greater than a value "x". For example, if the limiter is adjusted to 0.4, then the system must use only rules with a weight greater than 0.4. This limiter must be adjustable according to the posed situation. For example, if there are no results, the system can decrease the limiter value to use rules with a lower weight, and then the system can give results.

4   Implementation

We have implemented the rule management system for AD. We adopt that there is four phases in AD. That's mean we must have four PhRB and then three PDV. In this implementation we used different software modules. Eclipse (juno version 4.2.0) was used to build a Java software and interfaces. Jess (Java Expert System Shell) and NRC FuzzyJ Toolkit (Fuzzy Jess) was used to create a Java program that encodes rules and fuzzy rules. Finally, we have used xml files to store rules.

To add a new rule, user has to write it in Jess syntax and check if it doesn't contain an error because wrong rule cannot be added. If an error occurs, a complete description of the error will be presented in the check text area. Then, he must choose between (i) adding a rule in fuzzy way, by choosing one of the four membership function, and (ii) adding a rule with weight, by choosing "No Function". Once user hits the "Save rule" button, rule will be saved in xml files according to all its parameters inserted in the "add Jess rule" interface. Figure 5 presents an example of adding a rule in a fuzzy way using a trapezoidal membership function. User must specify all membership function parameters and give the three PDV which separate between phases.

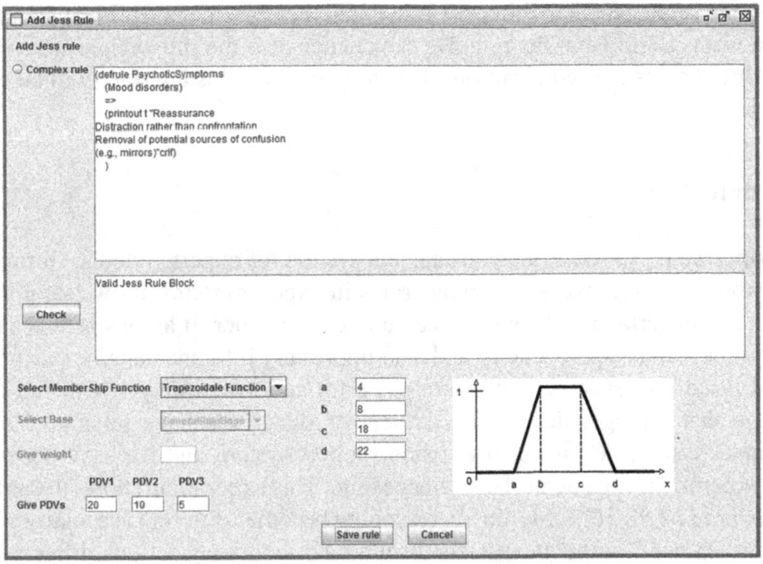

Fig. 5. Adding rule in fuzzy way

To edit a rule, user has to select the appropriate rule. Text and all parameters will be presented in the interface to be modified. User can also change the location of the rule by modifying the base containing the rule. Figure 6 presents an example of editing and moving a rule stored in $PhRB_1$. First, user must select the base from where we want to edit or move a rule. Then, he can modify the rule text and recheck the rule syntax by hitting the check button. If there are errors in rule syntax then the edited rule cannot be

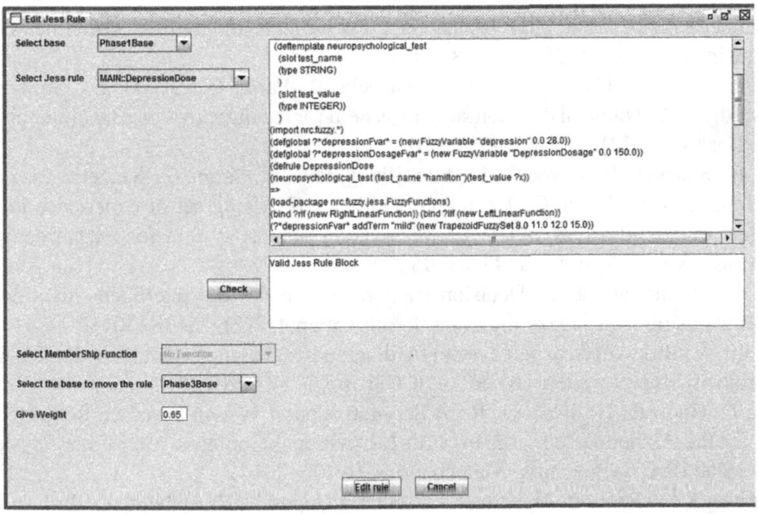

Fig. 6. Editing and moving a weighted rule

saved. Finally, if he wants to move the rule to another base he must select another base where he wants to move in the rule. He can change also the rule weight if he needs to do that. Once he hits the "edit" button all changes made in this interface will take effects in xml files.

5 Conclusion

In this paper we proposed a rule management system for expert systems. Our new way to store, to edit, to remove and to move rules in expert systems allows supporting all fuzziness and uncertainty present in disease like Alzheimer. It allows better managing medical information heterogeneity and also improving inference engine performance. This will make inference systems more powerful and faster since it will run only on knowledge that are specific to every patient enrolled. We have integrated our rule management system in a medical decision support system and we have performed a primary experiment to evaluate the inference time. This experiment shows that inference time is reduced by 38.56 %. Moreover, reasoning becomes different especially when we used the inference limiter. In fact, the results and recommendations differ from one patient to another. Even for the same patient, when we decrease or increase the limiter, the results change according to rule weight, because decreasing the limiter means using low weight.

In our future work, we aim to experiment our new system in a real medical case to test the feasibility of the proposed improvement. The results of these experimentations will constitute the object of a forthcoming paper.

References

1. Alzheimer's Association: 2012 Alzheimer's disease facts and figures. Alzheimer's Dement. J. Alzheimer's Assoc. **8**(2), 131–168 (2012)
2. Sanchez, E., Toro, C., Carrasco, E., Bonachela, P., Parra, C., Bueno, G., Guijarro, F.: A knowledge-based clinical decision support system for the diagnosis of Alzheimer disease. In: IEEE Healthcom 2013, pp. 351–357 (2011)
3. Toro, C., Sanchez, E., Carrasco, E., Mancilla-Amaya, L., Sanín, C., Szczerbicki, E., Graña, M., Bonachela, P., Parra, C., Bueno, G., Guijarro, F.: Using set of experience knowledge structure to extend a rule set of clinical decision support system for Alzheimer's disease diagnosis. Cybern. Syst. Int. J. **43**, 81–95 (2012)
4. Obi, J.C., Imainvan, A.A.: Decision support system for the intelligent identification of Alzheimer using neuro fuzzy logic. Int. J. Soft Comput. **2**(2), 25–38 (2011)
5. Al-Hajji, A.: Rule-based expert system for diagnosis and symptom of neurological disorders -Neurologist Expert System (NES). In: ICCIT, pp. 67–72 (2012)
6. Zekri, F., Ghorbel, H., Bouaziz, R.: A decision support system based on fuzzy specialized rules for the Alzheimer disease. In: 11th International Conference on Fuzzy Systems and Knowledge Discovery, China, Xiamen, pp. 490–496 (2014)
7. Alzheimer's Association: Alzheimer's association report 2015 Alzheimer's disease facts and figures. Alzheimer's Dement. **11**, 332–384 (2015)

8. Sanchez, E.: Semantically steered clinical decision support systems. Thesis Submitted to the Department of Computer Science and Artificial Intelligence in Partial Fulfillment of the Requirements for the Degree of Doctor of Philosophy, University of the Basque Country Donostia - San Sebastian, p. 122 (2014)

9. Dr. Sarazin M.: La maladie d'Alzheimer, Fiche technique, décembre 2009

10. Ghorbel, H., Maalej, S., Bahri, A., Bouaziz, R.: Un framework pour la génération semi-automatique d'ontologies floues. Technique et Science Informatiques (TSI) **32**(6), 671–698 (2011)

11. Cox, E.: Fuzzy fundamentals. IEEE Spectr. **29**, 58–61 (1992)

12. Zadeh, L.: The concept of a linguistic variable and its application to approximate reasoning. Int. J. Inf. Sci. **4**(4), 301–357 (1975)

Stability Analysis of Switched Systems

Jinjing Zhang, Fan Li, Xiaobin Yang, and Li Li$^{(\boxtimes)}$

School of Computer and Information Science,
Southwest University, Chongqing 400715, China
1476509610@qq.com, lily@swu.edu.cn

Abstract. A switched system is a hybrid system that is composed of a family of continuous-time and discrete-time subsystems with a specific rule orchestrating the switching among the subsystems. In this paper, switched systems are grouped into two broad categories based on the switch strategy. They are time-driven switched systems and event-driven switched systems, respectively. The sufficient conditions for exponentially asymptotic stability of the above two switched systems are discussed. We then perform both state responses and phase planes simulation of those systems in MATLAB. The stability of the subsystems and the switched system as a whole is further analyzed based on the simulation results.

Keywords: Switched systems · Simulation · Switch strategy · Exponentially asymptotic stability

1 Introduction

In general, the most widely studied dynamic systems in the literature can be classified into two groups: continuous and discrete switched systems [1]. But in practice, the discrete and continuous dynamics constitute of the fundamental attribute of dynamic systems [2], such systems are called hybrid dynamic system. Switched system is a kind of important hybrid dynamic system which is studied from the aspects of Systematics and Cybernetics, and it is made up by continuous-time evolving subsystems, discrete-event driven subsystems which interact with each other and switch strategy that decides which of the subsystem is active at each moment(switch strategy is also called switch rate, switch law, switching signal etc.). A main problem of switched systems which is always inherent in all dynamical systems is the presence of uncertainties and external disturbances [3], the attribute of switched systems is not simple superposition of all subsystems'. Namely, even if each subsystem is not stable, by structuring switch strategy, it can ensure that switched systems are tended to be asymptotically stable, and even if each subsystem is stable, selecting inappropriate switch strategy, switched system may not have stability.

Switched systems can depict actual system more scientifically, and as an international cutting-edge direction of the research of hybrid system theory, it gets extensive attention on the field of control and computer. There are many applications of switched systems, such as in control of mechanical systems, traffic

© Springer International Publishing AG 2016
F. Lehner and N. Fteimi (Eds.): KSEM 2016, LNAI 9983, pp. 478–488, 2016.
DOI: 10.1007/978-3-319-47650-6_38

management and many other practical control fields. The motivation of studying switched systems is from the fact that many practical systems are inherently multi-models [3], and as a special simplified model of hybrid systems, its analysis and design method is easier to generalize to general hybrid dynamic system. So switched system draws extensive attention on the field of control and computer. At present, switched systems will be one of the most research directions and strategic important topic.

1.1 Research Background of System Simulation

Simulation is the imitating for real things, and it, which is based on the theory of similarity theory, cybernetics and computer information processing technologies, etc., is a multi-disciplinary crossing technology in order to research real systems by the tool of computers and physical devices to build mathematical models. Simulation is an important scientific mean to gradually know complex systems and an efficient mean of optimization for hybrid systems. It can help realize the essence of systems and accurately analyze the performance of systems, it also deepens the understanding of complex system and gains a set of system theory which can strengthen the control of systems. At present, where there is scientific research, there must have the support of simulation. Simulation becomes one of the main research methods in the field of scientific research after experiment and theory.

1.2 Introduction of Simulator

Simulator can intuitively model for the systems of realistic life and improve and efficiency of modeling through user friendly graphical interface. Based on the theory of switched systems, the simulation of systems can be implemented under the MATLAB environment more easily and conveniently with the framework [2]. So this article mainly simulates several types of switched systems under the design environment of MATLAB. Based on the MATLAB simulation, it can not only facilitate the modeling of complex systems, but also can be more understanding of interaction of each subsystem and the nature of complex system.

1.3 Main Work of This Paper

Most of the previous papers studied dynamic systems which are classified into continuous or discrete switched, linear or non-linear. Such as, discrete-time linear switched systems [4,5], continuous-time systems [6], and so on. And most consider the construction of Lyapunov function to study stability of systems [7], which is hard to get the conditions. But in this paper, we divide dynamic systems into event-driven and time-driven switched systems according to the switch strategy. We can simply simulate the stability of all kinds switched systems and the relationship with their subsystems. The main work of this paper as following.

The rest of the paper is organized as follows. In Sect. 1, we describe the related works about incentive schemes. Section 2 analyses and simulates time-driven switched system including periodically switched systems and impulsive switched systems. Event-driven switched system is analyzed in Sect. 3, in which subsystems are designed based on event driven switched system of user requirement.

2 Stability of Time-Driven Switched System and Its Simulation

2.1 Theory of Stability of Periodically Switched Systems

Periodically switched systems are timely carried out in accordance with the fixed time, namely periodically switched systems have a periodic switching rule. In a period, each subsystem switches based on the order of switch law in a fixed time. In each period, subsystems cycle successively and switch in sequence.

In accordance with the definition of periodically switched system, we establish the mathematical model of periodically switched systems, as the following Eq. 1:

$$\dot{x}(t) = A_{\sigma(t)}x(t)$$
$$x(t_0) = x_0 \tag{1}$$

In which, it can be easily seen that each subsystem $\dot{x}(t) = A_{\sigma(t)}x(t)$ is active for $\triangle t_k$ seconds within each Period [11]. If the period of periodically switched system is T, the system matrix appears periodical satisfying $A_i \epsilon R^{n \times n}$ and $A(t) = A(T + t)$, $x(t) \epsilon \Re^{n \times n}$ is state vector of switched systems and is the right continuous piecewise constant function, $\sigma(t) : [0, +\infty] \to M = 1, 2, \ldots m$ is switched signal, $\sigma(t) = i$ means that the i-th subsystem is activated. Switching sequence is $\{x_0; (i_0, t_0)(i_1, t_1)\ldots(i_k, t_k)|i_k \epsilon M, k \epsilon N\}$, In which, t_0 is initial time, (i_k, t_k) represents that the i_k-th subsystem of switched system is activated, when $t_k \leq t \leq t_{k+1}$, and at the time of t_{k+1}, i_k-th subsystem finishes and i_k-th subsystem is activated. Therefore, in the time of $[t_k, t_{k+1}]$, the motion trajectory of periodically switched system (1) is determined by the i_k-th subsystem. And periodical switching sequence is

$$\Sigma_r = \{x_0; (i_0, t_0)(i_1, t_1)\ldots(m, t_{m-1})$$
$$(1, T_0 + T)(2, t_1 + T)\ldots(m, t_{m-1} + t)\ldots$$
$$(1, t0 + lT)(2, t1 + lT)(m, tm + lT)\ldots \tag{2}$$
$$|T > 0, l = 0, 1, 2, 3\ldots\}$$

In which, switching period of periodically switched system is $T = t_m - t_0$, we define $\triangle t_k = t_k - t_{k-1} k = 1, 2 \ldots m$ as the running time of k-th subsystem in each period, which is equivalent to the dwell time of each subsystem in each period.

We know that if system matrix of a subsystem is stable, subsystem has stability. The main results about the stability of subsystem are as following Lemma 1.

Lemma 1. *If the system matrix A_i, $i = 1, 2 \ldots m$ of each subsystem is stable matrix, all eigenvalues of each system matrix A_i, $i = 1, 2 \ldots m$ has negative real part or negative real number.*

Lemma 1 is suitable for each kind of switched system, including impulsive switched system in Sect. 2.3 and designed event-driven switched system in Sect. 3.

2.2   Simulation Example of Periodically Switched Systems

Example 1. Supposed periodically switched system is $\begin{cases} \dot{x}(t) = A_1 x(t) \\ \dot{x}(t) = A_2 x(t) \\ \dot{x}(t) = A_3 x(t) \end{cases}$, which

has three subsystems. The system matrix of first subsystem is $A_1 = \begin{bmatrix} -3 & -1 & 0 \\ -4 & -3 & 0 \\ 0 & 0 & -3 \end{bmatrix}$, the system matrix of second subsystem is $A_2 = \begin{bmatrix} 0 & 0 & -3 \\ -2 & 1 & 0 \\ 0 & -4 & 1 \end{bmatrix}$,

the system matrix of third subsystem is $A_3 = \begin{bmatrix} -2 & -3 & 0 \\ 0 & -1 & 0 \\ 0 & 3 & -1 \end{bmatrix}$, initial value of

system is $x(0) = 1, y(0) = 2, z(0) = -2$, and period is 6 s.

Example 2.1 of periodically switched system switches according to the switched law in Fig. 1(a), we draw the system state response diagram in Fig. 1(b) and phase plane diagram of switched system in Fig. 1(c). According to Lemma 1, there exists a positive definition matrix P_1 which makes system matrix A_1 of

first subsystem have $P_1^{-1} A_1 P_1 = \begin{bmatrix} -1 & 0 & 0 \\ 0 & -5 & 0 \\ 0 & 0 & -1 \end{bmatrix}$, the eigenvalues of A_1 are -1,

-1 and -5 which are all negative, so first subsystem is stable; there exists a positive definition matrix P_2 which makes system matrix A_2 of second subsystem

have $P_2^{-1} A_2 P_2 = \begin{bmatrix} -2.2593 & 0 & 0 \\ 0 & 2.1296 + 2.4673i & 0 \\ 0 & 0 & 2.1296 - 2.4673i \end{bmatrix}$, the eigenvalues

of A_2 are -2.2593, $2.1296 + 2.4673$ and $2.1296 - 2.4673i$ which don't all have negative real part, so second subsystem is unstable; similarly, there exists a positive definition matrix P_3 which makes system matrix A_3 of third subsystem

have $P_3^{-1} A_3 P_3 = \begin{bmatrix} -2 & 0 & 0 \\ 0 & -1 & 0 \\ 0 & 0 & -1 \end{bmatrix}$, eigenvalues A_3 of are -2, -1 and -1 which

are all negative, so third subsystem has stability. We can get first and third subsystem are stable and second subsystem is unstable in Fig. 2 which is exactly consistent with above reasoning. Although subsystems of periodically switched system of Example 2.1 are not all stable, but the whole periodically switched system of Example 2.1 is tended to globally exponential stability in Fig. 1(b) under switch law in Fig. 1(a).

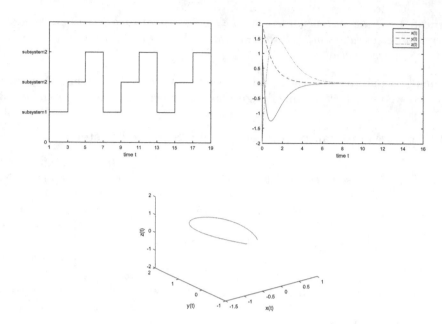

Fig. 1. (a) Switch law of Example 2.1; (b) state response curve of switched system; (c) phase plane of switched system

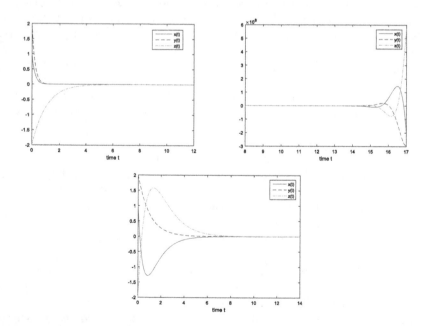

Fig. 2. State response of subsystems. (a) State response curve of the first subsystem; (b) state response curve of the second subsystem; (c) state response curve of the third subsystem

2.3   Theory and Stability of Impulsive Switched Systems

Impulsive dynamical systems are a type of hybrid systems consisting of three elements: an impulsive law being used to determine when the impulses occur; a continuous differential equation, which governs the continuous evolution of the system between impulses; and a difference equation governing the way the system states are changed at impulse times [9]. In other words, impulsive switched system is put a pulse function or impulse switch strategy into switched system, and the states of systems will sudden change at certain instants of switching which will change the value of state variable of some subsystems and which leads to switched system structure changed. Different switch law may lead to different switched results. Within the past several years, there is an increasing interest in the qualitative theory of impulsive switched systems. The reason is that impulsive switched systems can model nonlinear systems which exhibit not only impulsive dynamical behaviors but also switching phenomena [10].

Supposed there are m subsystems, we establish the mathematical model of impulsive switched systems, as the following Eq. 3:

$$\begin{cases} \dot{x}(t) = A_{i_k} x(t), t \in (t_k, t_{k+1}), i_k + 1, 2...m, k = 1, 2, ...m \\ \Delta x(t_k) = x_k^+ - x_k^- = B_k x(t_k), k = 1, 2...m \\ x_0^+ = x_0 \end{cases} \tag{3}$$

In which $x(t) \in \Re^n$ is input state vector, t_k is on behalf of switched state of k-th subsystem in the moment of t_k and in this moment adding a pulse control to switched system, $k = 1, 2 \ldots m$, t_k^- and t_k^+ respectively represent a moment before and after the moment of t_k and have

$$x(t_k) = x(t_k^-) = \lim_{h \to 0^+} x(t_k - h)$$
$$x(t_k) = x(t_k^+) = \lim_{h \to 0^-} x(t_k - h)$$

A_i is the system matrix of impulsive switched system, B_i is impulsive control matrix which is adding a pulse effect to switched system in a moment.

2.4   Simulation Example of Impulsive Switched Systems

Example 2. Supposed impulsive switched system is $\begin{cases} \dot{x}(t) = A_{i_k} x(t) \\ \Delta x(t_k) = B_k x(t_k) \end{cases}$ in

which system matrix of switched system is $A_1 = \begin{bmatrix} -7 & 0 & -1 \\ -9 & -8 & 0 \\ -8 & -3 & -1 \end{bmatrix}$ and impulsive

control matrix is $B_1 = \begin{bmatrix} -1 & 6 & -8 \\ 0 & -1 & -2 \\ 0 & 0 & -1 \end{bmatrix}$, initial value is $x(0) = [1; 0; -2]$.

Example 2.2 switches according to the switch law in the following Fig. 3(a). Then using MATLAB, we can draw the system state response diagram and phase plane diagram in Fig. 3(b) and in Fig. 3(c), respectively.

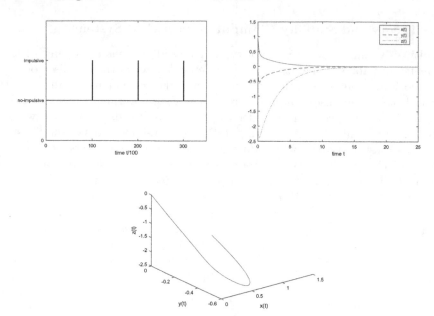

Fig. 3. (a) Switch law of Example 2.2; (b) state response curve of switched system; (c) phase plane of switched system

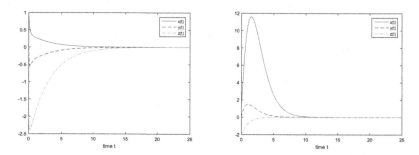

Fig. 4. State response of subsystems. (a) State response curve of switched system subsystem; (b) state response curve of impulsive control subsystem

Because in a moment the time of impulsive control matrix is too short, so the time axis of switch law of Example 2.2 in Fig. 5(a) expends 50 times.

According to computing, the eigenvalues of system matrix A of example 2.3 are -9.7190, -5.9526 and -0.3284, the eigenvalues of impulsive control matrix B are all -1 which is triple root. So all subsystems are stable. According to the state response and phase plane curve of switched system in Figs. 4(a) and (b), respectively. We can also see subsystem and impulsive control subsystem are tended to globally exponential stability. And we can gain the whole impulsive switched system is tended to globally exponential stability in Fig. 3(b) under the switch strategy in Fig. 3(a).

3   Stability of Event-Driven Switched System and Its Simulation

3.1   Theory of Stability of the Designed Switched Systems

Subsystem is designed based on event driven switched system of user requirement, switched systems carry out based on switch strategy related to system state, and it switches to the corresponding subsystem by judging the switching condition based on the change of switch law at any time. Supposed there are m subsystems, we establish the mathematical model of designed switched systems, as the following Eq. 4:

$$\begin{cases} \dot{x}(t) = f_{p(x)}(x(t)) = A_i x(t) \\ y(t) = g_{p(x)}(x(t)) \\ p(t^+) = \delta_{p(x)}(x(t), \sigma(t)) \end{cases} \tag{4}$$

In which, $p(x) \in \{1, 2, 3 \ldots n\}$ is discrete state variables in actual process, when it satisfies the switch strategy related to some system state, it switches to a subsystem, namely it indicates logic state of first subsystem, second subsystem etc. $\dot{x}(t) \in \Re^n$ presents the collection of subsystem of switched system which is a continuous set of input state. $y(t) \in \Re^n$ is output function, $\delta()$ function depends on piecewise constant function of state of subsystems $x(t)$ which is discrete input.

Above all, if we need determine whether a switched system is tended to asymptotically stable, we must examine each subsystem, if eigenvalues of subsystem matrix are all negative, the subsystem is stable. If subsystems are not all stable, we just need to find positive definite symmetric matrix P, if there exists such matrix, the system is stable, otherwise the system is unstable.

3.2   Simulation Example of Designed Switched Systems

Example 3. Supposed the following event driven switched system:

$$\frac{dx}{dt} = f(x) + y \quad f(x) = \begin{cases} -4x & x > 0 \\ -2x & -1 \le x \le 0 \\ -x - 3 & x < -1 \end{cases}$$
$$\frac{dy}{dt} = -x$$

This switched system includes three subsystem, when switch law $x > 0$, defining the first system is $\frac{dx}{dt} = -4x + y$, the general solution is $\frac{dy}{dt} = -x$

$$\begin{cases} x(t) = C1 \times (2 + \sqrt{3})e^{-(2+\sqrt{3})t} + C2 \times (2 - \sqrt{3})e^{(-2+\sqrt{3})t} \\ y(t) = C1e^{-(2+\sqrt{3})t} + C2 \times e^{(-2+\sqrt{3})t} \end{cases} \tag{5}$$

When switch law $-1 \le x \le 0$, define the second subsystem is $\frac{dx}{dt} = -2x + y$, $\frac{dy}{dt} = -x$ the general solution is

$$\begin{cases} x(t) = (C1 + C2 \times (t - 1))e^{-t} \\ y(t) = (C1 + C2 \times t)e^{-t} \end{cases} \tag{6}$$

When switch law $x < -1$, defining third subsystem is $\begin{array}{l} \frac{dx}{dt} = -x-3 \\ \frac{dy}{dt} = -x \end{array}$, the general solution is

$$\begin{cases} x(t) = \frac{1}{2}C1 \times e^{-\frac{1}{2}t}(\sin(\frac{\sqrt{3}}{2}t) - \sqrt{3}\cos(\frac{\sqrt{3}}{2}t)) \\ + \frac{1}{2}C2 \times e^{-\frac{1}{2}t}(\sqrt{3}\sin(\frac{\sqrt{3}}{2}t) + \cos(\frac{\sqrt{3}}{2}t)) \\ y(t) = C1 \times e^{-\frac{1}{2}t}\cos(\frac{\sqrt{3}}{2}t) + C2 \times e^{-\frac{1}{2}t}\sin(\frac{\sqrt{3}}{2}t) + 3 \end{cases} \qquad (7)$$

According to the general solution in Eqs. 5–7, when $t \to \infty$, $x(t)$, $y(t)$ are all tended to 0. Therefore, when $t \to \infty$, three subsystems of Example 3.1 are all tended to globally exponential stability. Supposed initial value is $x(0) = -1, y(0) = 1$ and the value of t is from 0 to 100. Switch law of Example 3.1 is only related to the state of system, and it switches according to switch law in Fig. 5(a). Using MATLAB, we can draw the system state response and phase plane diagram in Figs. 5(b) and (c).

According to state response and phase plane curve of subsystems in Fig. 6, we can see that first and second subsystem are all tended to stability in Fig. 6(a) and in Fig. 6(b), third subsystem is unstable in Fig. 6(c), but the whole switched system is tended to globally exponential stability under the switch law in Fig. 5(c).

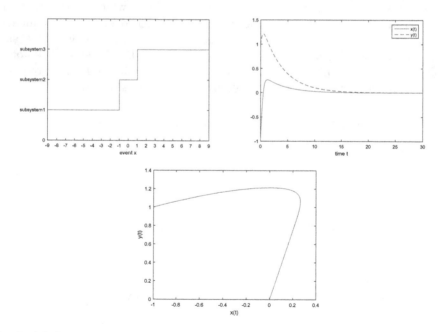

Fig. 5. (a) Switch law of Example 2.2; (b) state response curve of switched system; (c) phase plane of switched system

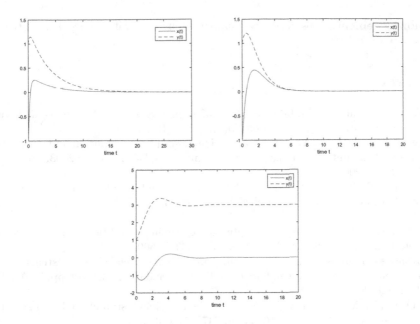

Fig. 6. State response of subsystems. (a) State response curve of the first subsystem; (b) state response curve of the second subsystem; (c) state response curve of the third subsystem

4   Conclusions

The article mainly expounds several kinds of switched system and their background, meaning and purpose of system simulation, establish mathematical model of each kind of switched system and analyzes stability of periodically switched system, impulsive switched system and designed event driven switched system whose switch strategy is related to the state of subsystem. Simultaneously, using MATLAB simulation software, we simulate and program those switched system and analyzes their stability from the whole switched system and their subsystems to study their relationship.

Stability is the basic condition of running a system, and is a challenging issue in the study of switched system [8]. So if you want to study the evolution process of a system, you must study the stable condition of system running. At present, many problems of switched systems still have miraculous differences and arguments in the academic which include the uncertainty and emergent of hybrid system, and there have no necessary and sufficient conditions to prove the stability of one switched system. So now, switched systems, scientific theories about simulation and method systems need to be further thought and developed in order to better apply to reality and solve more problems in more field of reality.

Acknowledgement. This work was supported by Natural Science Foundation of China (No. 61170192).

References

1. Yang, C., Zhu, W.: Stability analysis of impulsive switched systems with time delays. J. Math. Comput. Model. **50**(7), 1188–1194 (2009)
2. Fenghua, H., Jie, M., You, Y., Xia, Z.: The description simulation and verification for switched control systems. In: American Control Conference 2003. vol. 4, pp. 2791–2796 (2003)
3. Singh, H.P., Sukavanam, N.: Simulation and stability analysis of neural network based control scheme for switched linear systems. J. ISA Trans. **51**(1), 105–110 (2012)
4. Athanasopoulos, N., Lazar, M.: Alternative stability conditions for switched discrete time linear systems. IFAC Proc. Vol. **47**(3), 6007–6012 (2012)
5. Athanasopoulos, N., Lazar, M.: Stability analysis of switched linear systems defined by graphs. In: 53rd IEEE Conference on Decision and Control, 2014, pp. 5451–5456. IEEE (2014)
6. Chitour, Y., Mason, P., Sigalotti, M.: On the marginal instability of linear switched systems. J. Syst. Control Lett. **61**(6), 747–757 (2012)
7. Rubagotti, M., Zaccarian, L., Bemporad, A.: A Lyapunov method for stability analysis of piecewise-affine systems over non-invariant domains. Int. J. Control **89**(5), 950–959 (2016)
8. Song, Y., Fan, J., Fei, M., et al.: Robust H_∞ control of discrete switched system with time delay. J. Appl. Math. Comput. **205**(1), 159–169 (2008)
9. Bo, W., Shi, P., Wang, J., Song, Y.: Novel LMI-based stability and stabilization analysis on impulsive switched system with time delays. J. Franklin Inst. **349**(8), 2650–2663 (2012)
10. Honglei, X., Teo, K.L.: Robust stabilization of uncertain impulsive switched systems with delayed control. J. Comput. Math. Appl. **56**(1), 63–70 (2008)
11. Gökcek, C.: Stability analysis of periodically switched linear systems using Floquet theory. J. Math. Prob. Eng. **2004**(1), 1–10 (2004)

Implicit and Explicit Trust
in Collaborative Filtering

Yuanxin Ouyang[1,2(✉)], Jingshuai Zhang[1,2], Weizhu Xie[1,2],
Wenge Rong[1,2], and Zhang Xiong[1,2]

[1] Engineering Research Center of Advanced Computer Application Technology,
Ministry of Education, Beihang University, Beijing 100191, China
[2] School of Computer Science and Engineering,
Beihang University, Beijing 100191, China
{oyyx,weizhushieh,w.rong,xiongz}@buaa.edu.cn, zhangjs_it@163.com

Abstract. Recommender Systems based on collaborative filtering could
provide users with accurate recommendation. However, sometimes due
to data sparsity and cold start of the input ratings matrix, this method
could not find similar users accurately. In the past, researchers used
implicit trust weight instead of the similarity weight to find similar users,
to improve the quality of recommendation [17]. And they often ignore the
role of explicit trust in the process of finding similar users. Therefore,
in this paper, we explore the calculation of implicit trust and explicit
trust. Then according to their role in the recommendation system, we
propose a method that combined trust and similarity to get a better recommendation. At last, by experimenting on FilmTrust [5] data set which
has the explicit trust matrix, the result showed that the method we proposed significantly improve the quality of recommendation, in addition,
implicit trust and explicit trust have a positive effect on the quality of
the results of recommendation.

Keywords: Recommender system · Collaborative filtering · Implicit
trust · Explicit trust

1 Introduction

Collaborative filtering is one of the most successful method in the recommendation field. In 1990s, researchers began to pay attention on this field, and collaborative filtering also has promoted the development of the whole research of
personalized recommendation. In 1992, Goldberg et al. was the first to apply
collaborative filtering to the recommendation system, and collaborative filtering
became the first algorithm in recommendation field. In the early twenty-first
Century, the world's top researchers organized the first meeting of the Recommender System ACM in the field of personalized recommendation. After that,
the field of recommendation system has been greatly developed. Until now, both
in academic research and commercial application, personalized recommendation
have been made many achievements, and are changing people's habit to make

© Springer International Publishing AG 2016
F. Lehner and N. Fteimi (Eds.): KSEM 2016, LNAI 9983, pp. 489–500, 2016.
DOI: 10.1007/978-3-319-47650-6_39

people easily find the useful information for them in the age of information explosion.

However, there are still some problems such as data sparsity and cold start in the traditional collaborative filtering technology, which has seriously caused some negative influences on the quality of the results of recommendation. In the following research work, in order to improve the quality of recommendation, the researchers proposed many solutions, such as reducing the dimension of rating matrix [20], adding social relation in matrix operation [9], combining the item similarity and user similarity to find similar users. Recently, researchers have started to introduce machine learning and deep learning to recommendation system and some of them have got better results [18,21], but these methods are not mature, most of these methods cost a lot. Now take these problems from another point of view, we have considered the influence of interpersonal relationships on the recommendation system.

In our life, people often take the advice of friends who is trusted when they make decisions. Through these suggestions, the correctness of the decision will be greatly improved [1]. So in the Internet age, this relationship has become particularly important. In the traditional recommendation system, it often ignores the relationship, and the user is divided into a separate individual to consider. Therefore, no matter how we improve the recommendation, we cannot solve the actual problem of recommendation. Application of trust network among users in recommendation system, on the one hand can be combined with the users own characteristics and their neighbors and better mining user preferences and interest information [13], improve the quality of the recommendation system. On the other hand, through the analysis of the trust network can provide additional information, such as trust propagation, we can alleviate the problem of data sparsity and cold start problems in traditional recommendation system in a certain extent. So it is very necessary to study the trust network among the recommender systems.

At present, in the research of the application of trust network in the recommendation system, the explicit trust and the implicit trust are separately studied. The explicit trust is often used in collaborative filtering of matrix factorization [8,12,14]. And on the contrary, the implicit trust is used in collaborative filtering based on user [19]. In fact, the combination of the above two kinds of trust on the recommendation system also has an important significance, and it can better improve the quality of the recommendation.

In this paper, in order to solve the problem of data sparsity and cold start, we introduce a more abundant trust network included implicit trust and explicit trust to the recommendation system. We combine user similarity and trust as the new weight to find similar users and to recommend. Through several sets of control experiments, we found that either explicit or implicit trust could improve the recommendation results. And this proved the validity of trust network. Before the combination, we firstly discussed the calculation of implicit trust. Then because the trust is propagable, the method of trust propagation are developed. After having more abundant social relations, we can find the similar users accurately and get more accurate rating prediction, and this model named TPKNN.

2   Related Work

In order to attain the information which is useful for us efficiently, the recommendation system is widely applied to recommend films, books and songs. Collaborative filtering is one of the most successful technologies that has been applied in recommendation field. It gets huge success in both e-commerce and radio website applications. There are two kinds of collaborative filtering, the recommendation method based on items and based on users. The collaborative filtering based on items generated in mathematical statistics and the collaborative filtering based on users came from habits of people, the latter is more explanatory.

Collaborative filtering based on users is one of the earliest proposed collaborative filtering method, and it has been recognized in various industry recommendations. The core of collaborative filtering is mining the relationship between users, between items and between user and item. From these historical data, we can find the users potential behavior rules and recommend new items to users. And this is exactly using the shopping habits of people. K nearest neighbor model is a typical representative of collaborative filtering, and its working principle is using historical data to find k neighbors who is the most similar to the target user, then use the neighbor set to predict the target user's preference for the item. This approach does not feature extraction for the users or items, so this method is suitable for the item which is not structured.

In 1996, for the first time, M. Blaze et al. proposed the concept of trust management in their study [2]. However their purpose is to solve the security problem of Internet service. After the application, the concept of trust was born and many researchers began to put the trust management mechanism apply to recommendation systems. Trust management mechanism in the application of recommendation method is becoming more and more mature after the formulation of the calculation and propagation of trust. Although there is no uniform method to get trust network, researchers are still finding a better method.

Paolo Massa et al. discussed the propagation mode of explicit trust [15], and John O'Donovan et al. proposed the calculation method of implicit trust [16]. Yang Bo et al. in the collaborative filtering based on matrix factorization discussed the relationship between trustee and truster in the recommendation, and gived a method to combine the front two [22], and Soude et al., in the matrix factorization, compared the difference to the result of recommendation between the explicit trust and the implicit trust [3]. Khosravani compared the different trust recommender models in collaborative filtering recommendation [11]. Recently, Zheng Yaoyao et al. had put the distrust into trust networks to form a new trust model [23]. And this model improved the quality of recommendation for the dataset which has the relationship of distrust. So in this paper, we proposed a new networks combined explicit trust and implicit on these basis to get better results of recommendation.

3   TPKNN: A Model Combined with Similarity and Trust

3.1   Basic Steps of Collaborative Filtering

The collaborative filtering based users is a method which is recommended by the rating similarity, is also the first proposed and widely used in the commercial recommendation system. The collaborative filtering based users need find the similar users firstly, then generates the rating prediction by the set of similar users. And the general process includes getting user similarity matrix, finding K neighbors and prediction [4].

Getting User Similarity Matrix

This step is mainly for data preprocessing. Firstly, user similarity refers to the degree of similarity between the two users in the evaluation of the item, and it is standard to select the neighbors. The original input data is rating matrix, and the column vector and row vector of the matrix represent the set of users who rates the item and the set of items which rated by the user, respectively. So we can use the method of calculating the similarity between the vectors to express the similarity between the users and the users.

At present, the score similarity calculation method can be divided into the following three categories Cosine Similarity, Adjusted Cosine Similarity, Pearson Correlation Coefficient [1]. Cosine Similarity is the most basic and simple method to get the rating similarity, which is expressed by the cosine of the angle between the two vectors. However, it doesn't take into account the different rating criteria from all kinds of users. So Adjusted Cosine Similarity takes a way to eliminate this difference, and Pearson Correlation Coefficient further optimizes the method of calculating the similarity between users. In order to get more accurate results, we use the Pearson Correlation Coefficient to get the user similarity matrix, the formula is as follows:

$$sim(i,j) = \frac{\sum_{t \in K_i^R \cap K_j^R}(r_{it} - \bar{r}_i) \bullet (r_{jt} - \bar{r}_j)}{\sqrt{\sum_{t \in K_i^R}(r_{it} - \bar{r}_i)^2} \bullet \sqrt{\sum_{t \in K_j^R}(r_{jt} - \bar{r}_j)^2}} \quad (1)$$

where K_i^R and K_j^R denote the set of items which rated by user i and user j, respectively, $\bar{r}_i$ and $\bar{r}_j$ denote the mean rating of user i and user j, respectively.

Finding K Neighbors

The K nearest neighbors of user is a set of K users who have the highest similarity with the target users. And they have similar interests with the target users. Although it is the easiest step, the selection of neighbors will also affect the quality of the recommended results. For the user u, we first get the similarity between user u and other users, and then sorted from largest to smallest, finally select the front K users as the neighbor users. The number of nearest neighbor K determines the recommended range of changes. Whether the K value is too large or too small, the accuracy of the recommendation will not be guaranteed. So in this paper, we will take several groups of experiments to observe the effect of different K on the recommended result, and the number of K is 5, 10, 20, 30, 40, 50.

Prediction

Prediction would get the recommended result. After the user similarity matrix and neighbor users are obtained, we can predict the rating for users. Similar to the method of calculating the similarity, we also need take into account the different rating criteria from all kinds of users. And the formula is as follows:

$$\widehat{r}_{ui} = \bar{r}_u + \frac{\sum_{v \in T_u \cap K_i^R} sim(u,v) \bullet (r_{vi} - \bar{r}_v)}{\sum_{v \in T_u \cap K_i^R} sim(u,v)} \tag{2}$$

where $sim(u,v)$ denotes the similarity between user u and user v, T_u denotes the neighbor users of user u, $\bar{r}_u$ and $\bar{r}_v$ denote the mean rating of user u and user v, respectively.

This recommendation method is widely applied in all kinds of fields because it is easy to interpret and it has better accuracy, we called it PKNN. However, data sparsity and cold start problems still exist, so in the following section, we introduce trust network to the collaborative filtering to solve these problems.

3.2   Trust Network

Trust network is a complex network composed of interpersonal relationships. From the way of acquisition, it is divided into explicit trust and implicit trust. Explicit trust can be obtained directly from the dataset, and it comes from the mutual evaluation between users. The value of explicit trust is generally 0 or 1, and 0 and 1 represent distrust and trust respectively, so when combined with the similarity we need a weight to change the explicit trust between 0 and 1.

However, up to now, the trust network is very sparse, so we need add implicit trust to supplement. Implicit trust is derived from the users historical rating matrix [7], and it can also mining the hidden trust relationship between users [16]. Since the correctness of the recommendation has indicated the difference of user preferences, the accuracy of the recommendation can be used as the standard to calculate the implicit trust between two users. When calculating the implicit trust of the user u and the user v, we put the user v as a single neighbor of the user u, then we use the formula (2) to calculate the prediction rating of user u on the item i, set $\widehat{r}_{ui}$. And we use $Correct(i)$ to denote the set of items which is correct prediction, the formula is as follows:

$$Correct(i) = \{i | |\widehat{r}_{ui} - r_{ui}| < e\} \tag{3}$$

where r_{ui} denotes the real rating to item i from user u, e is a fixed value that represents the range of allowable errors.

Let the $Total(i)$ denotes the total number of items which have been predicted, and the formula for calculating the implicit trust between user u and user v is as follows:

$$trust(u,v) = \frac{count(Correct(i))}{Total(i)} \tag{4}$$

We can get the implicit trust matrix by this method easily. In this formula, $count(Correct(i))$ is always less than $Total(i)$, in other word, the molecular is always less than the denominator, so the range of the implicit trust value is between 0 and 1, which is convenient to combine with the similarity.

However, the above method to calculate the trust network is not enough, the trust network is still sparse. To avoid the problem, we introduce the method of trust propagation to the trust network [6]. Moreover, the method of explicit trust propagation and implicit trust propagation is different.

In the process, we use $R(m, n)$ to show the path between the node m and node n, that is the path of trust. And $N_{R(m,n)}$ indicates that the number of nodes on the path. In the method of explicit trust propagation, let d denotes the maximum propagation distance, and the formula is as follows:

$$DT_{R(m,n)} = (d - N_{R(m,n)})/d \qquad (5)$$

In the method of implicit trust propagation, there are two rules to calculate the indirect trust.

Rule 1: There is only one shortest path in the paths of node m to node n, the formula is as follows:

$$DT_{R(m,n)} = trust(m, a_1) \times trust(a_1, a_2) \times \ldots \times trust(a_k, n) \qquad (6)$$

where a_i denotes the node in the path of trust.

Rule 2: There are more than one shortest path in the paths of node m to node n, the formula is as follows:

$$DT_{R(m,n)} = \frac{\sum_i^k DT_{R_i}}{k} \qquad (7)$$

where DT_{R_i} is got by the **Rule 1**.

At the last, because the attenuation of the trust propagation we set the maximum propagation distance d to 4 in our research. And this not only increases the speed of calculation but also does not reduce the accuracy of the recommendation.

3.3 The New Weight

According to the above method, we can get the new trust network, then combine trust and similarity to get a new weight for finding the neighbor users and predicting. When we combine one trust and similarity, the formula of the new weight is as follows:

$$W(u, v) = sim(u, v) * (1 - \alpha) + trust(u, v) * \alpha \qquad (8)$$

where the $trust(u, v)$ denotes the explicit trust or implicit trust, α denotes the weight of the explicit or implicit trust. And when we combine two trust and similarity, the formula of the new weight is as follows:

$$W(u, v) = sim(u, v) * (1 - \alpha - \beta) + etrust(u, v) * \alpha + mtrust(u, v) * \beta \qquad (9)$$

where $etrust(u,v)$ denotes the explicit trust, $mtrust(u,v)$ denotes the implicit trust, α denotes the weight of explicit trust, β denotes the weight of implicit trust.

After get the new weight, we can replace user similarity with the new weight, and the new formula of prediction is as follows:

$$\widehat{r}_{ui} = \bar{r}_u + \frac{\sum_{v \in T_u \cap K_i^R} W(u,v) \bullet (r_{vi} - \bar{r}_v)}{\sum_{v \in T_u \cap K_i^R} W(u,v)} \tag{10}$$

where $W(u,v)$ denotes the new weight got from the above.

In this section, we first introduced the steps of collaborative filtering, and then introduced the trust network in detail and the calculation of getting explicit trust and implicit trust, finally get a new weight combined similarity and trust and put the new weight to the formula of prediction to get the final recommendation results. The above is the steps of TPKNN to recommend, and next we will experiment to verify the validity of this method.

4 Experiments and Validations

4.1 Description of Dataset

The dataset used in our experiments is from the FilmTrust website which uses trust in web-based social networks to create predictive movie recommendations. The FilmTrust dataset is a small dataset crawled from the entire FilmTrust website in June, 2011 [5]. And the date contains rating matrix and trust matrix, so we can get the explicit trust easily. In FilmTrust dataset, the range of rating id between 0.5 and 4, the value of explicit trust matrix is 0 or 1, representing the distrust or trust relationship, respectively. This dataset contains 1642 users and 2071 items, the total number of ratings is 35497 and the total number of trust is 1853. So the dataset is very sparse, and its density is 1.04 % and 0.069 % in terms of ratings and trust relations, respectively.

4.2 Experimental Setup

Evaluate metrics. In the experiment, we use the classical evaluation metrics to test the prediction quality of the proposed model, which is the mean absolute error (MAE) and the root mean square error (RMSE) [10].

The formula of MAE is as follows:

$$MAE = \frac{\sum_{i,j} |R_{i,j} - \widehat{R}_{i,j}|}{N} \tag{11}$$

where $R_{i,j}$ represents the rating user i gives to item j, $\widehat{R}_{i,j}$ denotes the rating user i gives to item j, and N is number of test dataset.

The formula of RMSE is as follows:

$$RMSE = \sqrt{\frac{\sum_{i,j}((R_{i,j}) - \widehat{R}_{i,j})^2}{N}} \tag{12}$$

where $R_{i,j}$ represents the rating user i gives to item j, $\widehat{R}_{i,j}$ denotes the rating user i gives to item j, and N is number of test dataset.

Comparison Methods. In order to comparatively evaluate the performance of our proposed methods, we compare the following four methods, and the existing method is traditional recommender method which only use the user similarity. And the details are as follows:

PKNN: the traditional collaborative filtering recommendation which using Pearson Correlation Coefficient to get the user similarity matrix and set the user similarity as the weight for finding k neighbors and predicting [1].

ETPKNN: proposed in this paper which combined the user similarity and explicit trust as the new weight for finding k neighbors and predicting.

MTPKNN: proposed in this paper which combined the user similarity and implicit trust as the new weight for finding k neighbors and predicting.

TPKNN: proposed in this paper which combined the user similarity and all trust as the new weight for finding k neighbors and predicting.

Cross-Validation. In the model training and rating testing, we used cross validation to process the data set. In simple terms, cross validation divides the data set into k parts, and then selects the (a) parts as a training set, and the rest (K-a) parts as a test set. Because the FilmTrust data set is sparse, we use a 5-fold cross-validation. In each time we randomly divide the data into 5 parts, then select 1 part as the test set and the rest 4 parts as the training set. That is to say, there are 80 % of the data for the training set, 20 % of the data for the test set. We make a total of 5 times tests, and take the mean as the final result.

4.3   Results

In Eqs. 8 and 9, it can be seen that the weight of trust α and β is very important for recommendation, since it shows the proportion of user similarity, explicit trust and implicit trust, and only when we find the best value of α and β, we can get the best comparison results. So before comparing these models, we need first determine the optimal weight of trust. According to the experimental results, we first plot the change of MAE according to the weight of explicit and implicit trust, as shown in Fig. 1, where the number of neighbor is set to 20. And we can find that when the weight of explicit trust is 0.2 and the weight of implicit trust is 0.3, the MAE is minimum, which means the performance of recommendation is the best at that time.

So in the ETPKNN model, we set the weight to 0.2 and the weight set to 0.3 in the MTPKNN model. Similar to the above experiment, the weight of explicit

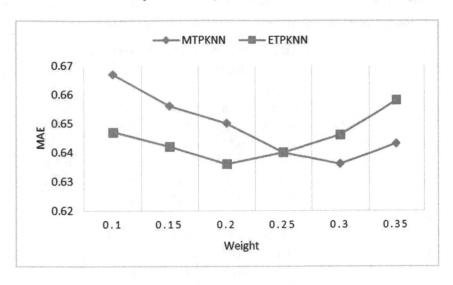

Fig. 1. Impact of weight

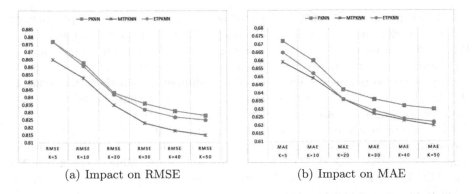

(a) Impact on RMSE   (b) Impact on MAE

Fig. 2. The comparison of recommendation performance

trust and implicit trust are both set to 0.18 in the TPKNN model. We could find that the weight of user similarity is the highest no matter which model is, and the weight of implicit trust is higher than explicit trust in single model.

By comparing the performance of ETPKNN, MTPKNN and PKNN to verify the effectiveness of the trust network in the recommendation. The experimental results are shown in Fig. 2. It shows that the performance of ETPKNN and MTPKNN is better than PKNN, so it proves that our theory about trust network and the propagation trust is reasonable, and this combination can improve the accuracy of the recommendation. But on the other hand, although the performance of ETPKNN does not improve much, it costs the same time as PKNN. On the contrary, the model of MTPKNN improves the accuracy of recommendation obviously, but it needs a lot of time to calculate the implicit trust.

Table 1. Comparison result against baselines in term of MAE and RMSE

The number of neighbor	Measure	PKNN	MTPKNN	ETPKNN	TPKNN
5	MAE	0.672	0.659	0.665	0.653
	RMSE	0.877	0.865	0.877	0.863
10	MAE	0.66	0.649	0.652	0.646
	RMSE	0.863	0.853	0.861	0.854
20	MAE	0.642	0.636	0.636	0.632
	RMSE	0.843	0.835	0.842	0.836
30	MAE	0.636	0.627	0.629	0.624
	RMSE	0.836	0.823	0.832	0.824
40	MAE	0.632	0.623	0.624	0.62
	RMSE	0.831	0.818	0.827	0.82
50	MAE	0.63	0.62	0.622	0.617
	RMSE	0.828	0.815	0.825	0.817

After verifying the validity of the trust network, we do another experiment to verify the performance of the TPKNN model and the results are shown in Table 1. We can find that the TPKNN performs the best of all, no matter how much the number of neighbors. On one hand, the more neighbors there are, the better the quality of the recommendation is. Until the number of neighbors increased to 40, the change of MAE is not obvious, which indicates that the recommended model has contained almost users who can be trusted, the increase of neighbors would not improve the accuracy of the recommendation. On the other hand, the MAE of TPKNN with 5 neighbors is less than the MAE of PKNN with 10 neighbors. However, the time of PKNN is the sum of the time of the ETPKNN and MTPKNN, so when TPKNN obtains more accurate results, it sacrifices other resources at the same time.

5 Conclusion

In this paper, we proposed a model combined user similarity and the trust networks which combined explicit trust and implicit trust. In the model, we not only use the user-item rating matrix, but also use the user-user trust matrix to get social relations. The experiment on Filmtrust dataset shows that the model we proposed model performs better compared with traditional method, so the validity of trust network is proved. The model also confirmed the validity of the way to get explicit trust and implicit trust, and the possibility of combination between user similarity and trust. So it could be a direction of the future study.

The model proposed in this paper showed the superiority in experiment, but it's definitely not the best. So in the future work, there are still several challenges for the researchers. Firstly, making the way of trust propagation more reasonable would make the trust network more abundant, then the recommendation result would be more accurate. Secondly, the way of combination between user similarity and trust could be more complex instead of a simple linear combination. Whats more, the success of the TPKNNN model would encourage more people to explore the calculation of trust networks.

Acknowledgements. This work was partially supported by the National Natural Science Foundation of China (No. 61472021), SKLSDE project under Grant No. SKLSDE-2015ZX-17 and the Fundamental Research Funds for the Central Universities.

References

1. Adomavicius, G., Tuzhilin, A.: Toward the next generation of recommender systems: a survey of the state-of-the-art and possible extensions. IEEE Trans. Knowl. Data Eng. **17**(6), 734–749 (2005)
2. Blaze, M., Feigenbaum, J., Lacy, J.: Decentralized trust management. In: 1996 IEEE Symposium on Security and Privacy, Proceedings, vol. 30, no. 1, pp. 164–173 (1996)
3. Fazeli, S., Loni, B., Bellogin, A., Drachsler, H., Sloep, P.: Implicit vs. explicit trust in social matrix factorization. In: Proceedings of the 8th ACM Conference on Recommender systems, pp. 317–320 (2014)
4. Ge, X., Liu, J., Qi, Q., Chen, Z.: A new prediction approach based on linear regression for collaborative filtering. In: Eighth International Conference on Fuzzy Systems and Knowledge Discovery, FSKD 2011, Shanghai, China, 26–28 July 2011, pp. 2586–2590 (2011)
5. Golbeck, J.: Generating predictive movie recommendations from trust in social networks. In: 4th International Conference, Trust Management, iTrust 2006, Proceedings, Pisa, Italy, 16–19 May 2006, pp. 93–104 (2006)
6. Guha, R.V., Kumar, R., Raghavan, P., Tomkins, A.: Propagation of trust and distrust. In: Proceedings of the 13th International Conference on World Wide Web, pp. 403–412 (2004)
7. Guo, G., Zhang, J., Thalmann, D., Basu, A., Yorke-Smith, N.: From ratings to trust: an empirical study of implicit trust in recommender systems. In: Proceedings of 29th ACM International Symposium on Applied Computing, pp. 248–253 (2014)
8. Guo, G., Zhang, J., Yorke-Smith, N.: TrustSVD: collaborative filtering with both the explicit and implicit influence of user trust and of item ratings. In: Proceedings of the Twenty-Ninth AAAI Conference on Artificial Intelligence, Austin, Texas, USA, 25–30 January 2015, pp. 123–129 (2015)
9. Guo, G., Zhang, J., Yorke-Smith, N.: A novel recommendation model regularized with user trust and item ratings. IEEE Trans. Knowl. Data Eng. **28**(7), 1607–1620 (2016)
10. Herlocker, J.L., Konstan, J.A., Terveen, L.G., Riedl, J.: Evaluating collaborative filtering recommender systems. ACM Trans. Inf. Syst. **22**(1), 5–53 (2004)
11. Khosravani, A., Farshchian, M., Jalali, M.: Introduction of synthetic and non-synthetic trust recommender models in collaborative filtering. In: International Congress on Technology, Communication and Knowledge, pp. 1–9 (2014)
12. Koren, Y., Bell, R.M., Volinsky, C.: Matrix factorization techniques for recommender systems. IEEE Comput. **42**(8), 30–37 (2009)
13. Ma, H., King, I., Lyu, M.R.: Learning to recommend with social trust ensemble. In: Proceedings of the 32nd Annual International ACM SIGIR Conference on Research and Development in Information Retrieval, pp. 203–210 (2009)

14. Ma, H., Yang, H., Lyu, M.R., King, I.: SoRec: social recommendation using probabilistic matrix factorization. In: Proceedings of the 17th ACM Conference on Information and Knowledge Management, pp. 931–940 (2008)
15. Massa, P., Avesani, P.: Trust-aware collaborative filtering for recommender systems. In: Meersman, R., Tari, Z. (eds.) OTM 2004. LNCS, vol. 3290, pp. 492–508. Springer, Heidelberg (2004). doi:10.1007/978-3-540-30468-5_31
16. O'Donovan, J., Smyth, B.: Eliciting trust values from recommendation errors. In: Eighteenth International Florida Artificial Intelligence Research Society Conference, Clearwater Beach, Florida, USA, pp. 289–294 (2005)
17. O'Donovan, J., Smyth, B.: Trust in recommender systems. In: Proceedings of the 10th International Conference on Intelligent User Interfaces, pp. 167–174. ACM (2005)
18. Portugal, I., Alencar, P., Cowan, D.: The use of machine learning algorithms in recommender systems: a systematic review. arXiv preprint arxiv:1511.05263 (2015)
19. Roy, F., Sarwar, S.M., Hasan, M.: User similarity computation for collaborative filtering using dynamic implicit trust. In: Khachay, M.Y., Konstantinova, N., Panchenko, A., Ignatov, D.I., Labunets, V.G. (eds.) International Conference on Analysis of Images, Social Networks and Texts, pp. 224–235. Springer, Heidelberg (2015)
20. Sarwar, B., Karypis, G., Konstan, J., Riedl, J.: Application of dimensionality reduction in recommender system-a case study. Technical report, DTIC Document (2000)
21. Wang, H., Wang, N., Yeung, D.Y.: Collaborative deep learning for recommender systems. In: Proceedings of the 21th ACM SIGKDD International Conference on Knowledge Discovery and Data Mining, pp. 1235–1244. ACM (2015)
22. Yang, B., Lei, Y., Liu, D., Liu, J.: Social collaborative filtering by trust. In: International Joint Conference on Artificial Intelligence, pp. 2747–2753 (2013)
23. Zheng, Y., Ouyang, Y., Rong, W., Xiong, Z.: Multi-faceted distrust aware recommendation. In: International Conference on Knowledge Science, Engineering and Management, pp. 435–446 (2015)

Neural Networks and Artificial Intelligence

A Subset Space Perspective on Agents Cooperating for Knowledge

Bernhard Heinemann$^{(\boxtimes)}$

Faculty of Mathematics and Computer Science,
FernUniversität in Hagen, 58084 Hagen, Germany
bernhard.heinemann@fernuni-hagen.de

Abstract. In this paper, we propose an additional application area of the subset space semantics of modal logic in terms of cooperating agents. While the original conception reflects both the knowledge acquisition process and the accompanying topological effect for a single agent, we show how a slight extension of that system can be utilized for modeling agents which, in a strict sense, cooperate for knowledge. In so doing, the agents will come in by means of so-called effort functions. These functions shall represent those of the agents' actions which are targeted at more knowledge of the whole group. Our investigations result in a particular multi-agent version of the well-known logic of subset spaces, which allows us to reason about qualitative aspects of cooperation like the dominance of a joint commitment over any individual effort. On the technical side, a soundness and completeness theorem for one of the logics arising from that will be proved.

Keywords: Reasoning about knowledge · Subset space semantics · Cooperating agents · Effort functions · Completeness

1 Introduction

The title of this paper contrasts with that of our contribution to the previous edition of KSEM in an obvious manner; see [12]. The research into the present object of study has, in fact, developed from a corresponding discussion at that meeting. The point here is that the suitability of the subset space semantics of modal logic for describing the joint knowledge of *cooperating agents* shall be examined.

Thus, we are concerned with the topic of *reasoning about knowledge*. The classic textbooks [6,16] give detailed information about the foundational concepts underlying this nowadays huge field of research. Accordingly, a binary *accessibility relation* R_A connecting *possible worlds* or *conceivable states of the world,* is associated with every instance A of a given finite group G of agents. The *knowledge of A* is then *defined* through the set of all formulas being valid at every state the agent considers possible at the actual one. This widespread and well-established way of thinking is complemented by Moss and Parikh's bi-modal

© Springer International Publishing AG 2016
F. Lehner and N. Fteimi (Eds.): KSEM 2016, LNAI 9983, pp. 503–514, 2016.
DOI: 10.1007/978-3-319-47650-6_40

logic of subset spaces, LSS (see [4,17], or Chap. 6 of [1]), of which the basic idea is reported in the following since our proceeding is based on it.

The *epistemic state* of an agent in question, i.e., the set of all those states that cannot be distinguished by what the agent topically knows, can be viewed as a *neighborhood U* of the actual state x of the world. The formulas of the language for LSS are then interpreted with respect to the resulting pairs x, U called *neighborhood situations.* Thus, both the set of all states and the set of all epistemic states constitute the relevant semantic domains as particular subset structures. The two modalities involved, K and $\square$, quantify over all elements of U and 'downward' over all neighborhoods contained in U, respectively. This means that K captures the notion of knowledge as usual (see [6] or [16] again), and $\square$ reflects a kind of *effort to acquire knowledge* since gaining knowledge goes hand in hand with a shrinkage of the epistemic state. In fact, knowledge acquisition is this way reminiscent of a *topological procedure* so that it is natural to ask for the actual topological significance of LSS (which, of course, has already been done; see, e.g., [7]). For some of the more recent developments pertaining to the topological and, respectively, the epistemic content of the logic of subset spaces, see [2,5,14,19].

In the paper [11], an attempt was made to obtain an appropriate multi-agent version of LSS. The key idea behind that approach is to incorporate the agents in terms of additional modalities. This leads to an essential modification of the logic (namely to a *hybridization* in the spirit of [3], Sect. 7.3), while the original semantics basically remains unchanged. The 'right' generalization of LSS to multiple agents has then been found by Wáng and Ågotnes; see [18]. Their *partition semantics,* however, assigns sets of sets of subsets to the agents, complicating the handling of the system considerably. Thus, one might still ask how far-reaching the 'naive' subset space semantics for multiple agents is. In this paper, we argue that the effect of agents being exclusively geared to *group knowledge* can satisfactorily be embodied on that basis.

In doing so, we follow a two-faced principle. The neighborhood components of any subset space are as of now (and here for the first time) viewed as knowledge states of the given group G of agents on the one hand, on the other hand, the agents are represented by the actions they are able to perform at any state; moreover, these actions are of such a nature that no decrease of the knowledge of G can result.[1] In other words, only (the improvement in) group knowledge is of interest to us and the individual knowledge of the agents is neglected here. In this way, the semantic problems with the latter can be avoided and a multi-agent logic of subset spaces can nevertheless still be brought about for an interesting special case.

Let us give a bit more precision to these ideas. In usual epistemic logic, the knowledge of a group of agents is differentiated into at least three kinds: every-one's knowledge, common knowledge, and (implicit or) distributed knowledge; see [6], Sect. 2.2. Each of these notions relies on the individual knowledge of the agents. As opposed to this, we take up the following position. We let the LSS-

---

[1] This is in accordance with the general setting in the context of subset spaces.

operator K become a *modality for group knowledge,* no matter whether its relation equals the union $\bigcup\limits_{A\in G} R_A$, the reflexive and transitive closure $\left(\bigcup\limits_{A\in G} R_A\right)^*$, or the intersection $\bigcap\limits_{A\in G} R_A$ of the individual relations (as related to the above cases), or something like that. Thus, we see it as less important how the interpretation of K has recourse to the relations for the individual agents. This 'coarsening of information', which is typical of LSS to a certain extent, is here justified since *cooperation* is the center of attention. To put it the other way round, it is only the group that counts. Several application examples relating to this immediately come into mind when thinking about that (e.g., *joint product development*). – We confine ourselves to such scenarios in this paper.

As indicated above, the idea of cooperation for group knowledge is realized by certain *actions,* which will formally be captured by agent-specific *functions* shrinking neighborhood situations. Since the latter requirement reminds one of the effect of the □-operator from above,[2] the term *effort functions* is used for them. And the fact that the set of all *group knowledge states* correspondingly decreases shows that the idea in question is adequately reflected by this means. In the logic, the effort functions will be appear as additional modalities.

Several examples of what can be expressed in the language resulting from this outline eventually will be given in Sect. 3. Both the concept of agent cooperation we have in mind and its distinction from a knowledge-competitive behavior of the agents as considered in the paper [12], are discussed in more detail there, too. However, that section primarily contains the formal design of our approach.

As a whole, this paper is organized as follows. In the next section, we recapitulate the language and the logic of subset spaces for single agents, for preparatory reasons. The content of Sect. 3 has just been mentioned. Section 4 contains the main technical contribution of this paper: the completeness of a new logic is proved there, which will deliberately be chosen from the system properties introduced in Sect. 3. In the final Sect. 5, we summarize, discuss some related issues, and point to a couple of topics for a future research.

Concerning the basic facts from modal logic needed here, the reader is referred to the standard textbook [3].

2   The Basic System

In this section, the starting point of our investigation is clarified on a technical level. To this end, the language for single-agent subset spaces, $\mathcal{L}$, is determined first. And second, the most important meta-results on the corresponding logic are recalled.

To begin with, we define the syntax of $\mathcal{L}$. Let Prop $= \{p, q, \dots\}$ be a denumerably infinite set of symbols called *proposition variables,* which shall represent the basic facts about the states of the world. Then, the set SF of all *subset formulas* over Prop is defined by the rule $\alpha ::= \mathsf{I} \mid p \mid \neg\alpha \mid \alpha \wedge \alpha \mid \mathsf{K}\alpha \mid \sqcup\alpha.$

---

[2] The part this operator plays in the novel system will become apparent later.

The missing boolean connectives are treated as abbreviations, as needed. The operators that are dual to K and $\Box$ are denoted by L and $\Diamond$, respectively. In view of our description in the previous section, K is called the *knowledge operator* and $\Box$ the *effort operator*.

We now fix the semantics of $\mathcal{L}$. For a start, we single out the relevant domains. We let $\mathcal{P}(X)$ designate the powerset of a given set X.

Definition 1 (Semantic Domains).

1. *Let X be a non-empty set (of* states*) and $\mathcal{O} \subseteq \mathcal{P}(X)$ a set of subsets of X. Then, the pair $\mathcal{S} = (X, \mathcal{O})$ is called a* subset frame.
2. *Let $\mathcal{S} = (X, \mathcal{O})$ be a subset frame. The set $\mathcal{N}_{\mathcal{S}} := \{(x, U) \mid x \in U \text{ and } U \in \mathcal{O}\}$ is then called the* set of neighborhood situations *of $\mathcal{S}$.*
3. *Let $\mathcal{S} = (X, \mathcal{O})$ be a subset frame. Under an $\mathcal{S}$-valuation we understand a mapping $V : \mathsf{Prop} \to \mathcal{P}(X)$.*
4. *Let $\mathcal{S} = (X, \mathcal{O})$ be a subset frame and V an $\mathcal{S}$-valuation. Then, $\mathcal{M} := (X, \mathcal{O}, V)$ is called a* subset space *(based on $\mathcal{S}$).*

Note that neighborhood situations denominate the semantic atoms of the bimodal language $\mathcal{L}$. The first component of such a situation indicates the actual state of the world, while the second reflects the uncertainty of the agent in question about it. Furthermore, Definition 1.3 shows that values of proposition variables depend on states only, which is in line with the common practice in epistemic logic; see again [6].

For a given subset space $\mathcal{M}$, the relation of *satisfaction*, $\models_{\mathcal{M}}$, is now defined between neighborhood situations of the underlying frame and formulas from SF. The notion of *validity* of formulas in subset spaces is then based on that. In the following, neighborhood situations are often written without parentheses.

Definition 2 (Satisfaction and Validity). *Let $\mathcal{S} = (X, \mathcal{O})$ be a subset frame.*

1. *Let $\mathcal{M} = (X, \mathcal{O}, V)$ be a subset space based on $\mathcal{S}$, and let $x, U \in \mathcal{N}_{\mathcal{S}}$ be a neighborhood situation of $\mathcal{S}$. Then*

$$
\begin{aligned}
&x, U \models_{\mathcal{M}} \top && \text{is always true} \\
&x, U \models_{\mathcal{M}} p && :\Longleftrightarrow x \in V(p) \\
&x, U \models_{\mathcal{M}} \neg\alpha && :\Longleftrightarrow x, U \not\models_{\mathcal{M}} \alpha \\
&x, U \models_{\mathcal{M}} \alpha \wedge \beta : \Longleftrightarrow x, U \models_{\mathcal{M}} \alpha \text{ and } x, U \models_{\mathcal{M}} \beta \\
&x, U \models_{\mathcal{M}} \mathsf{K}\alpha && :\Longleftrightarrow \forall y \in U : y, U \models_{\mathcal{M}} \alpha \\
&x, U \models_{\mathcal{M}} \Box\alpha && :\Longleftrightarrow \forall U' \in \mathcal{O} : [x \in U' \subseteq U \Rightarrow x, U' \models_{\mathcal{M}} \alpha],
\end{aligned}
$$

where $p \in \mathsf{Prop}$ and $\alpha, \beta \in \mathsf{SF}$. In case $x, U \models_{\mathcal{M}} \alpha$ is true we say that α holds in $\mathcal{M}$ at the neighborhood situation x, U.
2. *Let $\mathcal{M} = (X, \mathcal{O}, V)$ be a subset space based on $\mathcal{S}$. A subset formula α is called* valid *in $\mathcal{M}$ iff it holds in $\mathcal{M}$ at every neighborhood situation of $\mathcal{S}$.*

Note that the idea of knowledge and effort described in the introduction is made precise by Item 1 of this definition. In particular, knowledge is also here defined as validity at all states that are indistinguishable to the agent.

The next subject to be discussed is the *logic* of subset spaces, LSS. The subsequent axiomatization from [4], together with the standard modal proof rules, was proved to be *sound and complete for subset spaces* in Sect. 1.2 and, respectively, Sect. 2.2 there.

1. All instances of propositional tautologies
2. $\mathsf{K}(\alpha \to \beta) \to (\mathsf{K}\alpha \to \mathsf{K}\beta)$
3. $\mathsf{K}\alpha \to (\alpha \wedge \mathsf{KK}\alpha)$
4. $\mathsf{L}\alpha \to \mathsf{KL}\alpha$
5. $(p \to \Box p) \wedge (\Diamond p \to p)$
6. $\Box(\alpha \to \beta) \to (\Box \alpha \to \Box \beta)$
7. $\Box \alpha \to (\alpha \wedge \Box\Box\alpha)$
8. $\mathsf{K}\Box\alpha \to \Box\mathsf{K}\alpha$,

where $p \in \mathsf{Prop}$ and $\alpha, \beta \in \mathsf{SF}$.

The last schema is by far the most interesting one, as it displays the inter-relation between knowledge and effort. The members of this schema are called the *Cross Axioms (for* K *and* $\Box$*)* since [17].

Note that the axioms involving only proposition variables comply with the above remark on the third item of Definition 1. It ensues from these axioms that $\mathcal{L}$ can, in essence, only speak about the ongoing modification of knowledge (and, e.g., not about states changing over time).

Towards the final point in this section, let us take a brief look at the effect of the axioms from the above list within the relational framework of common modal logic. For this purpose, we consider bi-modal Kripke models $M = (W, R, R', V)$ satisfying the following four properties:

- the accessibility relation R of M belonging to the knowledge operator K is an equivalence,
- the accessibility relation R' of M belonging to the effort operator $\Box$ is reflexive and transitive,
- the composite relation $R' \circ R$ is contained in $R \circ R'$ (which is usually called the *cross property of* R, R'), and
- the valuation V of M is constant along every R'-path, for all proposition variables.

Such structures are called *cross axiom models* in the literature. It can be verified without much difficulty that LSS is also sound and complete with respect to the class of all cross axiom models, which was utilized for proving the *decidability* of LSS in [4], Sect. 2.3. This approach will play an important part in our discussion in Sect. 5 as well.

3   Subset Spaces for Cooperating Agents

Before we discuss some interesting attributes of cooperation, the formalism from the previous section will first be extended to the case of n agents, where $2 \leq n \in$

N. We again start with the logical language, containing n additional operators $C_1, \ldots, C_n$ as of now. Thus, the set nSF of all *n-subset formulas* over Prop is defined by the rule $\alpha ::= \top \mid p \mid \neg\alpha \mid \alpha \wedge \alpha \mid K\alpha \mid \Box\alpha \mid C_1\alpha \mid \cdots \mid C_n\alpha$. Obviously, SF is a subset of nSF. For $i = 1, \ldots, n$, the modality C_i is called the *cooperation operator associated with agent i*. The syntactic conventions from Sect. 2 apply correspondingly here. However, there is no need to consider the dual to C_i separately since C_i turns out to be a *functional* modality, thus being self-dual.

The additional semantic features are covered by the following definition, in particular.

Definition 3 (Subset Structures for Cooperating Agents). *Let $n \in \mathbb{N}$ be fixed as above, and let $\mathcal{S} = (X, \mathcal{O})$ be a subset frame.*

1. *For every agent $i \in \{1, \ldots, n\}$, let $f_i : X \times \mathcal{O} \to \mathcal{O}$ be a partial function satisfying the following two conditions for all $x \in X$ and $U \in \mathcal{O}$.*
   (a) The value $f_i(x, U)$ exists iff $x \in U$, and
   (b) f_i is contracting, i.e., if $f_i(x, U)$ exists, then $x \in f_i(x, U) \subseteq U$.
   Then, the triple
   $$\mathcal{S} = (X, \mathcal{O}, \{f_i\}_{1 \leq i \leq n})$$
   is called a subset frame for n cooperating agents (or an n-ca-subset frame for short), and the mapping f_i is called the effort function for agent i.
2. *The notions of neighborhood situation, $\mathcal{S}$-valuation and subset space for n cooperating agents (n-ca-subset space) are completely analogous to those introduced in Definition 1.*

Some comments on this definition seem to be appropriate. For a start, it is obviously guaranteed through the contraction requirement that the basic idea of the subset space semantics, as described in the introduction and made precise in Sect. 2, is preserved by the effort functions. These functions, considered after currying, depend on agents *and* states of the world. This is a (certainly debatable) question of system design, which at least results in the simplest ground logic.[3] On the other hand, taking *functions* for realizing the agents' epistemic actions is quite in conformity with the usual practice; cf. [6], Sect. 5.1. Finally, it should be mentioned that already the semantic domains represent the difference between the present approach and the one to knowledge-competitive agents proposed in the paper [12], since, apart from the appearance of the functions f_i, the set $\mathcal{O}$ remains undivided here; cf. Definition 3 there.

With regard to satisfaction and validity, we need not completely present the analogue of Definition 2 at this place, but may confine ourselves to the clauses for the new operators.

---

[3] If the effort functions shall depend on knowledge states alone, then *topological next-time logic,* see [9], would enter the field. This would lead to a somewhat more complicated but related system.

Definition 4 (Satisfaction). *Let* $\mathcal{S} = (X, \mathcal{O}, \{f_i\}_{1 \leq i \leq n})$ *be an n-ca-subset frame,* $\mathcal{M}$ *an n-ca-subset space based on* $\mathcal{S}$*, and* $x, U \in \mathcal{N}_\mathcal{S}$ *a neighborhood situation of* $\mathcal{S}$*. Then, for every* $i \in \{1, \ldots, n\}$ *and* $\alpha \in n\mathsf{SF}$*, we let*

$$x, U \models_\mathcal{M} \mathsf{C}_i \alpha : \iff x, f_i(x, U) \models_\mathcal{M} \alpha.$$

The *basic logic of n-ca-subset spaces*, CALSS_n, is yielded by adding the following list of axioms to those from Sect. 2 (and joining the necessitation rules for each of the C_i's with the proof rules for LSS).

9. $\mathsf{C}_i(\alpha \to \beta) \to (\mathsf{C}_i\alpha \to \mathsf{C}_i\beta)$
10. $\mathsf{C}_i \neg \alpha \leftrightarrow \neg \mathsf{C}_i \alpha$
11. $\mathsf{KC}_i\alpha \to \mathsf{C}_i\mathsf{K}\alpha$
12. $\Box\alpha \to \mathsf{C}_i\alpha$,

where $i \in \{1, \ldots, n\}$ and $\alpha, \beta \in n\mathsf{SF}$.

The effect of these axioms on the behavior of the operator C_i is as follows. Axiom 9 describes the usual distribution of a normal modality over implications. Axiom 10 captures the functionality of the accessibility relation associated with C_i; see, e.g., [8], Sect. 9 (for the operator *next* from temporal logic). In the present framework, it comes along with the case that those functions f_i have been assigned to the agents. The schema 11 is formally similar to the eighth one, thus comprising the *Cross Axioms for* K *and* C_i. And the fact that the effort functions, when defined, are *contracting*, is reflected by Axiom 11. With regard to the relational semantics, this last schema says that the accessibility relation for C_i is contained in that for $\Box$. It is also responsible, together with Axiom 10, for the absence of the counterpart of Axiom 5 for C_i[4].

The basic logic of n-ca-subset spaces might be regarded as not very interesting, since 'essentially new' relationships between the modalities are missing. In fact, it is known how to handle a system enriched by a set of functional modalities, particularly with regard to the typically desired meta-results. On the other hand, such relationships must not be too 'strong' in order to be technically manageable. In the following, we discuss a selection of candidates which do or might possibly constitute a good compromise in this respect. They all represent system properties that are worthy of discussion in the context of agents cooperating for group knowledge.

– There is some agent $j \in \{1, \ldots, n\}$, which contributes not less than all the other agents to the knowledge of the group.

This is to say that the knowledge state resulting from the effort function for j is (set-theoretically) equal to or smaller than those resulting from the effort functions for every $i \in \{1, \ldots, n\}$, which can formally be expressed by

$$\mathsf{C}_i\Box\alpha \to \mathsf{C}_j\alpha, \text{ where } i \in \{1, \ldots, n\} \text{ and } \alpha \in n\mathsf{SF}. \tag{1}$$

---

[4] In other words, the schema $(p \to \mathsf{C}_ip) \wedge (\mathsf{C}_ip \to p)$ is CALSS_n-derivable.

If this schema is added to CALSS_n, then the logic examined in the paper [13] is obtained. The understanding of the modality $\square$, playing the part of *missing effort operator* there, will be retained for the moment. However, the interpretations of both C_i as *cooperation action* and K as *group knowledge* are new here.

Concerning the relational semantics, it can be proved that a certain property of *lying in between* corresponds to the schema (1). In fact, (1) bears a structural resemblance to the schema for *weak density* from ordinary modal logic, $\square\square\alpha \to \square\alpha$; cf. [8], Sect. 1. This is the reason why that logic has turned out to have nice meta-properties like completeness and decidability.

The next item goes some way beyond this, namely by turning the lying-in-between criterion into a further *cross property*.

– By the *iterated effort* of any two agents i and j, at least as much can be achieved as by any single agent k :

$$\mathsf{C}_k\square\alpha \to \mathsf{C}_i\mathsf{C}_j\alpha, \text{ where } i, j, k \in \{1, \dots, n\} \text{ and } \alpha \in n\mathsf{SF}. \tag{2}$$

We will consider the system $\mathsf{CALSS}_n^+ := \mathsf{CALSS}_n + (2)$ in more detail below, in particular, with regard to the effect of the slight complication of the axioms.

Iterating the formal realizations of the agents' actions can be taken as a sequentialized modeling of *joint effort,* provided that the order in which the effort functions are applied does not play any role.[5] The latter postulation leads us to the schema

$$\mathsf{C}_i\mathsf{C}_j\alpha \to \mathsf{C}_j\mathsf{C}_i\alpha, \text{ where } i, j \in \{1, \dots, n\} \text{ and } \alpha \in n\mathsf{SF}, \tag{3}$$

since this one implies, together with Axiom 10, the 'Commutation Axioms' $\mathsf{C}_i\mathsf{C}_j\alpha \leftrightarrow \mathsf{C}_j\mathsf{C}_i\alpha$ needed for that.

In this connection it should be mentioned that one can also express that *every* effort gap can be closed by iteration, in other words that $\square$ represents the reflexive and transitive closure of the accessibility relation associated with $\mathsf{C}_1 \cup \cdots \cup \mathsf{C}_n$. As to the precise facts and circumstances regarding this, the reader is referred to the paper [10].

There is still another way of dealing with joint effort into cooperating, which stresses the aspect of parallelism a bit more.

– Any jointly acting agents i and j are more successful than one of them.

In other words, there is a *common refinement* of the outcomes of their respective effort functions:

$$\mathsf{C}_i\square\alpha \to \mathsf{C}_j\Diamond\alpha, \text{ where } i, j \in \{1, \dots, n\} \text{ and } \alpha \in n\mathsf{SF}. \tag{4}$$

Thus, $\square$ is now viewed as a *joint effort operator*. In usual modal logic, (4) means that the respectively assigned accessibility relations are *confluent* (also

---

[5] This point of view is derived by analogy with sequencing from the theory of parallel programming.

called *weakly directed;* cf. again [8], Sect. 1). In connection with subset spaces, the corresponding *closure under intersections* is a rather complicated issue, which shall not be considered further in this paper (except for the example following in a moment); cf. [20].

Each of the properties (1)–(4) captures a particular aspect of cooperation. Although this concept is the main focus in the present paper, it should be remarked that each of the corresponding schemata also tells us something about the (attainable) knowledge of the given group of agents. This is a fortiori the case when α is a knowledge formula. For instance, $C_i\Box K\alpha \to C_j\Diamond K\alpha$ says that the group knows α after a joint effort of the agents i and j.

The rest of the technical part of this paper deals with the logic CALSS_n^+. The intended semantic structures for this system are given by the next definition.

Definition 5 (n-Die-Subset Spaces). *Let* $\mathcal{M} = (X, \mathcal{O}, \{f_i\}_{1\leq i\leq n}, V)$ *be a subset space for n cooperating agents. Then, $\mathcal{M}$ is called an n-die-subset space (where 'die' means 'dominating iterated effort'), iff, for all $i, j, k \in \{1, \ldots, n\}$, $x \in X$, and $U \in \mathcal{O}$ containing x,*

$$f_j\left(x, f_i(x, U)\right) \subseteq f_k(x, U).$$

Finally in this section, it is proved that CALSS_n^+ is *sound* with respect to the class of all n-die-subset spaces.

Proposition 1. *Let* $\mathcal{M} = (X, \mathcal{O}, \{f_i\}_{1\leq i\leq n}, V)$ *be an n-die-subset space. Then, every axiom from the above list or from the schema (2) is valid in $\mathcal{M}$; moreover, every CALSS_n^+-rule preserves the validity of formulas.*

Proof. We confine ourselves to the schema (2). Let $C_k\Box\alpha \to C_iC_j\alpha$ be an instance of this, and let $x, U \models_{\mathcal{M}} C_k\Box\alpha$ be satisfied for any neighbourhood situation x, U of the frame underlying $\mathcal{M}$. According to Definition 4, this means that $x, f_k(x, U) \models_{\mathcal{M}} \Box\alpha$. Due to Definition 2, $x, U' \models_{\mathcal{M}} \alpha$ is valid for all $U' \in \mathcal{O}$ such that $x \in U' \subseteq f_k(x, U)$. From this we obtain that $x, f_j\left(x, f_i(x, U)\right) \models_{\mathcal{M}} \alpha$, since $\mathcal{M}$ is an n-die-subset space. Consequently, $x, f_i(x, U) \models_{\mathcal{M}} C_j\alpha$ is satisfied, again by Definition 4, which then gives us $x, U \models_{\mathcal{M}} C_iC_j\alpha$. This proves (the special case of) the proposition.

4   Completeness

This section is directed to the expert. It is aimed at proving the following main result of this paper.

Theorem 1. *The logic CALSS_n^+ is complete for the class of all n-die-subset spaces.*

Proof. Due to space limitations, only rough sketch can be drawn on the somewhat involved argument. Almost all technical details must therefore be omitted.

The overall structure of the proof consists of an infinite step-by-step model construction, as it is often the case with subset space logics.[6] Using such a procedure seems to be necessary, since subset spaces do not straightforwardly harmonize with the main modal means supporting completeness, viz *canonical models;* see [1], p. 321, for a more precise comment on this matter.

The canonical model of CALSS_n^+ will come into play nevertheless. So let us fix some notations concerning that model first. Let $\mathcal{C}$ be the set of all maximal CALSS_n^+-consistent sets of formulas from $n\mathsf{SF}$. Furthermore, let $\xrightarrow{\mathsf{K}}$, $\xrightarrow{\Box}$, and $\xrightarrow{\mathsf{C}_i}$ be the accessibility relations induced on $\mathcal{C}$ by the modalities K, $\Box$, and C_i, respectively, where $i \in \{1, \ldots, n\}$. And finally, let $\alpha \in n\mathsf{SF}$ be a formula which is *not* contained in CALSS_n^+. Then, we have to find a model for $\neg\alpha$.

This model is constructed by recursion in such a way that better and better intermediary structures are obtained, which means that more and more existential formulas of the form $\mathsf{L}\beta$, $\Diamond\beta$, or $\mathsf{C}_i\beta$ are realized. In order to ensure that the resulting limit structure behaves as desired, several requirements on those approximations have to be met at every stage. Describing the details of this construction and verifying the necessary properties each time, makes up the very core of the proof. The subsequent lemma is needed in the course of this.

Lemma 1. *Let $n \in \mathbb{N}$ be fixed as above, and let $i, j, k \in \{1, \ldots, n\}$. Suppose that $\Delta, \Theta, \Xi \in \mathcal{C}$ are maximal CALSS_n^+-consistent sets of formulas satisfying $\Delta \xrightarrow{\mathsf{C}_i} \Theta \xrightarrow{\mathsf{C}_j} \Xi$. Then, there is some $\Psi \in \mathcal{C}$ such that $\Delta \xrightarrow{\mathsf{C}_k} \Psi \xrightarrow{\Box} \Xi$.*

Proof. One can argue in a similar way as in the case of the Cross Axioms for the basic logic of subset sapces; cf. [4], Proposition 2.2. Of course, those axioms have to be replaced with the schema (2) here.

The previous lemma guarantees that the relational counterpart of the n-die-characteristic from Definition 5 is satisfied on the canonical model of CALSS_n^+, which marks one of the essential steps towards the success of our model construction. However, the decisive point here is that the otherwise applicable *end extensions* of the approximating subset structures must, in some cases, be substituted with suitable *inserting procedures,* which are technically feasible only because we are dealing with *functional modalities* in doing so.

For the reasons stated above, the rest of the proof has to be postponed to the full version of this draft.

5   Conclusion, Addenda and Final Comments

Summarizing the paper up to this point, it can be stated that its main contribution to the development of subset spaces (and, with this, to the field of knowledge science) comes from reinterpreting these domains as knowledge structures for a

---

[6] One or another proof of such a kind can be found in the literature; see, as regards a fully completed version for LSS, [4]. Note that the special circumstances of each individual case require an appropriate adjustment, which is most often non-trivial.

group of cooperating agents. In doing so, the agents are represented through their respective actions having an impact on the knowledge of the whole group. We discussed several conditions governing that cooperation in terms of the accompanying effort functions, particularly with regard to the idea of joint and iterated effort, respectively. Moreover, we introduced the basic logic CALSS_n for n agents cooperating for knowledge as well as the logic CALSS_n^+ for dominating iterated effort. The soundness (Proposition 1) and completeness (Theorem 1) of the latter system make up the main technical results of this paper.

As was to be expected, a corresponding soundness and completeness result can be proved for the logic CALSS_n as well. But we even obtain more in this case, as demonstrated in the following theorem.

Theorem 2. *The logic CALSS_n is sound and complete for the class of all n-ca-subset spaces. In addition, CALSS_n is a decidable set of formulas.*

Proof. The proof of the first part of this theorem is similar to the above proof for CALSS_n^+. As to the second assertion, cf. the proof of Theorem 2 in [13].

The question comes up why the decidability result for CALSS_n^+ is missing here. The reason for this lies in the fact that the additional cross property, being associated with the schema (2) from Sect. 3, cannot be dealt with on the basis of the methods established so far. One will, in particular, encounter problems when trying to apply *filtrations,* since none of the modalities involved here is S5 (like K; cf. Axiom 8 and Axiom 11 above). Thus, the solution of the decidability problem for CALSS_n^+ has to be postponed to a future research.

While the logic resulting from adding the schema (3) from Sect. 3 to CALSS_n can, even in combination with (2), be handled in a similar way to CALSS_n^+, the schema (4) is more difficult to manage. Thus the latter is, together with the treatment of further facets of cooperation (like the technically more promising 'complement' of (4), $\Diamond \mathsf{C}_i \alpha \rightarrow \Box \mathsf{C}_j \alpha$, or a combination of cooperating and competitive agents (see [12] once again)), a future topic as well.

Our final remark concerns the exact computational complexity of the satisfiability problem for the logics considered in this paper. It is to be expected that the corresponding bounds can be determined not until solving that problem for the usual logic of subset spaces. Regarding the latter, the following partial results are known. (1) The LSS-satisfiability problem is hard for PSPACE. (2) The LSS$^-$-satisfiability problem is complete for PSPACE, where LSS$^-$ is the Cross Axiom-free fragment of LSS. The first of these findings is contained in the paper [15] and the second in [2].

References

1. Aiello, M., Pratt-Hartmann, I.E., van Benthem, J.F.A.K.: Handbook of Spatial Logics. Springer, Dordrecht (2007)
2. Balbiani, P., Ditmarsch, H., Kudinov, A.: Subset space logic with arbitrary announcements. In: Lodaya, K. (ed.) ICLA 2013. LNCS, vol. 7750, pp. 233–244. Springer, Heidelberg (2013). doi:10.1007/978-3-642-36039-8_21

3. Blackburn, P., de Rijke, M., Venema, Y.: Modal Logic, Cambridge Tracts in Theoretical Computer Science, vol. 53. Cambridge University Press, Cambridge (2001)
4. Dabrowski, A., Moss, L.S., Parikh, R.: Topological reasoning and the logic of knowledge. Ann. Pure Appl. Logic **78**, 73–110 (1996)
5. Ditmarsch, H., Knight, S., Özgün, A.: Arbitrary announcements on topological subset spaces. In: Bulling, N. (ed.) EUMAS 2014. LNCS (LNAI), vol. 8953, pp. 252–266. Springer, Heidelberg (2015). doi:10.1007/978-3-319-17130-2_17
6. Fagin, R., Halpern, J.Y., Moses, Y., Vardi, M.Y.: Reasoning about Knowledge. MIT Press, Cambridge (1995)
7. Georgatos, K.: Knowledge theoretic properties of topological spaces. In: Masuch, M., Pólos, L. (eds.) Logic at Work 1992. LNCS, vol. 808, pp. 147–159. Springer, Heidelberg (1994). doi:10.1007/3-540-58095-6_11
8. Goldblatt, R.: Logics of Time and Computation. CSLI Lecture Notes, 2nd edn., vol. 7. Center for the Study of Language and Information, Stanford (1992)
9. Heinemann, B.: Topological nexttime logic. In: Kracht, M., de Rijke, M., Wansing, H., Zakharyaschev, M. (eds.) Advances in Modal Logic 1, vol. 87, pp. 99–113. CSLI Publications, Kluwer, Stanford, CA (1998)
10. Heinemann, B.: A PDL-like logic of knowledge acquisition. In: Diekert, V., Volkov, M.V., Voronkov, A. (eds.) CSR 2007. LNCS, vol. 4649, pp. 146–157. Springer, Heidelberg (2007). doi:10.1007/978-3-540-74510-5_17
11. Heinemann, B.: Topology and knowledge of multiple agents. In: Geffner, H., Prada, R., Machado Alexandre, I., David, N. (eds.) IBERAMIA 2008. LNCS (LNAI), vol. 5290, pp. 1–10. Springer, Heidelberg (2008). doi:10.1007/978-3-540-88309-8_1
12. Heinemann, B.: Subset spaces modeling knowledge-competitive agents. In: Zhang, S., Wirsing, M., Zhang, Z. (eds.) KSEM 2015. LNCS (LNAI), vol. 9403, pp. 3–14. Springer, Heidelberg (2015). doi:10.1007/978-3-319-25159-2_1
13. Heinemann, B.: Augmenting subset spaces to cope with multi-agent knowledge. In: Artemov, S., Nerode, A. (eds.) LFCS 2016. LNCS, vol. 9537, pp. 130–145. Springer, Heidelberg (2016). doi:10.1007/978-3-319-27683-0_10
14. Heinemann, B.: Topological facets of the logic of subset spaces (with emphasis on canonical models). J. Logic Comput. (2016). 22 pp. doi:10.1093/logcom/exv087
15. Krommes, G.: A new proof of decidability for the modal logic of subset spaces. In: Ten Cate, B. (ed.) Proceedings of the Eighth ESSLLI Student Session, Vienna, Austria, pp. 137–147, August 2003
16. Meyer, J.J.C., van der Hoek, W.: Epistemic Logic for AI and Computer Science, Cambridge Tracts in Theoretical Computer Science, vol. 41. Cambridge University Press, Cambridge (1995)
17. Moss, L.S., Parikh, R.: Topological reasoning and the logic of knowledge. In: Moses, Y. (ed.) Theoretical Aspects of Reasoning about Knowledge (TARK 1992), pp. 95–105. Morgan Kaufmann, Los Altos (1992)
18. Wáng, Y.N., Ågotnes, T.: Multi-agent subset space logic. In: Proceedings 23rd IJCAI, pp. 1155–1161. AAAI (2013)
19. Wáng, Y.N., Ågotnes, T.: Subset space public announcement logic. In: Lodaya, K. (ed.) ICLA 2013. LNCS, vol. 7750, pp. 245–257. Springer, Heidelberg (2013). doi:10.1007/978-3-642-36039-8_22
20. Weiss, M.A., Parikh, R.: Completeness of certain bimodal logics for subset spaces. Stud. Logica **71**, 1–30 (2002)

Context-Aware Tree-Based Convolutional Neural Networks for Natural Language Inference

Zhao Meng[1,2], Lili Mou[1,2], Ge Li[1,2(✉)], and Zhi Jin[1,2(✉)]

[1] Key Laboratory of High Confidence Software Technologies, Peking University,
Ministry of Education, Beijing, China
doublepower.mou@gmail.com, {lige,zhijin}@sei.pku.edu.cn
[2] Software Institute, Peking University, Beijing, China
zhaomeng.pku@outlook.com

Abstract. Natural language inference (NLI) aims to judge the relation between a premise sentence and a hypothesis sentence. In this paper, we propose a context-aware tree-based convolutional neural network (TBCNN) to improve the performance of NLI. In our method, we utilize tree-based convolutional neural networks, which are proposed in our previous work, to capture the premise's and hypothesis's information. In this paper, to enhance our previous model, we summarize the premise's information in terms of both word level and convolution level by dynamic pooling and feed such information to the convolutional layer when we model the hypothesis. In this way, the tree-based convolutional sentence model is context-aware. Then we match the sentence vectors by heuristics including vector concatenation, element-wise difference/product so as to remain low computational complexity. Experiments show that the performance of our context-aware variant achieves better performance than individual TBCNNs.

Keywords: Context-awareness · Tree-based convolutional neural network · Natural language inference

1 Introduction

Natural language inference (NLI), also known as *recognizing textual entailment*, is an important task in natural language processing (NLP), and has profound impact on other NLP applications [1,2], including paraphrase detection [3], question answering [4], and automatic summarization [5]. Formally, NLI aims to judge the relation between two sentences, called a premise sentence and a hypothesis sentence, respectively. The objectives of NLI are `Entailment`, `Contradiction`, and `Neutral`, where `Entailment` means the hypothesis sentence can be entailed from the premise sentence, `Contradiction` means the two sentences are contradictory to each other, and `Neutral` means the two sentences are logically independent to each other [2]. Examples are illustrated in Table 1.

Traditional approaches to NLI mainly focus on feature engineering [4] and formal reasoning [6]. Recently, neural networks have become one of the mainstream

© Springer International Publishing AG 2016
F. Lehner and N. Fteimi (Eds.): KSEM 2016, LNAI 9983, pp. 515–526, 2016.
DOI: 10.1007/978-3-319-47650-6_41

approaches to almost every NLP task, including natural language inference. Researchers have applied various neural models, e.g., recurrent neural networks (RNNs) [7,8], to capture sentence-level meanings; then heuristic matching or word-by-word attention mechanisms are used to classify the relation between the premise and the hypothesis. While attention mechanisms typically yield higher performance, they are more computationally intensive than heuristic matching in terms of complexity order: $\mathcal{O}(n^2)$ versus $\mathcal{O}(n)$, where n is the number of words in a sentence. Therefore, heuristic matching is still a hot research topic in the NLP community, especially when complexity is a major concern.

In our previous work [9], we apply a tree-based convolutional neural network (TBCNN) as the underlying sentence model and then match the premise and the hypothesis by heuristics like vector concatenation, element-wise product, and element-wise difference. Such approach has achieved state-of-the-art performance in the complexity of $\mathcal{O}(n)$, justifying the rationale of using TBCNN as the underlying sentence model.

However, the main shortcoming of this model is that the premise and the hypothesis are modeled independently, that is, when extracting features of the hypothesis by tree-based convolution, the model is unaware of the information of the premise. Evidence in the literature shows that context-awareness may be important in sentence pair modeling, which means that it is important to concern the information of the other sentence when we are modeling on one of the sentences: Rocktäschel et al. [8] propose a single-chain RNN that runs through both sentences, and achieve higher performance than two separate RNNs. This could be intuitively thought of as such that, in Table 1 for example, it is valuable to know the phrase *drinking orange juice* in the premise, when we model *drinking juice* in the hypothesis. Likewise, the phrase *An older man* provides a useful hint of the contradiction to *Two women*. Hence, we are curious whether such context-awareness can benefit our TBCNN model.

Table 1. Examples in the Stanford Natural Language Inference (SNLI) dataset. The classification objectives of NLI are `Entailment`, `Contradiction`, and `Neutral`. The `Neutral` class indicates two irrelevant sentences.

Premise	
An older man is drinking orange juice at a restaurant	
Hypothesis	Label
A man is drinking juice	Entailment
Two women are at a restaurant drinking wine	Contradiction
A man in a restaurant is waiting for his meal to arrive	Neutral

In this paper, we propose a context-aware tree-based convolutional neural network to improve the performance of NLI. Our idea is to summarize the premise's knowledge as fixed-size vectors, which are fed to the tree-based convolutional layer when we model the hypothesis. Then two sentences' information is matched by heuristics as in [9]. In this way, the underlying sentence model is context-aware, but the overall complexity remains low, i.e., $\mathcal{O}(n)$, in contrast to word-by-word attention mechanisms. We conduct our experiments on a large open dataset, the Stanford Natural Language Inference (SNLI) Corpus [10], and achieve better performance than our previously published TBCNN model did.

2   Related Work

In past years, researchers have mainly focused on feature-based approaches to natural language inference. For example, Harabagiu et al. [4] use linguistic knowledge and lexical alignment to decide the extent to which a sentence can be entailed from another. Bos et al. [6], on the other hand, use formal reasoning by combining shallow (word overlap) and deep (semantic parsing) NLP methods. While reasoning approaches can search for a proof in logical forms for entailment recognition, their scope and accuracy are highly limited.

Recent advances in neural networks bring new methods to NLI, which can be viewed as a task of sentence pair modeling, that is, the goal is to determine the relation between a pair of sentences (the premise and the hypothesis). Typically, these approaches involve two steps: sentence modeling and matching.

2.1   Sentence Modeling

In this step, the goal is aimed at capturing the meaning of a sentence. Kalchbrenner et al. [11] and Kim [12] use convolutional neural networks (CNNs) to model sentences; in CNNs, a sliding window extracts features of neighboring words. Recurrent neural networks (RNNs) iteratively pick up words in a sentence by keeping one or a few hidden states [13]. Socher et al. [14] propose recursive neural networks—which utilize a tree structure and—by propagating information recursively from leaf nodes to the root to summarize a sentence as a vector. In our previous work [15], we propose a tree-based convolutional neural network (TBCNN), which combines the merits of CNNs and recursive nets: it is structure-sensitive as recursive nets and has short propagation paths like CNNs. We have achieved state-of-the-art performance in several sentence classification tasks with TBCNN, showing its effectiveness. Based on the above sentence models, many studies build sentence pair models upon RNNs [7,8], CNNs [16,17], etc.

2.2   Sentence Matching

To determine the relation between a pair of sentences, Zhang et al. [17] and Hu et al. [16] concatenate the vectors of each sentence; He et al. [18] use Euclidean distance, cosine, and element-wise absolute difference as features. Other researchers compute word-by-word similarity matrices [3,7].

Recently, Rocktäschel et al. [8] demonstrate that context-awareness is important in sentence matching. They propose several methods including single-chain

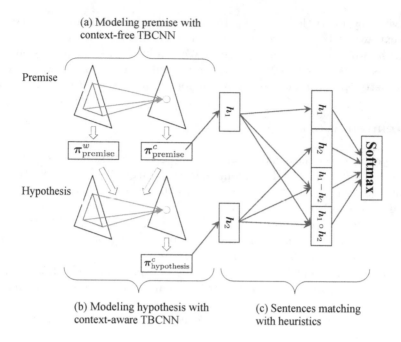

Fig. 1. The context-aware TBCNN model.

RNNs, static attention, and word-by-word attention, outperforming separate underlying sentence models. Wang et al. [19] further improve the performance by developing more elegant attention methods.

Although word-by-word attention and similarity matrices usually outperform simple matching heuristics, they are of higher overall complexity order, i.e., $\mathcal{O}(n^2)$. In this paper, we focus on $\mathcal{O}(n)$ methods: we enhance the TBCNN model with context-awareness, but remain in low complexity.

3 Our Approach

In this section, we describe our approach in detail. Subsection 3.1 provides an overview of our approach. Subsection 3.2 introduces the context-free tree-based convolutional neural network (TBCNN), which serves as the base model. We propose the context-aware TBCNN variant in Subsect. 3.3. Then we present matching heuristics and the training objective in Subsects. 3.4 and 3.5, respectively.

3.1 Overview

Figure 1 depicts the overview of our model. Concretely, our model has three main components:

- First, we apply TBCNN to capture the meaning of the premise (Fig. 1a). This part is essentially the same as the original context-free (i.e., individual) TBCNN model in [9].
- Then, we design another TBCNN to model the hypothesis (Fig. 1b). Contrary to previous work, the tree-based convolution here is aware of the premise's information. This is accomplished by summarizing the premise as fixed-size vectors, which are fed to the convolutional layer to interact with the hypothesis.
- After sentence modeling, we match the two sentences' vectors by heuristics including concatenation, element-wise product and difference. Finally, we use a softmax layer for classification. (See Fig. 1c.)

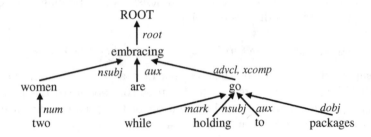

Fig. 2. Dependency parse tree

3.2   Tree-Based Convolution

The tree-based convolutional neural network (TBCNN) is proposed to model the parse trees of both programming languages [20] and natural languages [15]. In this part, we elaborate the process of tree-based convolution without context awareness, which is used to model the premise sentence in the NLI task.

First, we apply the Stanford parser[1] [21] to convert a sentence to a dependency tree, illustrated in Fig. 2. In our notations, an edge $a \xrightarrow{r} b$ refers to a being governed by b with the dependency type r. In total, we have approximately 30 different types (e.g., `nsubj`, `dobj`); other rare relations are mapped to a special type `default` in our study.

Then we perform the tree-based convolution over the dependency tree. We use pretrained word embeddings [22] as input signals. A subtree-based sliding window extracts structural features as the output of the convolution (denoted as y). Formally, given a parent node p and its child nodes $c_1, \cdots, c_n$, we have

$$ y = f\left(W_p p + \sum_{i=1}^{n} W_{r[c_i]} c + b \right) \tag{1} $$

---

[1] http://nlp.stanford.edu/software/lex-parser.shtml.

as detected features of the tree-based convolution at a certain position. Here, bold letters p and c_i refer to word embeddings of corresponding nodes; b is the bias vector. f is a nonlinear function; we used rectified linear units (ReLU) in our experiments, given by

$$f(x) = \begin{cases} x, & \text{if } x > 0 \\ 0, & \text{otherwise} \end{cases} \tag{2}$$

Notice that we have the same number of detected features y's and words in the original sentence, and that the number varies in different data samples. Hence, we apply dynamic pooling to summarize the features extracted by convolution. Specifically, a max-pooling operator takes the maximum value in each dimension of all features along the dependency tree. Suppose the tree-based convolutional layer extracts m features of n_c dimensions, the pooling layer outputs a vector π_{premise}^c, its j-th dimension being

$$\pi_{\text{premise}}^c[j] = \max\left\{ y_1[j], y_2[j], \cdots, y_m[j] \right\}, \quad 1 \leqslant j \leqslant n_c \tag{3}$$

The superscript c indicates that the features are pooled from those after convolution; the subscript suggests the features correspond to the premise. In all, π_{premise} provides summarized information of the premise and is used for sentence matching, described in Subsect. 3.4.

3.3   Context-Awareness for Tree-Based Convolution

As has been discussed in previous sections, context-awareness is important to sentence pair modeling, including the task of natural language inference. In this subsection, we propose a context-aware TBCNN variant to model the hypothesis sentence, that is to say, when we extract features of the hypothesis by tree-based convolution, we are equipped with the information of the premise sentence so that the second sentence model may focus on more relevant information rather than merely extract generic features of the sentence.

Concretely, we apply dynamic pooling to summarize the knowledge of the premise as fixed-size vectors and feed the vectors to the tree-based convolutional layer when modeling the hypothesis. More specially, we leverage two sets of features, which are of different abstraction levels, from the premise sentence:

- **Convolution-level features.** In Subsect. 3.2, we have summarized the premise's information as a vector π_{premise}^c. It is natural to use such knowledge in the context-aware TBCNN model.
- **Word-level features.** In addition to the above convolution-level features, we also pool the raw word embeddings in the premise sentence as a vector, given by

$$\pi_{\text{premise}}^w[j] = \max\left\{ w_1[j], w_2[j], \cdots, w_m[j] \right\}, \quad 1 \leqslant j \leqslant n_e \tag{4}$$

Similar to Eq. (3), we have m vectors of word embeddings $w_1, \cdots, w_m$, and n_e denotes embeddings' dimension. The superscript w indicates that this set of features is of word level.

After obtaining the convolution-level and word-level features, π^c_{premise} and π^w_{premise}, respectively, we feed them to the tree-based convolutional layer where we extract the hypothesis's features. We modify the convolution formula (1) as follows.

$$y = f\left(W_p p + \sum_{i=1}^{n} W_{r[c_i]} c_i + W^c \pi^c_{\text{premise}} + W^w \pi^w_{\text{premise}} + b\right) \qquad (5)$$

In the above equation, W_p, $W_{r[c_i]}$, and b are the same weights and bias as in Eq. (1) because the tree-based convolution operator detects general structural information of sentences. However, during the interaction, we also provide feature detectors (convolution operators) with the premise's information π^c_{premise} and π^w_{premise}, linearly transformed by the weight matrices W^c and W^w, respectively.

After the context-aware convolution process, we obtain a set of features over the dependency tree of the hypothesis sentence. Then we use max pooling to summarize them as a vector $\pi^c_{\text{hypothesis}}$ in a same way as Eq. (3). (Details are not repeated here.) The vector $\pi^c_{\text{hypothesis}}$, along with that of the premise sentence π^c_{premise}, is used for sentence matching, as will be described in the next subsection.

3.4 Matching Heuristics

Before matching the two sentences' vectors, we transform them by a fully-connected hidden layer

$$h_{\text{premise}} = f\left(W_h \pi^c_{\text{premise}} + b_h\right) \qquad (6)$$

$$h_{\text{hypothesis}} = f\left(W_h \pi^c_{\text{hypothesis}} + b_h\right) \qquad (7)$$

This hidden layer is designed empirically and not the main point of this paper.

Then we apply several heuristics proposed in our previous work [9] to match the premise and the hypothesis:

– Concatenation of the two sentence vectors:

$$concat = [h_{\text{premise}}; h_{\text{hypothesis}}]$$

– Element-wise difference:

$$diff = [h_{\text{premise}} - h_{\text{hypothesis}}]$$

– Element-wise product:

$$prod = [h_{\text{premise}} \circ h_{\text{hypothesis}}]$$

Then, they are further concatenated as the final features m

$$m = [concat; diff; prod]$$

which are fed to the softmax output layer. (In the above equations, semicolons refer to vector concatenation.)

In this way, we manage to integrate the premise's information to the hypothesis, but remain a low overall complexity of $\mathcal{O}(n)$. The computational complexity lies in the underlying (both context-free and context-aware) sentence models.

3.5 Training Objective

Finally, we feed the sentence matching vector m to a softmax layer as the output. We use standard cross-entropy loss as our cost function. Let m be the number of data samples in the training set and n_l be the number of labels. Suppose further $t^{(i)}$ is the one-hot ground truth and $y^{(i)}$ is the output of the softmax layer for the i-th data sample. The j-th element in $t^{(i)}$ is on ($= 1$) if the sample belongs to the j-th class. The training objective is

$$J = -\sum_{i=1}^{m}\sum_{j=1}^{n_l} t_j^{(i)} \log\left(y_j^{(i)}\right)$$

The network is trained by mini-batched stochastic gradient descent with backpropagation and regularized by dropout.

4   Evaluation

In this section, we describe the dataset of our experiment in Subsect. 4.1. We present our hyperparameters and settings in Subsect. 4.2. We compare our model with other models and analyze different context-aware TBCNN variants in detail in Subsect. 4.3.

Table 2. Statistics of the SNLI dataset.

Train	Validation	Test
550,152	10,000	10,000

4.1   Dataset

We use the Stanford Natural Language Inference (SNLI)[2] [10] to evaluate our context-aware TBCNN model. SNLI is a large dataset of more than 550k samples. All samples in SNLI are human-written sentences and are labeled manually. As illustrated in Table 1, SNLI has three categories of labels: Entailment, Contradiction, and Neural. Entailment means the hypothesis can be inferred

---

[2] http://nlp.stanford.edu/projects/snli/.

from the premise, while `Contradiction` means the two sentences have contradictory meanings. `Neural`, however, indicates that the premise and the hypothesis are irrelevant to each other. The labels are roughly equal-distributed in the dataset. We apply the official split for train/validation/test, which is listed in Table 2.

4.2 Experimental Settings

In this subsection, we details the experimental settings for our context-aware TBCNN. All layers including the word embeddings are 300-dimensional. Embeddings are pretrained on the Wikipedia corpus and fine-tuned during training. We use mini-batch stochastic gradient descent and set the mini-batch size to 50. The above values are chosen empirically mainly following [9]. We tune the following hyperparameters on the validation test: learning rate is chosen from $\{3, 1, 0.3, 0.1\}$. Power decay of learning rate is chosen from $\{1x, 0.9x, 0.3x\}$, which is the residual of learning rate after one epoch; intuitively, they can be thought of as no, slow, or fast decay. We do not add ℓ_2 penalty for convenience and simplicity. Instead, we use dropout [23] to regularize our model. The dropout rate is chosen from $\{0, 0.1, 0.2, 0.3, 0.4\}$. For efficiency of hyperparameter tuning, we do not conduct meaningless settings (e.g., a larger dropout rate when the model has ready been underfitting). Our context-aware TBCNN model reaches its peak performance when the learning rate is 0.3, the power decay is 0.9, and the dropout rate is 0.3. In the following part of our paper, we report the test accuracy that corresponds to the highest performance on the validation test.

In order to have a better understanding of the role of convolution-level features and word-level features when we model the hypothesis, we have an additional variant of context-aware TBCNN, where we only leverage the convolution-level features. That is to say, we only feed $\pi_{premise}^c$ to the tree-based convolutional layer when modeling on the hypothesis. Word-level features are simply ignored in this variant. Thus, the output vector of the hypothesis's convolutional layer is:

$$y = f\left(W_p \boldsymbol{p} + \sum_{i=1}^{n} W_{r[c_i]} \boldsymbol{c_i} + W^c \boldsymbol{\pi}_{\text{premise}}^c + \boldsymbol{b}\right) \qquad (8)$$

The above equation is different from Eq. (5) in that we do not have the $W^w \boldsymbol{\pi}_{\text{premise}}^w$ term. By contrast, the full context-aware TBCNN proposed in Sect. 3 has two levels of abstraction of the premise.

4.3 Performance

We present the performance of our models in comparison with previously published results in Table 3. As we can see, our full context-aware TBCNN model outperforms the competing approaches which are of $\mathcal{O}(n)$ overall complexity,

Table 3. Context-awared TBCNN compared with other models.

Overall complexity	Model	Test accuracy (%)
$\mathcal{O}(n)$	Unlexicallized features [10]	50.4
	Lexicallized features [10]	78.2
	Vector sum + MLP [10]	75.3
	Vanilla RNN + MLP [10]	72.2
	LSTM RNN + MLP [10]	77.6
	CNN + cat [9]	77.0
	GRU w/skip-thought pretraining [24]	81.4
	Single-chain LSTM RNN [8] + two-way attention [8]	81.4
		82.4
	Non-context-aware TBCNN [9]	82.1
	Full context-aware TBCNN	**82.7**
$\mathcal{O}(n^2)$	LSTM + word-by-word attention [8]	83.5
	mLSTM [19]	86.1

Table 4. Test accuracies of TBCNN variants with different levels of context awareness.

Variants of model	Test accuracy (%)
TBCNN w/o context information [9]	82.1
TBCNN w/$\pi_{premise}^{c}$	82.5
TBCNN w/$\pi_{premise}^{c}$ and $\pi_{premise}^{w}$	82.7

including two feature-based models (either unlexicalized or lexicalized), and several neural network-based models including RNNs and CNNs. Moreover, the proposed context-aware TBCNN model also outperforms the previous context-free TBCNN variant.

We compare TBCNN models of different levels of context awareness in Table 4 so as to have an in-depth analysis of context-awareness. If we only feed the premise's convolution-level features to the TBCNN model of hypothesis, we have an accuracy improvement of 0.4%. This provides consistent evidence of the effectiveness of context-awareness. By furthering feeding the word-level features, we improve the model by another 0.2%, indicating that more awareness of the premise results in higher performance.

We have to concede that our context-aware TBCNN model does not outperform LSTM models with intensive attention mechanisms. However, our complexity is lower than those in order, which is important is retrieval-and-reranking systems [25]. Our result is even comparable to one word-by-word attention model with LSTM-RNNs [8], showing the high performance of our model. Experiments show that context-awareness does improve TBCNN models in the NLI task. Intuitively, context-awareness helps the model find some relevant parts of the premise and the hypothesis when the model is judging the relation between the two sentences. For example, in Table 1, the model is aware of *a man* and *drinking orange juice* of the premise sentence when modeling on the hypothesis, which contains the information of *two women* and *drinking wine*. Hence it is easier for the model to judge that the two sentences are contradictory to each other, or, the hypothesis sentence does not logically follow the premise sentence [2].

5   Conclusion

In this paper, we proposed a context-aware TBCNN model for NLI. The model can leverage different levels of abstraction from the premise when modeling on the hypothesis. Such abstraction includes convolution-level features and word-level features. Our experiments have shown that context-awareness is helpful when applied on TBCNN model. Moreover, the overall complexity of our context-aware TBCNN model remains low despite the newly added mechanism of context-awareness.

Acknowledgments. We would like to thank anonymous reviewers for insightful comments. This research is supported by the National Basic Research Program of China (the 973 Program) under Grant No. 2015CB352201 and the National Natural Science Foundation of China under Grant Nos. 61232015, 91318301, 61421091, and 61502014.

References

1. MacCartney, B.: Natural language inference. Ph.D. thesis, Stanford University (2009)
2. Bowman, S.R.: Modeling natural language semantics in learned representations. Ph.D. thesis, Stanford University (2016)
3. Socher, R., Huang, E.H., Pennin, J., Manning, C.D., Ng, A.Y.: Dynamic pooling and unfolding recursive autoencoders for paraphrase detection. In: Advances in Neural Information Processing Systems, pp. 801–809 (2011)
4. Harabagiu, S., Hickl, A.: Methods for using textual entailment in open-domain question answering. In: Proceedings of the 21st International Conference on Computational Linguistics and the 44th Annual Meeting of the Association for Computational Linguistics, pp. 905–912 (2006)
5. Harabagiu, S., Hickl, A., Lacatusu, F.: Negation, contrast and contradiction in text processing. In: Proceedings of the 20th AAAI Conference on Artificial Intelligence, pp. 755–762 (2006)
6. Bos, J., Markert, K.: Combining shallow and deep nlp methods for recognizing textual entailment. In: Proceedings of the First PASCAL Challenges Workshop on Recognising Textual Entailment, Southampton, UK, pp. 65–68 (2005)
7. Wan, S., Lan, Y., Guo, J., Xu, J., Pang, L., Cheng, X.: A deep architecture for semantic matching with multiple positional sentence representations. arXiv preprint arXiv:1511.08277 (2015)
8. Rocktäschel, T., Grefenstette, E., Hermann, K.M., Kočiský, T., Blunsom, P.: Reasoning about entailment with neural attention. In: Proceedings of the International Conference on Learning Representations (2015)
9. Mou, L., Men, R., Li, G., Xu, Y., Zhang, L., Yan, R., Jin, Z.: Natural language inference by tree-based convolution and heuristic matching. In: Proceedings of the 54th Annual Meeting of Association for Computational Linguistics (2016)

10. Bowman, S.R., Angeli, G., Potts, C., Manning, C.D.: A large annotated corpus for learning natural language inference. In: Proceedings of the Conference on Empirical Methods in Natural Language Processing (2015)
11. Kalchbrenner, N., Grefenstette, E., Blunsom, P.: A convolutional neural network for modelling sentences. arXiv preprint arXiv:1404.2188 (2014)
12. Yin, W., Schütze, H.: Convolutional neural network for paraphrase identification. In: Proceedings of the Conference of the North American Chapter of the Association for Computational Linguistics: Human Language Technologies, pp. 901–911 (2015)
13. Xu, Y., Mou, L., Li, G., Chen, Y., Peng, H., Jin, Z.: Classifying relations via long short term memory networks along shortest dependency paths. In: Proceedings of the Conference on Empirical Methods in Natural Language Processing, pp. 1785–1794 (2015)
14. Socher, R., Perelygin, A., Wu, J.Y., Chuang, J., Manning, C.D., Ng, A.Y., Potts, C.: Recursive deep models for semantic compositionality over a sentiment treebank. In: Proceedings of the Conference on Empirical Methods in Natural Language Processing, pp. 1631–1642 (2013)
15. Mou, L., Peng, H., Li, G., Xu, Y., Zhang, L., Jin, Z.: Discriminative neural sentence modeling by tree-based convolution. In: Proceedings of the Conference on Empirical Methods in Natural Language Processing, pp. 2315–2325 (2015)
16. Hu, B., Lu, Z., Li, H., Chen, Q.: Convolutional neural network architectures for matching natural language sentences. In: Advances in Neural Information Processing Systems, pp. 2042–2050 (2014)
17. Zhang, B., Su, J., Xiong, D., Lu, Y., Duan, H., Yao, J.: Shallow convolutional neural network for implicit discourse relation recognition. In: Proceedings of the Conference on Empirical Methods in Natural Language Processing, pp. 2230–2235 (2015)
18. He, H., Gimpel, K., Lin, J.: Multi-perspective sentence similarity modeling with convolutional neural networks. In: Proceedings of the Conference on Empirical Methods in Natural Language Processing, pp. 1576–1586 (2015)
19. Wang, S., Jiang, J.: Learning natural language inference with LSTM. In: Proceedings of the Conference of the North American Chapter of the Association for Computational Linguistics: Human Language Technologies, pp. 1442–1451 (2016)
20. Mou, L., Li, G., Zhang, L., Wang, T., Jin, Z.: Convolutional neural networks over tree structures for programming language processing. In: Proceedings of the 30th AAAI Conference on Artificial Intelligence (2016)
21. de Marneffe, M.C., MacCartney, B., Manning, C.D.: Generating typed dependency parses from phrase structure parses. In: Proceedings of the Language Resource and Evaluation Conference, pp. 449–454 (2006)
22. Mikolov, T., Sutskever, I., Chen, K., Corrado, G.S., Dean, J.: Distributed representations of words and phrases and their compositionality. In: Advances in Neural Information Processing Systems, pp. 3111–3119 (2013)
23. Srivastava, N., Hinton, G., Krizhevsky, A., Sutskever, I., Salakhutdinov, R.: Dropout: a simple way to prevent neural networks from overfitting. J. Mach. Learn. Res. 15, 1929–1958 (2014)
24. Vendrov, I., Kiros, R., Fidler, S., Urtasun, R.: Order-embeddings of images and language. arXiv preprint arXiv:1511.06361 (2015)
25. Yan, R., Song, Y., Wu, H.: Learning to respond with deep neural networks for retrieval based human-computer conversation system. In: Proceedings of the 39th International ACM SIGIR Conference on Research and Development in Information Retrieval (2016)

Learning Embeddings of API Tokens to Facilitate Deep Learning Based Program Processing

Yangyang Lu, Ge Li[✉], Rui Miao, and Zhi Jin[✉]

Key Lab of High-Confidence Software Technology,
Ministry of Education, Peking University, Beijing, China
{luyy,lige,miaorui,zhijin}@pku.edu.cn

Abstract. Deep learning has been applied for processing programs in recent years and gains extensive attention on the academic and industrial communities. In analogous to process natural language data based on word embeddings, embeddings of tokens (e.g. classes, variables, methods etc.) provide an important basis for processing programs with deep learning. Nowadays, lots of real-world programs rely on API libraries for implementation. They contain numbers of API tokens (e.g. API related classes, interfaces, methods etc.), which indicate notable semantics of programs. However, learning embeddings of API tokens is not exploited yet. In this paper, we propose a neural model to learn embeddings of API tokens. Our model combines a recurrent neural network with a convolutional neural network. And we use API documents as training corpus. Our model is trained on documents of five popular API libraries and evaluated on a description selecting task. To our best knowledge, this paper is the first to learn embeddings of API tokens and takes a meaningful step to facilitate deep learning based program processing.

Keywords: API tokens · Embeddings · Program processing · Deep neural networks

1 Introduction

Deep learning has achieved significant breakthroughs in a number of fields, such as image processing [9,11], speech recognition [5] and natural language processing [2,19]. The mainstream models of deep learning, *deep neural networks (DNNs)*, can extract complex features from raw data with little human engineering knowledge. Recently, the advantages of deep learning have also been exploited in program processing and gains more and more attention on the academic and industrial communities [1,15,22].

When processing programs with DNNs, it's usually required to represent tokens (e.g. classes, variables, methods, etc.) in programs as real-value embeddings so that DNNs could accept programs as inputs. This is analogous to processing natural language sentences and paragraphs based on word embeddings [4,10]. A few works have been proposed to learning embeddings of tokens in

© Springer International Publishing AG 2016
F. Lehner and N. Fteimi (Eds.): KSEM 2016, LNAI 9983, pp. 527–539, 2016.
DOI: 10.1007/978-3-319-47650-6_42

programs, such as learning embeddings of identifiers based on programs' abstract syntax trees [15] or learning embeddings of keywords in programs for software document retrieval [21].

However, learning embeddings of API[1] tokens has not been exploited yet, while they appear in lots of real-world programs. API tokens here refer to the program artifacts related to API libraries, like classes, interfaces, methods etc. Nowadays more and more programs rely on API libraries for implementation. Numbers of API tokens are used to leverage functions provided by API libraries. These API tokens indicate important semantics of programs and play a vital role in processing and analyzing the corresponding programs.

Figure 1 shows an example of a code snippet *SimpleHostConnectionPool.java* of the *Astyanax* project. It implements an internal method to wait for a connection on the available connection pool. As shown in Fig. 1(a), API tokens are related to core functions of this program, such as obtaining the start time, trying to get a free connection, throw timeout exception and interrupting threads of waiting for connections. We can infer that these API tokens have more important influences than other components (Fig. 1(b)) of programs. Learning embeddings of API tokens is meaningful in processing programs that rely on API libraries.

In this paper, we propose a neural model to learn embeddings of API tokens to facilitate deep learning based program processing. Our model is composed of a recurrent neural network with a convolutional neural network to learn embeddings of API tokens from API documents. API documents provide detailed descriptions of functions for API tokens. Our dataset contains documents of five popular Java API libraries to train our model. Finally, we use a description selecting task to evaluate learnt embeddings. To the best of our knowledge, our work is the first to learn embeddings of API tokens and takes a meaningful step on processing programs with deep learning.

The rest of this paper is organized as follows. Section 2 introduces related work of deep learning based program processing. Section 3 describes our model for learning embeddings of API tokens with deep neural networks. Section 4 illustrates the dataset information and evaluation results. Section 5 presents the conclusion.

2   Related Work

Deep learning has been applied to program processing of different application scenarios. In this section, we first introduce the most relevant work to ours, then discuss other related work.

Mou et al. [15–17] proposed a tree-based convolutional neural network (TBCNN) for classifying programs with different algorithm labels. TBCNN was constructed based on programs' abstract syntax trees (ASTs) to capture programs' structural features. Also, TBCNN learned embeddings for identifiers of AST trees and program vectors. The dataset was composed of C programs which

---

[1] https://en.wikipedia.org/wiki/Application_programming_interface.

```
private Connection<CL> waitForConnection(int timeout) throws ConnectionException {
    Connection<CL> connection = null;
    long startTime = System.currentTimeMillis();
    try {
        blockedThreads.incrementAndGet();
        connection = availableConnections.poll(timeout, TimeUnit.MILLISECONDS);
        if (connection != null)
            return connection;

        throw new PoolTimeoutException("Timed out waiting for connection")
            .setHost(getHost())
            .setLatency(System.currentTimeMillis() - startTime);
    }
    catch (InterruptedException e) {
        Thread.currentThread().interrupt();
        throw new InterruptedOperationException("Thread interrupted waiting for connection")
            .setHost(getHost())
            .setLatency(System.currentTimeMillis() - startTime);
    }
    finally {
        blockedThreads.decrementAndGet();
    }
}
```

(a) API tokens are notated in blue.

```
private Connection<CL> waitForConnection(int timeout) throws ▇▇▇▇▇▇▇▇ {
    Connection<CL> connection = null;
    long startTime = ▇▇▇▇▇▇▇▇;
    try {
        ▇▇▇▇▇▇▇▇;
        connection = ▇▇▇▇▇▇▇▇(timeout, ▇▇▇▇▇▇▇▇);
        if (connection != null)
            return connection;

        throw new PoolTimeoutException("Timed out waiting for connection")
            .setHost(getHost())
            .setLatency(▇▇▇▇▇▇▇▇ - startTime);
    }
    catch (▇▇▇▇▇▇▇▇ e) {
        ▇▇▇▇▇▇▇▇;
        throw new InterruptedOperationException("Thread interrupted waiting for connection")
            .setHost(getHost())
            .setLatency(▇▇▇▇▇▇▇▇ - startTime);
    }
    finally {
        ▇▇▇▇▇▇▇▇;
    }
}
```

(b) API tokens are hidden in grey.

Fig. 1. Code from SimpleHostConnectionPool.java of the Astyanax project (Color figure online)

were written by students to solve algorithm assignments. However, TBCNN is inappropriate for processing real-world programs that contain numbers of API tokens. API tokens will be transferred to identifiers of AST trees, which leads to the lost of important semantics in such programs. We propose to learn embeddings of API tokens based on its corresponding descriptions in API documents, which can capture the tokens' semantics.

Ye et al. [21] split code and text into sets of keywords, then adapted Skipgram model to learn embeddings of keywords. The learnt embeddings were used to compute document similarities for software document retrieval tasks. They took documents as bag-of-words and used word-based equations to measure similarities between documents. API documents were used as a part of training datasets. Differently, we take API tokens and their descriptions as word sequences, capture the sequential context rather than bag-of-words and output embeddings for API tokens completely.

Despite the above work, deep learning was also used to predict programs' execution results or generate programs. Zaremba et al. [22] presented a character-level recurrent neural network to predict outputs of short python code. Allamanis et al. [1] proposed an attentional convolutional neural network to generate short, descriptive function name-like phrases for given code snippets. Ling et al. [13] presented a neural network architecture to generate programs for two card games with a mixed natural language and structured specification Gu et al. [6] modified the sequence-to-sequence network of machine translation [19] to generate API call sequences for given queries.

3   Learning Embeddings of API Tokens

3.1   Overview

We use API documents as training corpus to learn embeddings of API tokens. Because API documents provide mappings of API tokens and descriptions, which contain useful semantic information about API tokens' function.

Table 1 shows a mapping example. As for the API token part, we involve the package prefix ("java.awt.color") as a part of the entire API token. Because in real-world programs, packages should be imported before using API tokens. As for the description part, it provides a natural language summary of the corresponding API token's function, which is useful for program processing.

Table 1. An example of API tokens-description mappings

Token	Description
java.awt.color.ColorSpace	This abstract class is used to serve as a color space tag to identify the specific color space of a Color object or, via a ColorModel object, of an Image, a BufferedImage, or a GraphicsDevice

Here we explain several pre-processing steps for the actual inputs of our neural model first. We process the raw data of tokens and descriptions into word sequences. First, we remove punctuations and split the text into sequences by whitespaces. Then we decompose the words of the CamelCase[2] format

---

[2] https://en.wikipedia.org/wiki/CamelCase.

(e.g. names of classes, interfaces, etc.) into componential words by breaking at the positions of the capital letter or the underline (Sect. 4.1). There are two reasons for this operation. First, CamelCase rules use word combinations for token naming. The chose componential words usually help to express semantics of the entire token. Second, API tokens under the same package often share componential words as the common prefix, which indicates important lexical context of API tokens. For example, "InputMethod","InputMethodContext" and "Input-MethodDescriptor" under the package "java.awt.im.spi" have the common componential words "Input" and "Method". The CamelCase decomposition helps to capture the common prefix and makes the semantic space to be respectively dense. After that, we do the lowercase and stemming operations on each word. Finally, we get word sequences of the token and description. They are notated as *tSeq* and *dSeq* respectively.

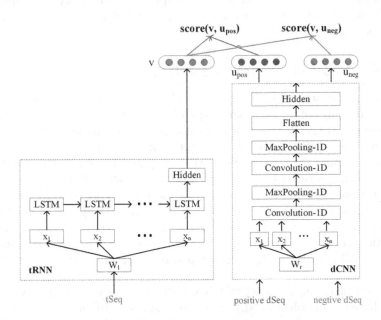

Fig. 2. Overview of the model architecture

The overview of the model architecture is shown as Fig. 2. Our neural model consists of a recurrent neural network (notated as tRNN) and a convolutional neural network (notated as dCNN). (The basic knowledge of networks is introduced in Sect. 3.2). tRNN accepts tokens as inputs and outputs embeddings of tokens. Correspondingly, dCNN accepts descriptions as inputs and outputs embeddings of descriptions. To capture the semantic mapping relation between tokens and descriptions, we use a *score* function to compute the similarity between tRNN's outputs and dCNN's outputs. Then to train the model in an unsupervised manner, we introduce noise-contrastive estimation [7] into our model, which is widely used in natural language modeling [14] and proved to be fast and effective on model training [4,23]. Its basic idea is to train a logistic regression classifier to discriminate between samples from the data distribution (i.e. positive samples) and samples from some "noise" distribution (i.e. negative/noisy samples). In this paper, the positive sample is the correct token-description pair and the noisy pair is a wrong description paired with a given token. Then positive and negative samples share dCNN in our architecture. Finally, the objective function of our model is to minimize the max-margin loss J as Eq. 1.

$$J(\theta) = \sum_{(i,j) \in P} \sum_{k \in S_i} \max(0, \Delta - score(v_i, u_j) + score(v_i, u_c)) \tag{1}$$

In Eq. 1, v_i is the learnt embedding for the token part. u_j and u_c are the learnt embeddings for the correct and noisy descriptions respectively. P is the set of all matched API token-description pairs (i,j) and S is the set of unmatched descriptions for i-th token.

3.2 Basic Knowledge of RNNs and CNNs

Recurrent neural networks (RNNs) are widely used to process sequential data across a wide range of applications in speech recognition [5] and NLP [18]. For a given sequence, RNNs do the same operation for every element in it, which means "recurrent". Usually, the basic RNNs [18] read one element at one time step, then send both the output of the previous time step and the current element input into the recurrent layer at the next step. The basic RNNs suffer from the gradient vanish problem in training, which leads to the lost of long history information. Then gated recurrent units, like LSTM [8] and GRU [3] cells, are proposed to memorize longer dependencies. They use neural gates that could read or forget information via internal memory states and have been successfully applied to numbers of NLP tasks, such as machine translation [19] and answer selecting [20].

Convolutional neural networks (CNNs) are firstly used in image classification [12] and have been applied to NLP tasks successfully [4,10] in recent years. CNNs contain two core operations: convolution and pooling. The convolution operation uses a filter with a fixed window size to extract local features of the input data. The pooling operation deals with the variant size of the input data and tailors them into the same size with given functions like maximal or average. CNNs usually process inputs with groups of convolutional layers and pooling layers, then use a layer to flatten all the feature maps into one fixed-size embedding for the supervised classification tasks.

3.3 tRNN

Here we first explain the reason for choosing the LSTM-based recurrent neural network to learn embeddings for the token part. The token part keep a hierarchical structure, which indicates different levels of concept abstraction. It is shown in the package prefix. Then after decomposing tokens of the CamelCase format into words, there's also a hierarchy among these componential words. Since RNNs have been proved powerful on capturing sequential context [18] and LSTM cells could memorize long sequential context better [8], we use a LSTM-based recurrent neural network (tRNN) to capture tSeq's hierarchical context and learn the embedding of the token part.

Then we introduce the process of learning embeddings for tokens via tRNN. As shown in Fig. 2(left), tRNN reads words in a direction from the general package prefix snippet to the specific word after CamelCase decomposition. Each word of the *tSeq* is transferred to an embedding via a look-up matrix W_l, then sent to a *LSTM* layer. When tRNN reads one word, it moves one time step. At each time step, the *LSTM* layer accepts current word's embedding and the output of the previous time step, then sends its output to the next step. After reading the last word, tRNN sends the output of the *LSTM* layer in the last time step into a fully connected hidden layer, which outputs a d-dimensional embedding for the token part.

3.4 dCNN

Here we also explain why we use a convolutional neural network to encode the description part first. According to our observations and statistics on API documents of five API libraries (shown in Sect. 4.1), descriptions usually contain a long sentence, even a small paragraph. Then we note that the core function is usually related to several local phrases in the long descriptions. Since CNNs are good at extracting local features across words without any syntactic parsing operations, we apply CNNs to learn the semantics of the description into embeddings.

The architecture of dCNN is shown in Fig. 2 (right). Words of the *dSeq* are transferred to embeddings via a look-up matrix W_r first. Then we process the sequence of word embeddings via two groups of 1D-convolutional layers and max-pooling layers. Here we use the 1D-convolutional layer in [10] to extract local features of phrases. It contains several convolution kernels. Each kernel slides a window of m words on the entire word sequence and outputs a filtered feature map of the input sequence. Then max-pooling layers are arranged after the 1D-convolutional layer and used to remain maximal values of each dimension based on the given window, so that the features of key phrase can be captured. After that, a flatten layer is used to concat features of different feature maps into one fixed-size embedding. Finally, dCNN sends the flatten layer's output into a fully connected hidden layer and gets the d-dimensional vector of the description part.

3.5 Tree-Based Negative Sampling

In this section, we illustrate our negative sampling method. To train a model via noise-contrastive estimation, negative samples are needed as the input. Since API documents only provide positive pairs, which are matched token-description

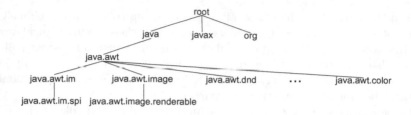

Fig. 3. A multiway tree constructed from package prefixes of Java SE 8

pairs, we need to extract negative samples by ourselves. For a given API token, we expect to find appropriate noisy descriptions. If they were too hard to distinguish on lexical information, it would take a long time for our model to converge to the optimal point. If they were too easy via identifying just from the type information, the robustness of our model would be weak.

Here we present a tree-based negative sampling method. We note that the package prefixes in tokens hold hierarchical structures that indicate different abstraction levels. Descriptions under the same package prefix share more common words than those under different package prefix. Then descriptions of API tokens with different types, such as *Class,Exception, Enum*, are usually easier to be distinguished from the sentence pattern that those with same types. So construct a multiway tree based on all the package prefixes of the corpus to sample noisy descriptions, like the one shown in Fig. 3. Each node presents a part of the package prefix. A virtual root node is added to connect all the package prefixes into a universal tree. Token-description mappings are saved under leaf nodes. For tokens or descriptions under different leaf nodes, we assume that the more common ancestor nodes they share, the more similar they are. So for a given API token T, we trace from the leaf node L of its package prefix, and go up with h depth. Then we find the find the ancestor A with h depth far from the leaf node L. Now we collect descriptions saved under the leaf nodes of the ancestor A except for the subtrees containing L. Then we filter descriptions holding the same type of T as candidates. Finally, we sample k descriptions randomly from the collected candidates as noisy descriptions for the given API token.

4   Evaluation

4.1   Datasets

In this paper, we use API reference documents of five Java API libraries as datasets: Jave SE 8, Eclipse Platform 4.3, Spring 4.3, Lucene 6.1 and Java EE 7. We download API documents from their official websites and use an HTML parser to extract mappings of API tokens and descriptions in raw text.

Several pre-processing steps are done to transfer these raw text of mapping into word sequences, so that they can be fed into the model in Sect. 3. First we remove punctuations and numerical digits, then split the text into word sequences by whitespaces. Then we do CamelCase decomposition operations on tokens of the CamelCase format, of which the reasons have been explained in Sect. 3.1. We break tokens of the CamelCase format into words at the position of the first capital letter followed by a lowercase letter or the position of the underline symbol. For example, "ICC_ColorSpace" is transferred to a word sequence of "ICC", "Color" and "Space". Finally, we do lowercase transformation and stemming on all the words.

Table 2 shows the basic statistical information of the five datasets. We remove the mappings of one-letter template classes and mappings of empty descriptions from the datasets, of which the number is tagged by "Filtered". Then we split the data into the training, dev, test subsets with the ratio of 8:1:1. "AvgLength" and "MaxLength" show the average and maximal length of the word sequences – tSeq and dSeq. "VocabSize" gives the vocabulary size for words in the token and description part.

Table 2. API documents for training and test

API library	Java SE 8	Eclipse Platform 4.3	Spring 4.3	Lucene 6.1	Java EE 7
#Sample (filtered/original)	4252/4457	3883/4001	3821/3875	2659/2672	2038/2134
#Train/#Dev/#Test	3401/425/426	3106/388/389	3056/382/383	2127/266/266	1630/204/204
#AvgLength (TOKEN/DESC)	5/14	7/13	7/14	8/11	5/15
#MaxLength (TOKEN/DESC)	11/69	15/55	21/57	20/73	13/93
#VocabSize (TOKEN/DESC)	1268/2268	885/1683	959/3229	1025/1838	729/1341

4.2 Training

We pre-process our datasets with python and implement our model by Keras 1.0.3 with the backend of Theano 0.8.3. All the weights in our model, including the word embeddings, are randomly initialized. The dimension is set to 256 for embeddings of words, tokens and descriptions. The number of units in *LSTM* layer is set to 256. The two 1D-convolution layers use convolutional kernels in the number of 32 and 64 respectively. The size of the convolutional window is set to 3 words for both convolutional layers. And the max-pooling window is set to be 2 words. Two kinds of score functions are used to measure the similarity of token-description pairs: cosine similarity and flipped Euclidean distance (i.e. negative values of the original euclidean distance). We train the model with 100 epochs for each dataset.

4.3 Results of Description Selecting

Here we explore the quantitative metrics to evaluate the learnt API embeddings of API tokens via our model. Since we expect to capture the mapping relations of API tokens and descriptions after learning embeddings for the two part, we propose a description selecting task.

The description selecting task is defined as follows. For each API token in the test set, we mix its original matched API description with k noisy descriptions sampled in Sect. 3.5 as candidates. Using the tRNN's output as embeddings of API tokens and the dCNN's output as embeddings of the descriptions, we compute the similarity *score* between the given API token and candidate descriptions. Then we rank their matching scores and compute the accuracy by identifying the correct description as the top one result.

We use Bag-of-Words (BoW) model as our baseline. BOW uses one-hot embeddings for words in tokens and descriptions and takes the average embeddings of all words in a sequence as the embedding of the entire sequence.

Table 3. Accuracy (%) of description selecting task

API library		Java SE 8	Eclipse Platform 4.3	Spring 4.3	Lucene 6.1	Java EE 7
Cos Score	BoW	87.90	88.43	90.19	90.20	86.15
	OurWork	93.45	93.35	91.71	91.43	87.69
EucScore	BoW	72.54	72.25	78.10	62.44	72.30
	OurWork	86.65	81.50	87.86	86.94	75.90

Table 3 shows the accuracy on five datasets. Here we mixed one correct description and three noisy descriptions as candidates. Our model runs 100-epoch training on each dataset. "Cos Score" means that the model is trained with the score function of cosine similarity and "EucScore" with the flipped Euclidean distance. Our work outperforms the baseline on both cosine score and Euclidean score. Then we can infer that using cosine similarity gets better results in all the datasets and more training samples lead to better results after the same training epochs. Also, more training samples the model is trained with, more improvement our model gets than the baseline.

We present results of several test cases in Table 4. The correct description is marked with * and bold fonts. *CaseA* shows a correct selecting case by both BoW and our work. Because the correct description share some keywords like "channel" and "write" with the given token, while other candidates do not. *CaseB* gives a case that can be identified correctly by our work while Bow cannot. In this case, there's no shared keywords between the given token and all the candidates. And our work could capture the semantics expressed by the token itself and its description effectively. *CaseC* shows a case failed by BoW and our work. Both models capture the semantics from keywords like "error", "factory", "configuration", but they both failed to identify the core semantics of the keyword "transformer".

All in all, the results of Tables 3 and 4 show that the learnt API embeddings could identify the correct description with high accuracies and our model could capture semantics of tokens and descriptions effectively.

Table 4. Results of several test cases

CaseA: **java.nio.channels.NonWritableChannelException**
*Unchecked exception thrown when an attempt is made to write to a channel that was not originally opened for writing
Checked exception thrown when an input character (or byte) sequence is valid but cannot be mapped to an output byte (or character) sequence
Runtime exception thrown when a file system cannot be found
Checked exception thrown when an input byte sequence is not legal for given charset, or an input character sequence is not a legal sixteen-bit Unicode sequence
CaseB: **java.nio.charset.Charset**
*A named mapping between sequences of sixteen-bit Unicode code units and sequences of bytes
A multiplexor of SelectableChannel objects
A selectable channel for stream-oriented connecting sockets
A token representing the membership of an Internet Protocol (IP) multicast group
CaseC: **javax.xml.transform.TransformerFactoryConfigurationError**
Thrown when a problem with configuration with the Schema Factories exists
Thrown when a problem with configuration with the Parser Factories exists
An error class for reporting factory configuration errors
*Thrown when a problem with configuration with the Transformer Factories exists

5   Conclusion

In this paper, we propose a neural model to learn embeddings of API tokens, which is important for processing API dependent programs with deep learning. Our model combines a recurrent neural network with a convolutional neural network and integrates a tree-based negative sampling method in training. The experimental results show that learnt embeddings of API tokens capture semantics expressed by API documents effectively.

Acknowledgements. This research is supported by the National Basic Research Program of China (the 973 Program) under Grant No. 2015CB352201 and the National Natural Science Foundation of China under Grant Nos. 61232015, 91318301, 61421091, and 61502014.

References

1. Allamanis, M., Peng, H., Sutton, C.: A convolutional attention network for extreme summarization of source code. arXiv preprint arXiv:1602.03001 (2016)
2. Bengio, Y., Ducharme, R., Vincent, P., Jauvin, C.: A neural probabilistic language model. J. Mach. Learn. Res. **3**, 1137–1155 (2003)

3. Chung, J., Gülçehre, C., Cho, K., Bengio, Y.: Gated feedback recurrent neural networks. CoRR, abs/1502.02367 (2015)
4. Collobert, R., Weston, J., Bottou, L., Karlen, M., Kavukcuoglu, K., Kuksa, P.: Natural language processing (almost) from scratch. J. Mach. Learn. Res. **12**, 2493–2537 (2011)
5. Graves, A., Mohamed, A.R., Hinton, G.: Speech recognition with deep recurrent neural networks. In: 2013 IEEE International Conference on Acoustics, Speech and Signal Processing, pp. 6645–6649. IEEE (2013)
6. Gu, X., Zhang, H., Zhang, D., Kim, S.: Deep API learning (2016)
7. Gutmann, M.U., Hyvärinen, A.: Noise-contrastive estimation of unnormalized statistical models, with applications to natural image statistics. J. Mach. Learn. Res. **13**, 307–361 (2012)
8. Hochreiter, S., Schmidhuber, J.: Long short-term memory. Neural Comput. **9**(8), 1735–1780 (1997)
9. Jia, Y., Shelhamer, E., Donahue, J., Karayev, S., Long, J., Girshick, R., Guadarrama, S., Darrell, T.: Caffe: convolutional architecture for fast feature embedding. In: Proceedings of the 22nd ACM International Conference on Multimedia, pp. 675–678. ACM (2014)
10. Kalchbrenner, N., Grefenstette, E., Blunsom, P.: A convolutional neural network for modelling sentences (2014)
11. Krizhevsky, A., Sutskever, I., Hinton, G.E.: Imagenet classification with deep convolutional neural networks. In: Advances in Neural Information Processing Systems, pp. 1097–1105 (2012)
12. Le, Q.V., Mikolov, T.: Distributed representations of sentences and documents. ICML **14**, 1188–1196 (2014)
13. Ling, W., Grefenstette, E., Hermann, K.M., Kocisky, T., Senior, A., Wang, F., Blunsom, P.: Latent predictor networks for code generation. arXiv preprint arXiv:1603.06744 (2016)
14. Mnih, A., Teh, Y.W.: A fast and simple algorithm for training neural probabilistic language models. arXiv preprint arXiv:1206.6426 (2012)
15. Mou, L., Li, G., Jin, Z., Zhang, L., Wang, T.: TBCNN: a tree-based convolutional neural network for programming language processing. arXiv preprint arXiv:1409.5718 (2014)
16. Mou, L., Li, G., Zhang, L., Wang, T., Jin, Z.: Convolutional neural networks over tree structures for programming language processing. In: Thirtieth AAAI Conference on Artificial Intelligence (2016)
17. Peng, H., Mou, L., Li, G., Liu, Y., Zhang, L., Jin, Z.: Building program vector representations for deep learning. In: Zhang, S., Wirsing, M., Zhang, Z. (eds.) KSEM 2015. LNCS (LNAI), vol. 9403, pp. 547–553. Springer, Heidelberg (2015). doi:10.1007/978-3-319-25159-2_49
18. Schmidhuber, J.: A local learning algorithm for dynamic feedforward and recurrent networks. Connect. Sci. **1**(4), 403–412 (1989)
19. Sutskever, I., Vinyals, O., Le, Q.V.: Sequence to sequence learning with neural networks. In: Advances in Neural Information Processing Systems, pp. 3104–3112 (2014)
20. Wang, B., Liu, K., Zhao, J.: Inner attention based recurrent neural network for answer selection. In: The Annual Meeting of the Association for Computational Linguistics (2016)

21. Ye, X., Shen, H., Ma, X., Bunescu, R., Liu, C.: From word embeddings to document similarities for improved information retrieval in software engineering. In: Proceedings of the 38th International Conference on Software Engineering, pp. 404–415. ACM (2016)
22. Zaremba, W., Sutskever, I.: Learning to execute. arXiv preprint arXiv:1410.4615 (2014)
23. Zhou, G., He, T., Zhao, J., Hu, P.: Learning continuous word embedding with metadata for question retrieval in community question answering. In: Proceedings of ACL, pp. 250–259 (2015)

Closed-Loop Product Lifecycle Management Based on a Multi-agent System for Decision Making in Collaborative Design

Fatima Zahra Berriche[1], Besma Zeddini[1(✉)],
Hubert Kadima[1], and Alain Riviere[2]

[1] Quartz Laboratory, EISTI, Avenue de parc, Cergy, France
{fbe,bzi,hk}@eisti.eu
[2] Quartz Laboratory, SUPMECA, 3, rue Fernand Hainaut, Saint Ouen, France
alain.riviere@supmeca.fr

Abstract. In the collaborative design environment, there is an increasing demand for information exchange and sharing to reduce lead time and to improve product quality and value. Software and communication technologies can be a relevant approach in this context, using for instance PLM (Product Lifecycle Management) systems. Each product lifecycle development phase generates knowledge, and managing this knowledge can be placed in a closed-loop. In this paper, we present a research in progress that exposes a collaborative architecture based on a multi-agent system which aims to support the knowledge management process in the closed-loop. This is a new strategic approach to manage the product lifecycle information efficiently in a distributed environment. The purpose of this paper is to illustrate the use of DOCK (Design based on Organization, Competence and Knowledge) methodology for the design of our multi-agent system and to demonstrate how to handle intelligent knowledge via a use case study.

Keywords: Product lifecycle management · Knowledge management · Intelligent knowledge · Multi-agent systems · Collaborative design

1 Introduction

Collaborative design of industrial products is the inevitable trend in the enterprises to improve the engineering activities of product lifecycle. The objective is to generate lower costs and to ensure higher efficiency of the product design and the manufacturing process. To remain competitive, a lot of industrial enterprises and networks increasingly need to collaborate with each other into collaborative design environment of industrials products [1–3]. In this context, several actors are involved with different views during different phases of the product lifecycle, while exchanging knowledge expressed in heterogeneous formats. This knowledge is related to different concepts such as product structure, usage, project history and design activities, parameters and constraints. Enterprises have now to differentiate themselves by extending their strategic business approach to the entire

© Springer International Publishing AG 2016
F. Lehner and N. Fteimi (Eds.): KSEM 2016, LNAI 9983, pp. 540–551, 2016.
DOI: 10.1007/978-3-319-47650-6_43

product lifecycle and by deploying knowledge management techniques [4]. The objective is to improve their knowledge management over the complete product lifecycle by implementing the approach of closed-loop PLM. The idea is that some information of one lifecycle stage could be useful for another stage. For instance, information of the production phase could be used at the recycling stage, to support deciding the most appropriate recycling option (especially to take decisions about remanufacturing and reuse). This information can also be combined with the recycling information and used as feedback for the production to improve the new generations of the product. Our research consists in the modeling and implementation of an intelligent knowledge-based system for collaborative design. The purpose of our system is to identify, capture, synthesize and reuse the professional knowledge in the closed-loop. To address this need, the multi-agent systems (MAS) paradigm provides a relevant solution framework to our research idea. MAS allows to design and implement such systems based on the interactions of different agents. The design of our system is based on the DOCK methodology (Design based on Organization, Competence and Knowledge) [5], a methodology to design intelligent knowledge-based systems, supporting the knowledge management process.

The present paper is organized as follows. Section 2 introduces closed-loop PLM for collaborative design. It describes how closed-loop PLM improves knowledge management over the complete product lifecycle and the use of MAS to handle intelligent knowledge in the context of closed-loop PLM. Section 3 describes the multi-agent system architecture that we propose and the main concepts of the DOCK methodology, which is used to the design of our intelligent knowledge-based system. Section 4 illustrates the first results. Section 5 concludes the paper and discusses some future improvements.

2   Literature Review

2.1   Closed-Loop PLM for Collaborative Design

Product Lifecycle Management (PLM) is primarily a business strategy [6], which aims to efficiently manage the product related information during the whole product lifecycle, for all internal and external actors involved in its creation. PLM most widely used definition is "a strategic business approach that applies a consistent set of business solutions in support of the collaborative creation, management, dissemination and use of product definition information across the extended enterprise from concept to end of life - integrating people, processes, business system, and information" [7]. The PLM approach is supported by software to create a collaborative work environment for all actors involved in product development cycle. It allows to orchestrate the progress of business processes (or workflow), in which the responsibilities, access rights and roles are clearly defined. The information recorded in the common database of PLM are structured using metadata involving individuals (objects), classes (sets of object types), attributes (object properties), relationships (links between objects) and events. Despite the benefits of PLM systems in the project design, the collaborative design of industrials products still remains complex.

To face the new industrial challenges and enable the factory of the future such as digital manufacturing, decision support technologies, etc., the information loops along the whole product lifecycle have to be closed. The concept of closed-loop PLM has appeared in 2007 [8] and is based on the new tracking technologies and monitoring products (Radio Frequency IDentification (RFID), chips, etc.). These new technologies allow to follow each product during the "beginning of life" (BOL), "middle of life" (MOL) and the "end of life" (EOL). They also allow to extract the necessary information for maintenance and recycling, and to improve the design and manufacturing of future products. With the closed-loop PLM, the information is not transmitted from phase to phase as in the PLM approaches. As shown in Fig. 1, these technologies involve closing the information loops throughout the lifecycle.

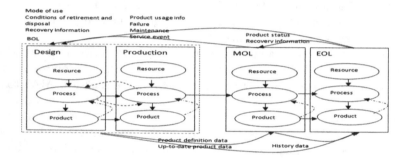

Fig. 1. Closed-loop PLM

In all these phases of closed-loop PLM, data, information and knowledge are created. Closed-loop PLM can improve the quality of product design and the efficiency of production. In this paper, we focus on the BOL phase (design and production), which is the phase where the product concept is generated and physically created. In fact, the creation of data, information and knowledge are supported by intelligent systems such as Computer aided design (CAD), Computer-aided manufacturing (CAM) and Computer-aided engineering (CAE) [9]. In the closed-loop PLM, designers will receive feedback about detailed product information such as product status, product usage, conditions of retirement and disposal. This feedback information is valuable for product design as it enables designers to exploit the expertise and know-how of other participants in the product lifecycle.

2.2   The Use of Multi-agent System to Handle Intelligent Knowledge

The term "agent" is loosely defined in the literature as it is used in very different contexts. In our context, an agent can be referred to as a component of software and/or hardware, which is an intelligent entity capable of acting like a human user in order to accomplish tasks. The agent paradigm provides the ability to model the distributed activities and exchanges between the business actors. A multi-agent system (MAS) [10] is a system composed of several agents interacting with each other. The interaction is usually done by messages. To successfully interact, agents require the ability to cooperate, coordinate, and negotiate with each others, much as people do. MAS have become an appealing paradigm for the design of computer systems based on autonomous cooperating software entities.

Knowledge management systems (KMS) [11] are complex systems especially when knowledge is distributed and heterogeneous. Our idea is to use a MAS to manage a KMS. Indeed, MAS offer several benefits such as interoperability with other existing systems, and heterogeneity management. This is possible because agents that are autonomous, cooperating and communicating to decompose problems into sub-problems and solve them more easily.

The development of a KMS based on the agent paradigm has been the subject of several research works such as (Vizcaino and al) [12] which recommended a multi-agent system to computer engineers to support the different stages of lifecycle of a KMS. (Wang and Xu) [13] have proposed a KMS based on MAS to help an organization in problem solving and decision-making activities. To manage and share knowledge of more applications (manufacturing, banking, medicine, entertainment), Sajja [14] followed a fixed topological modeling such as a relational database management system managed by a MAS. (Zhang et al.) [15] also proposed a MAS to manage heterogeneous and distributed knowledge but in the case of implementation of projects in collaboration via the Internet. While (Toledo et al.) [16] have handled this problem in the case of business process management based on the platform Jacamo (Boissier et al.) [17].

In the work presented in this paper, we are interested in multi-agent systems in the field of industrial design to manage tacit or explicit knowledge throughout the design process. The goal is to collect the knowledge based on the cooperation of intelligent agents and formalize them in ontologies. The ontologies describe the concepts used by the trades actors to complete a project or design a product. The created ontologies will be used by agents to achieve the extraction of knowledge in business applications, dissemination and reuse for future projects.

3 Architecture of Distributed Environment for Collaborative Design

In this section, we describe the objective and components of our architecture based on multi-agent system. The contribution of this proposal is in the context of the decision support in collaborative design. The application allows:

1. to define the design tasks for a product,
2. to coordinate the achievements of each task,
3. to assemble all of the collected data from the tasks to build a solution,
4. and finally to distribute these tasks between the involved designers.

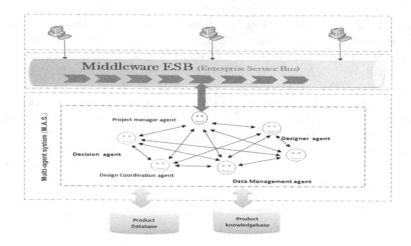

Fig. 2. Architecture of distributed collaborative environments

As illustrated in Fig. 2, the proposed architecture is composed of four layers:

- **Man/machine interface:** all hardware devices and software enabling the user to interact with the application.
- **Middleware ESB (Enterprise Service Bus):** enables the communication and the mediation between services and applications. It connects heterogeneous environments using a service oriented approach in order to orchestrate the different services
- **Application layer:** it is the process layer of our architecture. As mentioned before, in this part we propose our multi-agent system. A number of software agents have been designed. We describe the agents and their functions in the next section.
- **The data storage layer:** includes all the application data sources (product database + product knowledgebase).

3.1 The Design of Our Multi-agent System Using DOCK Methodology

To be reliable, our MAS must be designed and analyzed according to an organizational methodology oriented towards knowledge engineering applications called DOCK. As we have mentioned before, the main goal of the MAS is to allow designers to capitalize, share and reuse the knowledge generated throughout decision-making process. The DOCK approach is focused on the design of a knowledge-based system using the agent paradigm. This system must be responsive, proactive and flexible. To meet these requirements, the MAS has two types of agents:

- **Cognitive agents:** They are derived from modeling human organizations. These agents are continuously in touch with trades actors in order to provide them with proactive support in their activities.

- **Reactive agents:** they ensure the operation of knowledge management cycle, responding to external requests (from the cognitive agents or the trades actors).

This modeling of MAS with organizations and types of agents provides a flexible and adaptable system to different types of processes. The obtained system is composed of three organizational levels. The first is the human organization level, the second corresponds to the organization of cognitive agents and the third the reactive agents level.

To describe the modeling of human organization, DOCK uses organizational modeling methodology derived from KROM (Knowledge Reuse Organizational Model) [18], to expose the necessary and sufficient elements for the implementation of a knowledge management system. It highlights the skills of professional actors, description of the knowledge that they create, use and share, as well as the formalization of the organizational structure that they apply.

The chosen case study in this paper focuses on decision-making process in industrial product design projects. The MAS provides proactive assistance to users by capitalization and reuse of the knowledge generated throughout this process. These knowledge are created, used, shared and also reused in a closed-loop. To ensure the progress of the knowledge management process, four roles are formalized (see Fig. 3).

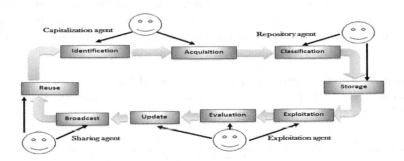

Fig. 3. The different types of agents and their roles in the knowledge management process

Following the DOCK approach, our system is composed of three organizations. The first concerns the human organization and represents the system user layer. In the second, cognitive agents interact with human actors and reactive agents to fulfill their missions (see Table 1).

In the latest organizational layer, reactive agents support the knowledge management process. The capitalization agent and repository agent work together to identify and store knowledge. Exploitation agent together with sharing agent are in charge of the evaluation, the update and the provision of the knowledge base. The Fig. 4 shows the general architecture of our MAS.

Table 1. The different types of reactive agents and their roles

Agents	Roles
Capitalization agent	Ensures the identification and formalization of the knowledge manipulated by the actors during the process of decision-making. The capitalization agent has no direct access to what the human actors do. It is the cognitive agents who request it by offering the knowledge to capitalize.
Repository agent	Provides classification and storage of knowledge in the knowledge base.
Exploitation agent	Ensures the use of the knowledge, as well as their evaluation and updating.
Sharing agent	Ensures the sharing and dissemination of the knowledge manipulated by other reactive agents. It is the only reactive agent that sends "informations, queries, knowledges" to cognitive agents or business actors

3.2   Descriptive Model of Cognitive Agents' Interactions

As shown in Fig. 5, the activity diagram describes the main functions of our system, the operating cycle and the interactions between different agents. There are five types of cognitive agents in our proposed system (project manager agent, data management agent, design coordination agent, design agent, designer agent). All of these agents work and react depending on their incoming messages. The main agents are briefly described in the following.

– **Project manager agent:** The main purpose of the project manager agent is to help the project manager to manage the design project. It allows to decompose a complex design into subtasks. Furthermore, it makes a schedule

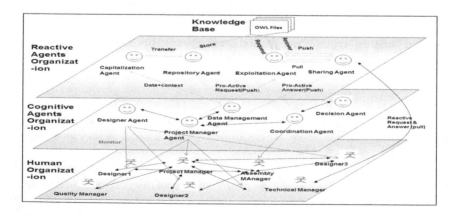

Fig. 4. The architecture of the MAS

for all the designers involved to realize their design tasks on time. It also allows to view a detailed vision on project progress (completed tasks, missed tasks, current tasks)

- **Designer agent:** The designer agent allows to provide support based on knowledge related to the design process driven by human designers. The interest of this agent is to identify and regroup all similar strategies adopted at the realization of the previous design tasks, from this cluster, the designer agent proposes to the involved designers a process model based on the different similar processes.
- **Data management agent:** This agent is responsible of collecting the data in the collaborative product design process.
- **Design coordination agent:** The design coordination agent is specially designed to manage the coordination between the design processes. Moreover, it alerts the designers involved in the design tasks of constraint conflicts and their respective reason.
- **Decision agent:** The decision agent is specifically designed to support decision-making in the design of industrial products. In our MAS, the information is distributed and controlled by each agent. Therefore, it is necessary to have a decision agent that controls and validates the solution proposed by the designer agent. It also runs the simulation and optimization of the process based on the key performance indicators.

4 Experimental Case

To validate our approach, we decide to implement a Web-based shared environment for experience feedback in design projects. Figure 6 shows the first user interface of the Web application. It is an environment that provides the following services:

- **List of current projects:** The list includes all projects driven with the platform (list of completed projects, list of current projects and list of future projects)

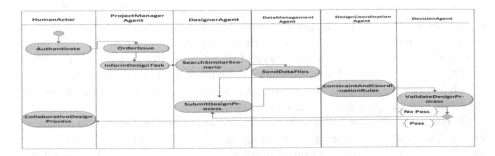

Fig. 5. Descriptive model of cognitive agents interactions

– **Management of project design:** This part is proposed to support designers in their design process. It focuses on the management and coordination of the collaborative processes using the knowledge and experiences of the designers. It also recovers all similar processes then discover a process model and displays it as workflows.
– **Mode RETEX:** This section provides an overview of the platform's projects, with the involved designers team. More precisely, it present a repository of knowledge in order to continuously improve the industrial design processes.

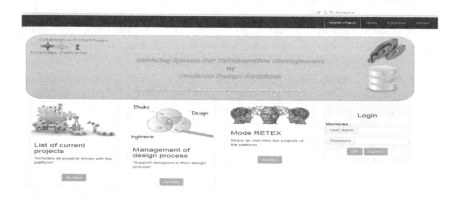

Fig. 6. User interface

The proposed system takes into account the knowledge management cycle linked to the process of decision making. To validate our approach, we decide to expose and explain how reactive agents interact with cognitive agents on a concrete example which is "designing a mechanical engine".

As we have already mentioned, reactive agents ensure the operation of knowledge management cycle. The first phase in this cycle is to identify the knowledge of the design project by the capitalization agent. These knowledge are identified from previous experience in similar projects and formalized to build an ontology [19]. The ontology provides a structured and semantic vocabulary for dealing with knowledge related to a specific area.

Before the creation of the ontology, the capitalization agent targets the business application to specify in what type of database it will be stored and therefore the storage format. Each ontology is identified by name. The Fig. 7 shows a simple example of an ontology built in the design of a mechanical engine. It describes the components of a mechanical engine.

When running a new process that is similar to our design process "mechanical engine", the capitalization agent is automatically activated to enrich the ontology. The backup and integration of ontology are guaranteed by the repository agent, this agent gathers the name of the ontology proposed by the capitalization agent. The created ontology is stored in an OWL (Web Ontology Language)

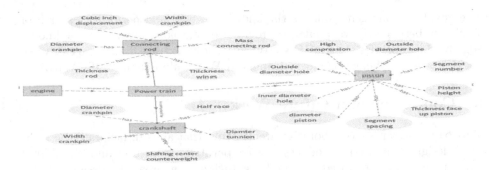

Fig. 7. Ontology of a mechanical engine

format [20] file in a knowledge base. The OWL file includes classes and "Object-Property".

The following example shows an example of the class "piston" built by the repository agent. This class is a subclass of class "Power train".

```
<owl:Class  rdf:ID=?piston?>
<rdfs:subClassOf rdf:resource="#Power train" />
</owl:Class>
```

The OWL file includes "DatatypeProperty" and restrictions on these properties. The following example presents a creation of the **property** "Outside diameter hole" with a **domain** that allows to specify in which class we can assign this particular property, as well as the **range** that allows to specify the type of property. In our example, the property contains a restriction that is specified by the syntax "hasValue". This restriction on the property requires that "outside hole diameter" should be equal to 15. So, only the pistons which have the outside hole diameter = 15 will be returned to the user as a result.

```
<owl:DatatypeProperty rdf:ID="Outside_diameter_hole">
<rdfs:domain rdf:resource="#piston"/>
<rdfs:range rdf:resource="&xsd;string"/>
</owl:DatatypeProperty>


<owl:Class rdf:ID="piston">
<rdfs:comment> the outside diameter hole must be equal to 15 cm
</rdfs:comment>
<rdfs:subClassOf>
<owl:Restriction>
<owl:onProperty rdf:resource="#Outside_diameter_hole"/>
<owl:hasValue rdf:resource="15"/></owl:Restriction>
</rdfs:subClassOf>
</owl:Class>
```

To exploit these knowledge, the cognitive agent "designer agent" sends a request to the "data management agent". The latter contacts the "Exploitation agent".

The "exploitation agent" queries the knowledge base and performs a search of similarity between the vocabularies provided by ontologies stored in the knowledge base. Ontologies are stored as OWL files, so to extract data more easily, we decide to use the Semantic Query-Enhanced Web Rule Language (SQWRL).

5   Conclusion and Future Work

This paper presents a collaborative architecture that aims to facilitate the technical design of industrial products in a cooperative, multi-user and distributed environment. This required the implementation of a proactive design scheme and an interdisciplinary approach that includes monitoring, decision and control execution process. More precisely, it allows designers to create, collect, search, share and reuse knowledge from the database. To achieve this goal, we have used multi-agent systems. This paradigm has the ability to manage complex systems such as heterogeneous knowledge management systems.

We use ontologies to solve the problem of management and knowledge reuse. These ontologies describe heterogeneous and distributed knowledge. They represent a semantic contribution through the concepts that they express.

The first experiment exposed on this paper, shows an overall view of how our agents interact to manage knowledge. We have chosen as a case study the designing of a mechanical motor. Our future task is to equip the agents with inference rules to provide relevant knowledge for designers. In addition, we plan to deploy and test our system on a real industrial case.

Acknowledgments. This paper reflects only the authors' views. The work reported is carried out by the PLACIS project, which has received research funding from French Ministry for Higher Education and Research.

References

1. Helms, R.: Product data management as enabler for concurrent engineering controlling the flow of preliminary information in product development. Ph.D. thesis, Technische Universiteit Eindhoven (2002) ISBN 90-386-1787-9
2. Saaksvuori, A., Immonen, A.: Product Lifecycle Management. Springer, Heidelberg (2003)
3. Stark, J.: 21st Century Product Realisation. Springer, London (2004)
4. Shen, W.: Knowledge sharing in collaborative design environments. Comput. Ind. **52**, 1–93 (2003)
5. Julien, G., Davy, M., Eric, B., Maggy, P.: An organizational approach to designing an intelligent knowledge-based system: application to the decision-making process in design projects. In: Advanced Engineering Informatics, p. 18 (2015)
6. Terzi, S.: Element of product lifecycle management: definitions, open issues and reference models. Ph.D. thesis, Universit Henry Poincar Nancy-I (2005)
7. CIMdata Inc.: Product Lifecycle Management, Empowering the future of business (2003)
8. Jun, H.-B., Kiritsis, D., Xirouchakis, P.: Research issues on closedloop PLM. Comput. Ind. **58**(8), 855–868 (2007)

9. Jun, B., Shin, H., Kiritsis, D., Xirouchakis, P.: System architecture for closed-loop PLM. Int. J. Comput. Integr. Manuf. **20**(7), 685–698 (2009)
10. Ferber, J.: Les Systmes multi-agents: Vers une intelligence collective. Dunod (1195)
11. Monticolo, D., Hilaire, V., Koukam, A., Gomes, S.: KATRAS: un systme multia-gents pour la gestion des connaissances lors des projets de conception mcanique. In: Journes Francophones Sur Les Systmes Multi-Agents, pp. 219–230 (2009)
12. Vizcaino, A., Soto, J.-P.: A multi-agent model to develop knowledge management systems. In: Proceedings of the 40th Annual Hawaii International Conference on System Sciences, pp. 296–299 (2007)
13. Wang, X., Xu, F.: Study on knowledge management system based on MAS. In: 2nd International Conference on Networking and Digital Society (ICNDS), pp. 395–398 (2010)
14. Sajja, P.: Multi-agent system for knowledge-based access to distributed databases. Interdisc. J. Inf. Knowl. Manag. **3** (2008)
15. Zhang, C., Tang, D., Liu, Y., You, J.: Multi-agent architecture, a for knowledge management system. In: Fifth International Conference on Fuzzy Systems and Knowledge Discovery (2008)
16. Toledo, C., Bordini, R., Chiotti, O., Galli, M.: Developing a knowledge management multi-agent system using JaCaMo. In: Dennis (2012)
17. Boissier, O., Bordini, R., Hbner, J., Ricci, A., Santi, A.: JaCaMo project [WWW Document] (2013)
18. Girodon, J., Monticolo, D., Bonjour, E., Perrier, M.: KROM: an organizational metamodel oriented to knowledge: a case from ophtalmic industry. In: Eighth International Conference on Signal Image Technology and Internet Based Systems (SITIS), Naples, pp. 845–851 (2012)
19. Astrova, I., Korda, N., Kalja, A.: Storing OWL Ontologies in SQL relational data-bases. In: World Academy of Science, Engineering and Technology (2007)
20. McGuinness, D., VanHarmelen, F.: OWL web ontology language overview. In: W3C Recommendation (2004)

Ontologies

A Benchmark for Ontologies Merging Assessment

Mariem Mahfoudh[1(✉)], Germain Forestier[2], and Michel Hassenforder[2]

[1] CNRS, LORIA, UMR 7503,
615 Rue du Jardin Botanique, 54506 Vandœuvre-lès-Nancy, France
mariem.mahfoudh@loria.fr
[2] MIPS EA 2332, Université de Haute Alsace,
12 Rue des Frères Lumière, 68093 Mulhouse, France
{germain.forestier,michel.hassenforder}@uha.fr

Abstract. In the last years, ontology modeling became popular and thousands of ontologies covering multiple fields of application are now available. However, as multiple ontologies might be available on the same or related domain, there is an urgent need for tools to compare, match, merge and assess ontologies. Ontology matching, which consists in aligning ontology, has been widely studied and benchmarks exist to evaluate the different matching methods. However, somewhat surprisingly, there are no significant benchmarks for merging ontologies, proving input ontologies and the resulting merged ontology. To fill this gap, we propose a benchmark for ontologies merging, which contains different ontologies types, for instance: taxonomies, lightweight ontologies, heavyweight ontologies and multilingual ontologies. We also show how the GROM tool (Graph Rewriting for Ontology Merging) can address the merging process and we evaluate it based on coverage, redundancy and coherence metrics. We performed experiments and show that the tool obtained good results in terms of redundancy and coherence.

Keywords: Ontologies merging · Benchmark · Graph rewriting · GROM tool

1 Introduction

In the two last decades, ontologies have become widely used in several domains such as semantic web, medicine, e-commerce and natural language processing. With this multitude of ontologies that represent and cover sometimes the same domain, there is a growing need to merge these ontologies [1].

Merging ontologies have the goal of *"creating a new ontology from two or more existing ontologies with overlapping parts, which can be either virtual or physical"* [2]. It starts with the identification of the overlapping part (the similarities) between the ontologies entities and based on this result, it merges them and creates a new one [3]. Merging ontologies is a challenging issue that depends

© Springer International Publishing AG 2016
F. Lehner and N. Fteimi (Eds.): KSEM 2016, LNAI 9983, pp. 555–566, 2016.
DOI: 10.1007/978-3-319-47650-6_44

on several factors. Among them, the most important are: (1) the quality of initials ontologies (consistent or inconsistent), (2) the quality of ontology alignment (the identification of the similarities could recognize syntactic and/or semantic and/or structural correspondences between the ontologies ?), and (3) the merging strategy and the quality of its results. To evaluate the merging process and compare the tools proposed by the community, it is important to: (1) check the consistency of the initials ontologies (it could be done with a reasoner such as Pellet, Hermit, etc.); (2) evaluate the ontologies alignment; (3) evaluate the merging result.

To evaluate ontologies alignment, researchers can use existing proposed benchmarks. This research field has now reached a significant maturity and there are specialized conferences that propose important benchmarks. As an example, the Ontology Matching[1] conference presents more than 20 datasets and publishes annually the results of their alignment. However, somewhat surprisingly, there are no significant benchmarks for merging ontologies [1,3]. To the best of our knowledge, the only work is the one of Raunich and Rahm [1]. It presents some taxonomies and the result of their merging using the ATOM (Automatic Target-driven Ontology Merging) tool [3]. This work has two main limitations. The first one is that it studies only taxonomies. Therefore, it can not test and evaluate the heavyweight ontologies (no property, no axioms contradiction, etc.). The second limit is that the benchmark is not published and therefore cannot be used.

We propose in this paper a benchmark for ontologies merging, which contains both lightweight and heavyweight ontologies. We show how our tool GROM (Graph Rewriting for Ontology Merging) [4] can address the merging process and we evaluate it based on coverage, redundancy and coherence metrics. All the ontologies, the result of their alignment and of their merging are available on the web for download.

The paper is structured as follows: Sect. 2 introduces the background of the work. It presents the merging process and the typed graphs grammars formalism. Section 3 defines our formalism and discusses how to use it to merge ontologies and resolve their inconsistencies. Section 4 presents a benchmark for ontologies merging and evaluates the GROM tool. Finally, a conclusion summarizes the presented work.

2   Background

2.1   Ontologies Merging

Ontologies are living objects that represent knowledge with an explicit and formal way [5]. They conceptualize a given domain by their concepts (classes, properties, individuals, axioms, etc.) in order to offer mechanisms of reasoning and inference. Given two or more ontologies, ontologies merging aims at producing a new ontology with their overlapping parts [2]. This process can be symmetric or

---

[1] http://www.ontologymatching.org.

asymmetric. Symmetric solutions aim at completely integrating all input ontologies with the same priority. Asymmetric approaches, by contrast, take one of the ontologies as the source and merge the other as a target. In this type of approach, the concepts of the source ontology are preserved and only the concepts of the target ontology, that not alter the consistency, are added [3].

Several approaches have been proposed in the literature, we briefly present in this section the approaches that are implemented with a proposed tool. The Table 1 summarizes these approaches according to: (1) the merging strategy (symmetric or asymmetric), (2) the ontology specification (OWL (Web Ontology Language), RDFS (Resource Description Framework Schema), Frame, etc.), (3) the tool, and (4) their specificities and the inconsistencies resolution (conflicts management). Note that because of the no-existence of the benchmark for merging ontologies, the researchers do not present details evaluation for their tools.

Table 1. Summary of some ontologies merging approaches.

Approach	Merge strategy	Specification	Tool	Specificity and conflicts management
Stumme et al. [6]	Symmetric	Frame	FCA-Merge	- Semi-automatic approach
				- Approach based on the formal concept analysis (FCA)
				- No conflicts management
Noy et al. [7]	Symmetric & Asymmetric	Frame	Prompt	- Semi-automatic approach
				- Conflicts detection
				- User intervention for the conflicts resolution
Kotis et al. [8]	Symmetric	–	HCONE	- Semi-automatic approach
				- Approach based on the Latent semantic analysis (LSA)
				- No conflicts management
Li et al. [9]	Symmetric	OWL	MOMIS	- Automatic approach
				- Management of some conflicts (structural and semantic)
Raunich et al. [3]	Asymmetric	OWL Taxonomy	ATOM	- Automatic approach
				- Delete the redundancy
				- Management of some structural conflicts

2.2   Quality Measures for Ontology Merging

Measuring the quality of ontology merging finds its origins in the field of conceptual schemas. Indeed, to evaluate the quality of an integrated schema, Duchateau et al. [10] proposed two measures: minimality and completeness. The minimality checks that no redundant concept appears in the integrated schemas. The completeness represents the percentage of concepts presented in the data sources that are covered by the integrated schemas. It is calculated as follows:

$$comp(Si_{tool}, Si_{exp}) = \frac{|Si_{tool} \cap Si_{exp}|}{Si_{exp}} \tag{1}$$

where Si_{exp} is the integrated schema proposed by an expert and the Si_{tool} is the integrated schema generated by a tool.

Rahm et al. [1] proposed the same metrics to evaluate the quality of the merging ontologies result and they called them: *coverage* (for the completeness) and *redundancy* (for the minimality). The coverage is then related to the degree of information preservation. It measures the share of input concepts preserved in the result and it depends on the type of approach symmetric or asymmetric one. With symmetric approaches (full merge) the coverage is equal to 1: all the concepts are preserved. For the asymmetric approaches, the concepts of the target ontology are preserved but only the concepts non-redundant of the source ontology are preserved (*i.e.* all the redundant concepts are removed).

Besides the coverage and the redundancy metrics, it is also important to check the *coherence* of the ontologies merging result. This metric checks if the ontology result is coherent or contains some conflicts. To ensure theses three metrics, we have used the typed graph grammar formalism for our approach of merging ontologies.

2.3   Typed Graph Grammars

Typed Graph Grammars (*TGG*) are a mathematical formalism that permits to represent and manage graphs. They are used in several fields of computer science such as software systems modelling and formal language theory [11]. Recently, they started to be used in the ontology field, in particular for the modular ontologies formalization [12] and consistent ontologies evolution [13,14]. A typed graph grammar is defined by $TGG = (G, TG, P)$ where:

- G is a start graph also called host graph.
- TG is a type graph that represents the elements type of the graph G.
- P is a set of production rules also called graph rewriting rules which are defined by a pair of graphs patterns (*LHS*, *RHS*) where: (1) *LHS* (Left Hand Side) represents the preconditions of the rewriting rule and describes the structure that has to be found in G; (2) *RHS* (Right Hand Side) represents the postconditions of the rule and must replaces *LHS* in G (see Fig. 1).

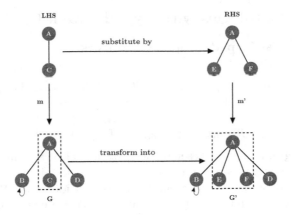

Fig. 1. The principle of graph transformation.

A rewriting rule can be extended with a set of negative application conditions ($NACs$). A NAC is another graph pattern such as: "if there exist a morphism from NAC to the host graph G, then, the rule cannot be applied". In this way, a graph transformation defines how a graph G can be transformed to a new graph G'. More precisely, there must exist a morphism that replaces LHS by RHS to obtain G'. To apply this replacement, different graph transformations approaches are proposed [15]. In this work, we use the algebraic approach [16] based on the *pushout* concept [17]. Given three objects (in our case graphs) G_1, G_2 and G_3 and two morphisms $f : G_1 \rightarrow G_2$ and $g : G_1 \rightarrow G_3$, the pushout of G_2 and G_3 consists of: (1) an object G_4 and two morphisms $f' : G_2 \rightarrow G_4$ and $g' : G_3 \rightarrow G_4$ where $f' \circ f = g' \circ g$; (2) for any morphisms $f'' : G_2 \rightarrow X$ and $g'' : G_3 \rightarrow X$ such that $f \circ f'' = g \circ g''$, there is a unique morphism $k : G_4 \rightarrow X$ such that $f' \circ k = f''$ and $g' \circ k = g''$. Algebraic approaches are divided into two categories: the *Single PushOut, SPO* [18] and the *Double PushOut, DPO* [19]. The DPO approach consists of two pushouts and requires an additional condition called the "dangling condition". This condition states that the transformation is applicable only if it does not lead to "dangling edges", i.e. an edge without a source or a target node. Indeed, in the SPO approach, one pushout is required and the dangling edges are removed which permits to write a wide variety of transformations not allowed by the DPO approach. Thus, in this work, we only consider the SPO approach. Applying a rewriting rule to an initial graph (G) with the SPO method consists in the following steps:

1. find a matching of LHS in G, i.e. find a morphism $m : LHS \rightarrow G$.
2. delete the sub-graph $m(LHS) - m(LHS \cap RHS)$ from G.
3. add the sub-graph $m(RHS) - m(LHS \cap RHS)$ to G to get G'.

3 Merging Ontologies with Typed Graph Grammars

3.1 Ontologies as Typed Graphs Grammars

In order to represent the ontologies and the ontology changes with the typed graph grammars (see Fig. 2), we use the $TGGOnto$ (Typed Graph Grammars for Ontologies) model [13,14]:

$$TGGOnto = \{TG_O, G_O, R_O\}, \text{where} :$$

TG_O is a type graph that represents the meta-model of an ontology, G_O is a host graph that represents an ontology and R_O is a set of rewriting rules that formalize the ontology changes. In our work, we consider the OWL ontologies. Therefore, the type graph (TG_O) represents the OWL meta-model. Thus, the vertices types of TG_O are:

$$V_T = \{Class(C), Property(P), ObjectProperty(OP), -$$
$$DataProperty(DP), Individual(I), DataType(D), Restriction(R)\}.$$

The edge types correspond to properties used to relate different entities. For example, $subClassOf$ is used to link nodes of the type $Class$.

$$E_T = \{subClassOf, equivalentTo, range, domain, ...\}.$$

An ontology change (CH) is formalized by rewriting rules and executed as graphs transformation using SPO algebraic method [14].

$$CH = r_i \in R_O = (NACs, LHS, RHS, CHDs), \text{where} :$$

- $NACs$ are graph patterns that define the conditions should not be satisfied to apply the change;
- LHS is a graph pattern that defines the preconditions that should be satisfied to apply an ontology change;
- RHS is a graph pattern that defines the change to apply in the ontology;
- $CHDs$ are derived changes added to the principal change (CH) for keeping the consistency of the modified ontology.

3.2 GROM Approach

To merge ontologies, we use the automatic and asymmetric approach GROM (Graph Rewriting for Ontologies Merging) which consists in three main steps that are briefly described below (for more details, we invite readers to refer to our previous work [4]).

1. Similarity search that identifies the correspondences between the ontologies entities based on syntactic, structural and semantic similarities. Given two ontologies (O_1 and O_2), we distinguish: (1) CN, the set of commons nodes between O_1 and O_2; (2) EN, the set of the syntactically equivalent nodes; (3) SN, the set of the synonyms nodes and (4) $IsaN$, the set of the nodes that share a subsumption relation.

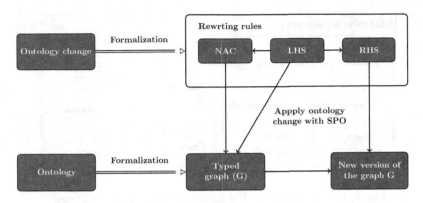

Fig. 2. The coupling between ontologies and typed graph grammars

2. *Ontologies merging* that represents ontologies with $TGGOnto = \{TG_O, G_O, R_O\}$ model and merges them based on the result of the similarity search step. The merging process consists in applying a set of consecutive rewriting rules with the SPO algebraic method in order to create a consistent global ontology (see Algorithm 1).

Inputs: two ontologies O_1, O_2
　　　　a set of correspondences: CN, EN, SN, $IsaN$
Outputs: a global ontology GO

for $N \in EN$ **do**
$\quad |\quad O_1' \leftarrow SPO\_RenameEntity\ (O_1,\ EN\{O_1\},\ EN\{O_2\}\ );$
end

$CN \leftarrow CN \cup EN\{O_1\}$;

$CO \leftarrow$ Create the common ontology;

$GO \leftarrow SPO\_MergeGraph\ (O_1',\ CO,\ O_2);$

```
/* Adapt the global ontology                                    */
```
for $N \in SN$ **do**
$\quad |\quad GO \leftarrow SPO\_AddEquivalentEntity\ (GO,\ SN\{O_1\},\ SN\{O_2\}\ );$
end
for $N \in IsaN$ **do**
$\quad |\quad GO \leftarrow SPO\_AddSubClass\ (GO,\ IsaN\{O_1\},\ IsaN\{O_2\}\ );$
end

Algorithm 1. Merging ontologies algorithm.

3. *Global ontology adaptation* step enriches the global ontology with the synonym (SN) and the subsumption ($IsaN$) relations identified in the step 1 (similarity search). As example, the rewriting rule of $AddEquivalentClasses(C_1, C_2)$ ontology change is presented in the Fig. 3.

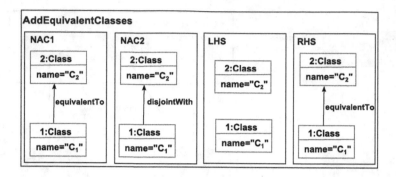

Fig. 3. Rewriting rule for the `AddEquivalentClasses` change.

The Rewriting rule (Fig. 3) preserves the ontology consistency thanks to the *NACs*. *NAC*1 ensures the no redundancy. It prohibits the adding of existing knowledge. *NAC*2 ensure that no contradictory axiom is added to the ontology by the application of the *AddEquivalentClasses* ontology change.

4   Experimental Results

4.1   Test Scenarios

We presented here two examples of ontologies (lightweight and heavyweight) to show and discuss the conflicts that can be found in the merging process.

Taxonomies case (Cars ontologies example [1]*):* The Fig. 4 shows an example of two taxonomies that represent the vehicle domain. They share commons concepts ("Automobile", "BMW", "Fiat") and subsumption relations (isA ("German_Car", "European_Car"), isA ("Italian_Car", "European_Car") and isA ("Mercedes", "German_Car")). Merging these taxonomies can cause the following conflicts and situations:

- *Data redundancy.* Considering that the ontologies share common concepts, their merge result can contain redundant elements, for example ("Automobile", "Automobile"), ("Audi", "Audi"), etc.
- *Sharing subsumption relations.* The ontologies could share subsumption relations, for example: isA("German_Car", "European_Car") and isA("Italian_Car", "European_Car").
- *Existence of cycles.* Adding the subsumption relations could provide cycles. For example, if we merge the ontologies we can obtain the cycle: isA("German_Car", "European_Car"), isA("German_Car", "Automobile"), isA("European_Car", "Automobile").

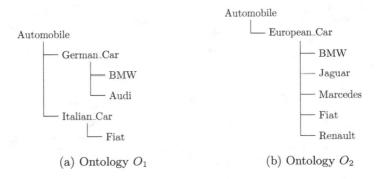

(a) Ontology O_1 (b) Ontology O_2

Fig. 4. European Cars ontologies [1].

OWL ontologies case (CCAlps ontologies example [14]): Fig. 5 presents extracts from the EventCCAlps and CompanyCCAlps ontologies developed in the frame of the European project CCAlps.

Merge OWL ontologies can cause the following situations and conflicts:

- *Data redundancy.* The concepts that have names syntactical close could be considered as redundant knowledge, for example: "hasPlace" and "has_place".
- *Synonyms concepts.* The ontologies could share synonyms relations, for example: "Individual" and "Person". It is important in this case, to link theses concepts by synonyms relations on the global ontology.
- *Axioms contradiction.* Merging OWL ontologies could cause several axioms contradictions relating to the disjunction, the equivalence, the restrictions, the subclass axiom, etc.

4.2   Benchmark and Results

To evaluate the ontologies merging process and test the GROM tool, we are based on the benchmark below, which is available for download in the supplementary material attached with this paper[2]:

- Cars ontologies [1] are taxonomies which are composed of classes and subsumption relations (already described in Sect. 4.1, Fig. 4). They are mainly included in the benchmark, for testing the hierarchical properties and for checking if a tool of ontologies merging can remove the cycles.
- CCAlps ontologies [13] are heavyweight OWL ontologies (described in Sect. 4.1, Fig. 5). These ontologies represent restrictions, properties, axioms, etc. and offer a good case study to check if an approach can manage or not the contradictory axioms and preserve or not the consistency of the global ontology.

---

[2] http://mariem-mahfoudh.info/ksem2016.

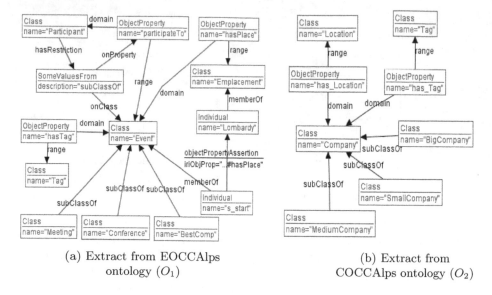

(a) Extract from EOCCAlps
ontology (O_1)

(b) Extract from
COCCAlps ontology (O_2)

Fig. 5. CCAlps ontologies [14].

- Lebensmittel (Google, web) [20] are multilingual ontologies that cover the food domain. `Google.Lebensmittel.owl` is represented in the English language and it is composed of 59 classes, 1306 individuals and 58 subClass axioms. `Web.Lebensmittel.owl` is represented in the German language and it is composed of 53 classes, 1566 individuals and 52 subClass axioms. The two ontologies have 15 synonyms classes that describe the same concepts but in two different languages.
- Freizeit (dmoz, Google) are multilingual ontologies that represent leisure. `Google.Freizeit.owl` is represented in the English language and it is composed of 71 classes. `Web.Freizeit.owl` is represented in the German language and it is composed of 76 classes. The two ontologies have 67 synonyms classes.
- Conference (ekaw, iasted, cmt, confof) ontologies [21] describe the conferences organization. They are datasets published by the OEAI (Ontology alignment evaluation initiative) to provide benchmark for the ontology alignment. We use the same ontologies to provide benchmark for the ontology merging.

   The Table 2 presents the set of ontologies that form the proposed benchmark and their result merge by GROM approach. It specifies: (1) the number of concepts of each ontology (nbC), (2) their similarities (the set of commons nodes (CN), the set of synonyms nodes (SN), the set of nodes that share isa relations ($IsaN$) and the set of nodes that are syntactically close (EN)), (3) the number of concepts of the merge result, (4) the number of redundant concepts in the merge result, (5) the number of inconsistencies and (6) the value of coverage (Cov).

Table 2. Merging result of some ontologies with GROM tool.

Ontologies	nbC	Similarities	Merge	Redundancy	Inconsistencies	Cov	Used in
Cars	7 .. 6	6 #CN=3, #IsaN =3	13	0	0	1.00	[1]
CCAlps (Event, Company)	12 .. 8	4 #CN=1, #EN=1, #SN=1, #IsaN=1	18	0	0	0.9	[4]
Lebensmittel (Google, web)	53 .. 59	15 #SN=15	112	0	0	1.00	[20]
Freizeit (dmoz, Google)	71.. 67	67 #EN=67	71	0	0	0.51	[20]
Conference (ekaw, iasted)	107 .. 182	10 #CN=5, #EN=3, #SN=2	280	0	1^3	0.97	[21]
Conference (cmt, confOf)	29 .. 38	9 #CN=9	58	0	0	0.86	[21]

[3] The detected inconsistency is from iasted ontology.

The Table 2 shows that all the output ontologies (the results of the process merging) don't contain any redundancy. Furthermore, GROM tool manages the inconsistencies and produces consistent ontologies.

5   Conclusion

In this paper, we presented a benchmark for ontologies merging that covers different ontologies types: taxonomies, lightweight ontologies, heavyweight ontologies and multilingual ontologies. We have ensured that the benchmark presents different pathological cases for merging process (data redundancy, existence of cycles, axioms contradiction, etc.) in order to assess the performance of ontologies merging tools. We also presented our tool, GROM (Graph Rewriting for Ontology Merging) and we described how it can automatically merge ontologies. Thanks to the graph grammar formalism, in particular the NAC (Negative Application conditions), GROM preserves the ontology consistency and removes the redundancies. To evaluate the merging process, we have used the coverage, the redundancy and the coherence metrics. The evaluation has shown that GROM obtained good results in terms of redundancy and coherence. All the ontologies are provided for download. In the near future, we plan to expand this benchmark with the results of other researchers in order to compare the different results of the related work and test the performance of the ontologies merging tools.

References

1. Raunich, S., Rahm, E.: Towards a benchmark for ontology merging. In: Herrero, P., Panetto, H., Meersman, R., Dillon, T. (eds.) OTM 2012. LNCS, vol. 7567, pp. 124–133. Springer, Heidelberg (2012). doi:10.1007/978-3-642-33618-8_20

2. Klein, M.: Combining and relating ontologies: an analysis of problems and solutions. In: IJCAI-2001, Workshop on Ontologies and Information Sharing, pp. 53–62 (2001)
3. Raunich, S., Rahm, E.: Target-driven merging of taxonomies with atom. Inf. Syst. **42**, 1–14 (2014)
4. Mahfoudh, M., Thiry, L., Forestier, G., Hassenforder, M.: Algebraic graph transformations for merging ontologies. In: Ait Ameur, Y., Bellatreche, L., Papadopoulos, G.A. (eds.) MEDI 2014. LNCS, vol. 8748, pp. 154–168. Springer, Heidelberg (2014). doi:10.1007/978-3-319-11587-0_16
5. Gandon, F.: Ontologies in computer science. In: Management and Design: Advanced Tools and Models (2010)
6. Stumme, G., Maedche, A.: FCA-MERGE: bottom-up merging of ontologies. In: Seventeenth International Joint Conference on Artificial Intelligence, pp. 225–230 (2001)
7. Noy, N.F., Musen, M.A.: Algorithm and tool for automated ontology merging and alignment. In: 17th National Conference on Artificial Intelligence, pp. 450–455. AAAI Press/The MIT Press (2000)
8. Kotis, K., Vouros, G.A.: The HCONE approach to ontology merging. In: Bussler, C.J., Davies, J., Fensel, D., Studer, R. (eds.) ESWS 2004. LNCS, vol. 3053, pp. 137–151. Springer, Heidelberg (2004). doi:10.1007/978-3-540-25956-5_10
9. Li, G., Luo, Z., Shao, J.: Multi-mapping based ontology merging system design. In: 2nd International Conference on Advanced Computer Control, pp. 5–11. IEEE (2010)
10. Duchateau, F., Bellahsene, Z.: Measuring the quality of an integrated schema. In: Parsons, J., Saeki, M., Shoval, P., Woo, C., Wand, Y. (eds.) ER 2010. LNCS, vol. 6412, pp. 261–273. Springer, Heidelberg (2010). doi:10.1007/978-3-642-16373-9_19
11. Ehrig, H., Montanari, U., Rozenberg, G., Schneider, H.J.: Graph Transformations in Computer Science. Geschäftsstelle Schloss Dagstuhl (1996)
12. d'Aquin, M., Doran, P., Motta, E., Tamma, V.A.: Towards a parametric ontology modularization framework based on graph transformation. In: WoMO (2007)
13. Mahfoudh, M., Forestier, G., Thiry, L., Hassenforder, M.: Consistent ontologies evolution using graph grammars. In: Wang, M. (ed.) KSEM 2013. LNCS (LNAI), vol. 8041, pp. 64–75. Springer, Heidelberg (2013). doi:10.1007/978-3-642-39787-5_6
14. Mahfoudh, M., Forestier, G., Thiry, L., Hassenforder, M.: Algebraic graph transformations for formalizing ontology changes and evolving ontologies. Knowl.-Based Syst. **73**, 212–226 (2015)
15. Rozenberg, G.: Handbook of Graph Grammars and Computing by Graph Transformation, vol. 1. World Scientific, Singapore (1999)
16. Ehrig, H., Pfender, M., Schneider, H.J.: Graph-grammars: an algebraic approach. In: Switching and Automata Theory, pp. 167–180. IEEE (1973)
17. Barr, M., Wells, C.: Category Theory for Computing Science, vol. 10. Prentice Hall, New York (1990)
18. Löwe, M.: Algebraic approach to single-pushout graph transformation. Theoret. Comput. Sci. **109**(1), 181–224 (1993)
19. Ehrig, H.: Introduction to the algebraic theory of graph grammars (a survey). In: Claus, V., Ehrig, H., Rozenberg, G. (eds.) Graph Grammars 1978. LNCS, vol. 73, pp. 1–69. Springer, Heidelberg (1979). doi:10.1007/BFb0025714
20. Peukert, E., Massmann, S., Koenig, K.: Comparing similarity combination methods for schema matching. In: GI Jahrestagung (1), pp. 692–701. Citeseer (2010)
21. OAEI: Ontology alignment evaluation initiative (2016). http://oaei.ontologymatching.org/2016/conference

TRSO: A Tourism Recommender System Based on Ontology

Yan Chu[1], Hongbin Wang[1(✉)], Liying Zheng[1(✉)],
Zhengkui Wang[2], and Kian-Lee Tan[3]

[1] Harbin Engineering Universtiy,
Nantong Street No.145, Nangang District, Harbin, China
{chuyan,wanghongbin,zhengLiying}@hrbeu.edu.cn
[2] InfoComm Technology, Singapore Institute of Technology, Singapore, Singapore
zhengkui.wang@singaporetech.edu.sg
[3] Department of Computer Science, National University of Singapore,
Singapore, Singapore
tankl@comp.nus.edu.sg

Abstract. In the era of information explosion, the Internet has become one of the most important tools for users to get information. As one of the main applications, most of the tourists, if not all, utilize the search engine to obtain the useful travelling information online which makes tourism recommender systems valuable. However, given a huge amount of online information, it still remains challenging to develop an effective tourism recommender system. To tackle this challenge, in this work, we propose TRSO, an ontology-based tourism recommender system by incorporating different techniques. First, we adopt the association rules to dig out the associated users from a large number of users. By doing so, users in the database are divided into two categories: related users and unrelated users. Second, for the related users, we propose a collaborative filtering algorithm by incorporating the time and evaluation factors. For the unrelated users, we utilize a different collaborative filtering algorithm, which integrates the time factor and the tourism attraction ontology information. Third, we further filter useless information according to the context information. Finally, we expand the tourism attraction with other tourism information such as shopping, eating and traveling based on a tourism ontology. The experimental results on the standard benchmark show that the proposed tourism recommendation algorithm can achieve satisfactory and comprehensive recommendation performance.

Keywords: Tourism recommendation · Association rules · Collaborative filtering · Context information · Ontology

1 Introduction

Nowadays, with the substantial improvement of people's living standard, tourism has become an increasing popular leisure activity for people. In addition, the

© Springer International Publishing AG 2016
F. Lehner and N. Fteimi (Eds.): KSEM 2016, LNAI 9983, pp. 567–579, 2016.
DOI: 10.1007/978-3-319-47650-6_45

Internet has become an important resource for those who are planning their trips to get tourism information. However, the huge amount information on the Internet always get people overwhelmed. Therefore, it is highly desired to develop an effective tourism recommender system, which is able to provide people the useful tourism information.

Existing recommendation technologies [1] can be roughly classified into four categories: content-based recommendation, collaborative filtering recommendation, knowledge-based recommendation, and mixed recommendation. Among these recommendation technologies, collaborative filtering recommendation has been considered as the most successful recommendation strategy. The basic idea of collaborative filtering recommendation is that if users have the same preferences in the past (such as browsing the same webpages or purchasing the same products), then they are more likely to have similar preferences and hence make the same choices in the future. Considering the same phenomenon arises in tourists behaviors, the user-based collaborative filtering recommendation has been widely used in the tourism recommendation as well [3].

Although existing recommendation algorithms have been adopted in the major e-commerce sites [2], it still remains challenging in developing an effective tourism recommendation algorithms. The first challenge is how to improve the accuracy of the recommendation results to meet users' need. Most of the recommended results remain at the landscape level recommendation, which ignores other tourism factors, such as cloth, food, accommodation and travel. The other challenge is how to develop one dynamic recommendation algorithm, which can take both the context and the personalization into the consideration. To tackle these challenges, this paper aims at proposing a new tourism recommender system which is able to generate accurate result and achieve dynamic recommendation by incorporating the context and personalization.

The main contribution of the paper are two-fold. First, we propose TRSO, a tourism recommender system based on attraction ontology. In particular, we first adopt the association rules to identify the related users which avoids the problem of sparse matrix in collaborative filtering recommendation and reduces the time costs. Meanwhile, we construct the attraction ontology to provide a comprehensive definition and relationships among tourism concepts. Furthermore, we incorporate one hybrid recommendation approach by integrating the results derived from different types of users by applying different algorithms using the time factor, evaluation factor and the ontology. Second, we conduct extensive experiments to evaluate proposed algorithms and the experimental results indicate the superiorities of them over the traditional algorithms.

The rest of the paper is organized as follows. Section 2 introduces the tourism attractions ontology constructions. In Sect. 3, we introduce the TRSO system design as well as all the proposed algorithms. Sections 4 and 5 present our experimental evaluation and conclusion respectively.

2   Construction of Tourism Attractions Ontology

The ontology concept originated from the philosophy domain, which refers to the description of the object in existence in the world [4]. Recently, ontology theory has gradually appeared in the field of artificial intelligence, and has been widely applied in various contexts, such as the semantic web [5], intelligent information retrieval [6], and digital libraries [7]. However, there is still no clear definition of ontology today. The most widely quoted definition, proposed by Gruber et al. [8], is that "ontology is a clear specification of the conceptual model." In a sense, ontology is used to describe concepts and the relationship among these concepts in a particular area or a more general area. By assigning reasonable, clear and unique definitions of concepts and relationships, ontology facilitates the communication between users and computers [9,15,16].

Tourism, an integrated industry including shopping, food, accommodation and travel, involves more complex and diverse information. Simply listing the flat tourism information cannot meet users' need well. Meanwhile, ontology does not only provide clear and rich concepts, but also is capable of characterizing the hierarchical relationship among concepts. Therefore, it is important to take the ontology into consideration to recommend hierarchical tourism information.

Based on existing tourism attraction ontology and the online travelling information, this paper constructs a more comprehensive tourism attraction ontology. It is worth noting that we standardize the names according to the specification. The construction of the tourism attractions ontology includes ten classes: accommodation, services, transportation, food, culture, activity, shopping, environment, natural landscape, and cultural landscape. This ontology is the basis of the whole tourism recommendation and plays a significant role in the subsequent recommendation.

3   The Ontology-Based Tourism Recommender System

In the context of tourism recommendation, similar users may share similar preferences. Due to the high cost and exclusive characteristics of tourism, collaborative filtering strategy can be applied for the tourism recommendation. However, in reality, it is inevitable to encounter the problem of sparse matrix by using the collaborative filtering recommendation algorithm. Towards this end, we propose to employ the association rules. Association rules have one important step - minimum support filtering, which guarantees the completeness of users. Considering that the contextual information plays an essential role in tourism, we further incorporate the context filters. Finally, based on the tourism attraction information, we expand the proposed ontology and hence provide comprehensive tourism recommendation.

3.1   Mining Association Rules

Association rules were originally designed to mine the products sales correlation in data mining, such as the example of the "beer diaper" [10]. It is worth mentioning that such a relationship is not based on the similarity between products.

This paper aims to take advantage of association rules to find the relationship between tourists instead of the tourism attractions. In particular, based on users' travel history, we aim to find the related tourists for a given tourist. In this way, we can effectively solve the problem of sparse matrix in collaborative filtering, and also improve the efficiency.

Apriori algorithm [17] is one of the most widely used association rules and initially designed to solve database problems. In particular, Apriori algorithm repetitively scans the database to obtain the frequent item sets. Although this algorithm is simple and easy to implement, it is both time and space consuming. Later, FP-Growth algorithm was developed based on the Apriori algorithm, which employs a tree structure to store the data and hence significantly reduces the time costs regarding the repetitive scanning and space costs for the candidate sets.

3.2   Collaborative Filtering with Time Factor and Evaluation Factor

Different to existing collaborative filtering algorithm, this paper handles different categories of users differently. Users are first divided into two categories by association rules. The first category consists of the users whose supports are larger than the minimum support, while the second consists of the users whose supports are less than the minimum support. To this end, we propose the TEUCF (user based collaborative filtering with time factor and evaluation factor) algorithm to recommend tourism attractions to users. Detailed algorithm is introduced as follows.

The phenomenon of user's dynamic interests that may change with time, is common in recommendation field. Users' recent behavior can reflect users' current interests better than that occurred long time ago. In other words, users' interests gradually decrease with time. This paper uses Ebbinghaus forgetting curve as the time factor to improve the recommendation performance. Ebbinghaus forgetting curve [11] can simulate the forgetting curve of human brain well, and hence better reflect the normal case of human forgetting. Ebbinghaus forgetting curve is where the horizontal axis represents time (in days) and the vertical axis represents the percentage of the amount of memory (total memory capacity of 100 %).

The Ebbinghaus forgetting curve simulates human memory retention in time. It can be seen from the figure that the amount of memory goes down as time goes on. The curve sharply decreases in the beginning and then gradually levels off. Overall, the curve is gradually approaching the horizontal axis. Due to the fact that the memory curve cannot be expressed by regular functions, several scholars [12] proposed the maintained memory function to fit the memory curve. Maintained memory means the amount of memory maintained in brain. The function of maintained memory curve is:

$$J(t) = \frac{20e^b}{(t + t_0)^c} \tag{1}$$

where t is the time variable (in days), e is the natural log base, b, c and t_0 are the constants to be determined. In particular, we empirically set $b = 0.42$, $c = 0.0225$, $t_0 = 0.00255$, to make the function most consistent with human memory curve. This paper normalizes the formula, and proposes time factor formula as follows.

$$T(t) = \begin{cases} 1 & t > 1 \\ \frac{e^b}{5(t+t_0)^c} & t \leq 1 \end{cases} \tag{2}$$

Apart from giving scores to tourism attractions, users also provide other evaluation forms, such as the textual evaluation comments. These comments not only illustrate the reasons for the scores but also affect the overall rating of the tourism attraction. If users feel the comments useful, they can give an "thumbs up" to express their agree with the comments. In particular, this paper introduces users' such "thumbs up" behaviors. The more "thumbs up" an evaluation comment harvests, the more accurate the corresponding score is and the higher weights this score deserves. This paper incorporates users' "thumb up" behavior as the evaluation factor into the collaborative filtering algorithm by using the below evaluation factor formula:

$$C = \begin{cases} 1 & i = 0 \\ \bigcup_{i=1}^{m} c_i & i > 1 \end{cases} \tag{3}$$

$$E(u) = 1 + \frac{C_u}{C} \tag{4}$$

where C_i and C are the number of reviews for comment i and the total number of reviews, and C_u and E are the number of comments for user U and the evaluation factor respectively. We incorporate the time factor and evaluation factor into the scoring matrix. In particular, we use Pearson correlation coefficient to measure the similarity. The Pearson correlation coefficient [13] is shown in formula 5.

$$sim_{uv} = \frac{\sum_{i \in I_{uv}} (r_{ui} - \bar{r}_u)(r_{vi} - \bar{r}_v)}{\sqrt{\sum_{i \in I_u} (r_{ui} - \bar{r}_u)^2} \sqrt{\sum_{i \in I_v} (r_{vi} - \bar{r}_v)^2}} = \frac{\sum_{i \in I_{uv}} R_{ui} R_{vi}}{\sqrt{\sum_{i \in I_u} R_{ui}^2} \sqrt{\sum_{i \in I_v} R_{vi}^2}} \tag{5}$$

Therefore, the Pearson correlation coefficient with time factor and evaluation factor is shown in formula 6.

$$R_{ui} = (r_{ui} - \bar{r}) \times T(ui) \times E(u) = \begin{cases} (1 + \frac{C_u}{C})(r_{ui} - \bar{r}) & t < 1 \\ \frac{e^b(1+\frac{C_u}{C})(r_{ui}-\bar{r})}{5(t+t_0)^c} & t \geq 1 \end{cases} \tag{6}$$

TEUCF algorithm procedure is shown as follows.

Algorithm 1. TEUCF Algorithm

Input: Target users; user-rating matrix M; Time factor; Evaluation factor;
Output: Top-N recommendation;
   01: FP-Growth algorithm is used to mine the association rules of matrix M.
   And get the related user set Ub, which consist of the related user - attraction
   rating matrix.
   02: Put time factor into rating matrix.
   03: Based of step two, put evaluation factor into new rating matrix constructed
   by step two.
   04: Use the new rating matrix which includes time factor and evaluation factor
   to calculate the similarity of related users to find out Top-N high similarity
   users, according to these users choice to generate recommendations.

3.3   Collaborative Filtering Based on Attractions Ontology

Certain tourists with a limited tourism attraction history would be filtered out by
the association rules. Many existing algorithms would simply remove these users
to achieve better performance which may result in incomplete result for such
filtered users. Differently, in this paper, we propose an ontology-based collaborate
filtering recommendation algorithm to deal with such users.

It is apparent that directly incorporate users with limited tourism attraction
history in the collaborate filtering may devastate the recommendation performance due to the sparse matrix. To solve this problem, we use attraction ontology to classify various attractions into different categories. And we generate a
new users-attractions class matrix, where the rating of each class is the average
rating of all attractions of this class.

In this part, time is also an important factor which affects users' interests.
Recent interests may reflect users current interests better. It is thus reasonable to
put time factor into this algorithm. However, it is worth noting that the method
to incorporate time factor is different from that in TEUCF. Normally, users may
become interested in visiting one type of attractions within one time period and
then change their interests after a while. But, it is likely that tourists will visit
the similar type of attractions as their previous touring. In other words, the
tourists' interests can be regained. On the basis of this observation, this paper
uses the last time of visiting one specific type of attractions as the time for the
whole continuous visiting for that type to avoid such regain behavior getting
overwhelmed by the huge amount of history data. By doing so, each continuous
visiting of one specific type of attractions is given one identified time. The rating
with time factor formula is shown as follows.

$$R_{u_{Li}} = (r_{u_{Li}} - \bar{r}) \times T(ui) = \begin{cases} (r_{u_{Li}} - \bar{r}) & t < 1 \\ \dfrac{e^b(r_{ui} - \bar{r})}{5(t+t_0)^c} & t \geq 1 \end{cases} \tag{7}$$

TCUCF algorithm procedure is shown as follow.

Algorithm 2.   TCUCF Algorithm
Input: no related users Ul; attractions ontology;
Output: Top-N recommendation;
01: Divide the old no related users-attractions rating matrix into the new related users-attractions types rating matrix based on attractions ontology.
02: Find out the last time of visiting this type of attractions. Put this time into the new related users-attractions types rating matrix.
03: Use the new rating matrix which includes time factor to calculate the similarity of related users to find out Top-N high similarity users, according to these users choice to generate recommendations.

3.4   Hybrid Algorithm

Hybrid algorithm is a common approach in recommender system, which is able to overcome the limitation of a single recommendation algorithm. Existing hybrid algorithms mainly have two kinds of hybrid modes: one for aggregating the results resulted in multiple individual algorithms and the other one for aggregating the algorithm processes. Due to the fact that this work separates users into two categories and handles them respectively, the first hybrid mode is preferred. In addition, the proposed tourism recommender system has two main branches. On branch takes the time factor as well as evaluation factor into consideration, while the other one considers the ontology. In particular, the second branch actually also achieves the hybrid of algorithm processes. The basic idea of the proposed hybrid algorithm is, by adopting the ontology information, to apply different algorithms for different category of tourists individually where the results are further aggregated together. Due to the space limitation and the simplicity of the algorithm, we will omit the detail here.

3.5   Multiple Context Information Filtering

Now, we are ready to introduce how the multiple context information filtering is applied in the system. Tourism is a field that is always influenced by context. The context may not only decide whether a tourism activity is feasible, but also affect the performance of tourism recommendation. In this work, the context information is obtained from the tourism attraction ontology. The proposed attraction ontology well fits the characteristics of tourism. The context information are set as 'season', 'location', 'weather' and 'time', which fully takes into account the characteristics of tourism.

To obtain the aforementioned context information, both the explicit and implicit approaches are applied. Specifically, the location and time can be explicitly collected, and the local weather conditions and season information are implicitly obtained online.

Adomavicius et al. [14] proposed two ways for context filtering in 2005, which are context pre-filter and context post-filter. Context pre-filter removes the irrelevant information about users' preferences regarding tourism attractions at first. And then traditional recommendation algorithm can be used to process the data set. In this way, the recommendation can fit well for users need and context need. Differently, context post-filter first conducts the recommendations by traditional recommendation algorithm. And then filter out the recommendations that do not fit for users' preferences.

Consider that the context pre-filter is much vulnerable to the context granularity. Coarse granularity may lead to the useless attractions in the recommendation, while too fine granularity may lead to the much sparse data set. Therefore, the post-filter method is adopted to ensure the recommendation performance.

3.6   Information Expansion Based on the Tourism Attractions Ontology

This work utilizes an ontology-based tourism recommendation algorithm which is able to find out the attractions which users may interested in. Furthermore, we use the context information (e.g., time, season, location and weather) to filter the recommendation results. This part uses a tourism attraction ontology to provide comprehensive tourism recommendation. The attraction ontology in this paper includes ten aspects: transportation, food, culture, activities, attractions, services, shopping, environment and accommodation. This ensures the diversity of tourism information and makes the tourism information recommendation being of more practical value.

4   Experimental Results and Analysis

In this section, we first introduce the dataset used for evaluation, and then present the method used for evaluation followed by the experimental results and analysis.

Datasets: In this work, the tourism information is crawled from the "Baidu tourism". In particular, we collect the following information about the attractions: names, attractions ratings, tourism time, weather and evaluation. After preprocessing, we store all these data in the database.

In total, we collected 1975 tourists' records which consists of about 2000000 attraction records. Each attraction has one rating, which can be five levels from low to high. Each record includes attraction, attraction type, location, time and "thumbs up" counts.

In the experiment, we randomly select 80 % rating data as the training set and the rest as the test set. This learning process is conducted in multiple rounds. The final result is calculated by averaging the value of all rounds.

Evaluation Method: The recommendation algorithm in this paper is developed on the basis of the traditional user-based collaborative filtering algorithm (UserCF), which consists of four steps as shown below:

- The first step is to find the association users whose supports are lager than 10 by the FP-Growth algorithm.

- The second step uses the UserCF algorithm to recommend attractions among related users. This algorithm is referred as FUCF algorithm.
- The third step first uses TEUCF and TCUCF to handle the related users and unrelated users, respectively, and then hybrids the recommendation results. We call this method as the MUCF algorithm.
- The fourth step takes the context information filter into consideration. This algorithm is called CMUCF algorithms.

4.1 Results and Analysis

First, we use the precision and recall to evaluate the quality of the recommendation. Precision represents the ratio of the intersection of the recommended attractions to the user and the actual tourism attractions of the user over the the actual tourism attraction of the user. Recall stands for the ratio of the intersection of the recommended attractions to the user and the user's actual viewing of the site over the recommended attractions to the user. Note that the larger the precision is, the more accurate the recommendation is. Recall follows the same manner. We change the number of K to do comparative experiments. Precision and recall are shown in Table 1 with different K-value.

Table 1. Precision comparison and Recall comparison

k	Pre(UserCF)	Pre(FUCF)	Pre(MUCF)	Rec(UserCF)	Rec(FUCF)	Rec(MUCF)
1	0.3678	0.4059	0.4785	0.1169	0.1236	0.1013
2	0.3029	0.3372	0.4119	0.1862	0.1967	0.1872
3	0.2651	0.2928	0.3779	0.2374	0.2459	0.2462
4	0.2359	0.2627	0.3477	0.2735	0.2836	0.2751
5	0.2157	0.2427	0.3296	0.3049	0.3177	0.3141
6	0.1996	0.2263	0.3149	0.3303	0.3451	0.3600
7	0.1869	0.2111	0.2971	0.3526	0.3633	0.3962
8	0.1767	0.2010	0.2840	0.3728	0.3856	0.4326
9	0.1689	0.1934	0.2709	0.3936	0.4087	0.4638
10	0.1621	0.1884	0.2610	0.4126	0.4356	0.4962

From Table 1, we can see that the precision of FUCF algorithm and MUCF algorithm significantly improved the recommendation performance compared with the traditional collaborative filtering algorithm. In terms of recall, when the number of recommended records is low (i.e., smaller than 6), all the three algorithms perform similarly. As the recommended number increases 6 and above, the MUCF significantly outperforms the other two, and the difference increases continuously. This indicates that the MUCF algorithm proposed in this paper is superior to the traditional algorithms in terms of the precision and recall.

We then employ MAE and RMSE to evaluate the recommendation quality. MAE and RMSE calculate the error between the predicted ratings and the actual users' ratings. The smaller the MAE and RMSE are, the more accurate the recommendation is. The performance of different algorithms regarding MAE and RMSE are shown in Table 2.

Table 2. MAE and RMSE comparison

k	MAE(UserCF)	MAE(FUCF)	MAE(MUCF)	RMSE(UserCF)	RMSE(FUCF)	RMSE(MUCF)
1	0.4561	0.4652	0.4611	0.6213	0.6165	0.6225
2	0.4648	0.4802	0.4553	0.6468	0.6308	0.6158
3	0.4523	0.4738	0.4544	0.6304	0.6326	0.6132
4	0.4623	0.4701	0.4526	0.6423	0.6275	0.4785
5	0.4690	0.4777	0.4494	0.6551	0.6386	0.6111
6	0.4700	0.4795	0.4516	0.6542	0.6426	0.6135
7	0.4701	0.4755	0.4508	0.6530	0.6407	0.6118
8	0.4673	0.4697	0.4515	0.6501	0.6335	0.6161
9	0.4704	0.4668	0.4502	0.6622	0.6313	0.6150
10	0.4724	0.4654	0.4495	0.6653	0.6296	0.6144

We can see that there is no clear advantages of FUCF over other algorithms regarding MSE when the recommended number is less than 8. The MAE of MUCF algorithm is lower than the other two algorithms only when k equals to 2. Overall, MUCF outperforms the other traditional recommendation algorithms.

Furthermore, we also evaluate the coverage rate as another metric in the recommendation to evaluate whether the personalization recommendation can be achieved. The coverage rate reflects the popularity of recommended attraction. The hot spots often appear on each tourism recommendation, thus the personalized recommendation is poor. The coverage rates are shown in Table 3.

Table 3. Coverage rate table

k	Cove(UserCF)	Cove(FUCF)	Cove(MUCF)
1	0.0629	0.0666	0.0988
2	0.1036	0.0962	0.1536
3	0.1480	0.1443	0.1762
4	0.1776	0.1813	0.1973
5	0.2072	0.2035	0.2518
6	0.1996	0.2263	0.2734
7	0.2960	0.2960	0.3506
8	0.3330	0.3404	0.3676
9	0.3478	0.3774	0.3854
10	0.3774	0.4218	0.4313

The corresponding coverage rate curves of different algorithms show that there is no significant difference between FUCF and UserCF in terms of coverage rate. But MUCF clearly outperforms UserCF which indicates the superiority of MUCF of handling the personalized recommendation.

Next, we evaluate the performance of context filtering algorithms. The context information is obtained from the attraction ontology which includes location, season, weather and time. In the experiment settings, we set the location as "Harbin", season as the "winter", time scheduled as the days in December. We obtain the weather conditions from the "www.tianqi.com". The sample weather conditions of Harbin in December is shown in Table 4.

Table 4. Part of December weather conditions in Harbin

Date	HighTemperature	LowTemperature	DayWeather	NightWeather	WindDirection	WindPower
1st	1	−4	Cloudy	Snow Shower	ES	Gentle Breeze
2nd	−10	0	Snow Shower	Snow Shower	ES	
3rd	−13	0	Snow Shower	Snow Shower	WS	
4th	−7	−13	Sunny	Snow	S	Gentle Breeze
5th	−7	−20	Snow Shower	Sunny	WN	Gentle Breeze
6th	−11	−21	Sunny	Sunny	WS	Breeze
7th	−10	−19	Sunny	Sunny	WS	Breeze
8th	−4	−15	Sunny	Cloudy	WS	Gentle Breeze
...	...	...	...	...	...	.
29th	−11	−24	Sunny	Smog	W	Breeze
30th	−15	−23	Smog	Smog	W	Breeze
31st	−11	−19	Smog	Smog	WS	Breeze

As shown in Table 4, there is no extreme snow or moderate gale in December. Only 30th and 31th had the smog weather. Therefore, it is reasonable to recommend indoor tourism activities from 30th to 31th due to the smog weather. And the rest of the month is fine for either indoor or outdoor tourism. We further calculate the precision and recall after the context filtering as shown in Table 5. As what Table 5 indicates, CMUCF achieves the best performance. This also shows that taking the context filtering into consideration can improve the precision and recall simultaneously.

Finally, instead of simply recommending tourism attraction, we expand the recommendation with other tourist attraction information based on the ontology which is able to meet users' basic necessities (e.g., cloth, food, accommodation and travel). For instance, we have set the location for the "Harbin", season for "winter", time for daytime, as the context. We thus based on the context information combined with the ontology to provide the final recommendation. For example, we can get the recommended attraction as "Snow World" and the nearby shopping malls such as "Wanda Plaza", nearby special restaurant such as "Manhattan restaurant" and nearby hotel such as "seven days inns". Moreover, considering the season context, we recommend the winter activity "Harbin International ice and snow festival".

Table 5. Precision and recall comparison

k	Pre(MUCF)	Pre(CMUCF)	Rec(MUCF)	Rec(CMUCF)
1	0.4785	0.4877	0.1013	0.1992
2	0.4119	0.4247	0.1872	0.2998
3	0.3779	0.3865	0.2462	0.3592
4	0.3477	0.3645	0.2751	0.4013
5	0.3296	0.3477	0.3141	0.4212
6	0.3149	0.3376	0.360	0.4415
7	0.2971	0.3303	0.3962	0.4598
8	0.2840	0.3243	0.4326	0.4737
9	0.2709	0.3190	0.4638	0.4856
10	0.2610	0.3161	0.4962	0.5019

5   Conclusions

Towards a better tourism recommendation service, we proposed TRSO, a tourism recommender system which adopts the ontology. Specifically, we introduced different techniques of identifying related/unrelated tourists using the association rules, constructing the attraction ontology, conducting the hybrid recommendation after applying different algorithms on different types of tourists and filtering the information based on the context information. The experimental results indicate the efficiency and accuracy of the proposed techniques.

Acknowledgments. The work is supported by the National Natural Science Foundation of China under Grant No. 61272185, the Natural Science Foundation of Heilongjiang Province of China under Grant No. F201340, the Science Foundation of Heilongjiang Province of China for returned scholars under Grant No. LC2015025, the Fundamental Research Funds for Central University under Grant No. HEUCF160602, and Harbin Special Fund for innovative talents of science and technology research under Grant No. 2013RFQXJ113.

References

1. Buettner, R.: Predicting user behavior in electronic markets based on personality-mining in large online social networks: a personality-based product recommender framework. Electron. Markets Int. J. Netw. Bus., 1–19 (2016)
2. Xu, W.H., Xiao, L.X., et al.: Comparison study of internet recommendation system. J. Softw. **10**, 350–362 (2009)
3. Jannach, D., Zanker, M.: Recommend System. Beijing University of Posts and Telecommunications Press, pp. 1–4 (2013)
4. Spiliopoulou, M.: The laborious way from data mining to the web mining. Int. J. Comput. Syst. Sci. Eng. **14**(2), 113–126 (1999). Special Issue on Semantics of the Web

5. Du, X., Li, M.: Summary of research on ontology learning. J. Softw. **17**, 1837–1848 (2006).

6. Strobbe, M., Van Leare, O.: Interest based selection of user generated content for rich communication services. J. Netw. Comput. Appl. **33**, 84–97 (2010)

7. Zghal, H.B., Moreno, A.: A system for information retrieval in a medical digital library based on modular ontologies and query reformulation. Multimedia Tools Appl. **72**, 2393–2412 (2010)

8. Gruber, T.R.: A translation approach to portable ontology specifications. Technical Report. 14, 2367–2456 (1993)

9. Ensan, F., Du, W.C.: A semantic metrics suite for evaluating modular ontologies. Inf. Syst. **38**, 745–770 (2013)

10. Baby, M., Idicula, S.M.: Apriori-based research community discovery in bibliographic database. Comput. Netw. Intell. Comput. **157**, 75–80 (2011)

11. Teyarachakul, S., Chand, S., Ward, J.: Effect of learning and forgetting on batch sizes. Prod. Oper. Manag. **20**, 116–128 (2011)

12. Zhiheng, J.: On the function of the past on the psychology of memory. J. Dyn. **3**, 3–23 (1988)

13. Orhan, E., Elvan, C., Yusuf, V.: A new correlation coefficient for bivariate time-series data. Phys. A Stat. Mech. Appl. **15**, 274–284 (2014)

14. Adomavicius, G., Sankaranarayanan, R., Sen, S., Tuzhilin, A.: Incorporating contextual information in recommender systems using a multidimensional approach. ACM Trans. Inf. Syst. **23**, 103–145 (2005)

15. Kang, W., Tung, A.K., Chen, W., Li, X., Song, Q., Zhang, C., Zhao, F., Zhou, X.: Trendspedia: an internet observatory for analyzing and visualizing the evolving web. In: ICDE 2014, Chicago, IL, USA, pp. 1206–1209 (2014)

16. Kang, W., Tung, A.K., Zhao, F., Li, X.: Interactive hierarchical tag clouds for summarizing spatiotemporal social contents. In: ICDE 2014, Chicago, IL, USA, pp. 868–879 (2014)

17. Agrawal, R., Srikant, R.: Fast algorithms for mining association rules in large databases. In: VLDB 1994, Santiago, Chile, pp. 487–499 (1994)

Generic Ontology Design Patterns: Qualitatively Graded Configuration

Bernd Krieg-Brückner[1,2]([⊠])

[1] CPS, BAALL, German Research Center for Artificial Intelligence (DFKI),
Bremen, Germany
Bernd.Krieg-Brueckner@dfki.de
[2] FB3 Mathematik und Informatik, Universität Bremen, Bremen, Germany

Abstract. For semantic modelling, ontologies are a good compromise between formality and accessibility to the layman, but lack sufficient methodological support and development tools. In particular, *Generic Ontology Design Patterns* are suggested as a domain independent methodological tool. *Qualitatively graded relations* combine semantic relations with qualitative valuations. *Qualitatively graded configuration* is illustrated in two application domains: mobility assistants for users with a variety of age-related impairments, and allowed food for users with diet restrictions.

1 Introduction

The objectives of this paper are twofold: to introduce *Generic Ontology Design Patterns* as a methodological tool in abstract semantic modelling, and to illustrate their use for *qualitatively graded configuration*.

Abstract Modelling. Mastering of complexity has traditionally been one of the major concerns in computer science, and is still a primary challenge in increasingly complex applications today. Apart from methods for separation of concerns (structuring of data, divide-and-conquer algorithms), *abstraction* is the key (cf. information hiding in Abstract Data Types), ranging from high level programming to abstract formal specification.

For abstract modelling, ontologies are a rather good compromise between data types (which may be too concrete and specific) and formal, e.g. algebraic, specifications (which may be too sophisticated). Data are abstracted into hierarchies of categories (*classes*); relations (*object properties*) abstractly capture semantics, modelling interrelations, but also operations and mere data access.

A particularly desirable feature of OWL-DL [5] is the decidability of abstract constraints expressed as axioms in Description Logic; thus modern reasoners provide effective execution of queries—indeed tools such as Stardog [7] combine modelling and data in one repository and make the implicit deduction of additional properties beyond the mere data appear as if it was one coherent data

© Springer International Publishing AG 2016
F. Lehner and N. Fteimi (Eds.): KSEM 2016, LNAI 9983, pp. 580–595, 2016.
DOI: 10.1007/978-3-319-47650-6_46

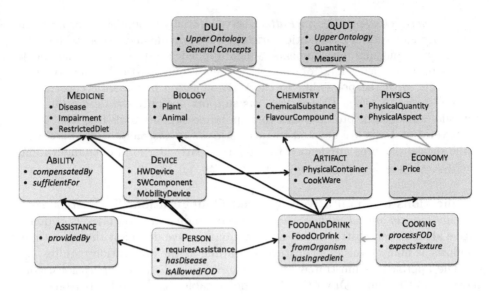

Fig. 1. Domain ontologies and import structure

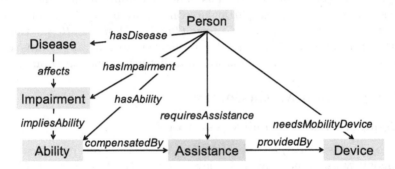

Fig. 2. Composite configuration for Person *needsMobilityDevice* MobilityDevice

base. Additional structuring can be achieved by modularisation of an ontology into a network of separate domain ontologies ("hyper-ontology"), cf. [8,15].

The challenge for modelling in OWL-DL is then to capture all relevant aspects of the application in Description Logic. However, the modelling should nevertheless be sufficiently structured and uncluttered to be understandable.

Generic Ontology Design Patterns. To further bolster the ontology development process, methodological support is provided in the form of *Generic Ontology Design Patterns* (see Sect. 3). While current *Ontology Design Patterns* are primarily intended for particular domains, these now act as generic tools, to be instantiated to a particular domain.

A *Generic Ontology Design Pattern* serves initially as guideline for a design with a particular form or semantic content; it also provides a constrained focus as a context for ongoing developments. When applied to a previous development, the pattern acts as a kind of "corset" to allow only particular development operations, with the possibility to check for violation of semantic constraints. Thus *Generic Ontology Design Patterns* support not only the design but also the safe development process, with corresponding support tools (see Sect. 4).

Several patterns are devoted to various kinds of *abstraction* (see Sect. 3.1), while special patterns are defined to support a particular kind of configuration, *qualitatively graded configuration* (see Sect. 3.2).

Applications of Qualitatively Graded Configuration. The methodology for *qualitatively graded configuration* supports grading at several levels of qualitative abstraction, abstracting away from quantitative detail; at the same time, it allows as many levels of sophisticated grading as the application requires.

The approach is illustrated by two significantly distinct applications: configuration of mobility DEVICEs to provide suitable assistance for persons with individual age-related impairments, and configuration of FOODANDDRINK for persons with individual diet restrictions. The first application is introduced in Sect. 2 as a motivation for the particular configuration approach, and the second in Sect. 3 while describing the *Generic Ontology Design Patterns*; thus both justify the abstraction from a particular application to general patterns and act as a "proof of concept" for instantiation in a distinct application domain.

2   Configuration Methodology

2.1   Configuration of Mobility Assistants

In the project *Assistants for Safe Mobility* (ASSAM) [2,14], multifaceted variants of mobility assistants have been developed on the basis of walkers, wheelchairs and tricycles. These are intended to compensate for individual age-related impairments: end-users may be afflicted by diseases leading to motoric impairments (e.g. loss of endurance, strength, visibility, or hearing) or cognitive impairments (e.g. disorientation). With appropriate hardware/software components, the mobility devices not only compensate for motoric impairments, but also provide orientation and navigation assistance, indoors and outdoors.

This variety presents a considerable challenge for assessment and configuration; a first attempt in [17,18] is elaborated here as a motivation for defining patterns for *Qualitatively Graded Configuration* (see Sect. 3.2).

Semantic Modelling. A structured hyper-ontology has been developed (comprising more than 10,000 hand-edited classes, object property relations, and axioms, cf. Fig. 1) consisting of 25 separate interlinked ontologies for general (e.g. MEDICINE) or specific application domains (e.g. ABILITY, FOODANDDRINK, or PERSON). They are based on Upper Ontologies for structuring (DUL, derived from DOLCE, cf. [3]) or quantity standards (QUDT, cf. [6]).

Fig. 3. Taxonomies of Disease and Impairment (excerpt)

Composite Configuration. The relation *needsMobilityDevice* (Fig. 2) is a step-wise composition of individual relations from Person to separately defined ontologies to manage complexity ("separation of concerns"). At first, individual Diseases of a particular Person are assessed, captured by the relation *hasDisease*. A Disease will, in general, result in some Impairment; which Impairment(s) are affected is modelled in the relation *affects*. Thus the relation *hasImpairment* already contains some entries when a further anamnesis of Impairments is made. In turn, an Impairment *impliesAbility* an Ability; an Ability may be *compensatedBy* some Assistance, which is *providedBy* some MobilityDevice. For (excerpts of) the classes Disease, Impairment, Ability, Assistance, MobilityDevice see Figs. 3, 4 and 5.

Thus the relation *needsMobilityDevice* is sequentially composed, where each relation is a "triangular" composition, e.g. for *needsMobilityDevice*:

requiresAssistance ∘ *providedBy* → *needsMobilityDevice*

There may be several such axioms. It is sufficient to model the individual relations *affects*, *impliesAbility*, up to *providedBy*, *separately*, and once and for all; the rest is deduced by OWL-DL reasoners by logical inference.

Such structuring is essential to master complexity. The introduction of Assistance is an example of an extra intermediate domain, modelled by *intermediate abstraction* (cf. the pattern in Fig. 11, Sect. 3.1; see also [13]).

Fig. 4. Taxonomies of Ability and Assistance (excerpt)

Example. Abel Muller is slightly afflicted by Asthma. As a consequence he needs a special walker. For the deduction consider the interrelation axioms[1]

Asthma $\sqsubseteq$ *affects* **some** EnduranceImpairment

EnduranceImpairment $\sqsubseteq$ *impliesAbility* **some** EnduranceAbility

EnduranceAbility $\sqsubseteq$ *compensatedBy* **some** UpHillForceAssistance

UpHillForceAssistance $\sqsubseteq$ *providedBy* **some** iWheelHWComponent

iWheelHWComponent $\sqsubseteq$ *isMobilityHWComponentOf* **some** eWalker

He also has a slight OrientationImpairment

OrientationImpairment $\sqsubseteq$ *impliesAbility* **some** WayfindingAbility

WayfindingAbility $\sqsubseteq$ *compensatedBy* **some** WayFindingAssistance

WayFindingAssistance $\sqsubseteq$ *providedBy* **some** WayfindingSWComponent

WayfindingSWComponent $\sqsubseteq$ *isDeviceOf* **some** eWalkerNavigationAid

thus an eWalkerNavigationAid with associated HW and SW is needed.

Complete Device Arrangements. The above relations capture the components that a device arrangement comprises. The modelling includes more applications of the respective *isHW/SWComponentOf* relations (cf. also Figs. 5 and 6).

---

[1] $X \sqsubseteq r$ **some** Y means: for each x in X, some y in Y exists, related by r. Because of the universal quantification of x and existential quantification of y, X should be a maximal and Y a minimal superclass, resp., for compositions $r \circ s$.

Fig. 5. MobilityDevice arrangements (excerpt)

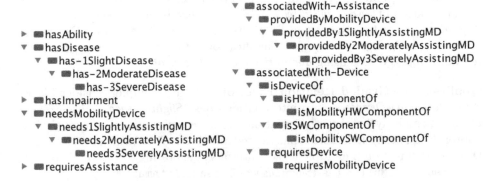

Fig. 6. Relations for Person, Assistance and Device (excerpt)

For example, a WayfindingSWComponent has a sister RoutePlanningSWCompo-
nent, and both are components of a NavigationAidiWheelSWComponent, have user
interface components; eWalkerNavigationAid has a hardware component SmartInteractionPC (such as a smartphone or tablet) to run the software components for
navigation and provide an appropriate interface, and so on. The component relations *isHWComponentOf* etc. (Fig. 6) have a transitive closure *isDeviceOf*; thus
the relation *providedBy* can be extended to cover all contributing components:

providedBy ∘ *isDeviceOf* → *providedBy*

Crosswise Requirements that a component requires another for proper functioning can be captured by a separate relation *requiresDevice*. It allows a proof by deduction, that such components are part of a complete device arrangement.

2.2   Qualitatively Graded Relations

This modelling captures the relationships from Disease to Assistance rather well. However, the assessment should be more subtle: is the person's impairment slight or severe? does the mobility device provide slight or moderate assistance?

Qualitative Abstraction. To achieve a graded valuation according to some qualitative abstraction of a semantic concept, we might introduce extra valuation domains with values such as Slight, Moderate, Severe (cf. the *value partitioning* design pattern in [1]), or some other (arbitrarily fine) qualitative metrics; the number of levels depends on the application. To combine a class, say C, with its valuations, resp., we would have to construct extra combined domains, say C-Slight, C-Moderate, C-Severe, one for each domain. Alternatively, we could introduce standard individuals for valuation combinations; but then each extra individual generated by some application would have to "inherit" the appropriate valuation or be made equivalent with such a standard individual, cf. [17]. This involves a very considerable overhead, is clumsy and ultimately error-prone.

Graded Composition. Instead, the novel approach described here (cf. [18] for an earlier attempt) uses *qualitatively graded relations* to encode such valuations. A *graded composition* then typically starts with an assessment of a Person (or other class in focus), i.e. a graded classification by relating the person to some relevant property class using *qualitatively graded relations*.

Qualitatively Graded Relation Example. As an example, consider Fig. 6. The relation *hasDisease* is split into relations *has-1SlightDisease*, ... *has-3Severe Disease*[2], similarly for the other relations with Person as domain (for a similar approach cf. Fig. 8). In a more sophisticated anamnesis Abel Muller clicks "Slight Disease Asthma" on a multiple choice interface, generating for the individual
    `Person_AbelMuller` *has-1SlightDisease* `Disease_Asthma`.
Note that the modelling for the relations *affects, impliesAbility, compensatedBy* stays as is, the associated grading is automatically transferred by the compositions (cf. the pattern for *Simplified Symmetric Configuration* in Fig. 12), e.g.:
    has-1SlightDisease ∘ *affects* → *has-1SlightImpairment*

Upgrading and Downgrading. Alternatively, we may wish to model some *upgrading* or *downgrading* influence. As an example, we split *affects* into several downgrading relations *affects0Neutrally, affects-1Worse, affects-2MuchWorse*; we replace the axiom for Asthma in Sect. 2.1 above by

---

[2] The strictly mnemonic names/numbers receive their semantics only through axioms.

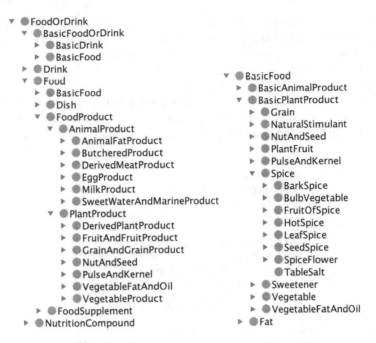

Fig. 7. FoodOrDrink and BasicFood taxonomies (excerpt)

Asthma $\sqsubseteq$ *affects-1Worse* **some** EnduranceImpairment

and the composition axioms, rather than relating the levels one-to-one, capture the effects of downgrading

has-$1SlightDisease \circ affects0Neutrally \rightarrow has$-$1SlightImpairment$

has-$1SlightDisease \circ affects$-$1Worse \rightarrow has$-$2ModerateImpairment$

has-$1SlightDisease \circ affects$-$2MuchWorse \rightarrow has$-$3SevereImpairment$

has-$2ModerateDisease \circ affects0Neutrally \rightarrow has$-$2ModerateImpairment$

has-$2ModerateDisease \circ affects$-$1Worse \rightarrow has$-$3SevereImpairment$

has-$3SevereDisease \circ affects0Neutrally \rightarrow has$-$3SevereImpairment$

Qualititative Assessment of the Target. The *graded composition* ends with a qualitative assessment of the target domain, in this case the device arrangements, cf. Fig. 6. The relation *providedBy* is split into relations *providedBy1SlightlyAssistingMD ... providedBy3SeverelyAssistingMD*, and the components of a device arrangement are classified as to their influence, resp., e.g.:

UpHillForceAssistance $\sqsubseteq$
providedBy2ModeratelyAssistingMD **some** iWheelHWComponent

Differentiated Qualitative Composition. As the modelling of Sect. 2.1 essentially stays the same, the effect for Abel Muller is not immediately apparent.

However, the achievements of a more differentiated modelling by qualitative composition emerge when we consider an additional VisualImpairment.

If it is slight, the corresponding restricted VisualPerceptionAbility can be compensated on the same channel by VisualPerceptionAssistance, provided, say, by large arrow symbols on a screen of the SmartInteractionPC.

If it is moderate, compensation on a different, e.g. the tactile, channel is required; VibrationFeedbackAssistance for direction instructions is provided by a VibrationOutputDevice in an additional iHandleBarHWComponent on the walker.

If it is severe, a different mobility assistant, a wheelchair with speech control and AutonomousDrivingAssistance with obstacle avoidance, will be selected.

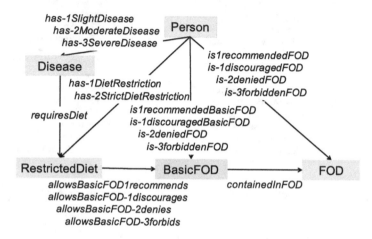

Fig. 8. Grading RestrictedDiet

2.3   Configuration of Food and Drink for Diet Restrictions

Before the abstraction to *Generic Ontology Design Patterns* in the next section, consider the other application domain in which these are to be instantiated.

Assume that you want to cook for several invited guests; it is quite likely, that they would have a variety of different diet restrictions or preferences to be considered. For this purpose (and other ongoing work to model the semantics of food and cooking), the target domain FOODANDDRINK has been extensively modelled, see Figs. 1 and 7. Figure 8 shows the overall structure of the *qualitatively graded configuration* (FOD occasionally abbreviates FoodOrDrink).

Note that the domains PERSON and MEDICINE are re-used and extended by appropriate class hierarchies etc., since the general configuration approach is the same: for mobility assistance, a Person has individual Diseases and age-related Impairments, for which adapted MobilityDevices are configured; analogously, a Person's individual Diseases (Fig. 3) require RestrictedDiets (Figs. 8 and 9), for which FoodOrDrink is *1recommended, -1discouraged, -2denied,* or *-3forbidden.* As an example of two different medical conditions consider:

GastroesophagealReflux ⊑ *requiresDiet* **some** LowFatDiet

Fig. 9. Taxonomies for MedicallyRestrictedDiet and PreferredDiet (excerpt)

GastroesophagealReflux ⊑ *requiresDiet* **some** LowSugarDiet
GastroesophagealReflux ⊑ *requiresDiet* **some** NoHotSpiceDiet
PregnancyCondition ⊑ *requiresDiet* **some** NoHotSpiceDiet
PregnancyCondition ⊑ *requiresDiet* **some** NoNaturalStimulantDiet
both require i.a. a NoHotSpiceDiet, which forbids any HotSpice in BasicFood:
NoHotSpiceDiet ⊑ *allowsBasicFOD-3forbids* **some** HotSpice
any food that contains[3] HotSpice (Fig. 7) is thus forbidden, e.g.:
HotSpice ⊑ *containedInFOD* **some** HotSauce
HotSpice ⊑ *containedInFOD* **some** HotSpiceMixture
In analogy to Impairments, a parallel assessment of individual preferences for
RestrictedDiets can be obtained; the union of these for a list of guests then leads
to overall FoodOrDrink restrictions to be taken as a basis for adapted recipes.

---

[3] Using an inverse relation *containsFOD* is OK as long as it is completely specified on
corresponding individuals. However, care has to be taken for axioms on the inverse:
seemingly equivalent axioms on the original relation do *not* hold in general and are
not deduced (similarly for *provides* as an inverse of *providedBy*, Fig. 2).

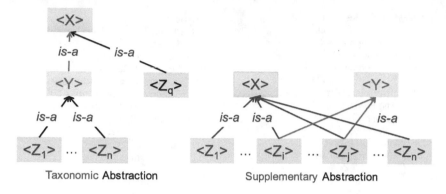

Fig. 10. Taxonomic and supplementary abstraction patterns

3   Generic Ontology Design Patterns

In the repository for *ontology patterns* (OPs) [4], most of the patterns treat Upper Ontologies, or design issues and structuring in specific application domains (cf. *Content Design Patterns* in [10, 16]). The OPs proposed in this paper fit better into the classification of [9] as "*Generic* domain independent *Ontology Design Patterns*". They abstract from application domains to a generic methodological level, and are instantiated to particular domains as part of an ontology engineering process. The patterns have initially been inspired by the work on configuration of mobility assistants (Sect. 2.1). They were then applied to modelling diet restrictions, food and drink, etc. (Sect. 2.3); it is this latter domain to which they shall be applied in this section for further illustration.

The notation $<X>$ shall denote a pattern variable X to be instantiated.

3.1   Abstraction Patterns

Taxonomic Abstraction. A class in a taxonomic hierarchy comprises the distinctive properties characterising this class, its subclasses and the constituent individuals. Axioms for the superclass hold for all subclasses and make an individual treatment for each subclass superfluous, an "added value" of abstraction.

The *taxonomic abstraction* pattern in Fig. 10 illustrates the introduction of a new class $<Y>$ as a superclass for $<Z_1>$ to $<Z_n>$, making $<Y>$ a subcategory of $<X>$; we assume, that previously $<Z_1>$ to $<Z_n>$ were direct subclasses of $<X>$ at the same level as $<Z_q>$. Common properties of $<Z_1>$ to $<Z_n>$ are *collected* at $<Y>$; $<X>$ now shares all properties of $<Y>$ and $<Z_q>$.

Tools supporting such patterns and associated development operations should automatically support a *re-distribution* of properties. Moreover, semantic constraints such as adherence to disjointness axioms must be checked: if $<Z_1>$ to $<Z_n>$ and $<Z_q>$ were previously stated to be disjoint, this axiom must be adapted to exclude $<Z_q>$; otherwise an inconsistency occurs, a common error.

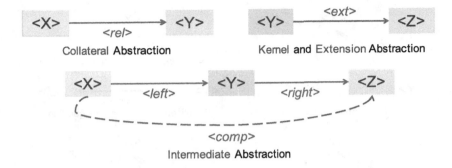

Fig. 11. Collateral, kernel extension and intermediate abstraction patterns

Supplementary Abstraction in Fig. 10 introduces $<Y>$ as a supplementary superclass, which collects the properties shared by $<Z_i>$ to $<Z_j>$.

This case occurs often, e.g. when defining BasicFood as a subclass of both, BasicFoodOrDrink and Food, cf. Fig. 7 and *kernel extension* below.

Similarly, HotSpice has been introduced in BasicFood as a *supplementary abstraction* for spices with spicy hot aromas, classified elsewhere: e.g. Mustard-SeedSpice is a subclass of SeedSpice. If the supplementary superclass can be characterised by relation axioms as a *defined class* (cf. recommendation in [1]), then shared subclasses can be deduced by reasoning, e.g.

HotSpice $\equiv$ Spice **and** (*has2EssentialFlavour* **some** SpicyHotSensationAspect)

Collateral Abstraction occurs when we perform some abstraction in two domains collaterally, see Fig. 11. As an instantiation of $<rel>$ consider the relation *fromOrganism*: it relates a BasicFoodOrDrink to the corresponding Organism in BIOLOGY. Since plants or animals often have local or colloquial names, such "anchoring" is indispensable for food to achieve a correct identification and translation in multilingual ontologies. *Taxonomic abstraction* is desirable on either side of the pattern, in a coordinated fashion, to relate superclasses, when shared properties correspond, e.g. GourdFruitVegetables *fromOrganism* Cucurbitaceae. However, CucurbitaMaxima (or CucurbitaMoschata) contains plants, whose fruits are in distinct GourdFruitVegetable categories w.r.t. their *culinary* properties having either a soft (squashes) *or* a hard rind (pumpkins) (cf. [13]).

Kernel Extension. A similar situation arises when a relation $<ext>$ extends a subdomain $<Y>$ to a domain $<Z>$ (Fig. 11).

The extension relation *isDeviceOf* in Sect. 2.1 for device arrangements shows a composition pattern $r \circ ext \rightarrow r$, such that r with the kernel as target now also holds for the extended domain.

Similarly, for *containedInFOD* (Sect. 2.3, Figs. 7 and 8), the focus of modelling can stay on the kernel BasicFoodOrDrink, which remains rather stable, while the extended domain FoodAndDrink will ever so often be augmented by a derived food product, in whose ingredients products from the kernel are contained.

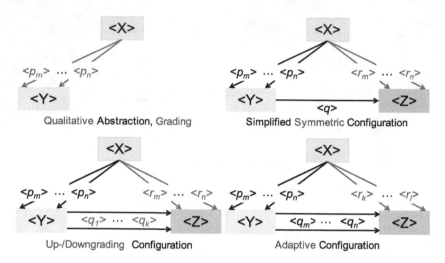

Fig. 12. Grading and configuration patterns

Intermediate Abstraction splits a relation $<comp>$ into a composition of $<left>$ and $<right>$ instead of defining it directly (Fig. 11).

Assistance is such an *intermediate abstraction* (Fig. 2): its class hierarchy may be structured according to need: "on the left" by defining *compensatedBy* from Ability, and "on the right" by defining *providedBy* to MobilityDevice. Similarly, DietRestriction and BasicFoodOrDrink are *intermediate abstractions* (Fig. 8).

A major methodological advantage (again: separation of concerns to master complexity) lies in the fact that $<left>$ and $<right>$ can be *separately* defined and focused on during the development. It is often helpful to introduce appropriate extra superclasses in $<X>$, $<Y>$, or $<Z>$ to gather common properties.

For example, HotSpice directly corresponds to NoHotSpiceDiet (Sect. 2.3).

3.2   Qualitatively Graded Configuration Patterns

Qualitative Abstraction and Grading has been described in Sect. 2.2. The first pattern in Fig. 12 shows the general situation: relations $<p_m>$ to $<p_n>$ are introduced, whose names convey some grading semantics.

A grading level often corresponds to an appropriate interval of quantitative properties. When the levels are ordered (the intervals are consecutive), it makes sense to let a higher grade subsume a lower one, mnemonically indicated by some numbering scheme. Then a higher relation *subsumes* a lower one (Figs. 6 and 8):
x *providedBy2ModeratelyAssistingMD* y → x *providedBy1SlightlyAssistingMD* y

The number of levels depends on the application. The support of a pattern for adding an *additional grade* ("step-by-step" pattern) will be tricky, since the previous grading for adjoining relations will most likely have to be revised.

Symmetric Configuration is the most common: the levels of the source and the target correspond. *Simplified symmetric configuration* (Fig. 12) takes advantage of this fact: only a single relation $<q>$ between $<X>$ and $<Y>$ is needed; composition axioms have the form $<p_i> \circ <q> \rightarrow <r_i>$; cf. *impliesAbility* or *compensatedBy* (Sects. 2.1 and 2.2, Fig. 2), and *requiresDiet* (Sect. 2.3, Fig. 8).

Up- and Downgrading Configuration (Fig. 12) is described in Sect. 2.2. In the composition axioms, the effects of the original graded relations $<p_i>$ and of the up/downgrading relations $<q_j>$ add up in the mnemonic index (or subtract, if j is negative) until the bounds of i; axioms have the form (for all i and j):

$$<p_i> \circ <q_j> \rightarrow <r_{i+j}>.$$

Adaptive Configuration allows for a general correspondence between grading levels m to n and k to l (Fig. 12). As an example consider Fig. 8: the axioms

    has-1SlightDisease ∘ *requiresDiet* → *has-1DietRestriction*
    has-2ModerateDisease ∘ *requiresDiet* → *has-2StrictDietRestriction*

are sufficient, since *has-3SevereDisease* is subsumed in *has-2ModerateDisease*.

However, depending on a positively or negatively formulated RestrictedDiet (Fig. 9) such as DietRichInFish or NoHotSpiceDiet, a corresponding BasicFOD category is *1recommended, -1discouraged*; in case of a strict diet restriction even *-2denied* or *-3forbidden*, depending on the *allowsBasicFOD...* classification (notice the non-continuity between 1 and negative subsumption):

has-1DietRestriction ∘ *allowsBasicFOD1recommends* → *is1recommendedBasicFOD*
has-1DietRestriction ∘ *allowsBasicFOD-1discourages* → *is-1discouragedBasicFOD*
has-2StrictDietRestriction ∘ *allowsBasicFOD-2denies* → *is-2deniedBasicFOD*
has-2StrictDietRestriction ∘ *allowsBasicFOD-3forbids* → *is-forbiddenBasicFOD*

4 Ongoing and Future Work

Development Support Tools. Further work on tools for patterns is under way, along with a formal definition of patterns and their semantic constraints. The patterns in Sect. 3 introduce a new class and all of its descendants, properties and axioms in one step. For actual tool support, "step-by-step" patterns would be preferable since they mirror the development situation more realistically: the pattern has already been previously applied and we want to add *one more* class and/or property/axiom guarded by the pattern. These should include associated development operations, (re-)distribution of properties, and constraint checking.

An *application domain expert* would expect a pattern-based editor for domain ontology development, with browsing, structural editing and semantic checking in the focus of an instantiated pattern; thus only the operations allowed by the corresponding pattern become possible. Similarly, a *semantic modelling expert* would expect a (meta) editor for ontology pattern development, with a toolbox for the definition of generic patterns, axioms, links to reasoners to prove semantic constraints, and similar facilities as for the pattern-based editor.

The above experts are allowed to change the modelling level of an ontology; an application interface for an *end-user* will only be able to browse (possibly in a variant of a pattern-based editor with a frozen model), and add or change (local) data (i.e. individuals/instances), such as person profiles, guest lists, new recipes, or shopping lists.

The potential for support is considerable. It cannot be expected that ontologies are ever "complete"; thus the development process and the correctness of changes on the resulting ontologies are a major concern.

Modelling. Modelling beyond the demonstration prototype stage will be pursued once appropriate tools are available. Indeed, the present modelling, although of considerable size and sophistication, can only be considered as preliminary until its contents are verified by domain experts, e.g. in MEDICINE.

An extension of the present hyper-ontology (Fig. 1) to cover various aspects of assistance in daily life, in particular *Ambient Assisted Living (AAL)*, is intended. Modelling personal impairments/abilities in AAL (Sect. 2.1), or diseases/diet restrictions (Sect. 2.3), could easily be extended to other medical conditions.

Configuration. The modelling approach for *qualitatively graded configuration* is very general and has promising perspectives in many application domains.

For mobility assistants it is presently extended to cover user-adapted space accessibility, such as suitable paths in appropriately configured routes.

Multimodal configuration for specific tactile, visual, or hearing impairments and their implications for interaction with dedicated input/output devices is being refined; switching from reinforcing *monomodal compensation* to *complementary multimodal compensation* (Sect. 2.2) will be discussed in a future paper.

The food modelling (Sect. 2.3, [13]) has been used in a shopping assistant [12]. The modelling of food flavours [11,19] is being considerably refined with *qualitatively graded configuration* for ingredients and flavours, now also covering abstract dosage and the influence of cooking operations on aroma persistence.

Dedication and Acknowledgements

This paper is dedicated to my old friend Prof. Lu Ruqian, Chinese Academy of Sciences, Beijing, China. He has been an active supporter (if not founder) of the KSEM conference series. I very much appreciate his suggestion for this presentation, and Martin Wirsing's invitation.

I am also very grateful for the contributions of Serge Autexier (i.a. requiresDevice as a separate relation, Sect. 2.1), Sidoine Ghomsi Nokam [11], Jana Köhler, Philipp Kolloge (initial diet restrictions), Jens Pelzetter, and Martin Rink (the idea for *up- and downgrading configuration*, cf. [17] and Sects. 2.2 and 3.2).

References

1. A Practical Guide To Building OWL Ontologies Using Protege 4 and CO-ODE Tools. http://mowl-power.cs.man.ac.uk/protegeowltutorial/resources/ProtegeOWLTutorialP4_v1_3.pdf
2. Assistants for Safe Mobility, ASSAM. www.assam-project.eu
3. DUL - DOLCE+DnS Ultralite ontology - Ontology Design Patterns (ODP). www.ontologydesignpatterns.org/ont/dul/
4. Ontology Design Pattern Types. http://ontologydesignpatterns.org/wiki/OPTypes
5. OWL Web Ontology Language - Use Cases and Requirements - W3C Recommendation 10. www.w3.org/TR/2004/REC-webont-req-20040210/
6. QUDT - Quantities, Units, Dimensions and Data Type Ontologies. www.qudt.org/
7. Stardog. http://stardog.com/
8. Bateman, J.A., Castro, A., Normann, I., Pera, O., Garcia, L., Villaveces, J.M.: OASIS Common hyper-ontological framework (COF). EU FP7 Project OASIS - Open architecture for Accessible Services Integration and Standardization Deliverable D1.2.1, Bremen University, Bremen, Germany, January 2010
9. Blomqvist, E., Sandkuhl, K.: Patterns in ontology engineering: classification of ontology patterns. In: Chen, C., Filipe, J., Seruca, I., Cordeiro, J. (eds.) ICEIS 2005, Proceedings of the Seventh International Conference on Enterprise Information Systems, Miami, USA, May 25–28, 2005, pp. 413–416 (2005)
10. Gangemi, A.: Ontology design patterns for semantic web content. In: Gil, Y., Motta, E., Benjamins, V.R., Musen, M.A. (eds.) ISWC 2005. LNCS, vol. 3729, pp. 262–276. Springer, Heidelberg (2005). doi:10.1007/11574620_21
11. Ghomsi Nokam, S.: A food ontology for the assistance of shopping and cooking. Master's thesis, Universität Bremen (2015) (in German)
12. Kozha, D.: Shopping assistance from the kitchen cabinet to the supermarket shelf. Master's thesis, Universität Bremen (2015) (in German)
13. Krieg-Brückner, B., Autexier, S., Rink, M., Nokam, S.G.: Formal modelling for cooking assistance, essays dedicated to martin wirsing. In: Nicola, R.D., Hennicker, R. (eds.) Software, Services and Systems, pp. 355–376. Springer International Publishing (2015)
14. Krieg-Brückner, B., Mandel, C., Budelmann, C., Martinez, A.: Indoor and outdoor mobility assistance. In: Wichert, R., Klausing, H. (eds.) Ambient Assisted Living. Advanced Technologies and Societal Change, pp. 33–52. Springer, Verlag (2015)
15. Mossakowski, T., Codescu, M., Neuhaus, F., Kutz, O.: The distributed ontology, modeling and specification language DOL. In: Koslow, A., Buchsbaum, A. (eds.) The Road to Universal Logic, vol. I, pp. 489–520. Birkhäuser (2015)
16. Presutti, V., Gangemi, A.: Content ontology design patterns as practical building blocks for web ontologies. In: Li, Q., Spaccapietra, S., Yu, E., Olivé, A. (eds.) ER 2008. LNCS, vol. 5231, pp. 128–141. Springer, Heidelberg (2008). doi:10.1007/978-3-540-87877-3_11
17. Rink, M.: Ontology Based Product Configuration Based on User Requirements. Master's thesis, Universität Bremen (2015) (in German)
18. Rink, M., Krieg-Brückner, B.: Wissensbasierte Konfiguration vonMobilitäts-Assistenten. In: VDE E.V. (ed.) Zukunft Lebensräume Kongress 2016 (ZL 2016), pp. 201–206. VDE Verlag, April 2016
19. Vierich, T.A., Vilgis, T.A.: Aroma — Die Kunst des Würzens, 3. Auflage. Stiftung Warentest, Berlin (2015)

Recommendation Algorithms and Systems

CUT: A Combined Approach for Tag Recommendation in Software Information Sites

Yong Yang[1], Ying Li[1,2(✉)], Yang Yue[1], Zhonghai Wu[1],
and Wenlong Shao[3]

[1] School of Software and Microelectronics, Peking University, Beijing, China
{yang.yong,li.ying,yue.yang,wuzh}@pku.edu.cn
[2] National Engineering Center of Software Engineering,
Peking University, Beijing, China
[3] VMware, Inc., Beijing, China
wshao@vmware.com

Abstract. Software information sites such as Stack Overflow and Ask Ubuntu allow programmers to post their questions and share knowledge online. Usually tags that describe the key content of the questions are required by the website. These tags play an important role in organizing and indexing user posts efficiently and provide accurate abstracts of complicated technical problems. Users attach tags to the questions according to their experience and knowledge. Due to the expression difference and lack of grasp of the software, choosing the accurate tags is not an easy job. In this paper, we propose CUT, an automatic tag recommendation approach which recommends appropriate tags after users post their questions. This approach incorporates code fragments, text content, users' preference to tags and tag relation in recommendation process. We evaluated CUT by conducting comparative experiments on the Stack Overflow dataset. The results show that CUT achieves 69.9 % and 81.6 % respectively for recall@5 and recall@10, which outperforms the latest relevant approach.

Keywords: Tag recommendation · Software information sites · Labeled LDA · Combined approach

1 Introduction

Tagging, which refers to labelling objects with pivotal short descriptions, usually a few of words, has been widely used in local knowledge management and online community-based sites, such as Quora and Flickr. Tagging for the text or multimedia objects on the websites not only enable users to organize and index their content efficiently [5], but also produce tremendous value for the websites. For example, with the well-tagged objects, websites can push interesting contents to users according to

Y. Li—The work is supported by Key Program of National Natural Science Foundation of China (Grant No. 61232005), and VMware UR project.

F. Lehner and N. Fteimi (Eds.): KSEM 2016, LNAI 9983, pp. 599–612, 2016.
DOI: 10.1007/978-3-319-47650-6_47

their focused tags to foster the development of websites. Meanwhile, with the boom in open source software, lots of websites for programmers' communication have arisen, such as Stack Overflow and Ask Ubuntu, which are called software information sites. Generally, an object of software information sites is mainly composed of tags, poster information, post content including text content and several code fragments. Posters are required to attach several tags to the questions they posted. Accurate tags help bridge the gap between social and technical aspects [4] and contribute to quick and professional answers.

Websites usually allow users to freely tag their posts with unlimited words, which makes it easier to tag objects. However, it also results in inaccurate tags and reduces the value of tags. On the one hand, tags differ a lot due to expression difference even if they describe the same question or object. Thus similar objects will be tagged with greatly different tags. On the other hand, compared to other websites, software information sites require more professional knowledge to tag the objects accurately. It's a tough job for newbies to choose the appropriate tags especially with the increasing scale and complexity of software systems. Therefore, a tag recommendation system for software information sites to recommend proper and accurate tags to users is necessary, which will greatly reduce the impact of expression difference and make it easier for users to choose the appropriate tags even if they don't have deep insights of the software system.

In this paper, we tackle the tag recommendation problems in software information sites by proposing CUT which incorporates *code* fragments, *user* profile, *text* content and tag relation to perform tag recommendation. Given the untagged object, CUT will analyze the object from different aspects and recommend accurate tags for it with the knowledge learned from historical objects. Besides, we respectively implemented a prototype of CUT and EnTagRec [2] — the state-of-art tag recommendation approach for software information sites, and conducted comparative experiments on Stack Overflow dataset to quantitatively evaluate CUT with *recall@k* which we will introduce in Sect. 4 in detail. What makes CUT different from other tag recommendation approaches is that it is the first approach to consider code fragment which is the key character that discriminates software information sites with other websites. Besides, content including code and text content, user, and tag relation factors are combined in CUT to produce a powerful tag recommendation approach.

The rest of the paper is organized as follows. Section 2 presents previous related works in tag recommendation. An overview and detailed description of CUT are given in Sect. 3. In Sect. 4, we introduce the datasets and metrics. Then CUT is evaluated and experiment results are shown. Section 5 draws some conclusions and explore the future directions.

2   Related Works

Tag has been a fertile research area and there exist various approaches to recommend or predict tags for different types of website, such as social networking sites [5–7, 9, 23, 24] and software information sites [1–3].

Van et al. [5] proposed an approach to recommend tags for photos on Flickr based on collective knowledge. Aiming at creating a more homogenous set of tags, Zangerle et al. [9] studied how to recommend suitable tags for a certain Twitter message. Wang [23] and Zhao [24] focused on tag recommendation for microblogs and they mainly applied topic models and collaborative filtering. Two tag recommendation algorithms for folksonomies, which are separately an adaptation of user-based collaborative filtering and a graph-based recommender, were proposed by Marinho [6]. Chirita et al. [7] used P-TAG, which combines the textual content of web pages and user profile, to automatically generate personalized tags for web pages when users browse web pages.

The synonym of tags in software information sites has also been studied. A similarity metric among software tags was proposed by Wang et al. [12] and then they created a taxonomy of tags based on their semantically related software terms detecting algorithm. Beyer et al. [11] analyzed the synonym tags in Stack Overflow and presented TSST to produce possible synonyms for each tag. Besides, Treude et al. [4] explored how tagging is used to bridge the gap between technical and social aspects of managing work items and showed that tagging has become a significant part of many informal processes. Thung et al. [10] proposed a solution to detect similar applications based on tags.

An object in software information sites is mainly composed of content including text content and code fragments, user who posts this object and tags. Previous works for tag recommendation in software information sites focus on different parts with various approaches. TagCombine [1] mainly considers the text content of objects with three core components—multi-label ranking component, similarity based ranking component and tag-term based ranking component. These three components generate tags and corresponding scores for the untagged objects respectively with TF-IDF (term frequency–inverse document frequency) [25], NB (Naïve Bayes) and affinity of tag to term. NetTagCombine [3] is the latest work. It expands TagCombine by adding a component which leverages bipartite graph between users and tags. It also improves similarity based ranking component of TagCombine by clustering the objects as communities based on the objects network. EnTagRec [2] achieves the best result among the three approaches. It not only applies LLDA (Labeled Latent Dirichlet Allocation) [8] in text content to produce tags and scores, but also builds tag network to take tag relation into consideration. Table 1 summarizes the key techniques and main difference of these approaches including CUT.

Table 1. Comparison of different tag recommendation approaches

		TagCombine	NetTagCombine	EnTagRec	CUT
Content	text-text	TF-IDF	TF-IDF + clustering	–	–
	text-tag	NB + affinity	NB + affinity	LLDA + frequency	LLDA
	code-code	–	–	–	Edit distance
User	user-tag	–	Bipartite graph	–	Frequency
Tag	tag-tag	–		Tag network	Frequency

3   Methodology

When performing tag recommendation for a new post in software information site, none of previous work attaches importance to code fragments of post. And a composite approach which combines content (code and text), user and tag relation is lacked. Based on this observation, we propose CUT. In this section, we first present an overview of CUT. Then we describe each component of CUT in detail. *Question* and *object* will not be distinguished in this and following part of this paper.

3.1   An Overview of CUT

CUT is composed of six components, as shown in Fig. 1, which are Preprocessing Engine(PE), Code Engine(CE), User Engine(UE), Text Engine(TE), Expanding Engine (EE) and Combining Engine(CoE) respectively. The workflow of CUT can be divided into two stages—offline training stage and online recommendation stage.

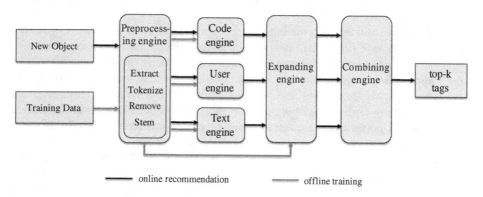

Fig. 1. An overview of CUT

In offline training stage, training data are processed by PE, then a code model, a user model, a text model and a tag relation model are generated based on the processed objects. Besides, key parameters for CoE are determined.

In online recommendation stage, an untagged question is processed by PE. Then CE, UE and PE will recommend limited tags to the question based on the trained models, which produce tags and corresponding scores according to the code fragments, user profile and text content separately. EE will expand and update tags and scores produced by the three components according to the trained tag relation model. Finally, CoE combines expanded results of CE, UE and TE and recommends k tags for the untagged question according to the final scores of tags.

3.2   Preprocessing Engine

At first, PE extracts code fragments, user information, text content and tags from raw objects. Then, tokenization, punctuation and number removal, stop words filtering and stemming are performed for text content. For code fragments, in order to preserve the code structure, we only remove number and common annotation such as text following //, /**. Code fragments are simply extracted from the question content according to the < code > </code > tag. Stop words are provided by NLTK Corpus [18] and NLTK implementation for Porter stemming algorithm [14] is adopted in PE.

3.3   Code Engine

Code fragments are treated as normal text [1–3] in previous work. In software information sites, posting questions with code fragments are common because code fragments usually depict problems more accurately compared to normal text. And questions with code fragments usually get quick and precise answers. We analyzed the proportion of questions that have code fragments of Stack Overflow from 2008 to 2015 by extracting the first 10,000 questions for each year. From Fig. 2 we can know that, with the development and maturity of Stack Overflow, more and more questioners tend to attach code fragments to their questions. Thus, tag recommendation will benefit a lot from taking code fragments into consideration.

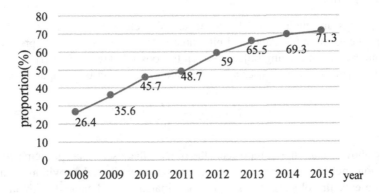

Fig. 2. The proportion (%) of questions with code fragments

We also analyzed the code fragments in questions and found that content of code fragments can be roughly categorized as *program code (including scripts), configuration file, software log* and *stack trace*. And questioners incline to post just a part of the initial code content. Therefore, although there are many approaches to analyzing each kind of code fragments, a general approach is absent. For example, abstract syntax tree (AST), which is widely used in program analysis, cannot be used in CE to generate possible tags according to the code fragments of a new question.

CE leverages the trained code model to generate tags and their scores. The score of a tag is determined by the similarity between code fragments of untagged objects and objects in training data. We denote training objects which have code fragments as OC, $OC = \{OC_1, OC_2, \ldots OC_n\}$. And correspondingly, their code fragments are denoted as C, $C = \{C_1, C_2, \ldots C_n\}$. The set of all possible tags are $T = \{t_1, t_2, \ldots t_m\}$. The untagged object and its code fragments are O^* and C^*. We use Levenshtein distance [13] to measure the code fragments similarity which is computed as follows:

$$Sim(C_i, C^*) = \frac{Dis(C_i, C^*)}{\max(|C_i|, |C^*|)}, 0 < i \leq n \tag{1}$$

Then the score of a tag t_i can be obtained from the most similar code fragments that has been tagged with t_i. The score of tag t_i is computed as follows:

$$Score_c(t_i) = \max(Sim(C_j, C^*)), C_j \in C^{ti} \tag{2}$$

In the above equation, C_{ti} means the set of code fragments that have been tagged with t_i. We summarize the workflow of CE as follows:

(1) If the untagged object doesn't have code fragment, return nothing
(2) For each tag t_i of T, calculate its score by (1) and (2)
(3) Sort tags by score and return the top-N tags and their scores.

3.4   User Engine

UE returns tags and their scores based on users' preference. Users are likely to frequently post content in the same or similar area with several tags. Users' preferences to tags are inferred from training objects. Given the possible tag set $T = \{t_1, t_2, \ldots t_m\}$ and user set $U = \{u_1, u_2, \ldots u_n\}$, user u_j' preference to tag t_i can be computed as follows:

$$Score_u(t_i) = \frac{|O_{uj,ti}|}{|O_{uj}|} \tag{3}$$

In the above equation, O_{uj} represents the objects that user u_j has participated in. The participation includes posting a question, giving answers and providing comments. $O_{uj,ti}$ represents the objects that u_j has participated in and tagged with tag t_i. We summarize the workflow of UE as follows:

(1) If the user of the untagged object doesn't appear in previous objects, return nothing
(2) For each tag t_i of T, calculate its score by (3)
(3) Sort tags by score and return the top-N tags and their scores.

3.5   Text Engine

TE outputs tags and their scores according to text content. TE is similar to the Bayesian Inference Component(BIC) of EnTagRec [2]. The major difference is that TE doesn't process code fragments which have been process in CE. We use topic models, in which topics are distributions of words and documents are distributions of topics, to represent questions and tags. Tags are mapped to topics and text contents of questions are mapped to documents. We adopt Stanford Labeled LDA(LLDA) [8, 17] implementation to build the topic model and use the model to produce tags and scores for the untagged objects. The workflow of TE can be summarized as follows:

(1)   For each tag, produce its score with the learned topic model according to the text content of the untagged object.
(2)   Sorted tags by score and return the top-N tags and their scores.

3.6   Expanding Engine

From the description of CE and UE we know that they will not produce tags and scores if untagged objects don't have code fragments or it's the first time for users to post content. And users' preferred tags usually are sparse when users participate in a few questions. Besides, complicated relations exist between tags such as hierarchical and synonym. Therefore, it's necessary to expand the limited tags produced by CE, UE and TE according to the tag relation.

In EE, we define the tag relation with asymmetric correlation. Each tag has a correlation score to all possible tags. For possible tags set $T = \{t_1, t_2, \ldots t_m\}$, a correlation matrix M with dimension $m \times m$ are learned from the training data. The correlation score can be computed with Eq. (4), in which $O_{t_i \cap t_j}$ represents the objects tagged with both t_i and t_j and O_{t_i} refers to the objects tagged with t_i.

$$Correlation(t_i, t_j) = \frac{|O_{t_i \cap t_j}|}{|O_{t_i}|} \qquad (4)$$

Given the recommended tags $T' = \{t_1, t_2, \ldots t_n\}$ from CE, UE or TE, EE expands recommended tags with the tag correlation matrix and update the scores of existed tags. For tag t_i and its current $Score_*(t_i)$, EE calculates expanding scores for each tag in T. The expanding score is denoted as $Score(t_i, t_j)$ and computed by Eq. (5).

$$Score(t_i, t_j) = Score_*(t_i) \times Correlation(t_i, t_j), \ t_i \in T', \ t_j \in T \qquad (5)$$

After expanding the recommended tag T', each tag of T has n expanding scores. Then, EE selects the maximum value as the scores for each tag in T. Finally, EE returns the top-N tags after sorting the tags by scores. If T' is an empty tag set, EE will generate top-N most frequently used tags and its frequency as the expanding results. The detailed step to expand and update recommended tags is shown in Algorithm 1.

Algorithm 1. Algorithm of expanding and updating tags

    Input: recommended tag set T' and their scores,

                 all possible tag set T, Correlation matrix M, parameter N

    Output: top-N tags and their scores

1: **if** T' is empty **then**

2:      **return** top-N most frequent tags and their frequency

3: **for** each tag t_i **in** T'

4:      **for** each tag t_j **in** T

5:          calculate $Score(t_i, t_j)$ by (5)

6: **for** each tag t_j **in** T

7:      $Score_*(t_j) = \max(\ Score(t_i, t_j)),\ t_i \in T'$

8: sort tag in T by $Score(t_j)^*$

9: **return** top-N tags and their scores

3.7 Combining Engine

CE, UE and TE separately generate the final N tags and their scores for the untagged object after expanding by EE. In order to combine the recommended tags by the three components, CoE assigns CE, UE and TE with different weights w_c, w_u and w_t. Then, for each tag t_i, the final score is calculated with Eq. (6).

$$Score(t_i) = w_c \times Score_c(t_i) + w_u \times Score_u(t_i) + w_t \times Score_t(t_i) \qquad (6)$$

To automatically find the w_c, w_u and w_t that generate the best recommendation results, we perform the grid search [19] which is also used in EnTagRec, TagCombine and NetTagCombine on a set of test questions. We provide the algorithm to find best w_c, w_u and w_t in Algorithm 2. The evaluation criteria in Algorithm 2 is *recall@k*. After obtaining the best w_c, w_u and w_t, we use them to combine the recommendation results of CE, UE and TE in other recommendation process. Finally, CoE will sort tags by the combined final score and return the first k tags as the final recommendation results.

Algorithm 2. Algorithm to generate best weights

Input: test objects O, possible tags set T,
       parameter N, parameter k

Output: w_c, w_u, w_t

1:   $w_c = w_u = w_t = 0$
2:   **for** each object O_j **in** O
3:      **for** each tag t_i in T
4:         calculate $Score_c(t_i)$, $Score_u(t_i)$, $Score_t(t_i)$
5:   **while** $w_c \leq 1.0$
6:      **while** $w_u \leq 1.0$
7:         **while** $w_t \leq 1.0$
8:            **for** each tag t_i **in** T
9:               calculate $Score(t_i)$ by (6)
10:           calculate $recall@k$ for w_c, w_u, w_t by (7)
11:           $w_t += 0.1$
12:        $w_u += 0.1$
13:     $w_c += 0.1$
14: **return** best w_c, w_u, w_t

---

4   Experiment and Evaluation

4.1   Datasets

To conduct comparative experiments with previous works of tag recommendation for software information sites, we use the common Stack Overflow datasets. We gain the datasets by extracting Stack Overflow data from Stack Exchange data dump on March 1,2016 [20], which almost contains all data of Stack Overflow from 2008 to March 1, 2016. We choose the first 50,000 questions and their answers, comments, users and tags, which we think their tags have stabilized and are relatively accurate. We perform the same filtering process of EnTagRec to filter out tags that appear less than 50 times. After the filtering, the real dataset is summarized in Table 2.

Table 2. Stack Overflow dataset information

	Question	Tags	Real question	Real tags
Number	50,000	9,233	47,750	438

4.2   Evaluation Metrics

Recall@k is a widely used metric in tag recommendation. It measures how many correct tags are recommended to the untagged objects. In this paper, we adopt *recall@k* to evaluate the recommendation result. k means the number of recommended tags.

Given the untagged objects set $O = \{O_1, O_2, \ldots O_n\}$, recommended tags $Tags_i^{top-k}$ and correct tags $Tags_i^{Correct}$ for O_i, the formula to calculate $recall@k$ is:

$$recall@k = \frac{1}{n}\sum_{i=1}^{n}\frac{|Tags_i^{top-k} \cap Tags_i^{Correct}|}{|Tags_i^{Correct}|} \tag{7}$$

As mentioned in Short's work [3], TagCombine may use the $recall\_Adj@k$ which only considers the filtered limited tags when performing recommendation and $recall@k$ calculation. In this paper, $recall@k$ means $recall\_Adj@k$. Then we show $recall@5$ and $recall@10$ of CUT and previous tag recommendation approaches.

4.3 Results and Discussion

We randomly divide the dataset into ten parts. We set $(w_c, w_u, w_t) = (1.0, 0.3, 0.3)$ as the default weights combination, which produces the best $recall@k$ on one part. And for each experiment, we perform ten-fold cross-validation to minimize the deviation.

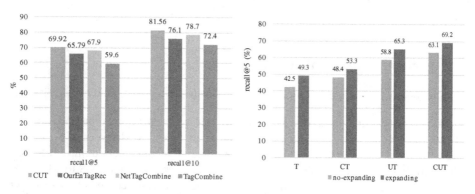

Fig. 3. Recall@k of different approaches

Fig. 4. Recall@5 of different components combination

4.3.1   Approach Comparison
We implement EnTagRec with Labeled LDA implementation of Stanford Topic Modeling Toolbox [17] and Stanford Log-linear Part-Of-Speech Tagger [15, 16] as a baseline for CUT. Parameter N is set to 30 as default. We name it OurEnTagRec.

Figure 3 shows the recall@5 and recall@10 of CUT, OurEnTagRec, NetTagCombine and TagCombine. CUT achieves 69.92 % and 81.56 for recall@5 and recall@10, which outperforms OurEnTagRec 4.13 % and 5.46 % respectively. CUT also outperforms NetTagCombine 2.02 % and 2.86 % respectively.

Besides, we find that OurEnTagRec significantly performs worse than the cited EnTagRec results. The difference is shown in Table 3. Beyesian and Frequentist are the two components of EnTagRec. The deviation may result from the different implementation of Labeled LDA which is not specified in the paper [2]. The deviation may also be related with the dataset or slight difference of preprocessing. However, we are only interested in how code fragments and a combined approach can improve the tag recommendation results, so we added CE and UE of CUT to EnTagRec and named it ExEnTagRec. After our expanding, ExEnTagRec obtains 2.6 % and 2.4 % improvement for recall@5 and rcall@10 compared to OurEnTagRec.

Table 3. Performance of claimed EnTagRec and our implementation of EnTagRec

	EntagRec/OurEnTagRec	Beyesian/OurBeyesian	Frequentist/OurFrequentist	ExEnTagRec
recall@5	80.5 %/65.8 %	56.5 %/47.6 %	59.3 %/46.2 %	68.4 %
recall@10	86.8 %/76.1 %	67.1 %/58.4 %	69.1 %/57.2 %	78.5 %

4.3.2 Component Effect Analysis

We use T to represent our CUT considering only text content, CT/UT to represent considering only text content with code fragments/user profile. Expanding means using EE to expand recommended tags and no-expanding means the opposite. Fig. 4 shows the component effect analysis result. It is known that TE and UE are the most important components and EE also brings obvious improvement. CE, which leverages code fragments, makes a little effect on the *recall@5*. The reason is that in order to keep consistent with TagCombine, NetTagCombine and EnTagRec, the first 50,000 questions and their corresponding answers, comments and tags which generated in 2008 are used. As shown in Fig. 2, the proportion of questions with code fragments was small in 2008, which impedes the advantage of CE. With the increasing proportion of questions with code fragments, CE will play a more important part in CUT.

4.3.3 Parameter Sensitivity

CE, UE, TE and EE produce top-N tags according to their scores. For simplicity, the unified N values is used by these four components. In order to investigate how the performance of CUT is affected by N, we varied N in a certain range and calculated *recall@5* and *recall@10* of CUT.

Figure 5 shows the impact of N on CUT. The minimum value of N is set to 10 to calculate *recall@10*. *Recall@10* increases with a larger N value ($10 \leq N \leq 30$). When N succeeds 30, N has little influence on the *recall@10* of CUT. We notice that *recall@5* of CUT is hardly affected by various N values. The experiment proves that CUT has a stable performance with various N values ($10 \leq N \leq 110$).

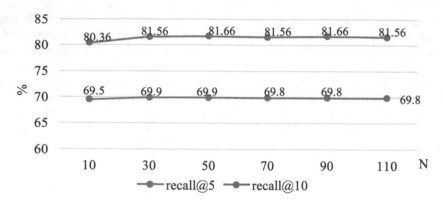

Fig. 5. Recall@k of CUT with various N

5   Conclusion and Future Work

In this paper, we propose CUT, which is the first attempt to leverage code fragments and combine post content, user and tag relation, to recommend proper tags for users when they post content in software information sites. We evaluated the performance of CUT on the common dataset—Stack Overflow dataset, which is also used by TagCombine, EnTagCombine and NetTagCombine. We conducted comparative experiments with previous approaches and showed that CUT outperforms TagCombine, NetTagCombine and our implementation of EnTagRec. The deviation between our EnTagRec and claimed EnTagRec may result from several aspects, but we show that EnTagRec will obtain improvement after combining the CE and UE components of CUT. Finally, we analyzed the importance of different components of CUT and performance sensitivity to parameter N, which shows CUT has a stable performance with different N values.

But there are still some limitations for CUT. Firstly, the current adopted method—Levenshtein distance cannot make full use of the characters of various kind of code fragments. Secondly, it is not quite proper to directly map tags to topics in topic model, because tags attached to post objects are usually limited and the relation of tags are complicated. For example, Stack Overflow allow users to attach no more than five tags to a question. In the future, we will search for more effective and efficient methods to leverage code fragments to improve the performance. We also plan to employ the LDA [21] or HDP [22], unsupervised methods, in CUT to replace the Labeled LDA used in text engine. Besides, we will also explore other metrics to evaluate our approach.

References

1. Xia, X., Lo, D., Wang, X., et al.: Tag recommendation in software information sites. In: Proceedings of the 10th Working Conference on Mining Software Repositories, pp. 287–296. IEEE Press (2013)

2. Wang, S., Lo, D., Vasilescu, B., et al.: EnTagRec: an enhanced tag recommendation system for software information sites. In: 2014 IEEE International Conference on Software Maintenance and Evolution (ICSME), pp. 291–300. IEEE (2014)
3. Short L, Wong C, Zeng D. Tag recommendations in stackoverflow. 2014
4. Treude, C., Storey, M.A.: How tagging helps bridge the gap between social and technical aspects in software development. In: Proceedings of the 31st International Conference on Software Engineering, pp. 12–22. IEEE Computer Society (2009)
5. Sigurbjörnsson, B., Van Zwol, R.: Flickr tag recommendation based on collective knowledge. In: Proceedings of the 17th International Conference on World Wide Web, pp. 327–336 ACM (2008)
6. Jäschke, R., Marinho, L., Hotho, A., Schmidt-Thieme, L., Stumme, G.: Tag recommendations in folksonomies. In: Kok, J.N., Koronacki, J., Lopez de Mantaras, R., Matwin, S., Mladenič, D., Skowron, A. (eds.) PKDD 2007. LNCS (LNAI), vol. 4702, pp. 506–514. Springer, Heidelberg (2007)
7. Chirita, P.A., Costache, S., Nejdl, W., et al.: P-tag: large scale automatic generation of personalized annotation tags for the web. In: Proceedings of the 16th International Conference on World Wide Web, pp. 845–854. ACM (2007)
8. Ramage, D., Hall, D., Nallapati, R., et al.: Labeled LDA: a supervised topic model for credit attribution in multi-labeled corpora. In: Proceedings of the 2009 Conference on Empirical Methods in Natural Language Processing, vol. 1. Association for Computational Linguistics, pp. 248–256 (2009)
9. Zangerle, E., Gassler, W., Specht, G.: Using tag recommendations to homogenize folksonomies in microblogging environments. In: Datta, A., Shulman, S., Zheng, B., Lin, S.-D., Sun, A., Lim, E.-P. (eds.) SocInfo 2011. LNCS, vol. 6984, pp. 113–126. Springer, Heidelberg (2011)
10. Thung, F., Lo, D., Jiang, L.: Detecting similar applications with collaborative tagging. In: 2012 28th IEEE International Conference on Software Maintenance (ICSM), pp. 600–603. IEEE (2012)
11. Beyer, S., Pinzger, M.: Synonym suggestion for tags on stack overflow. In: Proceedings of the 2015 IEEE 23rd International Conference on Program Comprehension, pp. 94–103. IEEE Press (2015)
12. Wang, S., Lo, D., Jiang, L.: Inferring semantically related software terms and their taxonomy by leveraging collaborative tagging. In: 2012 28th IEEE International Conference on Software Maintenance (ICSM), pp. 604–607. IEEE (2012)
13. Toutanova, K., Klein, D., Manning, C.D., et al.: Feature-rich part-of-speech tagging with a cyclic dependency network. In: Proceedings of the 2003 Conference of the North American Chapter of the Association for Computational Linguistics on Human Language Technology, vol. 1. Association for Computational Linguistics, pp. 173–180 (2003)
14. Porter, M.F.: An algorithm for suffix stripping. Program 14(3), 130–137 (1980)
15. Levenshtein, V.I.: Binary codes capable of correcting deletions, insertions and reversals. Sov. Phys. Dokl. 10, 707 (1966)
16. Log-linear Part-of-Speech Tagger. http://nlp.stanford.edu/software/tagger.shtml
17. Labeled LDA. http://nlp.stanford.edu/software/tmt/tmt-0.4/
18. English stop words. http://www.nltk.org/api/nltk.corpus.html
19. Bergstra, J., Bengio, Y.: Random search for hyper-parameter optimization. J. Mach. Learn. Res. 13, 281–305 (2012)
20. Stack Exchange Data Dump, 6 March 2016. https://archive.org/details/stackexchange
21. Blei, D.M., Ng, A.Y., Jordan, M.I.: Latent dirichlet allocation. the. J. Mach. Learn. Res. 3, 993–1022 (2003)

22. Teh, Y.W., Jordan, M.I., Beal, M.J., et al.: Hierarchical dirichlet processes. J. Am. Stat. Assoc. **101**(476), 1556–1581 (2012)
23. Wang, Y., Qu, J., Liu, J., Chen, J., Huang, Y.: What to tag your microblog: hashtag recommendation based on topic analysis and collaborative filtering. In: Chen, L., Jia, Y., Sellis, T., Liu, G. (eds.) APWeb 2014. LNCS, vol. 8709, pp. 610–618. Springer, Heidelberg (2014)
24. Zhao, F., Zhu, Y., Jin, H., et al.: A personalized hashtag recommendation approach using LDA-based topic model in microblog environment. Future Gener. Comput. Syst. **65**, 196–206 (2015)
25. Salton, G., Buckley, C.: Term-weighting approaches in automatic text retrieval. Inf. Process. Manage. **24**(5), 513–523 (1988)

CoSoLoRec: Joint Factor Model with Content, Social, Location for Heterogeneous Point-of-Interest Recommendation

Hao Guo[1], Xin Li[3], Ming He[1], Xiangyu Zhao[1], Guiquan Liu[1(✉)], and Guandong Xu[2]

[1] University of Science and Technology of China, Hefei, China
{guoh916,zxy1105}@mail.ustc.edu.cn, mheustc@gmail.com, gqliu@ustc.edu.cn
[2] University of Technology Sydney, Sydney, Australia
Guandong.Xu@uts.edu.au
[3] IFLYTEK Research, Hefei, China
xinli2@iflytek.com

Abstract. The pervasive use of Location-based Social Networks calls for more precise Point-of-Interest recommendation. The probability of a user's visit to a target place is influenced by multiple factors. Though there are several fusion models in such fields, heterogeneous information are not considered comprehensively. To this end, we propose a novel probabilistic latent factor model by jointly considering the social correlation, geographical influence and users' preference. To be specific, a variant of Latent Dirichlet Allocation is leveraged to extract the topics of both user and POI from reviews which is denoted as explicit interest. Then, Probabilistic Latent Factor Model is introduced to depict the implicit interest. Moreover, Kernel Density Estimation and friend-based Collaborative Filtering are leveraged to model user's geographic allocation and social correlation respectively. Thus, we propose CoSoLoRec, a fusion framework, to ameliorate the recommendation. Experiments on two real-word datasets show the superiority of our approach over the state-of-the-art methods.

Keywords: Location-based Social Network · Point-of-Interest recommendation · Topic model · Probabilistic latent factor model · Heterogeneous information

1   Introduction

In recent years, with rapidly development of Location-based Social Networks (LBSNs), boundary between the physical world and virtual networks is broken. As an interlink between these two worlds, Point-of-Interest (POI) refers to a place, such as restaurant that users may find useful or tend to visit and plays an essential role in LBSN thereby leading to an application - POI recommendation. This application can not only benefit merchants by increasing their

© Springer International Publishing AG 2016
F. Lehner and N. Fteimi (Eds.): KSEM 2016, LNAI 9983, pp. 613–627, 2016.
DOI: 10.1007/978-3-319-47650-6_48

revenue through virtual marketing but also benefit customers accrelating their decision-making by filtering out uninteresting places thus makes them satisfied.

Traditional recommender systems can be seamlessly applied by treating POI as an ordinary item, however there are several characteristics of POI recommendation that make it different from conventional recommendations thus if well considered, the performance would be improved in a significant margin.

- Tobler's Law of Geographical Influence. As Tobler [1] indicates, the aggregation of a user's check-ins depicts that users' check-in probability is inversely proportional to the geographical distance. Geographical influence can be denoted as a physical metric between the user and the POI.
- Homophily of Social Correlation. Homophily is one of the most important theories in sociology and also works in social network [2]. The depiction of homophily suggests that people tend to trust and have similar favorite as their friends psychologically, thus making social correlation a psychological metric between the user and the POI.
- Heterogeneous Information. A LBSN contains heterogeneous information, such as geographical location, social network, rating data and text reviews. Undoubtedly, utilizing heterogeneous data indubitably results in more precise user profiling and more personalized recommendations.

Despite the successes and improvements of the existing studies, the heterogeneous information is not comprehensively considered in one model. Generally speaking, users' behavior in choosing POIs can be influenced by multiple factors. So our recommendation will show more accurate and efficient if we consider more factors which will influence users' bahaviors. However, under some circumstance in reality, one or several pieces of information are not available, so real life applications demand for robust modeling. To this end, we propose a novel probabilistic latent factor model by considering the geographical location, social correlation and textual reviews simultaneously. Specifically, our model consists of four-fold measurements, i.e., physical distance, psychological distance, explicit interest and implicit interest. Firstly, physical distance denotes the distance between the user and the POI in real world. We apply Kernel Density Estimation (KDE) to estimate the visiting probability of a user at a target POI based on his or her historical visiting records in adherence to Tobler's Law. Secondly, for calculating psychological distance, friend-based Collaborative Filtering (CF) is simultaneously utilized to predict the visiting probability under the assumption called "the phenomenon of homophily" which means a user's visiting behavior is influenced by his or her friends. Thirdly, in order to leverage a user's explicit interest, we aggregate all the reviews written by him or her as a document and apply Latent Dirichlet Allocation (LDA) on it to derive the topic distribution of such user afterwards. The matching on the corresponding topics reveals preference for a user to a POI. Finally, under the belief that there are several factors that implicitly affect a user's decision-making, we apply the Latent Factor Model to depict the implicit interest thereby augmenting it to the explicit interest to form the entire interest in the model. The detail architecture of this model named as CoSoLoRec can be seen in Fig. 1.

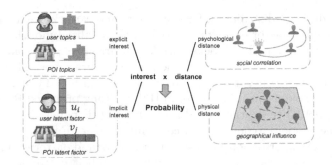

Fig. 1. The architecture framework of CoSoLoRec model

In summary, we have made several contributions in this paper:

- We propose a novel probabilistic latent factor model for heterogeneous Point-of-Interest recommendation by simultaneously considering the geographical location, social correlation and the user preference.
- Our model can keep its robustness due to its modularization. It means every heterogeneous information is embedded into the model as a module thus removing anyone of them won't affect the praticality but only decline the performance somewhat.
- We conduct extensive experiments on two real-world large scale datasets to evaluate both the efficiency and the effectiveness of our model. The results demonstrate that our approach outperforms other state-of-the-art methods.

The rest of the paper is organized as follows. We review the recent studies in Sect. 2. Problem is defined in Sect. 3.1 along with the model formation in Sect. 3.2 and the model description in Sect. 4. The experiment is demonstrated in Sect. 5. We conclude our work in Sect. 6.

2 Related Work

In recent years, POI recommendation has grown in popularity with the increasing demand of LBSNs. In reality, geographical information plays critical roles in influencing user's behaviors [4,5] since physical interactions are required in LBSNs which differs from other non-spatial situations totally. To exploit influence of geographical information in improving the quality of location recommendations, Some techniques model the distance between two locations visited by a user as a distribution for all users. In [4], Ye *et al.* employed power-law distribution (PD) to model user's checkin behaviors using naive bayesian method. Instead of making PD assumption, Cheng *et al.* [5] proposed the Multi-center Gaussian Model (MGM) to capture features of user's checkin behaviors. However, the geographical influence on individual user's visiting behaviors should be personalized rather than appearing as a common distribution. So Zhang *et al.* [17] used kernel density estimation (KDE) to model geographical influence as personalized distance distribution for each user. In our work, we adopt KDE to model geographical factor since its superior in personalized modeling.

According to homophily of social correlation, friends tend to share common interests which will make recommendation more accurate and efficient. H. Ma proposed social trust concept [12] summarized as friend-based Collaborative Filtering (CF) to explain social influence. In [13], H. Tong proposed Random Walk with Restart(RWR) to capture social relations. In this paper, we adopt friend-based CF since its lower computation cost and more accuracy.

Exploring text information can also better understand patterns in LBSNs and improve LBSNs services. Ye *et al.* [24] explored explicit patterns of individual places and implicit revelance among similar places through semantic annotation. Liu *et al.* [9] proposed TL-PMF model to consider both the extent to which a user interest matches the POI in terms of topic distribution and the word-of-mouth opinion of the POI. Kurashima *et al.* [19] proposed Geo-Topic Model which jointly estimates user's interests and activity areas. Yin [20] proposed LCA-LDA model by giving consideration to both personal interest and local preference. Zheng *et al.* [22] proposed a cross-region collaborative filtering method based on hidden topics about check-in records to recommend new POIs. Zhang *et al.* [23] distinguished the user preferences on the content of POIs from the POIs themselves and combined the predicted rating on content and location of POI.

Previous studies mainly focused on just one or two aspects which affects user's visiting behaviors. However, overall consideration on the joint effects of heterogeneous information can show great superiority in POI recommendation. Ye *et al.* [4] incorporated social and geographical influences into user-based CF framework by using linear interpolation. In [5], Cheng *et al.* considered geographical influence, check-in patterns, frequency and social networks in POI recommendation. Hu and Ester [6] considered spatial and textual aspects of posts published by mobile users and predicted user's willing locations. However, this work is more similar to a location prediction problem rather than POI recommendation. B. Liu [10] proposed the Geo-BNMF model to embrace geographical influence, popularity, text information and latent factor model. Based on this, They also proposed Geo-PFM [11] to combine geographical influence, latent factor model with latent region. Lian *et al.* [21] proposed a weighted matrix factorization method incorporating the modeling of the spatial clustering phenomenon. In this paper, we jointly fused heterogeneous information with latent factor model to describe users' behaviors.

3   Fusion Model with Heterogeneous Information

In this section, we define the problem of POI recommendation and introduce a fused probabilistic latent factor framework for heterogeneous information.

3.1   Problem Definition

The major task for POI recommendation is to recommend POIs which a user has not visited using heterogeneous information in LBSNs. Let $\mathcal{U} =$

$\{u_1, u_2, \ldots, u_i, \ldots, u_m\}$ be the set of users and $\mathcal{V} = \{v_1, v_2, \ldots, v_j, \ldots, v_n\}$ be the set of POIs. Each user u_i visited some POIs L_i historically and rated on these POIs.

User's visiting behavior may be not only influenced by geographical distance between his destination and visited POIs but also influenced by ratings made by user's friends F_i. We regard these factors as geographical influence and social influence which can also be regarded as physical distance and psychological distance respectively. Text information such as reviews may also reflects explicit interest of user u_i with user-topic distribution $\boldsymbol{\theta}_i$ in topic model.

The problem under investigation is essentially how to effectively and accurately estimate user's probability in visiting new POIs by employing information containing above three aspects. To attack this problem, in Sect. 3.2, We formulate a fused probabilistic latent factor model by incorporating these factors.

Table 1. Mathematical notations

Symbol	Size	Meaning
$\mathcal{U}$	m	Set of all users in one LBSN
$\mathcal{V}$	n	Set of all POIs in one LBSN
L_i	$|L_i|$	Set of locations visited by a user
X_i	$\binom{|L_i|}{2}$	Sample of distances between L_i
$\mathcal{F}_i$	$|\mathcal{F}_i|$	Friends of user i
$\boldsymbol{u}_i$	d	Preference of user i
$\boldsymbol{v}_j$	d	Affinity of POI j
$\boldsymbol{\theta}_i$	K	User i's review topic distribution
$\boldsymbol{\pi}_j$	K	POI j's review topic distribution

As for POIs, each POI v_j has its own location l_j labeled as vector $< lon_j, lat_j >$ in representing latitude and longitude respectively. Also each POI has its own textual profiles with its topic distribution $\boldsymbol{\pi}_j$ in topic model. For convenience, we term i and j as user u_i and location l_j respectively.

3.2 Fused Probabilistic Latent Factor Model

To make CoSoLoRec model concrete, we assume the follow factor representation: (1) each user i is associated with his or her interest $\eta(i,j)$ with respect to POI j. (2) each user has an intended visiting probability $p_f(i,j)$ with respect to POI j on the basis of friend-based CF. (3) geographical influence impels user i to estimate the probability he or she will visits POI j denoted as $p_l(i,j)$. We integrally consider user's interest, physical distance and psychological distance. Finally we got a joint model with these three factors:

$$p(i,j) \propto \eta(i,j)((1-\lambda)p_l(i,j) + \lambda p_f(i,j)) \tag{1}$$

The recommend process of user i for POI j can be represented in a generative way. For user's preference, $\eta(i,j)$ can be represented as a linear combination of latent factor $\mathbf{u_i^T v_j}$ and function of user's and POI's observable properties which can be expressed as topic distribution of user i and POI j named as θ_i, π_j. We denote these two parts as implicit interests and explicit interests of a user respectively. We use $\eta_1(i,j)$ and $\eta_2(i,j)$ to notate them. Also, user's rating $y(i,j)$ can be influenced by his visiting probability. Here we adopt Possion distribution to describe this relation. So fused probabilistic latent factor model can be expressed as follows:

1. Draw a user interest
   (a) Generator user latent factor $u_{iw} \sim Gamma(\alpha_U, \beta_U)$
   (b) Generator item latent factor $v_{jw} \sim Gamma(\alpha_V, \beta_V)$
   (c) user's explicit interest $\eta_1(i,j) = \theta_i^T \pi_j$, implicit interest $\eta_2(i,j) = \mathbf{u_i^T v_j}$
   (d) user's interest $\eta(i,j) = \eta_1(i,j) + \eta_2(i,j)$
2. $y(i,j) \sim P(p(i,j))$ where
   $p(i,j) = (\eta_1(i,j) + \eta_2(i,j))((1-\lambda)p_l(i,j) + \lambda p_f(i,j))$

4   Model Specification

In this section, we introduce detailed model specifications and present our fusion model called **CoSoLoRec**.

4.1   Geographical Influence

We aim to exploit geographical Influence by measuring the distance from a user's visited POIs to an unvisited POI. Thus we employ Kernel Density Estimation (KDE) to model the geographical influence of POIs on users' visiting behaviors.

Like MGM, KDE is also a widely-adopted method to estimate geographical influence. What's more, it shows superior to other methods which model geographical influence in considering visited POIs. We can evaluate general influence of all POIs using the following method:

$$p_l(i,j) = P\left(\bigcup_{t=1}^{|L_i|}(c_t \rightarrow c_0)\right) = 1 - P\left(\bigcap_{t=1}^{|L_i|}\overline{c_t \rightarrow c_0}\right) = 1 - \prod_{t=1}^{|L_i|}(1 - P(c_t \rightarrow c_0)) \quad (2)$$

From the Eq. (2), we can discover that before fetching geographical influence of locations $p_l(i,j)$, our task is to learn the probability that event $c_t \rightarrow c_0$ occurs. Here we use the same algorithm proposed in [17] to learn it.

$$P(c_t \rightarrow c_0) = \frac{1}{|X_i|}\sum_{x \in X_i} K\left(\frac{z_t - x}{\delta}\right) = \frac{1}{\sqrt{2\pi}|X_i|}\sum_{x \in X_i} e^{-\frac{(z_t - x)^2}{2\delta^2}} \quad (3)$$

Here, z_t is the distance between user visiting POI c_0 and each of user's historical POIs, which can be used to derive the probability of c_0. $K(\cdot)$ represents kernal function. Also δ is a smoothing parameter which is called the bandwidth. We use optimal bandwidth [3] $\delta \approx 1.06\hat{\delta}|X_i|^{-1/5}$. However, the computational complexity grows rapidly with the increment of L_i. So we use efficient approximation algorithm [17] to measure $p_l(i, j)$.

Eventually, by combining Eqs. (2) and (3), we can exploit *geographical influence of locations* by using $p_l(i, j)$. However, only the geographical information is not sufficient. Thus social correlation is introduced.

4.2 Friend-Based Collaborative Filtering

With the exponential growth of online social network, social relationship plays an important role in influencing users' behaviors. Friends usually have similar behaviors due to the phenomenon that sociologists call homophily [4].

Aiming to predict the probability of user i to a POI j, we adopt the user-based collaborative filtering(CF) by regarding all of i's friends as neighbors. In order to determine the probability in interval $[0, 1]$, We devise the calculation as:

$$p_f(i, j) = \frac{\sum_{i' \in \mathcal{F}_i} sim(i, i') r_{i'j}}{\sum_{i' \in \mathcal{F}_i} sim(i, i')} \cdot \frac{1}{r_{max}} \quad (4)$$

Here, r_{max} can ensure $p_f(i, j)$ is normalized. $sim(i, j)$ refers to similarity between user i and user j. In our study, we choose cosine similarity.

4.3 Probabilistic Latent Factor Model

Probabilistic Factor Model (PFM) [14] is a generative probabilistic model. The notations involved in PFM are defined in Table 1. Here, $\hat{f}_{ij}$ is assumed to follow Possion Distribution, the mean is $\hat{y}_{ij}$. Also, $\boldsymbol{u}_i$ and $\boldsymbol{v}_j$ are given certain distributions as priors. Here, $\boldsymbol{u}_i = (u_{i1}, u_{i2}, ..., u_{iw}, ..., u_{id})$ and $\boldsymbol{v}_j = (v_{j1}, v_{j2}, ..., v_{jw}, ..., v_{jd})$.

Therefore, the process of Probabilistic Factor Model is as follows:

1. for all w, generate $u_{iw} \sim p(u_{iw}|\Phi_{u_{iw}})$
2. for all w, generate $v_{jw} \sim p(v_{jw}|\Phi_{v_{jw}})$
3. generate $\hat{f}_{ij}$ from user i to location j with equation $\hat{f}_{ij} = \sum_{w=1}^{d} u_{iw} v_{jw} = \boldsymbol{u}_i \boldsymbol{v}_j$
4. generate $\hat{y}_{ij} \sim P\left(\hat{f}_{ij}\right)$

Here, $\Phi_{u_{iw}}$ and $\Phi_{v_{jw}}$ are hyperparameter lists respect to u_{iw} and v_{jw}. We assume latent factors are non-negative in real situations. So u_{iw} and v_{jw} are given Gamma distributions as empirical priors [14,15]. The gamma distribution of U and V can be represented as functions:

$$p(U|\alpha_U, \beta_U) = \prod_{i=1}^{m} \prod_{w=1}^{d} \frac{u_{iw}^{\alpha_U - 1} \exp(-u_{iw}/\beta_U)}{\beta_U^{\alpha_U} \Gamma(\alpha_U)} \quad (5)$$

$$p\left(V|\alpha_V,\beta_V\right) = \prod_{j=1}^{n} \prod_{w=1}^{d} \frac{v_{jw}{}^{\alpha_V-1} \exp\left(-v_{jw}/\beta_V\right)}{\beta_V{}^{\alpha_V} \Gamma\left(\alpha_V\right)} \tag{6}$$

where $u_{iw}, v_{jw}, \alpha_U, \beta_U, \alpha_V, \beta_V > 0$, $\Gamma\left(.\right)$ is the Gamma function. Given user latent factor $\boldsymbol{u_i}$ and item latent factor $\boldsymbol{v_j}$, the The possion distribution of y_{ij} given f_{ij} is

$$P\left(\hat{y_{ij}}|\hat{f_{ij}}\right) = (\boldsymbol{u_i v_j})^{y_{ij}} \frac{\exp\left(-\boldsymbol{u_i v_j}\right)}{\hat{y_{ij}}!} \tag{7}$$

Expressed in matrix of the above equation with $F = UV^T$. With the method of maximum a posterior(MAP), posterior distribution of Y can be modeled as

$$p\left(U,V|Y,\alpha_U,\beta_U,\alpha_V,\beta_V\right) \propto P\left(Y|F\right)p\left(U|\alpha_U,\beta_U\right)p\left(V|\alpha_V,\beta_V\right) \tag{8}$$

Finally, we can infer parameters with stochastic gradient descent (SGD) method.

4.4   Textual Analysis

In order to extract users' explicit interest, we use an aggregated LDA model. The model has two latent variables with corresponding super parameters as priors: (1) document-topic distributions Θ. (2) topic-word distributions Φ. In order to learn users' interests, we aggregate all the reviews written by each user into a document. Thus, user and document are interchangable in reflecting user's interest.

In this way, we build an aggregated Topic Model. Each user in LBSN is associated with topics following a multinomial distribution, denoted as $\boldsymbol{\theta}$. Also, each topic is associated with textual items according to a multinomial distribution. As we have sampled Θ and Φ of Topic Model of users. Obviously, the dimension for document-topic distribution of both Topic Model should be the same. Using gibbs sampling, the topic distribution for POI j and document-topic distribution θ of users is:

$$\pi_{js} = \frac{n_j^{(s)} + \alpha}{\sum\limits_{s=1}^{K} n_j^{(s)} + K\alpha} \qquad \theta_{is} = \frac{n_i^{(s)} + \alpha}{\sum\limits_{s=1}^{K} n_i^{(s)} + K\alpha} \tag{9}$$

Here, $n_j^{(s)}$ is the topic observation count for POI j. $n_i^{(s)}$ is the topic observation counts for user i(document d). V and K are the number of unique words and topics. α and β are hyperparameters in corresponding to topic model.

4.5   Learning and Inference

Parameter Estimation. Let us denote all parameters as $\Lambda = \{U, V\}$ and let $\Omega = \{\alpha_U, \beta_U, \alpha_V, \beta_V\}$ be the hyperparameters. Hyperparameters are all apriori given. Given the observed data collection $\mathcal{D} = \{p\left(i,j\right)\}^{I_{ij}}$ where $p\left(i,j\right)$ is the user visiting probability, and $I_{ij} = 1$ when user i visited POI j, and $I_{ij} = 0$ otherwise.

To estimate the parameters Λ, we use maximum likelihood estimation (MLE) method and sampling algorithm to learn all the parameters. So the postprior probability can be expressed as the following:

$$P\left(\Lambda|\mathcal{D},\Omega\right) \propto \prod_{\mathcal{D}} P\left(y\left(i,j\right)|U,V,\Omega\right)^{I_{ij}} \times P\left(U|\alpha_U,\beta_U\right) P\left(V|\alpha_V,\beta_V\right) \tag{10}$$

For simplicity, we use logarithmic form of posterior distribution instead. We express this as follows:

$$\mathcal{L}\left(U,V;\mathcal{D}\right) = \sum_{i=1}^{M}\sum_{j=1}^{N} I_{ij}\left(y\left(i,j\right)\log p\left(i,j\right) - p\left(i,j\right)\right) + \sum_{i=1}^{M}\sum_{w=1}^{d}\left(\left(\alpha_U - 1\right)\log u_{iw} - \frac{u_{iw}}{\beta_U}\right)$$

$$+ \sum_{j=1}^{N}\sum_{w=1}^{d}\left(\left(\alpha_V - 1\right)\log v_{jw} - \frac{v_{jw}}{\beta_V}\right) \tag{11}$$

Here, $p\left(i,j\right) = \left(\boldsymbol{u}_i^T\boldsymbol{v}_j + \boldsymbol{\theta}_i^T\boldsymbol{\pi}_j\right)\left(\left(1-\lambda\right)p_l\left(i,j\right) + \lambda p_f\left(i,j\right)\right)$ In order to approximate actual value of $\boldsymbol{u}_i$ and $\boldsymbol{v}_j$, we use stochastic gradient descent (SGD) method to optimize them and update parameters iteratively using all training samples, ξ_i and ξ_j are learning rates.

$$u_{iw} \leftarrow u_{iw} + \xi_i \times \frac{\partial \mathcal{L}}{\partial u_{iw}} \qquad v_{jw} \leftarrow v_{jw} + \xi_j \times \frac{\partial \mathcal{L}}{\partial v_{jw}} \tag{12}$$

4.6   Recommendation

After we learn the parameters Λ, CoSoLoRec model predicts the ratings of a user for a given POI using $\mathbb{E}\left(y\left(i,j\right)\right) = \left(\mathbf{u_i^T v_j} + \boldsymbol{\theta}_i^T\boldsymbol{\pi_j}\right)\left(\left(1-\lambda\right)p_l\left(i,j\right) + \lambda p_f\left(i,j\right)\right)$. We adjust hyperparameters in training process and adjust parameter λ to balance physical and psychological influence in making decisions. Our model learns the latent factors by SGD effectively. Therefore we make POI recommendation via our trained model.

5   Experiment

In this section, we conduct several experiments based on CoSoLoRec model and baselines to evaluate the performance of our proposed approach empirically. All experiments are conducted on two real-world datasets in LBSNs, collected from Yelp and Foursquare.

5.1   Dataset

Yelp dataset. Yelp is a famous website which provides reviews and ratings for restaurants and other business places [16]. To make the experimental results more convinced, we manually choose two cities named "Phoenix" and "Las Vegas" which consist of 80 % of original datasets to evaluate our approaches. We filter out users who have more than 300 friends to avoid spam generated by brushed from robots and less than 20 friends to make our datasets dense. With the same reason, we filter out users who have more than 300 reviews and less than 20 reviews. Finally, we obtain a dataset consists of 3059 users, 26446 business along with 180755 review records.

Foursquare dataset. Besides the Yelp dataset, we also evaluate our models on Foursquare (4sq). The dataset includes POIs distributed in the United States. With the same reason in handling Yelp dataset, we filter out users with more than 500 friends and less than 18 friends. Similarly, we filter our ratings given by more than 500 users and less than 20 users. We finalize a dataset of 6895 users for 13208 POIs with 166989 ratings. Table 2 indicates the data statistics for Yelp and Foursquare.

Table 2. Data description

	Yelp	Foursquare
Number of users	366715	571700
Number of locations	61184	8318919
Review items	1569265	5550203
User-location matrix density	6.99×10^{-5}	1.17×10^{-6}
Number of cities	10	50

To unify ratings with probability to visit a POI, we normalize the discrete rating using $f(x) = \frac{x}{max\{x\}}$, where $max\{x\}$ represents the largest rating value [7].

5.2   Evaluation Metrics

We present each user with N POIs sorted by the predicted probability and evaluated based on which of these POIs were actually visited by users.

rPrecision. To unify evaluation of a universal baseline and baseline with region-based attached, we introduce relative precision for evaluation. We assume C as the candidate POIs. The precision in a top-K list of a random recommendation system is $\frac{|S_{visited}|}{|C|}$. Then, the relative precision [10,25] is defined as:

$rPrecision@N = \frac{|S_{N,rec} \cap S_{visited}| \cdot |C|}{|S_{visited}| \cdot N}$

RMSE. We also use Root Mean Square Error (RMSE) for evaluation. $RMSE = \sqrt{\frac{1}{N} \sum_{(u,i) \in E} (r_{ui} - \hat{r_{ui}})^2}$. where E is the test dataset. r_{ui} and $\hat{r_{ui}}$ represent the observed and predicted performance used for user i on business j respectively. Smaller value of RMSE implies better performance in our recommendation.

5.3 Baseline Comparison

In this section, we introduce baselines and parameters involved in experiments and evaluate our model with baselines in a number of experiments.

Comparative Approaches. In this section, in order to show the effectiveness of our CoSoLoRec model, we compare our model with the following baselines:

① **Probabilistic Matrix Factorization (PMF).** PMF [7] is a recommendation method widely used for different recommendation tasks.

② **Non-negative Matrix Factorization (NMF).** NMF [8] is a method used in recommender system which constrains the factors to be non-negative.

③ **Bayesian Non-Negative Matrix Factorization (BNMF).** BNMF [18] is an unsupervised learning method using Markov chain Monte Carlo sampling method based on Non-negative Matrix Factorization.

④ **Geographical-Topical Bayesian Nonnegative Matrix Factorization (GT-BNMF)** [9]. This is a new method combining geographical information, textual information with other aspects based on BNMF.

⑤ **Geographical Probabilistic Factor Model (Geo-PFM)** [11]. This is a fusion method using latent factor model considering geographical information with no textual information and social correlation.

To further understand the benefits using different forms of implicit interest and distinction between implicit interest and explicit interest, we implement three modifications of CoSoLoRec model.

⑥ **CoSoLo-PMF (C-PMF)** is a modification of CoSoLoRec model using PMF to depict implicit interest.

⑦ **CoSoLo-NMF (C-NMF)** is a modification of CoSoLoRec model using NMF to depict implicit interest.

⑧ **CoSoLo-BNMF (C-BNMF)** is a modification of CoSoLoRec model using BNMF to depict implicit interest.

We divide data as training set and test set on the ratio of 7:3 with the review time order. For CoSoLoRec model, we set $\alpha_U = 5$ and $\beta_U = 0.2$ as hyper parameters of prior of U. Also, we set $\alpha_V = 20$ and $\beta_V = 0.2$ as hyper parameters of prior of V. We initialize $\lambda = 0.5$ for comparing our model with baselines. For PMF, we set hyper parameters $\sigma_U = 0.05$ and $\sigma_V = 0.05$ as priors of U and V. Also, we set $\sigma = 0.2$ as prior of $y(i,j)$. For textual aspects, we initialize the topic number $K = 30$, $\alpha = 40/K$, $\beta = 0.3$. We set the number of regions $|R| = 50$ for foursquare which is the number of regions according to all the states in USA(expect Alaska). For yelp, we set $|R| = 2$ for data from two cities.

Expermental Results. To reflect how the models actually outperform a universal baseline and baseline with region-based attached, we use relative performance. Figure 2(a) and (b) indicates CoSoLoRec model outperforms all the basslines including classical baseline models (PMF,NMF,BNMF) as well as recent proposed model (GT-BNMF,Geo-PFM) in both datasets. Furthermore,

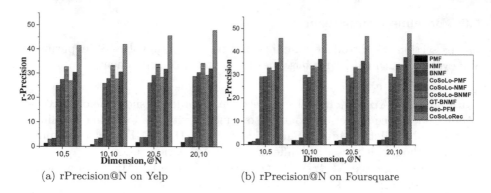

(a) rPrecision@N on Yelp   (b) rPrecision@N on Foursquare

Fig. 2. r-Precision with different latent factor dimension and @N

CoSoLo-PMF, CoSoLo-NMF and CoSoLo-BNMF show almost equivalent performance in precision. However, they performs better than classical beselines but no obvious better performance than recent proposed model. This phenomenon indicates heterogeneous information can reflect user's interests accurately. Also, probabilistic latent factor model can reflect user's implicit interest better. GT-BNMF model considers heterogeneous information such as geographical information, popularity and user's interests. However, our model considers further more about these factors. Geo-PFM model use Possion factor model which can guarantees a rigorous probabilistic generative process. However, our model performs better since heterogeneous information fused.

Table 3. Performance comparison in different dimensions

	D	Metrics	PMF	NMF	BNMF	C-PMF	C-NMF	C-BNMF	GT-BNMF	Geo-PFM	CoSoLoRec
Yelp	10	RMSE	0.8225	0.7644	0.766	0.7639	0.7824	0.7769	0.7241	0.7076	**0.6692**
		Improve	18.64 %	12.45 %	12.64 %	12.40 %	14.47 %	13.86 %	7.58 %	5.43 %	
	20	RMSE	0.8455	0.7502	0.7564	0.7365	0.7716	0.7672	0.7573	0.6881	**0.6693**
		Improve	20.84 %	10.78 %	11.52 %	9.12 %	13.26 %	12.76 %	11.62 %	2.73 %	
4sq	10	RMSE	0.8792	0.8515	0.8624	0.8335	0.8612	0.8454	0.8282	0.7815	**0.7476**
		Improve	14.97 %	12.20 %	13.31 %	10.31 %	13.19 %	11.57 %	9.73 %	4.34 %	
	20	RMSE	0.8763	0.8498	0.85	0.8019	0.8509	0.8334	0.8132	0.7739	**0.7319**
		Improve	16.48 %	13.87 %	13.89 %	8.73 %	13.99 %	12.18 %	10.00 %	5.43 %	

In order to measure the difference between estimated rating values and real rating values in test datasets, RMSE is introduced. We conduct experiments on different latent factor dimensions. From Table 3, we can conclude that our model achieves less mean square error than baselines with different data divisions. So it is obvious that our model can achieve more accurate recommendation for user's interestd places. Also, three modified model based on CoSoLoRec achieves less mean square error than classical baselines and performs almost equivalently with recently proposed models. This phenomenon indicates heterogeneous can ensure more precise prediction in recommendation system.

5.4   Parameters Sensitivity

As mentioned, both geographical influence and social influence play important roles in estimating user's interest on unvisited POIs. So in the following part, we respectively set λ as 0.2, 0.4, 0.6, 0.8 to detect importance of geographical and social factors.

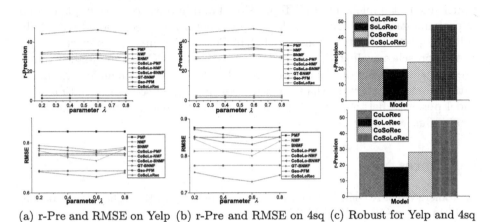

(a) r-Pre and RMSE on Yelp   (b) r-Pre and RMSE on 4sq   (c) Robust for Yelp and 4sq

Fig. 3. Experimantal results in parameter sensitivity and model robust

From Fig. 3(a) and (b), we can find that rPrecision rises first and falls later while RMSE shows a reverse trend. Since PMF,NMF,BNMF,GT-BNMF and Geo-PFM do not consider relation between geographical distance and psychological distance, results show no change with different parameter λ. We can find both geographical and social influence play comparative roles. Our model outperforms baselines in every value of λ while three modifications show almost equivalent performance with Geo-PMF and GT-BNMF.

5.5   Impact of Geographical, Social and Textual Information

In some situations, it is ideal that all aspects are covered in our model. So it is necessary to evaluate our model if not all the data is present. In this part, we choose three models which are based on CoSoLoRec model: (1) **CoLoRec:** social correlation removed. (2) **CoSoRec:** geographical factor removed. (3) **SoLoRec:** textual information removed. All the experiments are conducted in $K = 20, N = 10$.

Figure 3(c) shows results comparing the above three models with CoSoLoRec model. We conclude that above three models perform worse than CoSoLoRec model since user preference can not be completely described if one of factors in CoSoLoRec removed. However, relevant indicators show not much decline comparing with CoSoLoRec model. In particular, CoLoRec and CoSoRec models show not much decline compared to SoLoRec model which shows users' text information contributes greater than geographical factor and social correlation.

6 Conclusion

In this paper, we proposed CoSoLoRec model fusing heterogeneous information like geographical factor, social correlation and text information. We incorporate user preference with geographical factor realized by KDE and social correlation realized by friend-based CF. Further more, we devide user interest into explicit interest implemented by Topic Model and implicit interest implemented by probabilistic latent factor model. Experimental result conducted with Yelp and foursquare dataset demonstrated that CoSoLoRec model is superior to all other approaches evaluated, such as PMF, NMF, GT-BNMF and Geo-PFM and three different forms of CoSoLoRec model. Also we can conclude text information is more important than geographical factor and social correlation. Our model performs better in any different combinations between geographical and social influence.

References

1. Tobler, W.R.: A computer movie simulating urban growth in the Detroit region. Econ. Geogr. **46**, 234–240 (1970)
2. McPherson, M., Smith-Lovin, L., Cook, J.M.: Birds of a feather: homophily in social networks. Ann. Revi. Sociol. **27**, 415–444 (2001)
3. Dehnad, K.: Density estimation for statistics and data analysis. Technometrics **29**(4), 495–495 (1987)
4. Ye, M., et al.: Exploiting geographical influence for collaborative point-of-interest recommendation. In: Proceedings of ACM SIGIR. ACM (2011)
5. Cheng, C., Yang, H., et al.: Fused matrix factorization with geographical and social influence in location-based social networks. In: AAAI (2012)
6. Hu, B., Ester, M.: Spatial topic modeling in online social media for location recommendation. In: Proceedings of the 7th ACM Recsys. ACM (2013)
7. Mnih, A., Salakhutdinov, R.: Probabilistic matrix factorization. In: NIPS (2007)
8. Lee, D.D., Seung, H.S.: Learning the parts of objects by non-negative matrix factorization. Nature **401**(6755), 788–791 (1999)
9. Liu, B., Xiong, H.: Point-of-interest recommendation in location based social networks with topic and location awareness. In: SDM, vol. 13 (2013)
10. Liu, B., et al.: Learning geographical preferences for point-of-interest recommendation. In: Proceedings of ACM SIGKDD. ACM (2013)
11. Liu, B., et al.: A general geographical probabilistic factor model for point of interest recommendation. TKDE **27**, 1167–1179 (2015)
12. Ma, H., Lyu, M.R., King, I.: Learning to recommend with trust and distrust relationships. In: Proceedings of ACM Recsys. ACM (2009)
13. Tong, H., Faloutsos, C., Pan, J.-Y.: Fast random walk with restart and its applications. In: Proceedings of IEEE ICDM. IEEE Computer Society (2006)
14. Ma, H., et al.: Probabilistic factor models for web site recommendation. In: Proceedings of ACM SIGIR. ACM (2011)
15. Chen, Y., et al.: Factor modeling for advertisement targeting. In: NIPS (2009)
16. Anderson, M., et al.: Learning from the crowd: regression discontinuity estimates of the effects of an online review database. Econ. J. **122**(563), 957–989 (2012)

17. Zhang, J., Chow, C.-Y., et al.: iGeoRec: a personalized and efficient geographical location recommendation framework. IEEE Trans. Serv. Comput. **8**, 701–714 (2015)
18. Schmidt, M.N., Winther, O., Hansen, L.K.: Bayesian non-negative matrix factorization. In: Adali, T., Jutten, C., Romano, J.M.T., Barros, A.K. (eds.) ICA 2009. LNCS, vol. 5441, pp. 540–547. Springer, Heidelberg (2009). doi:10.1007/978-3-642-00599-2_68
19. Kurashima, T., Iwata, T., Hoshide, T., Takaya, N., Fujimura, K.: Geo topic model: joint modeling of user's activity area and interests for location recommendation. In: Proceedings of ACM WSDM. ACM (2013)
20. Yin, H., Sun, Y., Cui, B., Hu, Z., Chen, L.: Lcars: a location-content-aware recommender system. In: Proceedings of ACM SIGKDD. ACM (2013)
21. Lian, D., Zhao, C., Xie, X., Sun, G., Chen, E., Rui, Y.: GeoMF: joint geographical modeling and matrix factorization for point-of-interest recommendation. In: Proceedings of ACM SIGKDD. ACM (2014)
22. Zheng, N., Jin, X., Li, L.: Cross-region collaborative filtering for new point-of-interest recommendation. In: Proceedings of WWW Companion (2013)
23. Zhang, C., Wang, K.: POI recommendation through cross-region collaborative filtering. KIS **46**, 369–387 (2016)
24. Ye, M., Shou, D., Lee, W.-C., et al.: On the semantic annotation of places in location-based social networks. In: Proceedings of ACM SIGKDD. ACM (2011)
25. Yin, P., Luo, P., Lee, W.-C., Wang, M.: App recommendation: a contest between satisfaction and temptation. In: Proceedings of ACM WSDM. ACM (2013)

Evidential Item-Based Collaborative Filtering

Raoua Abdelkhalek$^{(\boxtimes)}$, Imen Boukhris, and Zied Elouedi

LARODEC, Institut Suprieur de Gestion de Tunis,
Université de Tunis, Tunis, Tunisia
abdelkhalek_raoua@live.fr, imen.boukhris@hotmail.com,
zied.elouedi@gmx.fr

Abstract. Recommender Systems (RSs) in particular the collaborative filtering approaches have reached a high level of popularity. These
approaches are designed for predicting the user's future interests towards
unrated items. However, the provided predictions should be taken with
restraint because of the uncertainty pervading the real-world problems.
Indeed, to not give consideration to such uncertainty may lead to unrep-
resentative results which can deeply affect the predictions' accuracy as
well as the user's confidence towards the RS. In order to tackle this issue,
we propose in this paper a new evidential item-based collaborative filtering approach. In our approach, we involve the belief function theory
tools as well as the Evidential K-Nearest Neighbors (EKNN) classifier
to deal with the uncertain aspect of items' recommendation ignored by
the classical methods. The performance of our new recommendation approach is proved through a comparative evaluation with several traditional
collaborative filtering recommenders.

Keywords: Recommender systems · Collaborative filtering · Belief
function theory · Uncertainty · Evidential K-Nearest Neighbors

1 Introduction

Recommender systems (RSs) are considered as a powerful tool to guide users in
their decision making process [8]. That is why, much research has been recently
devoted to the development of RSs aiming to enhance the accuracy and the
performance of the recommendations. One of the most promising recommendation approaches is the collaborative filtering which is considered as a leading
approach in the RSs field due to its straightforwardness and its high accuracy
[2]. It tends to predict users' preferences of items not yet rated. For example,
when a user is browsing a movie website in order to get an idea about the new
releases, the system recommend him movies that he had not watched yet by
predicting a rating for each movie. The question that arises here is how far
the system can assume that the computed outputs are certain? In one way or
another, the provided predictions are not perfect and they involve uncertainty
which should not be ignored. In this case, it is obvious that this approach exhibits
some weakness related to its unability to deal with the uncertainty involved in

© Springer International Publishing AG 2016
F. Lehner and N. Fteimi (Eds.): KSEM 2016, LNAI 9983, pp. 628–639, 2016.
DOI: 10.1007/978-3-319-47650-6_49

the predictions. That is why, we are oriented to the improvement and the extension of the traditional version of the collaborative filtering in order to deal with this kind of problems. In fact, a little attention has been paid to the problem of managing uncertainty in the field of RSs, either by means of fuzzy set theory [3,4], probability theory [5] or possibility theory [6]. However, the uncertainty that is generally pervading in the prediction results has not been considered. Consequently, ignoring this important point can decrease the user's confidence towards the system which can deeply affect the quality of recommendations as well as their reliability. This fact prompted us resorting to a powerful theory which offers a flexible tool to deal with imperfect information namely the belief function theory [7]. This theory is appropriate to cope with uncertainty in classification problems within several machine learning techniques, out of which the K-Nearest Neighbors (KNN). Indeed, an extension of the classical KNN based on the belief function framework has been proposed by [18]. Such classifier allows the objects to belong to not only a specific class.

Thus, the idea behind our new recommendation approach is to take advantage of the belief function tools as well as the Evidential K-Nearest Neighbors (EKNN) in order to deal with the prediction uncertainty. The proposed method is able to provide an evidential representation of the ratings given by similar items as well as the aggregation of their contributions. In this paper, the EKNN formalism is used to represent both the interactions between similar items and the processes leading to a richer information content of the final recommendation. Besides, we show how incorporating uncertainty in the prediction process leads to more significant and accurate results.

The remainder of this paper is organized as follows: Sect. 2 provides a review of the Recommender Systems. Section 3 recalls the Evidential K-Nearest Neighbors. Our proposed recommendation approach is presented in Sect. 4. Section 5 exposes its experimental results conducted on a real world data set. Finally, the contribution is summarized and the paper is concluded in Sect. 6.

2   Recommender Systems

With the continual growth of the available information, RSs have sprung up as a suitable solution to provide the users with personalized recommendations [1]. Such systems start by collecting users' preferences and try to predict their future evaluation towards unrated items. Generally, RSs fall into three categories [10] namely the content-based recommender [9], the collaborative filtering recommender [11] and the hybrid approaches [12]. The first type consists of a matching process between the contents of the unrated items and those of the items in which the user has previously expressed an interest. In this category, a prior access to the item's features is required which cannot be always the case. Indeed, when items have a limit number of available features, no content-based recommender can provide suitable suggestions. Moreover, recommending items sharing the same features may lead to an overspecialization problem. That is to say, since content-based approach relies solely on items' descriptions, this latter cannot in any case find out different items that may please the user. Such phenomenon refers to serendipity which cannot be obtained by this approach unlike the collaborative filtering (CF). This second type relies on a matrix of user-item ratings rather than items' features to predict preferences. By making use of other users' ratings, this approach is able to generate recommendations even when items' descriptions are not available or hard to extract. Besides, it can deal with any kinds of content and help users find interesting items that they might not have discovered otherwise. Surprising the users and recommending different items even the ones that are dissimilar to those rated in the past is the key advantage of this approach. The CF is often characterized as either being model-based or memory-based. The model-based techniques learn a model to predict the future preferences based on the entire collection of users' ratings. The second category, refereed as neighborhood-based, computes the similarity between users (user-based [14]) or items (item-based [15]) and select the most similar ones for recommendation. In some applications, hybrid approaches have also emerged combining two or more recommendation techniques to increase their performance while leveling out the weakness of each one. However, the CF in particular the memory-based has remained the most popular and commonly implemented approach in RSs field due to its simpleness, robustness and its success in real-world applications [13]. That is why the new method will be centered around the memory-based CF approach.

3   Evidential K-Nearest Neighbors

In this section, we recall the basic concepts of the belief function theory as well as the Evidential K-Nearest Neighbors classifier.

3.1   Belief Function Theory

The belief function theory [16,17], also refereed to Dempster-Shafer Theory (DST), is among the most used theories for representing and reasoning with uncertainty. In this theory, a problem domain is represented by a finite set of elementary events called the frame of discernment and denoted by Θ. The belief committed to the elements of the frame of discernment Θ is expressed by a basic belief assignment (bba) which is a mapping function $m : 2^{\Theta} \rightarrow [0, 1]$ such that:

$$\sum_{A \subseteq \Theta} m(A) = 1 \qquad (1)$$

Each mass $m(A)$, called a basic belief mass (bbm), quantifies the degree of belief exactly assigned to the event A of Θ. An event A is called a focal element if $m(A) > 0$. The bba which has at most one focal element aside from the frame of discernment Θ is called simple support function. It is defined as follows:

$$m(X) = \begin{cases} w & \text{if } X = \Theta \\ 1 - w & \text{if } X = A \text{ for some } A \subseteq \Theta \\ 0 & \text{otherwise} \end{cases} \qquad (2)$$

where A is the focal element and $w \in [0, 1]$.

Given two *bba*'s m_1 and m_2 induced from two reliable and independent information sources, the evidence can combined using Dempster's rule of combination defined as:

$$(m_1 \oplus m_2)(A) = k.(m_1 \bigcirc m_2)(A) \tag{3}$$

$$where \quad (m_1 \bigcirc m_2)(A) = \sum_{B,C \subseteq \Theta : B \cap C = A} m_1(B) \cdot m_2(C) \tag{4}$$

$$and \quad k^{-1} = 1 - (m_1 \bigcirc m_2)(\varnothing) \text{ and } (m_1 \oplus m_2)(\varnothing) = 0 \tag{5}$$

To make decisions, beliefs can be represented by pigninstic probabilities defined as:

$$BetP(A) = \sum_{B \subseteq \Theta} \frac{|A \cap B|}{|B|} \frac{m(B)}{(1 - m(\varnothing))} \text{ for all } A \in \Theta \tag{6}$$

3.2   Evidential K-Nearest Neighbors

The Evidential K-Nearest Neighbors (EKNN) [18] is a pattern classification method based on the belief function theory. This latter improves the classification performance over the crisp KNN approach. It allows a credal classification of the objects which leads to a richer information content of the classifier's output.

Notations

- $\Theta = \{C_1, C_2, ..., C_M\}$: The frame of discernment containing the M possible classes in the system.
- $X_i = \{X_1, X_2, ..., X_n\}$: The object X_i belonging to the set of n distinct objects in the system.
- X: A new object to be classified.
- $N_K(X)$: The set of the K-Nearest Neighbors of X.

EKNN Procedure

The EKNN aims to classify a new instance X based on the information provided by the training set. A new pattern X to be classified should be assigned to one class of the $N_K(X)$ based on the selected neighbors. However, the knowledge that a neighbor X belongs to class C_q can be considered as a piece of evidence that raises the belief that the pattern X to be classified belongs to the class C_q. That is why, the EKNN technique treats each neighbor as a piece of evidence supporting a number of hypotheses regarding the class of the object X to be classified. Indeed, the more the distance between X and X_i is scaled down, the more the evidence is strong. This evidence can be represented by a simple support function with a *bba* verifying:

$$m_{X,X_i}(\{C_q\}) = \alpha_0 \exp^{-(\gamma_q^2 \times d(X,X_i)^2)} \tag{7}$$

$$m_{X,X_i}(\Theta) = 1 - \alpha_0 \exp^{-(\gamma_q^2 \times d(X,X_i)^2)}$$

where α_0 is a constant that has been heuristically fixed to 0.95 and $d(X, X_i)$ is the Euclidean distance between the object to be classified and the other objects in the training set. On the other hand, γ_q has been defined as a positive parameter assigned to each class C_q. It is considered as the inverse of the mean distance between all the training patterns belonging to the class C_q.

Once the different *bba*'s provided by the K-Nearest Neighbors are generated, they can be combined using Dempster's rule of combination.

$$m_X = m_{X,X_1} \oplus ... \oplus m_{X,X_K} \tag{8}$$

where $\{1, ..., K\}$ is the set containing the indexes of the K-Nearest Neighbors.

4   Evidential Item-Based Collaborative Filtering

The idea behind our contribution is to take into account the uncertain aspect of the predictions made by RSs. To ensure this task, we propose to rely on the EKNN which is a machine learning algorithm under the belief function framework. The whole process of our proposed evidential collaborative filtering approach is performed in three phases namely the initialization phase, the learning phase and the prediction phase.

Step1: Initialization Phase

The first step consists of assigning values to the two parameters α_0 and γ_{r_i} to be used in the next phase. This procedure starts by initializing the parameter α_0 and then exploits the user-item matrix in order to compute the second parameter γ_{r_i}. The parameter α_0 is initialized to the value 0.95 as mentioned in the EKNN formalism [18]. Note that the initialization of α_0 is executed only once while the γ_{r_i} computation is performed each time according to the current items' ratings. In order to ensure the γ_{r_i} computation, we should firstly find items having separately exclusive ratings. In other words, we extract the items having equal values of the provided ratings. According to the selected items, we assign a parameter γ_{r_i} to each rating r_i which will be computed as the inverse of the average distance between each pair of items i and j having the same ratings. This computation is performed based on the normalized Euclidean distance denoted by d(i,j) and defined as follows:

$$d(i,j) = \frac{\sum_{u \in u_i \cap u_j} (r_{u,i} - r_{u,j})^2}{|u_i \cap u_j|} \tag{9}$$

where $r_{u,i}$ and $r_{u,j}$ correspond to the rating of the user u for the items i and j. Moreover, u_i and u_j are the users u who have rated the items i and j.

Step2: Learning Phase

Once the two parameters α_0 and γ_{r_i} have been assigned, the second phase consists in the items' selection. This selection is performed according to a similarity computation as that proposed in [15] by isolating the co-rated items. In our method, we firstly consider users who have rated common items. Then, we compute for each item j in the database, its distance with the target item i. Given a target item, we have to spot to its K-most similar items, also referred to the neighbors, by picking out only the K items having the lowest distances that we denote by dist(i,j).

Step3: Prediction Phase

The prediction phase is the most important one in RSs since it provides users with their future evaluations regarding the unrated items. In this section, we shed light on the prediction process of our contribution. The two key procedures of this phase are respectively, the *bba*'s generation and the *bba*'s combination.

1. **The *bba*'s Generation**

   Traditional methods in this step, provide the user with a predicted rating that indicates a score of his future degree of satisfaction given an item. As we argued in the introduction, the RS cannot draw any certain inference about the future rating. Even if the ratings provided by the most similar items can increase our belief about the most probable one, we cannot admit that such knowledge is certain. In view of this assumption, we emphasize the presence of uncertainty facet in the predictions through the belief function theory. In such case, we maintain that different pieces of evidence which involve a particular hypothesis about the predicted rating can contribute to the final prediction. Hence, the evidence would be over the ratings provided by the K-most similar items. The main advantage under this representation is that the final prediction must be a basic belief assignment which reflects more credible results. We start by observing the ratings provided by the different pieces of evidences (i.e. the K-similar items). Accordingly, we generate the corresponding *bba*'s.

   Using the terminology of the belief function theory, we can define the frame of discernment corresponding to this situation as:

   $$\Theta = \{r_1, r_2 \ldots r_n\} \tag{10}$$

   where n denotes the number of the possible ratings r that can be provided in the system.

   Since each item involves a particular hypothesis about the predicted rating, we generate a *bba* over each rating provided by the similar items as well as the whole frame of discernment Θ. According to the similarities computed in the learning phase as well as the two parameters α_0 and γ_{r_i} initially assigned, we can represent this *bba* as a simple support function defined as following:

   $$m_{i,j}(\{r_i\}) = \alpha_{r_i}$$
   $$m_{i,j}(\Theta) = 1 - \alpha_{r_i} \tag{11}$$

   where i is the target item and j is its similar item such that: $j = \{1, .., K\}$, $\alpha_{r_i} = \alpha_0 \exp^{-(\gamma_{r_i}^2 \times dist(i,j)^2)}$, α_0 and γ_{r_i} are the two parameters assigned in the initialization phase and $dist(i, j)$ is the distance between the items i and j computed in the learning phase.

In this situation, each neighbor of the target item has two possible hypotheses. The first one corresponds to the value of its provided rating while the rest of the committed belief is allocated to the frame of discernment Θ. Thereupon, the focal elements of the belief function are the ratings provided by the K-similar items and Θ. By treating the K-most similar items as independent sources of evidence, each one is represented by a basic belief assignment. Hence, K different bba's can be generated for each item.

2. **The bba's Combination**

In the previous step, we showed how to generate bba's for each similar item. Now, we describe how to aggregate these bba's in order to synthesize the final belief about the rating of the target item. Using the belief function theory, such bba's can be combined using Dempster's rule of combination. Therefore, the resulting bba encodes the evidence of the K-Nearest Neighbors regarding the rating that should be provided to the target item.

$$m_{Target\ item} = m_{Target\ item, Item\ 1} \oplus \dots \oplus m_{Target\ item, Item\ K} \quad (12)$$

This final bba is obtained as follows:

$$m(\{r_i\}) = \frac{1}{R}(1 - \prod_{i \in i_K}(1 - \alpha_{r_i})) \cdot \prod_{r_j \neq r_i} \prod_{i \in i_K}(1 - \alpha_{r_j}) \quad \forall r_i \in \{r_1, .., r_n\} \quad (13)$$

$$m(\Theta) = \frac{1}{R}\prod_{i=1}^{n}(1 - \prod_{i \in i_K}(1 - \alpha_{r_i}))$$

where $i_K = \{i_1, i_2 ..., i_K\}$ is the set containing the indexes of the K-nearest neighbors of the target item over the user-item matrix, n is the number of the ratings provided by the similar items, α_{r_i} is the belief committed to the rating r_i, α_{r_j} is the belief committed to the rating $r_j \neq r_i$, R is a normalized factor defined by:

$$R = \sum_{i=1}^{n}(1 - \prod_{i \in i_K}(1 - \alpha_{r_i})) \prod_{r_j \neq r_i} \prod_{i \in i_K,}(1 - \alpha_{r_j}) + \prod_{i=1}^{n}(\prod_{i \in i_K}(1 - \alpha_{r_j})) \quad (14)$$

Figure 1 illustrates the prediction phase.

5   Experimental Study

We conduct some experiments on the MovieLens[1] data set which is one of the widely used real word data sets in the field of CF. It contains in total 100.000

---

[1] http://movielens.org.

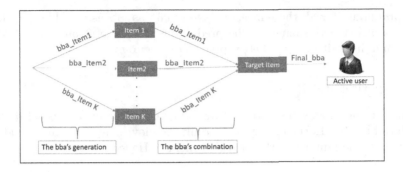

Fig. 1. The prediction phase

ratings collected from 943 users on 1682 movies. Indeed, a study of the choice of the appropriate metric to be used in CF has been performed in [20] using MovieLens data set. According to this study, the Euclidean distance achieves the best results in terms of accuracy. Otherwise, another experimental study [21] has proven that both Pearson and Cosine are the most appropriate similarity measures. Hence, the choice of the most suitable metric is still arguable in the CF framework. That is why, we propose to perform a comparative evaluation over our proposed method as well as the traditional one using these three different similarity measures commonly used in the memory-based CF category.

5.1 Evaluation Metrics

To evaluate our approach, we carry out experiments over three evaluation metrics namely the *Mean Absolute Error* (MAE) [22], the *Root Mean Squared Error* (RMSE) [23] and the *Distance criteron* (dist_crit) [24] defined by:

$$MAE = \frac{\sum_{u,i} |p_{u,i} - r_{u,i}|}{N}, \tag{15}$$

$$RMSE = \sqrt{\frac{\sum_{u,i} (p_{u,i} - r_{u,i})^2}{N}} \tag{16}$$

$$dist\_crit = \frac{\sum_{u,i} dist\_crit(i)}{N} \tag{17}$$

where

$$dist\_crit(i) = \sum_{i=1}^{n} (BetP(\{r_{u,i}\}) - \delta_i)^2 \tag{18}$$

- $r_{u,i}$ is the real rating for the user u on the item i and $p_{u,i}$ is the predicted value of the rating. N is the total number of the predicted ratings over all the users and n is the number of the possible ratings that can be provided in the system. δ_i is equal to 1 if $r_{u,i}$ is equal to $p_{u,i}$ and 0 otherwise.

It is appropriate for all these measures to remain as low as possible in order to achieve a higher performance of the predictions. Hence, a small value of MAE, RMSE and dist_crit means a better prediction accuracy.

5.2   Experimental Protocol

Working on the MovieLens data set, we follow the same experimental protocol introduced by [19]. In the first step, we rank the movies available in the data set according to the number of the provided ratings. Hence, we get:

$$Nb_{user}(m_1) \geq Nb_{user}(m_2) \geq ...Nb_{user}(m_{1682})$$

where $Nb_{user}(m_i)$ is the number of users who rated the movie m_i. From the original MovieLens data set, we extract 10 subsets each of which contains the ratings provided by the users for 20 movies. The selection of the subsets is performed by progressively increasing the number of the missing rates. In other words, since few ratings provided for the total number of items are available, each subset will contain a specific number of ratings leading to different degrees of sparsity. For each subset, we randomly extract 20 % of the available ratings as a test set and the remaining 80 % are used as a training set. We compute the MAE, the RMSE and the dist_crit for each subset by varying each time the value of the neighborhood size K.

5.3   Experimental Results

Our proposed approach is characterized by its ability to deal with the uncertainty pervaded in the prediction task. Hence, we will be concerned by showing how this contribution improves the accuracy of the predictions. That is why, we perform several experiments over the 10 subsets by varying each time K from 1 to 10.

Performance for Different Sparsity Degrees

Table 1 recapitulates results considering different sparsity degrees and recommendations' approaches in the two cases namely, certain case and uncertain case. We mention that the obtained results for each subset correspond to the average of 10 Nearest-Neighbors in order to have fair results over the four approaches. As can be seen, the evidential item-based CF has practically better MAE, RMSE and dist_crit values comparing to the three other approaches under a certain framework. For example, at a sparsity level of 75 %, the MAE of our proposed method (equal to 0.744) is lower than the MAE of Pearson CF (equal to 0.943) leading to a reduction of 20 % in the error rate. Similarly, it outperforms both Cosine (equal to 0.877) and Euclidean CF (equal to 0.851). Besides, the dist_crit corresponding to our approach remains the lowest over the different sparsity degrees. (e.g. 0.786 compared to 1.283, 1.353 and 1.444 at a sparsity of 95.9 %.)

Performance for Different Neighborhood Sizes

According to the neighborhood size, we can observe a variation of the MAE, the RMSE and the dist_crit corresponding to the four approaches. This variation

Table 1. Comparison result in term of MAE, RMSE and dist_crit

		Certain framework			Uncertain framework
Measure	Sparsity	Euclidean	Pearson	Cosine	Evidential
MAE	53%	0.815	0.839	0.824	**0.751**
RMSE		1.178	1.231	1.158	**1.089**
dist_crit		1.195	1.27	1.205	**0.859**
MAE	56.83%	0.886	0.936	0.87	**0.84**
RMSE		1.22	1.291	1.215	**1.158**
dist_crit		1.279	1.293	1.251	**0.875**
MAE	59.8%	0.853	0.863	0.825	**0.761**
RMSE		1.223	1.256	1.198	**1.135**
dist_crit		1.217	1.191	1.178	**0.795**
MAE	62.7%	0.858	0.905	0.876	**0.763**
RMSE		1.21	1.267	1.232	**1.092**
dist_crit		1.255	1.308	1.244	**0.859**
MAE	68.72%	0.914	0.990	1	**0.831**
RMSE		1.249	1.367	1.351	**1.184**
dist_crit		1.307	1.372	1.374	**0.862**
MAE	72.5%	0.915	0.976	0.917	**0.851**
RMSE		1.28	1.348	1.272	**1.184**
dist_crit		1.257	1.333	1.29	**0.858**
MAE	75%	0.851	0.943	0.877	**0.744**
RMSE		1.194	1.27	1.212	**1.187**
dist_crit		1.24	1.322	1.266	**0.858**
MAE	80.8%	0.792	0.927	0.848	**0.718**
RMSE		1.112	1.265	1.179	**1.079**
dist_crit		1.214	1.322	1.259	**0.837**
MAE	87.4%	0.889	0.958	0.978	**0.840**
RMSE		1.248	1.309	1.334	**1.18**
dist_crit		1.263	1.317	1.32	**0.856**
MAE	95.9%	0.98	**0.913**	1.13	0.991
RMSE		1.381	**1.217**	1.527	1.445
dist_crit		1.283	1.353	1.444	**0.786**

is illustrated in Figs. 2, 3 and 4. As seen, the four approaches have almost the same behavior over the different neighborhood sizes. However, we can observe that the curve of the evidential item-based CF remains always under those of the three traditional methods. According to these results, the evidential approach shows the greatest performance over all the traditional item-based CF methods.

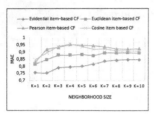
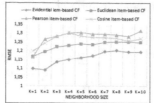
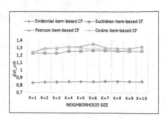

Fig. 2. The MAE results

Fig. 3. The RMSE results

Fig. 4. The dist_crit results

Thereupon, we can conclude that our new approach significantly improves the prediction accuracy.

6 Conclusion

In this paper, we have proposed a new recommendation approach based on the Evidential K-Nearest Neighbors. Our approach is based on the generation and the combination of the different *bba*'s corresponding to each similar item which allows an improvement over the crisp results related to the traditional methods. This solution may increase the user's confidence towards the system as well as the accuracy of the provided predictions which would be fully beneficial to the RSs field especially, nowadays where reliability becomes a crucial parameter to attend user's satisfaction. Moreover, the proposed evidential item-based collaborative filtering approach offers the users the possibility of having a global overview of their future interests which leads to a better decision making. As future works, we suggest to introduce uncertainty in the users' preferences which are considered as the inputs of our approach.

References

1. Bobadilla, J., Ortega, F., Hernando, A., Gutierrez, A.: Recommender systems survey. Knowl. Based Syst. **46**, 109–132 (2013)
2. Koren, Y., Sill, J.: Collaborative filtering on ordinal user feedback. In: International Joint Conference on Artificial Intelligence, pp. 3022–3026 (2013)
3. Zenebe, A., Norcio, A.F.: Representation, similarity measures and aggregation methods using fuzzy sets for content-based recommender systems. Fuzzy Sets Syst. **160**(1), 76–94 (2009)
4. Boulkrinat, S., Hadjali, A., Aissani Mokhtari, A.: Handling preferences under uncertainty in recommender systems. In: IEEE International Conference on Fuzzy Systems, pp. 2262–2269 (2014)
5. Yu, K., Schwaighofer, A., Tresp, V., Xu, X., Kriegel, H.-P.: Probabilistic memory-based collaborative filtering. IEEE Trans. Knowl. Data Eng. **16**(1), 56–69 (2004)

6. Slokom, M., Ayachi, R.: Towards a new possibilistic collaborative filtering approach. In: Second International Conference on Computer Science, Computer Engineering, and Social Media, pp. 209–216 (2015)

7. Smets, P.: The transferable belief model for quantified belief representation. In: Smets, P. (ed.) Quantified Representation of Uncertainty and Imprecision, pp. 267–301. Springer, Heidelberg (1998)

8. Ricci, F., Rokach, L., Shapira, B.: Introduction to recommender systems handbook. In: Ricci, F., Rokach, L., Shapira, B., Kantor, P.B. (eds.) Recommender Systems Handbook, pp. 1–35. Springer, Heidelberg (2011)

9. Gu, Y., Zhao, B., Hardtke, D., Sun, Y.: Learning global term weights for content-based recommender systems. In: Proceedings of the 25th International Conference on World Wide Web, pp. 391–400 (2016)

10. Isinkaye, F.O., Folajimi, Y.O., Ojokoh, B.A.: Recommendation systems: principles, methods and evaluation. Egypt. Inf. J. **16**(3), 261–273 (2015)

11. Su, X., Khoshgoftaar, T.M.: A survey of collaborative filtering techniques. Adv. Artif. Intell. **2009**, 1–19 (2009)

12. Burke, R.: Hybrid web recommender systems. In: Brusilovsky, P., Kobsa, A., Nejdl, W. (eds.) The Adaptive Web. LNCS, vol. 4321, pp. 377–408. Springer, Heidelberg (2007). doi:10.1007/978-3-540-72079-9_12

13. Zheng, Y., Ouyang, Y., Rong, W., Xiong, Z.: Multi-faceted distrust aware recommendation. In: Zhang, S., Wirsing, M., Zhang, Z. (eds.) KSEM 2015. LNCS (LNAI), vol. 9403, pp. 435–446. Springer, Heidelberg (2015). doi:10.1007/978-3-319-25159-2_39

14. Zhao, Z.D., Shang, M.S.: User-based collaborative-filtering recommendation algorithms on hadoop. In: Third International Conference on Knowledge Discovery and Data Mining, pp. 478–481 (2010)

15. Sarwar, B., Karypis, G., Konstan, J., Riedl, J.: Item-based collaborative filtering recommendation algorithms. In: International Conference on World Wide Web, pp. 285–295 (2001)

16. Dempster, A.P.: A generalization of bayesian inference. J. R. Stat. Soc. Ser. B (Methodol.) **30**, 205–247 (1968)

17. Shafer, G.: A Mathematical Theory of Evidence, vol. 1. Princeton University Press, Princeton (1976)

18. Denoeux, T.: A K-nearest neighbor classification rule based on Dempster-Shafer theory. IEEE Trans. Syst. Man Cybern. **25**(5), 804–813 (1995)

19. Su, X., Khoshgoftaar, T.M.: Collaborative filtering for multi-class data using bayesian networks. Int. J. Artif. Intell. Tools **17**(01), 71–85 (2008)

20. Sanchez, J., Serradilla, F., Martinez, E., Bobadilla, J.: Choice of metrics used in collaborative filtering and their impact on recommender systems. In: IEEE International Conference on Digital Ecosystems and Technologies, pp. 432–436 (2008)

21. Bobadilla, J., Hernando, A., Ortega, F., Bernal, J.: A framework for collaborative filtering recommender systems. Expert Syst. Appl. **38**(12), 14609–14623 (2011)

22. Pennock, D.M., Horvitz, E., Lawrence, S., Giles, C.L.: Collaborative filtering by personality diagnosis: a hybrid memory-and model-based approach. In: The Conference on Uncertainty in Artificial Intelligence, pp. 473–480 (2000)

23. Bennett, J., Lanning, S.: The netflix prize. In: KDD Cup and Workshop, vol. 2007, p. 35 (2007)

24. Elouedi, Z., Mellouli, K., Smets, P.: Assessing sensor reliability for multisensor data fusion within the transferable belief model. IEEE Trans. Syst. Man Cybern. Part B Cybern. **34**(1), 782–787 (2004)

Author Index

Printed in the United States
by Bookmasters

Printed in the United States
By Bookmasters